Husrev Taha Sencar, Sergio Velastin, Nikolaos Nikolaidis, Shiguo Lian (Eds.)

Intelligent Multimedia Analysis for Security Applications

Studies in Computational Intelligence, Volume 282

Editor-in-Chief
Prof. Janusz Kacprzyk
Systems Research Institute
Polish Academy of Sciences
ul. Newelska 6
01-447 Warsaw
Poland
E-mail: kacprzyk@ibspan.waw.pl

Further volumes of this series can be found on our homepage: springer.com

Vol. 261. Nadia Nedjah, Leandro dos Santos Coelho, and Luiza de Macedo de Mourelle (Eds.)
Multi-Objective Swarm Intelligent Systems, 2009
ISBN 978-3-642-05164-7

Vol. 262. Jacek Koronacki, Zbigniew W. Ras, Slawomir T. Wierzchon, and Janusz Kacprzyk (Eds.)
Advances in Machine Learning I, 2009
ISBN 978-3-642-05176-0

Vol. 263. Jacek Koronacki, Zbigniew W. Ras, Slawomir T. Wierzchon, and Janusz Kacprzyk (Eds.)
Advances in Machine Learning II, 2009
ISBN 978-3-642-05178-4

Vol. 264. Olivier Sigaud and Jan Peters (Eds.)
From Motor Learning to Interaction Learning in Robots, 2009
ISBN 978-3-642-05180-7

Vol. 265. Zbigniew W. Ras and Li-Shiang Tsay (Eds.)
Advances in Intelligent Information Systems, 2009
ISBN 978-3-642-05182-1

Vol. 266. Akitoshi Hanazawa, Tsutom Miki, and Keiichi Horio (Eds.)
Brain-Inspired Information Technology, 2009
ISBN 978-3-642-04024-5

Vol. 267. Ivan Zelinka, Sergej Celikovský, Hendrik Richter, and Guanrong Chen (Eds.)
Evolutionary Algorithms and Chaotic Systems, 2009
ISBN 978-3-642-10706-1

Vol. 268. Johann M.Ph. Schumann and Yan Liu (Eds.)
Applications of Neural Networks in High Assurance Systems, 2009
ISBN 978-3-642-10689-7

Vol. 269. Francisco Fernández de de Vega and Erick Cantú-Paz (Eds.)
Parallel and Distributed Computational Intelligence, 2009
ISBN 978-3-642-10674-3

Vol. 270. Zong Woo Geem
Recent Advances In Harmony Search Algorithm, 2009
ISBN 978-3-642-04316-1

Vol. 271. Janusz Kacprzyk, Frederick E. Petry, and Adnan Yazici (Eds.)
Uncertainty Approaches for Spatial Data Modeling and Processing, 2009
ISBN 978-3-642-10662-0

Vol. 272. Carlos A. Coello Coello, Clarisse Dhaenens, and Laetitia Jourdan (Eds.)
Advances in Multi-Objective Nature Inspired Computing, 2009
ISBN 978-3-642-11217-1

Vol. 273. Fatos Xhafa, Santi Caballé, Ajith Abraham, Thanasis Daradoumis, and Angel Alejandro Juan Perez (Eds.)
Computational Intelligence for Technology Enhanced Learning, 2010
ISBN 978-3-642-11223-2

Vol. 274. Zbigniew W. Raś and Alicja Wieczorkowska (Eds.)
Advances in Music Information Retrieval, 2010
ISBN 978-3-642-11673-5

Vol. 275. Dilip Kumar Pratihar and Lakhmi C. Jain (Eds.)
Intelligent Autonomous Systems, 2010
ISBN 978-3-642-11675-9

Vol. 276. Jacek Mańdziuk
Knowledge-Free and Learning-Based Methods in Intelligent Game Playing, 2010
ISBN 978-3-642-11677-3

Vol. 277. Filippo Spagnolo and Benedetto Di Paola (Eds.)
European and Chinese Cognitive Styles and their Impact on Teaching Mathematics, 2010
ISBN 978-3-642-11679-7

Vol. 278. Radomir S. Stankovic and Jaakko Astola
From Boolean Logic to Switching Circuits and Automata, 2010
ISBN 978-3-642-11681-0

Vol. 279. Manolis Wallace, Ioannis E. Anagnostopoulos, Phivos Mylonas, and Maria Bielikova (Eds.)
Semantics in Adaptive and Personalized Services, 2010
ISBN 978-3-642-11683-4

Vol. 280. Chang Wen Chen, Zhu Li, and Shiguo Lian (Eds.)
Intelligent Multimedia Communication: Techniques and Applications, 2010
ISBN 978-3-642-11685-8

Vol. 281. Robert Babuška and Frans C.A. Groen (Eds.)
Interactive Collaborative Information Systems, 2010
ISBN 978-3-642-11687-2

Vol. 282. Husrev Taha Sencar, Sergio Velastin, Nikolaos Nikolaidis, and Shiguo Lian (Eds.)
Intelligent Multimedia Analysis for Security Applications, 2010
ISBN 978-3-642-11754-1

Husrev Taha Sencar, Sergio Velastin,
Nikolaos Nikolaidis, and Shiguo Lian (Eds.)

Intelligent Multimedia Analysis for Security Applications

Springer

Dr. Husrev Taha Sencar
TOBB ETÜ Computer Engineering
Department
Sögütözü Cad.No:43, Sögütözü
Ankara, 06560
Turkey
E-mail: htsencar@etu.edu.tr

Dr. Nikolaos Nikolaidis
Department of Informatics
Aristotle University of Thessaloniki
Box 451
Thessaloniki, GR-54124
Greece
E-mail: nikolaid@aiia.csd.auth.gr

Dr. Sergio Velastin
Computing, Information Systems
and Mathematics
Kingston University London
Penrhyn Road
Surrey KT1 2EE
UK
Email: sergio.velastin@kingston.ac.uk

Dr. Shiguo Lian
France Telecom R&D (Orange Labs) Beijing
Raycom Infotech Park C
2 Science Institute South Road
Haidian District, Beijing, 100080
China
E-mail: shiguo.lian@orange-ftgroup.com

ISBN 978-3-642-11754-1 e-ISBN 978-3-642-11756-5

DOI 10.1007/978-3-642-11756-5

Studies in Computational Intelligence ISSN 1860-949X

Library of Congress Control Number: 2010924914

Typeset & Cover Design: Scientific Publishing Services Pvt. Ltd., Chennai, India.

Printed on acid-free paper

9 8 7 6 5 4 3 2 1

springer.com

Preface

The advances in the generation and processing of multimedia data (e.g. documents, images, video, audio, animations, etc.) have had an immense impact on multimedia applications and, as a result, multimedia has permeated almost every aspect of our daily lives. This development has also brought with it a whole host of issues and challenges which were either not as apparent before or were non-existent. Today, digital media is relied upon as primary news and information resource, as evidence in a court of law, as part of medical records or as financial documents. However, there is still lack of authoritative mechanisms to verify the origin and veracity of media data. Indeed, multimedia content has become an extremely valuable asset, and it is being both disseminated and consumed on a larger scale than ever before, but the issues concerning how the content owners and publishers should control the distribution of and access to their content have not been satisfactorily resolved yet.

There are various other issues related to use of multimedia that require further analysis and research. For example, it is a known fact that some criminal organizations communicate with its members by posting information embedded media to public forums and web-sites to evade surveillance by law enforcement. Conventional multimedia processing approaches do not provide sufficiently effective means for defending against such communication. At the same time, audio and video surveillance is becoming increasingly more ubiquitous, with systems deployed to detect and report abnormal events and to prevent criminal activity. Essentially, this leads to accumulation of insurmountable amount of media data for which appropriate analysis, indexing and search tools need to be designed. Similarly, speech and face recognition capabilities are now being required in many critical applications to automatically identify individuals. However, achieving these tasks with high accuracy still remains difficult due to the variety of operating conditions encountered. All these existing and emerging security-related problem domains require radically new approaches to dealing with multimedia data and attract the interest of many researchers in diverse application areas.

To the best of our knowledge, this is one of the very few books focused on analysis of multimedia data and newly emerging multimedia applications with an emphasis on security. The main objective of this project was to assemble as much research coverage as possible related to the field by defining the latest innovative technologies and providing the most comprehensive list of research references. The book includes sixteen chapters highlighting current concepts, issues and emerging technologies.

Distinguished scholars from many prominent research institutions around the world contributed to the book, shedding light on various aspects, including not only some fundamental knowledge and the latest key techniques, but also typical applications and open issues. Topics covered include dangerous or abnormal event detection, interaction recognition, person identification based on multiple traits, audiovisual biometric person authentication and liveness verification, emerging biometric technologies, sensitive information filtering for teleradiology, detection of nakedness in images, audio forensics, steganalysis, media content tracking authentication and illegal distributor identification through watermarking and content-based copy detection.

We believe that the comprehensive coverage of diverse disciplines in the field of intelligent multimedia analysis for security applications will contribute to a better understanding of all topics, research, and discoveries in this emerging and evolving field and that the included contributions will be instrumental in the expansion of the corresponding body of knowledge, making this book a reference source of information. It is our sincere hope that this publication and its great amount of information and research will assist our research colleagues, faculty members and students, and organization decision makers in enhancing their understanding for the concepts, issues, problems, trends, challenges and opportunities related to this research field. Perhaps this book will even inspire its readers to contribute to the current discoveries in this immense field.

Editors

Dr. Husrev Taha Sencar
TOBB University of Economics and Technology, Turkey

Dr. Sergio Velastin
Kingston University London, UK

Dr. Nikolaos Nikolaidis
Aristotle University of Thessaloniki, Greece

Dr. Shiguo Lian
France Telecom R&D (Orange Labs) Beijing, China

Acknowledgments

The editors would like to acknowledge the help of all people that were involved in the collation process of the book, without whose support the project could not have been satisfactorily completed. Deep appreciation and gratitude is due to the authors of all the chapters for their efforts to achieve a high quality project.

Special thanks go to the publishing team at Springer, whose contributions throughout the whole process from inception of the initial idea to final publication have been invaluable. In particular to Dr. Thomas Ditzinger, who continuously prodded via e-mail to help us keep the project on schedule and to Prof. Janusz Kacprzyk, whose enthusiasm motivated us to initiate the project.

And last but not least, a big and sincere expression of gratitude to our families, for their unfailing support and encouragement during the months it took to give birth to this book.

November 2009

Dr. Husrev Taha Sencar
Dr. Sergio Velastin
Dr. Nikolaos Nikolaidis
Dr. Shiguo Lian

Contents

Moving Pixels in Static Cameras: Detecting Dangerous Situations due to Environment or People

Simone Calderara[1], Rita Cucchiara[1], and Andrea Prati[2]

[1] D.I.I. - University of Modena and Reggio Emilia
[2] Di.S.M.I. - University of Modena and Reggio Emilia

Summary. Dangerous situations arise in everyday life and many efforts have been lavished to exploit technology to increase the level of safety in urban areas. Video analysis is absolutely one of the most important and emerging technology for security purposes. Automatic video surveillance systems commonly analyze the scene searching for moving objects. Well known techniques exist to cope with this problem that is commonly referred as "change detection". Every time a difference against a reference model is sensed, it should be analyzed to allow the system to discriminate among a usual situation or a possible threat. When the sensor is a camera, motion is the key element to detect changes and moving objects must be correctly classified according to their nature. In this context we can distinguish among two different kinds of threat that can lead to dangerous situations in a video-surveilled environment. The first one is due to environmental changes such as rain, fog or smoke present in the scene. This kind of phenomena are sensed by the camera as moving pixels and, subsequently as moving objects in the scene. This kind of threats shares some common characteristics such as texture, shape and color information and can be detected observing the features' evolution in time. The second situation arises when people are directly responsible of the dangerous situation. In this case a "subject" is acting in an unusual way leading to an abnormal situation. From the sensor's point of view, moving pixels are still observed, but specific features and time-dependent statistical models should be adopted to learn and then correctly detect unusual and dangerous behaviors. With these premises, this chapter will present two different case studies. The first one describes the detection of environmental changes in the observed scene and details the problem of reliably detecting smoke in outdoor environments using both motion information and global image features, such as color information and texture energy computed by the means of the Wavelet transform. The second refers to the problem of detecting suspicious or abnormal people behaviors by means of people trajectory analysis in a multiple cameras video-surveillance scenario. Specifically, a technique to infer and learn the concept of normality is proposed jointly with a suitable statistical tool to model and robustly compare people trajectories.

1 Introduction

Dangerous situations arise in everyday life and many efforts have been lavished in new technologies to increase the level of safety in urban areas. Everyday threats

H.T. Sencar et al. (Eds.): Intel. Multimedia Analysis for Security Appli., SCI 282, pp. 1–28.
springerlink.com

are of different nature and a reliable solution to prevent all them is a chimera. Nevertheless, even though a “solution” is not possible, an aid to ease the work of public officers is still technologically feasible. The aim of this aid is to pop up the attention on significant parts of the scene, to stores evidences of relevant events, to enhance the image/sound quality for investigation purposes, and so on.

Many media have been explored and effectively exploited to prevent criminal acts or to react to dangerous situations. Among these, it is worth to mention how sound or images can be exploited to successfully detect dangerous events in everyday life. Many applications exist, ranging from sound analysis to image retrieval, that are currently adopted for forensic purposes. For instance, several approaches to computational audio analysis have been proposed, mainly focused on the computational translation of psychoacoustics results, such as the so-called *computational auditory scene analysis* (CASA) [1], aimed at the separation and classification of sounds. However, video analysis is absolutely one of the most important and emerging technology for security purposes. There are several reasons that concur to make the video information so crucial; obviously a video footage is a rich source of information containing time-dependent visual features thus allowing a deep analysis of objects' motion and behavior. Although CCTV systems are widespread in many urban areas, automatic video surveillance is a relatively new task and recently many effort have been spent in creating “intelligent systems” capable of detect dangerous situations in real scenarios. The reasons for this reside in the complexity of video data which need to be efficiently processed to extract useful information. In particular, a challenging task in video analysis is to limit false positives. Reflexes, shadows, false motion (due, for instance, to camera vibrations), changes in the illumination, are some examples of the causes of false positives and false alarms.

For example, automatic video surveillance systems commonly analyze the scene searching for moving objects. Well known techniques exist to cope with this problem that is commonly referred as “change detection”. Every time a difference against a reference model is sensed, it should be analyzed to allow the system to discriminate among a usual situation or a possible threat. When the sensor is a camera, motion is the key element to detect changes and moving objects must be correctly classified according to their nature. In this context we can distinguish among two different kinds of threat that can lead to dangerous situations in a video-surveilled environment. The first one is due to environmental changes such as rain, fog or smoke present in the scene. This kind of phenomena are sensed by the camera as moving pixels and, subsequently as moving objects in the scene. This kind of threats shares some common characteristics such as texture, shape and color information and can be detected observing the features' evolution in time.

The second situation arises when people are directly responsible of the dangerous situation. In this case a "subject" is acting in an unusual way leading to an abnormal situation. From the sensor's point of view, moving pixels are still observed, but specific features and time-dependent statistical models should be adopted to learn and then correctly detect unusual and dangerous behaviors.

With these premises, this chapter will present two different case studies. The first one describes the detection of environmental changes in the observed scene and details the problem of reliably detecting smoke in outdoor environments using both motion information and global image features, such as color information and texture energy computed by the means of the Wavelet transform. The second refers to the problem of detecting suspicious or abnormal people behaviors by means of people trajectory analysis in a multiple cameras video surveillance scenario. Specifically, a technique to infer and learn the concept of normality is proposed jointly with a suitable statistical tool to model and robustly compare people trajectories.

2 State of the Art

2.1 State of the Art on Video-Based Smoke Detection

Smoke detection in video surveillance systems is still an open challenge for computer vision and pattern recognition communities. It concerns the definition of robust approaches to detect, as soon as possible, fast propagation of smoke possibly due to explosions, fires or special environmental conditions. These systems can replace standard smoke and fire sensors, which cannot be applied in large and open spaces. Moreover, detecting smoke by visual cues could allow fast and reactive alarms also in some specific situations, where smoke is growing in unconventional directions, so that the time-to-alarm of normal sensors could become unacceptable.

The video analysis tasks for smoke detection are not trivial due to the variability of shape, motion and texture patterns of smoke, which appearance is dependent on the luminance conditions, the background manifolds and colors of the scene. The smoke identification becomes more challenging in presence of other moving objects and shadows and whenever the background is variable too.

The problem of studying environmental effects on the scene have been deep investigated in literature. In particular several situations arise when weather condition affects the object visibility in the scene leading to particular and dangerous situations. All these cases belong to the macro category of event detection problems that may be studied to trigger alarms or signaling whether a dangerous events happens or not. Among the natural phenomena that may be visually analyzed we can annoverate for examples rain, fog and smoke due to fire or different sources.

Narasimhan and Nayar have extensively studied the visual manifestation of different weather conditions. In [46] they propose an interesting method, based on atmospheric optic, to recover the "clear-day" scene color from two or more images taken under different and unknown weather conditions. They also developed a method for depth-segmentation and extracting three dimensional scene structure. In order to do this they used two scattering model: Attenuation and

Table 1. Summary of Reference for several natural event detection techniques.

Event	Technique	Authors
Haze	Statistical model	Oakley et. al. [45]
	Airtight and Attenuation scattering model	Narasimhan and Nayar [46]
	Light polarization	Schechner et. al. [47]
Rain	Setting camera parameters	Garg et. al. [50]
	Intensity constrain, photometric constrain, spatio-temporal correlation	Garg et. al. [53]
Fire	Color information	Phillips et. al. [49]
	FFT and boundary analysis	Fastcom Tech.SA [52]
	FFT and shape analysis	C.B. Liu [51]
Smoke	non-self similarity and motion irregularities	Kopilovic et. al. [34]
	Chromatic analysis, growth-rate and disorder measure	Chen et. al. [36]
	Mean Crossing Rate	Xiong et. al. [38]
	Wavelet transform, energy analysis and shape analysis	Toreyin et. al. [41]

Airlight (scattering caused by fog or haze), with the constraint that both observer and the object observed must lie at the ground level. Schechner et al. [47] propose a method for haze removal from an image. His method is based on the fact that usually the natural environmental light scattered by the atmosphere is partially polarized. This approach does not rely of previous knowledge about the scattering model or knowledge about the illumination direction, but require only two independent images. Oakley et al. [45], instead use a statistical model for detecting Airlight . Then a linear dependency between the real pixel value and the distorted pixel value is issued for both monochromatic and color images. This approach does not distort images taken in clear weather condition and tested with video sequences, it presents color stability for subsequent frame. Regarding the visual effect due to the rain, Garg et al. [48] studied the appearance of a single raindrop, developing a photometric and geometric model for the raindrop refraction and reflection. He shows that a raindrop behaves like a wide-angle lens and although it is a transparent entity, its brightness is independent from the background brightness, because the drop has a large field of view and the background subtend only a small angle. Subsequently in [50] is proposed how to remove the rain from a video without post-processing and without altering the scene perception, in fact he derives the relationship between the properties of rain, camera exposure time, depth of field and scene brightness. He shows that the rain visibility increase with the square of the raindrop size, and decrease linearly with the brightness of the background. Finally in [53] authors developed an algorithm capable to detect rain in a video sequence. The detection of the rain is composed by several step: in the first step all the pixel that present a peak in intensity over a set of three subsequent frame are selected; in the second step the false positive are discarded using the photometric constrain; subsequently the spatio-temporal correlation and the direction of the rain fall are computed.

Although these method are all of some interest in many security applications the problem of fire and smoke detection is definitely crucial for improving people safety and represents a hard challenge to be solved using cameras sensor.

For this reason we focus our attention on the problem of smoke detection and how to fast and reliably detect dangerous smoke presence in the scene. The problem of smoke detection has been discussed in the past in some works where local features of pixels in the images or measures on the shape temporal variations are exploited. In an early work, Kopilovic et al. [34] took advantage of irregularities in motion due to non-rigidity of smoke. They computed optical flow field using two adjacent images, and then used the entropy of the motion directions distribution as key feature to differentiate smoke motion from non-smoke motion. Similarly, motion was exploited in [35] where local motions from cluster analysis of points in a multidimensional temporal embedding space are extracted. The goal was to track local dynamic envelopes of pixels, and then use features of the velocity distribution histogram to discriminate between smoke and various natural phenomena such as clouds and wind-tossed trees that may cause such envelopes. In this work, the presence of other moving objects, typical of video surveillance scenes, has not taken into account.

Recently, Chen, Yin et al. [36] present a smoke detection approach working on pixel-level classification after motion segmentation based on frame difference. Pixels can be initially classified as a smoke-pixel with a very simple chromaticity-based static decision rule; it is based on two thresholds in the color space assuming that smoke usually displays grayish colors. A Further dynamic decision rule is dependent on the spreading attributes of smoke: the ratio between the sums of circumferences of smoke regions segmented and the number of smoke-pixel extracted can give a measure of disorder in the segmented objects. Similarly other works evaluate the contours of the object that are candidate to be classified as smoke. In [38], smoke detection is based on four steps: background subtraction, flickering extraction, contour initialization, and contour classification using both heuristic and empirical knowledge about smoke. Background subtraction uses the Stauffer and Grimson algorithm [37]. Then a measure of flickering is provided. They state that flickering frequency of turbulent flame has shown experimentally to be around 10Hz and it could be as low as 2 or 3 Hz for slowly-moving smoke. The temporal periodicity can be calculated using Fast Fourier Transform (FFT), Wavelet Transform or Mean Crossing Rate (MCR). They adopt the Mean Crossing Rate (MCR). Finally as in [36], a measure of the shape complexity given by the ratio between edge length and area is provided to achieve classification. Also in this work, only qualitative measure are provided.

An interesting and robust approach has been defined by Toreyin et. al. [39] and further improved in [40] and [41]. They use the Collins background subtraction method to extract moving objects [42]. Then as in previous work a flickering analysis and a measure of turbulence is provided by evaluating the edge and texture variation using the Wavelet Transform. In each block of the sub-image resulting, after the wavelet decomposition, the variation of energy is computed. The energy is given by the sum of the high-frequency components in the wavelet domain. Finally two thresholds are given to measure an acceptable energy variation. The dynamism of the variation is modeled with a simple three state Random Markov Model (RMM), trained with smoke and non-smoke pixels. Finally, an analysis of smoke shape complexity is provided as in [36] and [38], based on the distance between the contour points and the center of mass of the shape. This approach is quite robust in the given examples, but a precise evaluation of the different features contributions is not provided. Many systems exist for detecting natural event in the scene and could be effectively used for detecting tempestively possible threat. In the following we will present a case study that deeply covers the problem of detecting whether a moving object in the scene is smoke or not using motion clues and texture.

2.2 State of the Art on People Path Analysis

Recent advances in computational resources and algorithms have made distributed video surveillance very appealing to both the academia and the industry. As a consequence, there exist in the literature many works addressing some of or all the steps related to distributed video surveillance: from motion detection and moving object segmentation [2, 3, 4], to object tracking with occlusion handling [5, 6, 7], to fusion among multiple cameras, with either overlapped [8, 3, 9] or disjoint views [10], to higher-level reasoning modules to analyze the behaviors and the interactions, or to detect and classify events [11, 12].

Despite this considerable amount of papers and techniques, the focus of these proposals has been mainly on proposing innovative solutions capable to handle the most complex situation possible, with few (or none) attention to the real-time constraints or to the computational requirements in general. However, even if complexity is a requirement in order to propose significant advances with respect to the state of the art, real-time alarming is often a must in this type of systems, since off-line processing does not guarantee a timely response to relevant events. Obviously, a careful tuning of the trade-off between efficiency and accuracy must be achieved in order to preserve as much as possible the flexibility and the applicability of the system to different contexts.

Several real-time video surveillance systems have been proposed in the past, but they basically proposed quite complex techniques for fusing single (low-level) algorithms from multiple cameras, such as moving object detection and tracking [13, 14]. Instead, the implementation of higher level tasks (e.g., trajectory/path classification) in real time has not been deeply explored. This section will mainly focus on related works in the field of (real-time) trajectory analysis, which is the main contribution of this chapter.

Trajectory analysis has been studied in depth over the last years, especially for its application in people surveillance. Morris and Trivedi in [15] proposed a recent survey on state-of-art techniques for modeling, comparing and classifying trajectories in video surveillance. The simplest way to define a similarity measure between trajectories is the adoption of Euclidean distance between spatial coordinates as proposed in [16], while the Hausdorff distance was adopted by Junejo *et al.* [17]. However, both these measures only perform point-to-point comparison on trajectories of the same length and, additionally, the Euclidean distance needs that the trajectories have the same length, while Hausdorff distance does not need same length, but cannot distinguish the opposite directions. Chen et al. in [18] presented a method to compare the trajectories after projecting them in a null-space to obtain a representation insensitive to projective transformation of the trajectories themselves.

The distance between correspondent points only can be affected by segmentation errors, noise, temporal shifts, or in general misalignments between trajectories. Thus, many inexact matching techniques have been extensively used both for trajectory analysis [19, 20] and for several different applications ranging from speech [21] to handwriting recognition [22]. Alignment technique like *Longest Common SubSequence* (LCSS) and *Dynamic Time Warping* (DTW) have been efficiently applied to compare trajectory shapes in sign language recognition and surveillance applications [21, 20].

The similarity measures with or without alignment are typically defined in a statistical framework. Mecocci and Panozzo in [23] suitably modified the iterative *Altruistic Vector Quantization* algorithm to robustly cluster trajectories by pure spatial observations obtaining representative prototypes. The anomaly detection is based on fitting a spatial Gaussian on each prototype and statistically checking for fitness of new trajectory samples. In [17], Junejo *et al.* applied graph cuts to cluster trajectories with the Hausdorff distance. In [9] a system for learning statistical motion patterns using a two-stage fuzzy k-means is presented. Porikli [24] proposed the use of a HMM-based similarity measure where each trajectory is modeled with a HMM and compared using the cross likelihood. The results are promising but, in general, a large amount of data is needed to avoid overfitting in the HMM training phase.

Our approach mutuates from the two most common approaches for trajectory comparison and clustering. On the one hand, it adopts an alignment-based distance measure to compare sequences of different lengths and, on the other hand, it employs a statistical measure to perform point-to-point comparison to deal with inaccuracies of the automatic video surveillance system that extracts the people trajectories. Additionally, a specific on-line distance measure has been developed to obtain a robust and efficient comparison among paths each time a new trajectory point is extracted by the system.

3 Environment Threat Detection: Video-Based Smoke Detection

Even though humans can quite easily identifying smoke with a joint use of vision and nose senses, finding smoke in digital video is a challenging problem since smoke shares several features with other objects but also results to be very transparent. This section will present a complete system for detecting smoke in difficult situations by only processing video streams.

Fig. 1. Example of discrete wavelet transform. The leftmost image is the original image. The right image is the transformed one. The components are: top left compressed image C_t, top right horizontal coefficient image H_t, bottom left vertical coefficient image V_t and bottom right diagonal coefficient image D_t.

3.1 Smoke Detection for Foreground Object Classification

The proposed model evaluates the joint contribution coming from the graylevel image energy and color intensity attenuation to classify an object as possible smoke. We assume that when smoke grows and propagates in the scene its image energy is attenuated by the blurring effect of smoke diffusion.

We firstly detect possible candidate objects by means of a motion segmentation algorithm. When a new foreground object is detected we analyze its energy using the Wavelet Transform coefficients and evaluate its temporal evolution. The color properties of the object are analyzed accordingly to a smoke reference color model to detect if color changes in the scene are due to a natural variation or not. The input image is then divided in blocks of fixed sized and each block is evaluated separately. Finally a Bayesian approach detect whether a foreground object is smoke.

Energy analysis using the direct wavelet transform

An efficient way to evaluate the energy variation of an intensity image is the discrete wavelet transform DWT [43].

The DWT is obtained convolving the image signal with several banks of filters obtaining a multiresolution decomposition of the image. Given the input image I_t the decomposition produces four subimages, namely the compressed version of the original image C_t, the horizontal coefficient image H_t, the vertical coefficient image V_t and the diagonal coefficient image D_t. An example decomposition is computed with the algorithm proposed in [43] is shown in Fig. 1

The energy is evaluated blockwise dividing the image in regular blocks of fixed size and summing up the squared contribution coming from each coefficient image:

$$E(b_k, I_t) = \sum_{i,j \in b_k} V_t^2(i,j) + H_t^2(i,j) + D_t^2(i,j) \tag{1}$$

where b_k is the kth block in the input image I^t.

The energy value of a specific block varies significantly over time in presence or absence of smoke, Fig. 2.

When the smoke covers part of the scene the edges are smoothed and the energy consequently lowered. This energy drop can be further emphasized computing the ratio $r(B_k)$ between the image energy of the current input frame and the one of the background model. The energy ratio has the advantage of normalizing the energy values and allowing a fair comparison between different scenes where the block energy itself can vary significantly. The ratio of the block b_k is given by:

$$r(b_k, I_t, Bg_t) = \frac{E(b_k, Bg_t)}{E(b_k, I_t)} \tag{2}$$

where Bg_t is the background model up to time t and I_t is the input frame.

The analysis of the energy ratio is performed in two different context to account for both global and local energy drops. Firstly the image energy variation is computed frame by frame to bias the detection using global information. Several clips containing a smoke events have been analyzed and the global energy ratio of the scene computed by sum the block energy. The *Parzen window* technique is adopted to build a non parametric distribution from global energy ratio values computed on several clips. The parzen window method is a kernel density estimator that computes a non parametric distribution from a set of iid samples $X = \{x_i \mid i = 1 \ldots N\}$ of a random variable x. Adopting a specific kernel distribution the approximated pdf is computed summing the kernel for all the sampled values:

$$\widehat{f} = \frac{1}{N\,h} \sum_{i=1}^{N} K(x - x_i) \tag{3}$$

using a standard Gaussian kernel function $K = \frac{1}{2\pi} e^{-\frac{1}{2}x^2}$.

Secondly each block is then locally evaluated to capture the temporal evolution of the energy ratio. When an energy drop is observed for a significant period

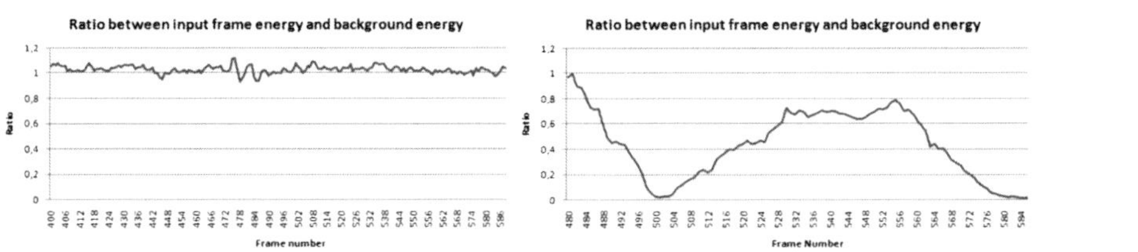

Fig. 2. Left figure: the energy ratio trend of a non smoke block. Right figure: the energy ratio trend of a smoke block. In presence of smoke the energy ratio is subjected to gradual drops in its value.

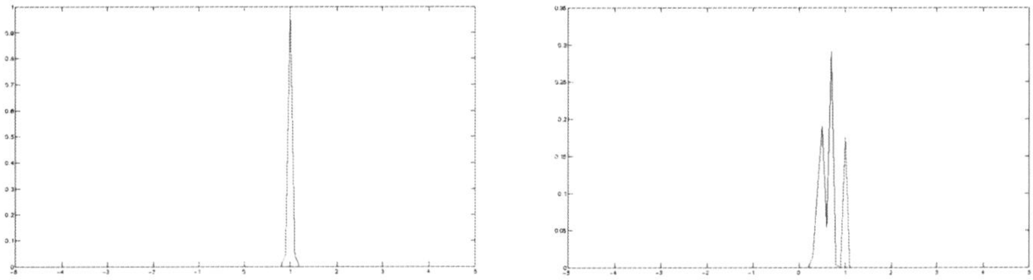

Fig. 3. Gaussian mixtures obtained observing energy ratio values at a single block. The left plot shows the mixture when there is no smoke in the block. The right plot shows how the mixture changes when smoke is in the scene. It is worth noting that when a block is covered by smoke the mixture components mean values move gradually towards 0.

of time an edge smoothing process occurs. The edge smoothing process can be affected by noise due to light variation in the scene. A Mixture of Gaussian model is adopted to improve the analysis robustness.

The MoG has the great advantage to correctly catch variations for multimodal distributions. To compute the probability for each frame the on-line expectation maximization algorithm proposed in [44] is used. In detail, for all blocks b_k of the image I_t at time t the value $r(b_k, I_t, Bg_t)$ is computed and the MoG of block b_k updated using a selective update method.

This process has a main advantage. The mixture component reweighting process is able to catch slow and gradual variations of energy ratio. Values that do not occur frequently are filter out and assigned to the least probable Gaussian of the mixture. This property is helpful for evaluating the gradient intensity lowering process of smoking regions that has the peculiarity of being slow and continuous in time, Fig. 3.

To capture the time variation of the energy ratio the Gaussian Mixture Model was preferred to a Hidden Markov model (HMM). Although HMMs are widely adopted to classify and model temporal stochastic processes, the data values sequence is crucial to obtain a good classification. Instead, as previously stated, the block energy ratio is subject to strong fluctuations of energy values due to noise and natural scene lighting. This reason makes the lowering sequence unpredictably variable in different setups; thus the specific energy drop trajectory can produce misleading results. On the contrary is interesting to analyze the global trend.

Color analysis to detect blended smoke regions

When a smoke event occurs, scene regions covered by smoke change their color properties. The smoke can either be completely opaque or partially transparent. In the former case the covered region changes completely its color while in the latter case the color of the covered region appears to be blended with the smoke color.

This simple observation remains valid in all the observed cases and intuitively suggests a hint to characterize the color of a smoke region. The proposed model simply adopts an evaluation based on a blending function mutuated from computer graphics. A reference color model is chosen in the RGB color space to represent the color of the smoke in the scene. The model is selected by analyzing the different color tones produced combusting different materials. For explanatory purposes is possible to concentrate the analysis to the case of a light gray color model as the smoke in the leftmost image of Fig. 1. Each pixel $I_t(i,j)$ of the input frame at time t is then checked against the smoke model and the background model Bg_t to evaluate the reference color presence computing the blending parameter bl using equation 4. The evaluation takes into account the case where the scene color and the smoke color are mixed together.

$$bl(i,j,I_t,Bg_t,S) = \frac{I_t(i,j) - Bg_t(i,j)}{S - Bg_t(i,j)} \tag{4}$$

where Bg_t is the current background model at time t and S is the smoke reference color model.

To filter out the errors and possible measurements inaccuracy the blending value is computed for each image block as the average of bl values in the block:

$$\beta_{b_k}(I_t,Bg_t,S) = \frac{1}{N^2} \sum_{i,j \in\ ,b_k} \frac{I_t(i,j) - Bg_t(i,j)}{S - Bg_t(i,j)} \tag{5}$$

where block size is $N \times N$

In conclusion the β measure quantifies how much each block globally shares chromatic properties with the reference color model.

A Bayesian approach for classification

Previously the blockwise energy ratio measure r and the color blending measure β have been presented as possible discriminant features to identify a smoke region in the scene. A Bayesian formulation has been chosen to identify whether a block b_k is likely to belong to a smoke region. For each block the posterior probability of smoke presence, the event $f = 1$, considering the block b_k is defined:

$$P(f=1|b_k) \propto P(b_k|f=1)P(f=1) \tag{6}$$

The likelihood value is obtained by combining both the contributions coming from energy ratio and color information. These terms are considered probabilistically independent to simplify the treatment.

$$P(b_k|f=1) = P(r_{b_k}, \beta_{b_k}|f=1) = P_r(b_k|f=1) \cdot P_\beta(b_k|f=1) \tag{7}$$

The likelihood contribution due to energy ratio decay is obtained by summing the weighted Gaussians of the MOG having mean value below a considered threshold th_1 computed empirically observing the mean energy ratio value in smoke regions.

$$P_r(b_k|f=1) = \sum_{i=1}^{K} w_i N(r(b_k, I_t, Bg_t)|\mu_i \sigma_i) \tag{8}$$

when the ith Gaussian mean value $\mu_i < th_1$.

The color contribution to the likelihood value is directly computed as the block color blending measure β_{b_k} according to equation 5.

$$P_\beta(b_k|f=1) = B_k(I_t, Bg_t, S) \tag{9}$$

The classification is biased making use of prior knowledge acquired observing several clips containing smoke. The prior probability of a smoke event in the current frame is directly related to the mean energy ratio value of the scene and computed using the non parametric distribution obtained by equation 3.

$$P(f=1) = \widehat{f}(\frac{1}{M} \sum_{\forall b_k \in I_t} r(b_k, I_t, Bg_t)) \tag{10}$$

where I_t is composed by M blocks.

The posterior probability value is thresholded to identify a candidate smoke block. The test for smoke presence is performed after foreground object segmentation. For any segmented object in the scene the number of candidate blocks intersecting the object's blob is computed. Finally an object is classified as smoke when the 70% of its area overlays candidated smoke blocks.

3.2 Experimental Validation of the Model

The proposed smoke detection system can be used in conjunction with a whichever video surveillance system providing moving object segmentation using a background model. The background model should be updated regularly but smoke regions should not be included in the background. This can be achieved choosing a slow background update rate and avoiding updating the background model areas where a smoke object is detected. The tests were performed using both the Stauffer and Grimson background model with selective update [37] and the SAKBOT median background model with knowledge based update proposed in [32]. Although the results did not vary significantly changing the background model and object detection technique, the second method has been preferred since discriminates the presence of possible shadows objects too.

In all the tests carried out the learning rate α, [37], of the MoGs used to model the energy ratio decay was set to 0.1. Although changing this parameter does not have major effects on the system performance, a fast learning rate is preferable to detect energy ratio variations rapidly. The system was tested on 50 clips of varying length in both indoor and outdoor setups where moving objects such as people or vehicles were present in the scene during the smoke event. Each clip contained a smoke event. Part of the dataset is publicly available at website *http://imagelab.ing.unimore.it/visor*. Each likelihood term was evaluated separately to measure the impact on the system performance.

Table 2. System results on 15 reference clips. The detection rate was evaluated using energy and color component respectively and their joint contribution. Time to detect(TtD) and False positives rate(FP) are reported for each considered contribution.

			Temporal Analysis		Color Analysis		Global analysis	
Clip	**Frame No**	**Type**	**TtD**	**FP**	**TtD**	**FP**	**TtD**	**FP**
Clip1	165	Outdoor	22	-	1	-	1	-
Clip2	210	Indoor	18	-	1	-	1	-
Clip3	2200	Outdoor	28	-	34	-	20	-
Clip4	3005	Indoor	212	-	273	-	285	-
Clip5	1835	Indoor	87	-	100	3	52	-
Clip6	2345	Outdoor	129	-	161	-	116	-
Clip7	2024	Indoor	57	3	99	-	35	-
Clip8	2151	Outdoor	88	2	88	-	42	-
Clip9	1880	Outdoor	59	-	56	-	45	-
Clip10	2953	Outdoor	457	-	498	-	300	-
Clip11	1485	Indoor	62	-	x	5	62	-
Clip12	499	Outdoor	43	-	8	-	16	-
Clip13	195	Indoor	53	-	23	-	27	-
Clip14	1226	Outdoor	77	-	370	-	69	-
Clip15	109	Outdoor	29	-	x	1	3	-

The table Tab. 2 summarizes the results obtained on 15 reference clips. The first column of the table reports the video type and its frame-length. The average clips framerate is 25fps. The remaining columns report the results obtained using each likelihood term separately and finally the results of the whole system. The detection time after the smoke event occurs is reported for all the test clips. The table clearly shows that the likelihood term due to temporal analysis (eq.8) is effective in most of the observed cases. The main problem is the long detection time. This is caused by the time based statistics used to capture the energy ratio decay. Although the likelihood contribution due to color blending has the advantage of speed up the detection process it tends to detect much false positives if used alone. See seventh column of Tab. 2. Observing the last two columns of Tab. 2 we can state that the complete approach is fast and reliable enough even in situations where each likelihood contribution fails. The overall system results on the 50 clips used for testing purposes report a detection rate of 77% 3 seconds later the smoke event occurs, 98.5% 6 seconds later and finally 100% 10 seconds later with an average true positive rate of 4%. Fig. 4 shows some snapshots of the system working on different conditions.

In conclusion the proposed case study underlines another important application of pattern recognition in video surveillance contexts for a deep understanding of the monitored scene and event detection.

Fig. 4. Snapshots of the proposed system working on several clips in different conditions. The blue area in the images is detected as smoke.

4 People Threat Detection: Abnormal Path Detection

Recent advances in computational resources and algorithms have made distributed video surveillance very appealing to both the academia and the industry. As a consequence, there exist in the literature many works addressing some of or all the steps related to distributed video surveillance: from motion detection and moving object segmentation [2, 3, 4], to object tracking with occlusion handling [5, 6, 7], to fusion among multiple cameras, with either overlapped [8, 3, 9] or disjoint views [10], to higher-level reasoning modules to analyze the behaviors and the interactions, or to detect and classify events [11, 12].

This section will describe a complete and full-working system for real-time detection of abnormal paths in real multi-camera scenarios. In fact, in most of the surveillance scenarios "abnormal" is often synonymous of "dangerous". The first step consists in the extraction of the points composing the trajectory/path and it requires to segment moving objects in all the cameras and to track them both in each single camera and across adjacent cameras.

Moving object detection and tracking from a single static camera is a well-known and almost-solved problem. Our system makes use of the approach proposed by us in [3]. The approach is based on background suppression using a suitable modification of the median background model that improves both the initialization and the update phases using a knowledge based reasoning scheme. Proper techniques for background bootstrapping, ghost suppression, and object validation have also been introduced to improve the accuracy of the segmentation in cluttered outdoor setup. The objects detected as moving are then tracked in each single view by means of an appearance-based algorithm proposed in [5]. The algorithm uses a classical predict-update approach, but it takes into account not only the status vector containing position and speed, but also the memory appearance model and the probabilistic mask of the shape.

Once each camera has processed the video stream and obtained the object tracks, there is the need to render the track labels/ids consistent among the different cameras: this step is crucial to keep track of the object when it moves across the fields of view of the different cameras and thus to obtain longer and more stable trajectories. This problem is also known as *consistent labeling* and has been fully studied in the literature [3]. We borrowed the approach, proposed in [3], valid for cameras with partially-overlapped fields of view (FoVs), which adopts a geometric approach that exploits cameras' FoV relationships and constraints to impose identities consistency. In detail, when cameras partially overlaps the shared portion of the scene is analyzed and people identities are matched geometrically, by exploiting ground-plane homographies and pairwise epipolar geometry.

Once that the people trajectories (on the ground plane) are obtained, we can develop a method for comparing trajectories analyzing different characteristics: trajectories shape and trajectories positions in a given scene. The shape analysis is important when unfrequent or particular behaviors must be extracted without the knowledge of where and when the event of interest occurs. Conversely, positional analysis is useful when a specified portion of the scene should be analyzed and scene properties, such as entry or exit zones, can be deduced directly from people activities.

A typical reference application is shown in Fig. 5. Here it is depicted how the system could be useful for instance in forensic application. First, a query is performed on the trajectory shape, Fig. 5.a; second, several exemplars having

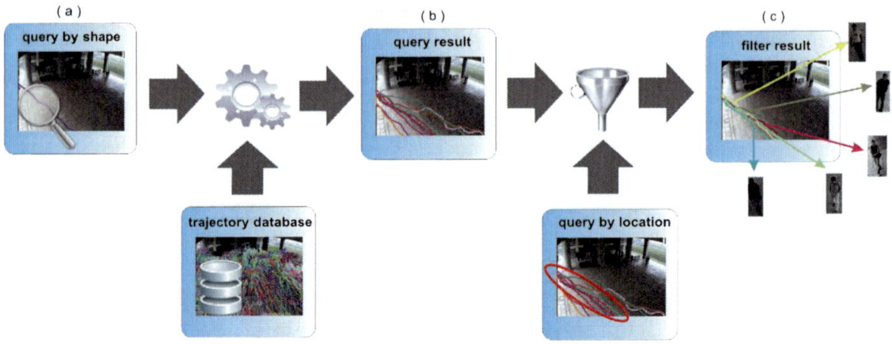

Fig. 5. Example of a possible application scenario of the trajectory analysis framework.

the desired shape are shown to the user. It is then possible to choose a specific example, according to its position in the scene, and the system will retrieve all the trajectories similar to the desired one (Fig. 5.c). Finally, it is possible to retrieve people snapshots and trajectories information (such as creation time) that could be of interest during the investigation process.

4.1 Trajectory Model

As stated in the previous Section, people trajectories can be modelled by means of either the sequence of spatial locations or the sequence of directions.

Spatial Model for Positional Analysis

The people trajectory projected on the ground plane is a very compact representation based on a sequence of 2D data ($\{(x_1, y_1), \cdots, (x_n, y_n)\}$ coordinates), often associated with the motion status, e.g. the punctual velocity or acceleration.

When large data are acquired in a real system they should be properly modeled to account for tracking errors, noise in the support point extraction and inaccuracies due to the multi-camera data fusion module. Positional trajectories must then be correctly extracted by the tracking system and analyzed in order to discriminate or aggregate different kinds of people behaviors.

When observing a video surveillance scenario some paths are considerably more common than others, and this can be very meaningful in suurveillance applications. Different path frequencies are mainly due to two factors. First, the structure of the environment may condition significantly the way people move. Second, according to the scenario, people tend to reproduce frequent behaviors.

Given the kth rectified trajectory projected on the ground plane $T_k = \{\mathbf{t_{1,k}} \dots \mathbf{t_{n_k,k}}\}$, where $\mathbf{t_{i,k}} = (x_{i,k}, y_{i,k})$ with n_k the number of points of trajectory T_k, a bi-variate Gaussian centered on each data point $\mathbf{t_{i,k}}$ (i.e., having

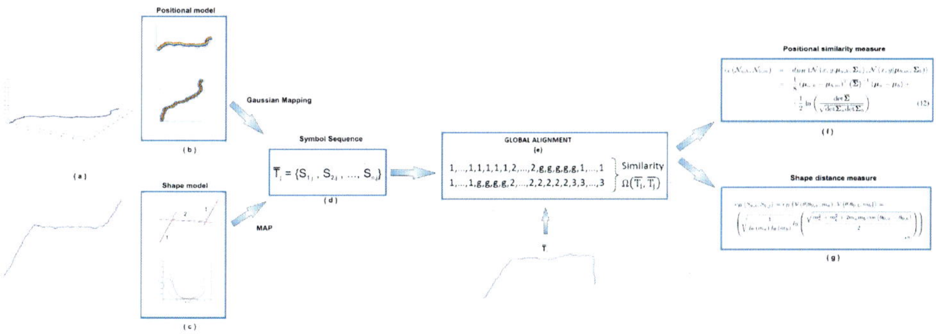

Fig. 6. Example of the trajectory model.

the mean equal to the point coordinates $\boldsymbol{\mu_{i,k}} = (x_{i,k}, y_{i,k}))$ and with fixed covariance matrix $\boldsymbol{\Sigma}$ can be defined as:

$$\mathcal{N}_{i,k} = \mathcal{N}(x, y \mid \boldsymbol{\mu_{i,k}}, \boldsymbol{\Sigma}) \tag{11}$$

An example of the fitting of Gaussians onto the trajectory points is shown in Fig. 6, where (a) shows an exemplar trajectory, (b) the 3D plot of the superimposed Gaussians and the x-y projection.

The main motivation for this modeling choice relies in the fact that when comparing two points belonging to different trajectories small spatial shifts may occur and trajectories never exactly overlap point-to-point. Using a sequence of Gaussians, one for each point, allows to build an envelope around the trajectory itself, obtaining a slight invariance against spatial shifts.

After assigning a Gaussian to each trajectory point, the trajectory can be modeled as a sequence of symbols corresponding to Gaussian distributions $\overline{T}_j = \{S_{1,j}, S_{2,j}, ..., S_{n_j,j}\}$, where each symbol $S_{i,j}$ is modeled as in equation 11.

Angular Model for Shape Analysis

Using a constant frame rate, the sequence of (x, y) coordinates can be easily converted in directions/angles, in order to model the single trajectory T_j as a sequence of n_j directions θ, defined in $[0, 2\pi)$:

$$T_j = \{\theta_{1,j}, \theta_{2,j}, \ldots, \theta_{n_j,j}\} \tag{12}$$

In order to analyze its shape, *circular* or *directional statistics* [25] is a useful framework for the analysis. We propose to adopt the von Mises distribution, that is a special case of the von Mises-Fisher distribution [26, 27]. The von Mises distribution is also known as the *circular normal* or the *circular Gaussian*, and it is particularly useful for statistical inference of angular data. When the variable is univariate, the probability density function (pdf) results to be:

$$\mathcal{V}(\theta|\theta_0, m) = \frac{1}{2\pi I_0(m)} e^{m \cos(\theta - \theta_0)} \tag{13}$$

where I_0 is the modified zero-order Bessel function of the first kind, defined as:

$$I_0(m) = \frac{1}{2\pi} \int_0^{2\pi} e^{m \cos\theta} d\theta \tag{14}$$

and represents the normalization factor. The distribution is periodic so that $p(\theta + M2\pi) = p(\theta)$ for all θ and any integer M.

Von Mises distribution is thus an ideal pdf to describe a trajectory T_j by means of its angles. However, in the general case a trajectory is not composed only of a single main direction; having several main directions, it should be represented by a multi-modal pdf, and thus we propose the use of a mixture of von Mises (MovM) distributions:

$$p(\theta) = \sum_{k=1}^{K} \pi_k \mathcal{V}(\theta|\theta_{0,k}, m_k) \tag{15}$$

As it is well known, EM algorithm is a very powerful tool for finding maximum likelihood estimates of the mixture parameters, since the mixture model depends on unobserved latent variables (defining the "responsibilities" of a given sample with respect to a given component of the mixture). The EM algorithm allows the computation of the parameters for the K components of the MovM. A full derivation of this process can be found in [28].

Each direction $\theta_{i,j}$ is encoded with a symbol $S_{i,j}$ with a MAP approach, that, assuming uniform priors, can be written as:

$$S_{i,j} = \arg\max_{r=1,\ldots,K} p(\theta_{0,r}, m_r|\theta_{i,j}) = \arg\max_{r=1,\ldots,K} p(\theta_{i,j}|\theta_{0,r}, m_r) \tag{16}$$

where $\theta_{0,r}$ and m_r are the parameters of the rth components of the MovM. With this MAP approach each trajectory T_j in the training set is encoded with a sequence of symbols $\overline{T}_j = \{S_{1,j}, S_{2,j}, \ldots, S_{n_j,j}\}$.

4.2 Sequence Similarity Measure

In order to cluster or classify similar trajectories, a similarity measure $\Omega(\overline{T}_i, \overline{T}_j)$ is needed. Due to acquisition noise, uncertainty and spatial/temporal shifts, exact matching between trajectories is unsuitable for computing similarity. Thus, two sequences of symbols can be compared by using an inexact matching technique. The main motivation resides in the fact that trajectories are never equal both in number and position of points. Small changes can occur between two similar sequences: for example, there may be some time stretches that result in sequences having different lengths; additionally, sequences may be piecewise-similar, sharing some common parts, but they can be different in other parts. In choosing the similarity measure it is desirable to gain control on the amount of common points that two sequences must share in order to be considered "similar".

For these motivations, the best way to compare two sequences is to identify the best alignment of the sequence data, based on a given point-to-point distance metrics. Point-to-point comparison can be made either directly on the data or by selecting a data representation which assigns a symbol (with a given "meaning") to each data and performing a symbol-to-symbol comparison. However, the trivial model that simply performs a point-wise comparison in the rectified Euclidean plane will result extremely imprecise. We decided to adopt a model that employs statistics to model data points sequences, being consequently robust against measurement errors and data uncertainties, but imposing some constraint and limitation to achieve real-time performance. As stated in the introduction, this permits to achieve a good trade-off between efficiency and accuracy.

Once a sequence of data/symbols is achieved, we can borrow from bioinformatics the method for comparing DNA sequences in order to find the best inexact matching between them, also accounting for gaps. Then, we propose to adopt the *global alignment*, specifically the well-known Needleman-Wunsch algorithm [29] for comparing sequences of probability distributions. A global alignment (over the entire sequence) is preferable over a local one, because preserves both global and local shape characteristics. Global alignment of two sequences $\overline{T}_i$ and $\overline{T}_j$ is obtained by first inserting spaces, either into or at the ends of the sequences so that the length of the sequences will be the same; by doing this, every symbol (or space) in one of the sequences is matched to a unique symbol (or space) in the other.

The algorithm is based on the concept of "modification" to the sequence (analogous to the mutation in a DNA sequence). The modifications to a sequence can be due to *indel* operations (insertion or deletion of a symbol) or to *substitutions*. By assigning different weights/costs to these operations it is possible to measure the degree of similarity of the two sequences. Unfortunately, this algorithm can be very onerous in terms of computational complexity if the sequences are long. For this reason, *dynamic programming* is used to reduce computational time to $O(n_i \cdot n_j)$, where n_i and n_j are the lengths of the two sequences. Dynamic programming overcomes the problem of the recursive solution to global alignment by not comparing the same subsequences for more than one time, and by exploiting tabular representation to efficiently compute the final similarity score. Each element $V(a, b)$ of the table contains the alignment score of the symbol $S_{a,i}$ of sequence $\overline{T}_i$ with the symbol $S_{b,j}$ of sequence $\overline{T}_j$. This inexact matching is very useful for symbolic string recognition and theoretically could be used on whichever data have been organized in a sequence. However, we do not adopt it directly on the data since they can be affected by measurement noise, but on the pdf corresponding to trajectory data. Thus, the one-to-one score between symbols can be measured statistically as a function of the distance between the corresponding distributions. If the two distributions result sufficiently similar, the score should be high and positive, while if they differ significantly, the score (penalty) should be negative.

The alignment is simply achieved by arranging the two sequences in a table, the first sequence row-wise and the second column-wise, starting from the base conditions:

$$\begin{aligned} V(a,0) &= \Omega\left(S_{a,i},-\right) \\ V(0,b) &= \Omega\left(-,S_{b,j}\right) \end{aligned} \tag{17}$$

where Ω represents a suitable similarity measures and $-$ indicates a zero-element or gap.

This is due to the fact that the only way to align the first k elements of the sequence $\overline{T}_i$ with zero elements of the sequence $\overline{T}_j$ (or viceversa) is to align each of the elements with a space in the sequence $\overline{T}_i$.

Starting from these base conditions, the alignment is performed exploiting the recurrent equation of global alignment that computes the best alignment score for each subsequence of symbols:

$$V(a,b) = \max \begin{cases} V(a-1,b-1) + \Omega\left(S_{a,i},S_{b,j}\right) \\ V(a-1,b) + \Omega\left(S_{a,i},-\right) \\ V(a,b-1) + \Omega\left(-,S_{b,j}\right) \end{cases} \tag{18}$$

with $1 \leq a \leq n$ and $1 \leq b \leq m$ and where $V(a,b)$ is the score of the alignment between the subsequence of $\overline{T}_i$ up to the ath symbol and the subsequence of $\overline{T}_j$ up to the bth symbol.

Assuming that two distributions are sufficiently similar if the coefficient is above 0.5 and that the score for perfect match is +2, whereas the score (penalty) for the perfect mismatch is -1 (that are the typical values used in DNA sequence alignments), we can write the general score of alignment between two symbols/distributions as follows:

$$\Omega\left(S_i,T_j\right) = \begin{cases} 2\cdot(c_B) & \text{if } c_B \geq 0.5 \\ 2\cdot(c_B - 0.5) & \text{if } c_B < 0.5 \\ 0 & \text{if } S_i \text{ or } T_j \text{ are gaps} \end{cases} \tag{19}$$

where c_B represents the cost of aligning two symbols. The following Section will report the proposed way for computing this cost in the two cases of spatial and angular data.

4.3 Statistics Symbol-to-Symbol Distance Metrics

Distance in the case of Spatial Model

In the case of symbol sequences that represent spatial-Gaussian probability distributions, a proper symbol-to-symbol similarity measure must be defined in order to perform the global alignment. Among the possible metrics to compare probability distributions we chose to employ the Bhattacharyya coefficient as in the case of shape model, to measure the distance between the two normal distributions $\mathcal{N}_{a,k}$ and $\mathcal{N}_{b,m}$ corresponding to a^{th} and b^{th} symbols of sequences $\overline{T}_k$ and $\overline{T}_m$, respectively:

$$c_b\left(\mathcal{N}_{a,k},\mathcal{N}_{b,m}\right) = d_{BH}\left(\mathcal{N}\left(x,y|\boldsymbol{\mu}_{a,k},\boldsymbol{\Sigma}_a\right),\mathcal{N}\left(x,y|\boldsymbol{\mu}_{b,m},\boldsymbol{\Sigma}_b\right)\right)$$
$$= \frac{1}{8}\left(\boldsymbol{\mu}_{a,k}-\boldsymbol{\mu}_{b,m}\right)^T\left(\overline{\boldsymbol{\Sigma}}\right)^{-1}\left(\boldsymbol{\mu}_a-\boldsymbol{\mu}_b\right)+ + \frac{1}{2}\ln\left(\frac{\det\overline{\boldsymbol{\Sigma}}}{\sqrt{\det\boldsymbol{\Sigma}_a\det\boldsymbol{\Sigma}_b}}\right) \tag{20}$$

where $2\cdot\overline{\boldsymbol{\Sigma}} = \boldsymbol{\Sigma}_a + \boldsymbol{\Sigma}_b$. Since in our case $\boldsymbol{\Sigma}_a = \boldsymbol{\Sigma}_b = \boldsymbol{\Sigma}$, we can rewrite the distance as:

$$c_b\left(\mathcal{N}_a^k,\mathcal{N}_b^m\right) = \frac{1}{8}\left(\boldsymbol{\mu}_a-\boldsymbol{\mu}_b\right)^T\boldsymbol{\Sigma}^{-1}\left(\boldsymbol{\mu}_a-\boldsymbol{\mu}_b\right) \tag{21}$$

As previously performed for the angular model this coefficient can be injected in equation 19 to obtain the symbol to symbol similarity measure used in the alignment process.

Distance in the case of Angular Model

When the data sequences is modeled using the Mixture of Von Mises Model, Section 4.1, one possible symbol-to-symbol distance between the univariate pdf associated to each symbol, following the scheme of Fig. 6, is the Bhattacharyya coefficient between Von Mises distribution,[30]. We can derive the *Omega* score for the Mixture of Von Mises Model; specifically, we measured the distance between distributions p and q using the Bhattacharyya coefficient:

$$c_B\left(p,q\right) = \int_{-\infty}^{+\infty}\sqrt{p\left(\theta\right)q\left(\theta\right)}d\theta \tag{22}$$

following the derivation in [28] for two univariate Von Mises distribution the analytic form of the coefficient results:

$$c_B\left(S_i,T_j\right) = c_B\left(\mathcal{V}\left(\theta|\theta_{0,i},m_i\right),\mathcal{V}\left(\theta|\theta_{0,j},m_j\right)\right) = \left(\sqrt{\frac{1}{I_0\left(m_u\right)I_0\left(m_b\right)}}I_0\left(\frac{\sqrt{m_i^2+m_j^2+2m_im_j\cos\left(\theta_{0,i}-\theta_{0,j}\right)}}{2}\right)\right) \tag{23}$$

where it holds that $0\leq c_B\left(S_i,T_j\right)\leq 1$.

4.4 Experimental Results and Discussion

Once a proper similarity measure is available, sequences can be compared according to either their position (Section 4.3) or their shape (Section 4.3). In particular, it could be of interest to retrieve all the sequences similar to a given exemplar (*query problem*) or the most or least frequent sequence sharing shape or position characteristics (*clustering problem*). In forensic and video surveillance applications, this could be of undoubtful utility; sequences can be retrieved according to their shape and then filtered according to their position or vice-versa.

Most common paths can also be extracted to synthesize a clear picture of normal and frequent (abnormal and unfrequent) behaviors in a specific scenario. To group together paths sharing some common characteristics we choose to adopt the k-medoids [31] clustering algorithm using the similarity measures introduced in section 4.2. Hereinafter we use interchangeably the trajectory and its symbolic representation in the Ω measure to keep the notation light, but the similarity measure is obviously computed, as previously stated, on the symbolic representation of the trajectory and consequently on the chosen probability density funtction.

The adopted clustering algorithms, K-medoids, is a suitable modification of the well-known k-means algorithm which has the appreciable characteristic to compute, as prototype of the cluster, the element that minimizes the sum of intra-class distances. In other words, let us suppose to have a training set $TS = \{T_1, \cdots, T_N\}$ composed of N trajectories and set $i = 0$ and $k(0) = N$. As initialization, each trajectory is chosen as prototype (medoid) of the corresponding cluster. The k-medoids algorithm iteratively assigns each trajectory T_j to the cluster $C_{\widetilde{m}}$ at the minimum distance d, i.e. given $k(i)$ clusters $C_1, \cdots, C_{k(i)}$ and the corresponding medoids $M_1, \cdots, M_{k(i)}$, $\widetilde{m} = \underset{m=1,\cdots,k(i)}{\arg\min}\ d\left(T_j, T_{M_m}\right)$, where T_{M_m} is the trajectory corresponding to the medoid M_m. Once all the trajectories have been assigned to the correct cluster, the new medoid M_s for each cluster C_s is computed as that one which minimizes the intra-cluster distances, i.e. $T_{M_s} \equiv T_{\widetilde{p}} = \underset{\forall T_p \in C_s}{\arg\min} \sum_{\forall T_r \in C_s} d\left(T_p, T_r\right) = \underset{\forall T_p \in C_s}{\arg\max} \sum_{\forall T_r \in C_s} \Omega\left(T_p, T_r\right)$. However, one of the limitations of k-medoids (as well as k-means) clustering is the choice of k. For this reason, we propose to use an *iterative k-medoids* algorithm. Then, the following steps are performed:

- Step 1: Run k-medoids algorithm with $k(i)$ clusters
- Step 2: If there are two medoids with a similarity greater than a threshold Th, merge them and set $k(i+1) = k(i) - 1$. Increment i and go back to step 1. If all the medoids have a two-by-two similarity lower than Th, stop the algorithm

In other words, the algorithm iteratively merges similar clusters until convergence. In this way, the "optimal" number $\widetilde{k}$ of medoids is obtained.

Performing the clustering on a given corpus of trajectories leads to two main advantages. First, after the clustering, clusters cardinality naturally represents by definition how often a specific path occurs, thus allowing to classify the paths in abnormal (unfrequent) and normal (frequent). Second, when the dataset grows dramatically in number of exemplars, the one-to-many approach that consists of comparing a query trajectory with all the trajectories previously stored, can be extremely onerous in term of computational time. Adversely, the adoption of clustering allows the classes to be represented by their prototype, reducing the number of comparisons in the case of query.

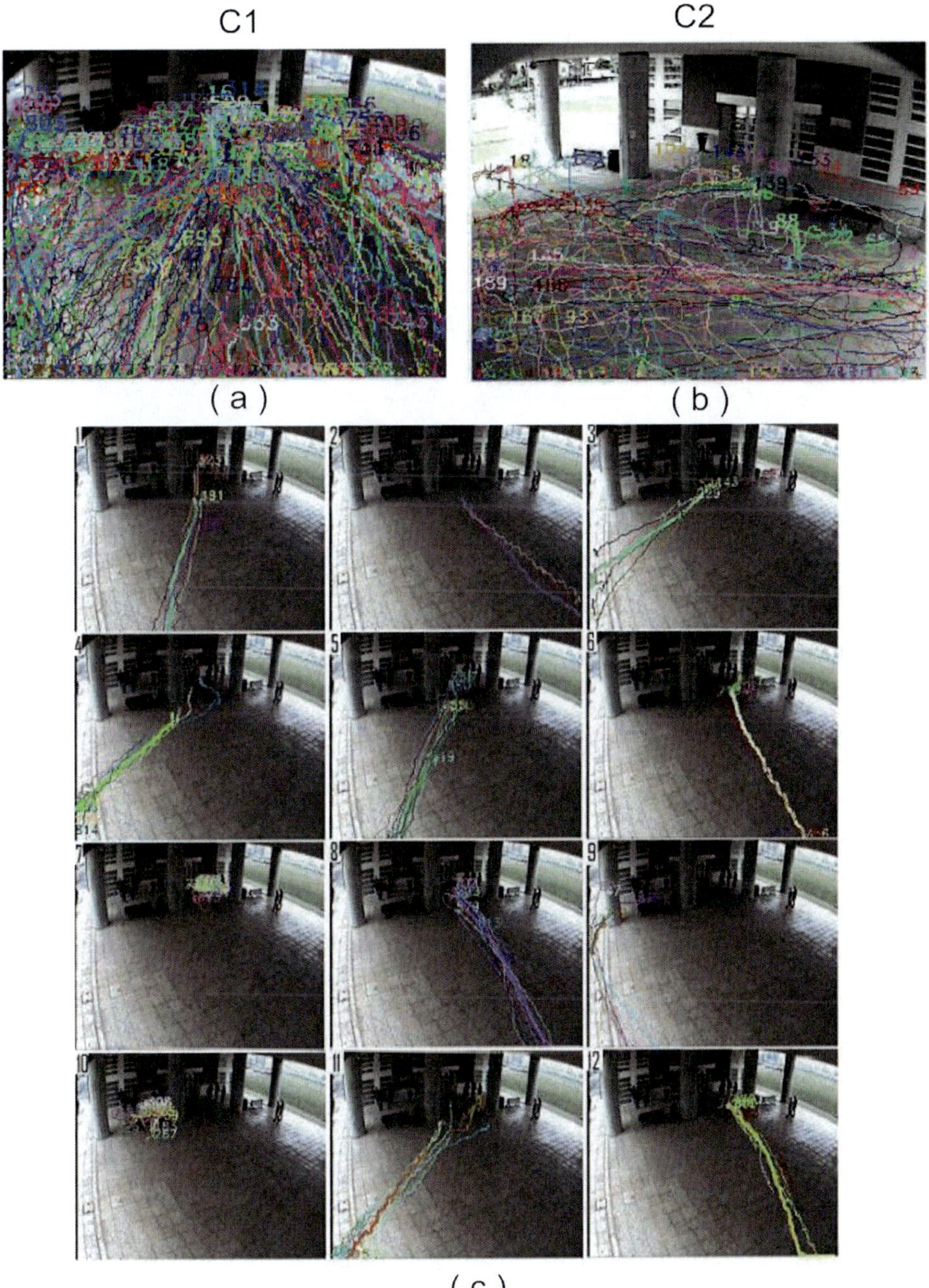

Fig. 7. In (a) and (b) is shown the training set used during the learning stage. (c) shows the obtained most frequent behaviors projected on the $C1$ view.

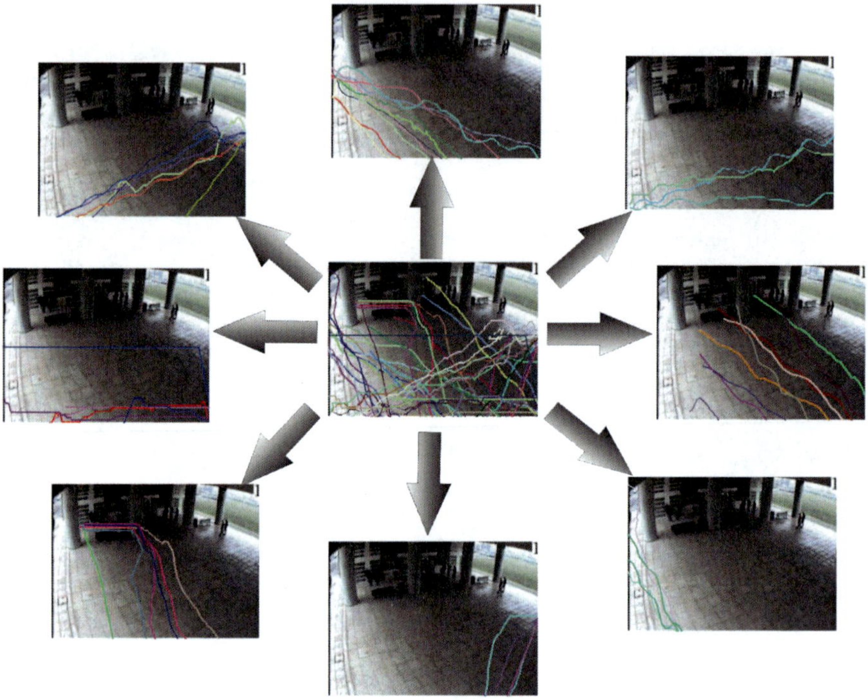

Fig. 8. The center figure shows the training set used for trajectories shape clustering. The remaining figure shows the most frequent behaviors according to their shape.

To keep this approach consistent when new data are presented to the system, the clusters must be updated every time a new sequence is classified. More operatively, we can define the maximum similarity between the new trajectory T_{new} and the set of clusters $\mathbf{C}$ as $\Omega_{max} = \Omega\left(C_{\widetilde{j}}, T_{new}\right)$, where:

$$\widetilde{j} = \arg\max_{j=1,\ldots,\widetilde{k}} \Omega\left(C_j, T_{new}\right) \tag{24}$$

If this value is below a given threshold Th_{sim} a new cluster $C_{\widetilde{k}+1}$ should be created with T_{new}. The cardinality $\mathcal{C}$ of each class (which represents the prior for a classification normal/abnormal) is updated to take into account the increased number of samples assigned to the cluster:

$$\begin{aligned} &C_{\widetilde{k}+1} = T_{new}\ ;\ \mathcal{C}\left(C_{\widetilde{k}+1}\right) = \frac{1}{N+1} \\ &\forall i = 1, ..., \widetilde{k} \Rightarrow \mathcal{C}_{new}\left(C_i\right) = \mathcal{C}_{old}\left(C_i\right)\frac{N}{N+1} \\ &k = k+1\ ;\ N = N+1 \end{aligned}$$

where N is the current number of observed trajectories.

Conversely, if the new trajectory is similar enough to one of the current medoids, the trajectory is assigned to the corresponding cluster C_j:

$$T_{new} \in C_j \ ; \mathcal{C}_{new}\left(C_{\widetilde{k}}\right) = \frac{\mathcal{C}_{old}\left(C_{\widetilde{k}}\right) \cdot N + 1}{N+1}$$

$$\forall i = 1, ..., \widetilde{k}, i \neq j \Rightarrow \mathcal{C}_{new}\left(C_i\right) = \mathcal{C}_{old}\left(C_i\right) \frac{N}{N+1}$$

$$N = N + 1$$

Moreover, if the average similarity of the new trajectory with respect to other medoids is smaller than the average similarity of the current medoid C_j, T_{new} is a better medoid than C_j since it increases the separability with other clusters. Consequently, T_{new} becomes the new medoid of the cluster. Finally, to avoid old and rare trajectories afffecting our model, clusters witth small cardinality and with no new trajectories assigned for a fixed-length time window are dropped.

We tested our system in a two cameras setup at our campus. People are extracted and tracked across camera streams using the multi-camera tracking system described in [32, 33]. Once the trajectories are reliably obtained, we first performed the clustering described above on a dataset of 900 trajectories acquired during an ordinary working day. In this way the most frequent behaviors in the chosen scenario, as shown in Fig. 7, can be extracted according to their position. In this case trajectories sharing similar shape and location are clustered together and it is possible to easily detect the most frequent activity zones of the scene, for example benches where people use to stop. In Fig. 8 trajectories are clustered according to their shape only. In this case it is possible to extract similar trajectories, and most frequent ones as shown in the figure, that share common directions and shape properties independently on where they occur in the scene.

5 Conclusions

This chapter tackles the problem of moving pixel detection from an original perspective: two quite different applications, namely vision-based smoke detection and abnormal path detection, are treated as methods for threat detection (due to the environment and the people behavior, respectively) based on similar concepts. In fact, both these applications start from the detection and analysis of moving pixels and require to effectively distinguish from the two cases. Smoke, in fact, shares some common characteristics such as texture, shape and color information with moving pixels due to objects.

The techniques reported in this chapter demonstrated to be very robust in both cases and can be easily included in advance video surveillance systems to provide both these functionalities.

References

1. Bregman, A.: Auditory Scene Analysis: The Perceptual Organization of Sound. MIT Press, London (1990)
2. Haritaoglu, I., Harwood, D., Davis, L.: W4: real-time surveillance of people and their activities. IEEE Trans. Pattern Anal. Mach. Intell. 22(8), 809–830 (2000)
3. Calderara, S., Prati, A., Cucchiara, R.: Hecol: Homography and epipolar-based consistent labeling for outdoor park surveillance. Comput. Vis. Image Underst. 111(1), 21–42 (2008)
4. Hu, W., Tan, T., Wang, L., Maybank, S.: A survey on visual surveillance of object motion and behaviors. IEEE Trans. Syst. Man Cybern. C 34(3), 334–352 (2004)
5. Vezzani, R., Cucchiara, R.: Ad-hoc: Appearance driven human tracking with occlusion handling. In: Proc. of First Int. Workshop on Tracking Humans for the Evaluation of their Motion in Image Sequences, THEMIS 2008, in conjunction with BMVC 2008 (2008)
6. Yilmaz, A., Javed, O., Shah, M.: Object tracking: A survey. ACM Comput. Surv. 38(4), 13 (2006)
7. Zhang, Z., Piccardi, M.: A review of tracking methods under occlusions. In: Proc. of Int. Conf. on Mach. Vis. Appl., pp. 146–149 (2007)
8. Khan, S., Shah, M.: Consistent labeling of tracked objects in multiple cameras with overlapping fields of view. IEEE Trans. Pattern Anal. Mach. Intell. 25(10), 1355–1360 (2003)
9. Hu, W., Xiao, X., Fu, Z., Xie, D., Tan, T., Maybank, S.: A system for learning statistical motion patterns. IEEE Trans. Pattern Anal. Mach. Intell. 28(9), 1450–1464 (2006)
10. Madden, C., Cheng, E.D., Piccardi, M.: Tracking people across disjoint camera views by an illumination-tolerant appearance representation. Mach. Vis. Appl. 18(3-4), 233–247 (2007)
11. Parameswaran, V., Chellappa, R.: View invariance for human action recognition. Int. J. Comp. Vis. 66(1), 83–101 (2006)
12. Lavee, G., Khan, L., Thuraisingham, B.M.: A framework for a video analysis tool for suspicious event detection. Multimedia Tools Appl. 35(1), 109–123 (2007)
13. Javed, O., Rasheed, Z., Alatas, O., Shah, M.: Knight: a real time surveillance system for multiple and non-overlapping cameras. Proc. of Int. Conf. Multimedia and Expo. 1, 649–652 (2003)
14. Zhao, T., Aggarwal, M., Kumar, R., Sawhney, H.: Real-time wide area multi-camera stereo tracking. In: Proc. of Int. Conf. Comp. Vis. Pattern Recognit., vol. 1, pp. 976–983 (2005)
15. Morris, B., Trivedi, M.: A survey of vision-based trajectory learning and analysis for surveillance. IEEE Trans. Circuits Syst. Video Technol. 18(8), 1114–1127 (2008)
16. Ding, H., Trajcevski, G., Scheuermann, P., Wang, X., Keogh, E.J.: Querying and mining of time series data: experimental comparison of representations and distance measures. In: Proc. VLDB Endow., vol. 1(2), pp. 1542–1552 (2008)
17. Junejo, I., Javed, O., Shah, M.: Multi feature path modeling for video surveillance. In: Proc. of Int. Conf. Pattern Recognit., vol. 2, pp. 716–719 (2004)
18. Chen, X., Schonfeld, D., Khokhar, A.: Robust null space representation and sampling for view invariant motion trajectory analysis. In: Proc. of IEEE Int. Conf. Comp. Vis. Pattern Recognit. (2008)
19. Piciarelli, C., Foresti, G.: On-line trajectory clustering for anomalous events detection. Pattern Recognit. Lett. 27(15), 1835–1842 (2006)

20. Buzan, D., Sclaroff, S., Kollios, G.: Extraction and clustering of motion trajectories in video. In: Proc. of Int. Conf. Pattern Recognit., vol. 2 (2004)
21. Keogh, E.J., Pazzani, M.J.: Scaling up dynamic time warping for datamining application. In: Proc. of ACM SIGKDD Int. Conf. Knowl. Discov. Data Min., pp. 285–289 (2000)
22. Qiao, Y., Yasuhara, M.: Affine invariant dynamic time warping and its application to online rotated handwriting recognition. Proc. of Int. Conf. Pattern Recognit. 2, 905–908 (2006)
23. Mecocci, A., Pannozzo, M.: A completely autonomous system that learns anomalous movements in advanced videosurveillance applications. Proc. of IEEE Int. Conf. Image Process. 2, 586–589 (2005)
24. Porikli, F., Haga, T.: Event detection by eigenvector decomposition using object and frame features. In: Proc. of Comp. Vis. Pattern Recognit. Workshop, vol. 7, pp. 114–121 (2004)
25. Mardia, K., Jupp, P.: Directional Statistics. Wiley, Chichester (2000)
26. Fisher, R.: Dispersion on a sphere. Proc. Roy. Soc. London Ser. A 217, 295–305 (1953)
27. Bishop, C.: Pattern Recognit. Mach. Learn. Springer, Heidelberg (2006)
28. Prati, A., Calderara, S., Cucchiara, R.: Using circular statistics for trajectory shape analysis. Proc. of Comp. Vis. Pattern Recognit. (2008)
29. Needleman, S., Wunsch, C.: A general method applicable to the search for similarities in the amino acid sequence of two proteins. J. Mol. Biol. 48(3), 443–453 (1970)
30. Kailath, T.: The divergence and Bhattacharyya distance measures in signal selection. IEEE Trans. Commun. Tech. COM-15(1), 52–60 (1967)
31. Reynolds, A., Richards, G., Rayward-Smith, V.: The Application of K-Medoids and PAM to the Clustering of Rules. In: Yang, Z.R., Yin, H., Everson, R.M. (eds.) IDEAL 2004. LNCS, vol. 3177, pp. 173–178. Springer, Heidelberg (2004)
32. Cucchiara, R., Grana, C., Piccardi, M., Prati, A.: Detecting moving objects, ghosts and shadows in video streams. IEEE Trans. Pattern Anal. Mach. Intell. 25(10), 1337–1342 (2003)
33. Calderara, S., Cucchiara, R., Prati, A.: Bayesian-competitive Consistent Labeling for People Surveillance. IEEE Trans. Pattern Anal. Mach. Intell. 30(2), 354–360 (2008)
34. Kopilovic, I., Vagvolgyi, B., Sziranyi, T.: Application of panoramic annular lens for motion analysis tasks: surveillance and smoke detection. In: Proc. of 15th Int. Conf. Pattern Recognit., vol. 4, pp. 714–717 (2000)
35. Vicente, J., Guillemant, P.: An image processing technique for automatically detecting forest fire. Int. J. Therm. Sci. 41(12), 1113–1120 (2002)
36. Chen, T.-H., Yin, Y.-H., Huang, S.-F., Ye, Y.-T.: The Smoke Detection for Early Fire-Alarming System Base on Video Processing. In: Proc. of Int. Conf. Intell. Inf. Hiding and Multimedia, pp. 427–430 (2006)
37. Stauffer, C., Grimson, W.E.L.: Adaptive Background Mixture Models for Real-Time Tracking. In: Proc. of IEEE Conf. Comp. Vis. Pattern Recognit., pp. 246–252 (1999)
38. Xiong, Z., Caballero, R., Wang, H., Finn, A., Lelic, M.A., Peng, P.: Video-based Smoke Detection: Possibilities, Techniques, and Challenges Suppression and Detection Research and Applications. In: A Techn. Working Conf., SUPDET (2007)
39. Toreyin, B.U., Dedeoglu, Y., Cetin, A.E.: Flame detection in video using hidden Markov models. In: Proc. of IEEE Int. Conf. Image Proc. (2005)

40. Toreyin, B.U., Dedeoglu, Y., Cetin, A.E.: Wavelet based real-time smoke detection in video. In: EUSIPCO (2005)
41. Toreyin, B.U., Dedeoglu, Y., Cetin, A.E., Fazekas, D., Chetverikov, A.T., Kiryati, N.: Dynamic texture detection, segmentation and analysis. In: Proc. of ACM Conf. Image Video Retr., pp. 131–134 (2007)
42. Collins, R.T., Lipton, A.J., Kanade, T.: A system for video surveillance and monitoring. In: Proc. of 8th Int. Top Meet on Robot and Remote Syst. (1999)
43. Mallat, S.G.: A theory for multiresolution signal decomposition: The wavelet representation. IEEE Trans. Pattern Anal. Mach. Intell. 11(7), 674–693 (1989)
44. Sato, M.: Fast learning of on-line EM algorithm. Technical Report TR-H-281, ATR Human Information Processing Research Laboratories
45. Oakley, J.P., Bu, H.: Correction of Simple Contrast Loss in Color Images. IEEE Trans. Image Proc. 16(2), 511–522 (2007)
46. Narasimhan, S.G., Nayar, S.K.: Vision and the atmosphere. Int. J. Comput. Vis. 48(3), 233–254 (2002)
47. Schechner, Y.Y., Narasimhan, S.G., Nayar, S.K.: Polarization-based vision through haze. Appl. Opt. 42(3) (2003)
48. Garg, K., Nayar, S.K.: Photometric Model of a Rain Drop. Technical Report, Department of Computer Science, Columbia University (2004)
49. Wilfred, P., Shah, M., Lobo, N.V.: Flame Recognition in Video. Pattern Recognit. Lett. 23(1-3), 319–327 (2002)
50. Garg, K., Nayar, S.K.: When Does a Camera See Rain? In: Proc. of IEEE Int. Conf. Comput. Vis., vol. 2, pp. 1067–1074 (2005)
51. Liu, C.B., Ahuja, N.: Vision Based Fire Detection. In: Proc. of Int. Conf. Pattern Recognit., vol. 4 (2004)
52. Fastcom Tech. SA, Blvd. de Grancy 19A, CH-1006 Lausanne, Switzerland: Method and Device for Detecting Fires Based on Image Analysis. In: Patent Coop. Treaty (PCT) Appl. No: PCT/CH02/00118, PCT Pubn. No: WO02/069292
53. Garg, K., Nayar, S.K.: Detection and Removal of Rain from Videos. In: Proc. of Comput. Vis. Pattern Recognit., vol. 1, pp. 528–535 (2004)
54. Casella, G., Berger, R.: Statistical Inference, 2nd edn. Duxbury Press, Boston (2002)

Recognizing Interactions in Video

Murtaza Taj and Andrea Cavallaro

Queen Mary University of London
murtaza.taj@elec.qmul.ac.uk, andrea.cavallaro@elec.qmul.ac.uk

Summary. Detection and tracking algorithms generates useful information in the form of trajectories from which the behaviors and the interactions of moving objects can be inferred through the analysis of spatio-temporal features. Interactions occur either between a dynamic and a static object, or between multiple dynamic objects. This chapter presents an interaction modeling framework formulated as a state sequence estimation problem using time-series analysis. Bayesian network-based methods and their variants are studied for the analysis of interactions in videos. Moreover, techniques such as Coupled Hidden Markov Model are also discussed for more complex interactions, such as those between multiple dynamic objects. Finally, the interaction modeling is demonstrated on real surveillance and sport sequences.

1 Introduction

Event recognition and behavior analysis are desirable yet demanding functionalities of video analytics. These functionalities can help CCTV operators to focus more effectively on cameras that are observing activities of interest. They can also help in automated analysis of large volumes of videos for indexing and retrieval applications, such as multimedia databases and video surveillance. Video activity detection can be decomposed into three main steps: (i) extraction of the *features* that best describe the activities of interest; (ii) learning *models* that describe the various activities given the static and dynamic contextual information and, finally, (iii) *recognizing* the events of interest. Figure 1 shows a generic flow diagram for activity recognition.

Depending upon the type of activity to be detected, different types of *features* can be extracted. For activities such as crowd behavior that affect the entire scene (scene-level activities), motion vectors can be used [1]. For activity such as road crossing or running that may be related to single objects only (object-level activities), information about an object location and its temporal evolution can be used [2, 3]. For object-level activities, object detection [4, 5, 6, 7, 8, 9] and tracking [5, 10, 11, 12, 13, 14, 15, 16] information form part of the features [17, 18]. Derived features such as velocity or spatio-temporal cuboids of patches inside object detection area over a certain number of frames can also be used [3, 19, 20].

H.T. Sencar et al. (Eds.): Intel. Multimedia Analysis for Security Appli., SCI 282, pp. 29–57.
springerlink.com

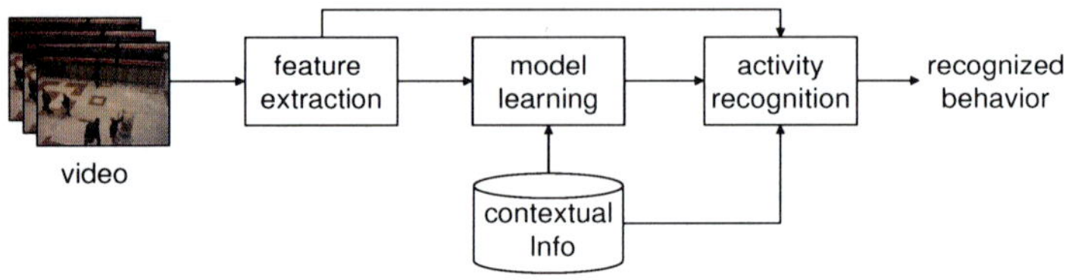

Fig. 1. Flow diagram of a generic activity recognition system.

An activity can be decomposed into a set of atomic events or states through which an object passes over time. An event can be defined as an observable occurrence or a distinguished occurrence that can be explained using a set of rules. For example, a running event can be defined as a detected dynamic object whose speed is larger than the average walking speed, where the average walking speed is defined either statistically, using the labeled interval of a walking event, or based on domain knowledge. Given the rules, an event can be recognized using thresholds on the values of the extracted features. For example, a common event in video analytics is a *tripwire* [21] event. A tripwire event occurs whenever an object crosses a line or a boundary defined by the user.

In probabilistic terms, an event can be defined as a set of outcomes to which a probability is assigned. Under this definition, an observable measurement or underlying state can be represented by a distribution such as a Gaussian, a Multivariate Gaussian or a Mixture of Gaussians. Most approaches [3, 19] involving a *learning step* require the estimation of the parameters of these distributions [22] or the boundaries employed by the set of rules defining the event or state [3, 19].

Once an atomic event or state is defined, the next step is *to model a certain activity*. An activity may be represented using a 3D shape such as motion history volumes [2], an action cylinder [23] or a quantized vocabulary of local spatio-temporal (ST) volumes (or cuboids), and a quantized vocabulary of spin-images [20]. However, these approaches are generally computationally expensive, have high memory requirements and are sensitive to noise in the extracted features. Alternatively, an activity can be modeled as an evolution of states over time. Certain activities can be represented as temporal templates that define the occurrence of various states or events in a specific order where the goal is to recognize the temporal template in the extracted features accumulated over time [24, 25, 18].

Graphical methods are well suited to represent such structures as they have inherently capability of encoding the various rules that govern the transition between states [26]. Examples of graphical models used for event detection are Petri-nets [24], Bayesian Networks [27] and Hidden Markov Models (HMMs) [28, 29, 18].

In this chapter, we focus on video event detection using HMMs and in particular on the recognition of object interactions. We propose a solution based on activity analysis on both the image and the ground plane to understand objects' behaviors and events happening in the monitored scene. The solution is based on object-level feature extraction through detection and tracking. In the learning step, we give a probabilistic representation of states as a set of distributions. Event recognition is based on the estimation of the most likely object state sequence using information about the temporal evolution of objects in the scene (tracks) and contextual information (e.g., zone boundaries and events of interest). We model the problem using time-series analysis and solve the task of state sequence estimation using the Viterbi algorithm for Hidden Markov Model (HMM) decoding. We evaluate the proposed framework on standard surveillance datasets, including sequences from i-Lids (Imagery Library for Intelligent Detection Systems) [30], ETISEO (Video Understanding Evaluation) [31], PETS (Performance Evaluation of Tracking and Surveillance) [32] and CAVIAR (Context Aware Vision using Image-based Active Recognition) [33]. These sequences represent different scenarios, such as a metro station, an airport, a building entrance and a road.

This chapter is organized as follows: related work on event detection is discussed in Section 2. Section 3 formalizes the event detection problem, whereas different solutions are discussed in Section 4. Finally, Section 5 draws the conclusions.

2 Related Work

Events are generally composed of specific sequences of actions and therefore HMMs are appropriate for their modeling [34]. HMMs are also used for abnormal activity detection in crowds [1] by modeling normal motion paths using a single HMM [1] or a Mixture of Gaussian HMMs [35]. The main limitation of the aforementioned HMM-based techniques is the use of a *evaluation* strategy to obtain the sequences of events, as this result in a dependence on the selected pattern. Moreover, the above techniques model events involving single objects only. As objects move simultaneously based on their intentions, their actions depend also upon the behaviors of other objects. For this reason, monitoring a single target separately [36, 27, 37] may not provide the complete information about its state.

Target interactions can be modeled using either predefined heuristics (rules) [38] or by using graphical models [39, 40, 22, 29]. *Heuristics-based methods* generally define rules for interactions based on each target spatio-temporal features, such as speed and distance, to compute the probability of interaction [38]. Events can be assumed to be composed of sub-events and graphs can be used to model the conditional dependency between them. Normalized cuts can be used to partition such a dependency graph to extract complex events as a highly correlated chain of sub-events. *Probabilistic graphical models* [26] provide a simple way to visualize the structure of a probabilistic model, insight about their conditional dependence and a way to solve complex computations as graphical manipulation. A common type of graphical model is the *directed acyclic graph* (DAG) [41]. In a DAG, one can move from one node to another along the links but cannot reach the initial node again. DAGs of stochastic processes are called Dynamic

Bayesian Networks (DBN) [42]. The simplest type of DAGs are HMMs [26], which are characterized by one discrete hidden node (state) and one discrete or continuous observed node (output or emitting symbol) at each time stamp.

HMMs and their variants [43] are used to address the problem of Interaction Event Recognition (IER) due to their inherent capability of modeling uncertain temporal information. Interactions can happen between a static object and a dynamic object or between dynamic objects. Static objects can either be temporally static (such as a bag or a car) or permanently static such as structural objects. HMMs were used to model interactions with such static objects in [17] where object trajectories were used to obtain a 4D features vector accounting for object position and size. This feature vector is then used inside a continuous distribution HMM using multi-variate Gaussian for estimation of emission probabilities for detection of interactions associated to static objects. The approach is further extended by modeling the duration of each state which imposes a practical constrain that objects should take a transition from one state to another after certain interval, Hidden-Semi Markov models (HSMMs) were used to model such duration-related transitions.

HSMMs have been shown to better model temporal evolution of object behavior. However, they are limited in modeling interactions with static objects only. Coupled Hidden Markov Models (CHMMs) have received significant attention [39, 40, 22, 29], to model group activity among multiple dynamic objects, as they allow modeling the full coupling between the processes and can be solved in polynomial time using dynamic programming [39]. Other methods for modeling such interactions include Multi-Observation-Mixture+Counter Hidden Markov Models (MOMC-HMM), which allows representation of multiple observations of different objects at each state [44] similar to Variable-length HMM (VLHMM) [45]. The extension to VLHMM is proposed in [25]. This method is made immune to noise by abstracting the continuous variables (tracks) into discrete space while preserving the underlying behavioral patterns and applying them in modeling interaction among cars on highways. However, it should be noted that such discretization may result in losing some relevant details and therefore may not be suitable for some applications.

3 Problem Formulation

Let an object detector generate at each time k a set of N_o objects $O_k = \{o_k^1, o_k^2, \cdots, o_k^{N_o}\}$. Let a tracker associate object instances between consecutive frames to establish the track $O_{k_0:k}^r = \{o_{k_0}^r, o_{k_1}^r, \ldots, o_k^r\}$, up to time k, for each rth object (for simplicity the superscript r will be dropped whenever the discussion is related to a single object).

Event detection can be modeled as a random process that is segmental in nature, as the piecewise stationarity assumption of HMMs [28] is well suited for time-series analysis. Let $\lambda = \{A, B, S, s_0\}$ be a continuous distribution first-order Hidden Markov Model, where $S = \{s_1, s_2, \cdots, s_{N_S}\}$ represents a set of N_S discrete events (states) to be detected (the actual state at time k is denoted as a time indexed discrete variable $s_k \in S$), $A = \{a_{ij}\} = \{P(s_k = s_j|s_{k-1} = s_i)\}$ represents the state transition probabilities where $P(.)$ represents the probability of transition from state s_i to state s_j from time $k-1$ to k; $B = \{b_{jk}\}$ are the emission probabilities, with $b_{jk} = P(o_k|s_k = s_j)$ and s_0 is the known initial state at time $k = 0$. The emitting symbols of each state are provided by the features extracted from the objects.

The observed objects interact with the environment as well as with other objects. The problem of interaction event detection can therefore be divided into two categories of state estimation, namely dynamic object with respect to static objects (see Fig. 4 and Fig. 5) and dynamic objects with respect to each other (see Fig. 11). More formally, the problem of interaction event detection can be defined as follows.

- **Dynamic-static interaction.** Given the model λ and the sequence of observations $O^p_{k_0:k}$ and $O^q_{k_0:k}$ from a dynamic and a static object, it is desired to generate the sequence in which they interact and have gone through various states in S.
- **Dynamic-dynamic interaction.** Given the model λ, the observations $O^p_{k_0:k}$ and $O^q_{k_0:k}$ for two interacting objects p and q, and set of interaction event templates $\{E_1, \cdots, E_{N_E}\}$, it is desired to find the interaction between the objects.

Interaction event analysis will be performed using the track information and the available contextual information. The next section discusses various approaches for solving these problems.

4 Interaction Event Recognition

4.1 Bayesian Networks for Interaction Event Modeling

Let $\lambda = \{A, B, S, s_0\}$ be a continuous distribution first-order Hidden Markov Model. The associated optimal state sequence can be obtained by applying Bayes' rule

$$\hat{S} = \arg\max_S P(S|O) = \arg\max_S \frac{P(O|S)P(S)}{P(O)}, \tag{1}$$

where $\hat{S}$ is the estimated state. To estimate the posterior probability $P(S|O)$, the likelihood $P(O|S)$, the prior $P(S)$ and the probability $P(O)$ needs to be calculated. $P(O)$ is a normalizing constant that can be estimated as

$$P(O) = \sum_{c=1}^{N_C} \sum_{i=1}^{N_S} P(o_c|s_i)P(s_i), \tag{2}$$

where N_C is the total number of interacting chains. Using the Markov property, the likelihood $P(O|S)$ is given as

$$\begin{aligned} P(O|S) &= P(o_1, o_2, \cdots, o_K | s_1, s_2, \cdots, s_K) \\ &= P(o_1|s_1)P(o_2|s_2)\cdots P(o_K|s_K) \\ &= \prod_{k=1}^{K} P(o_k|s_k), \end{aligned} \tag{3}$$

where K is the total number of time stamps. Similarly, the prior $P(S)$ can be computed using the Markov property as

$$\begin{aligned} P(S) &= P(s_1, s_2, \cdots, s_k) \\ &= P(s_1)P(s_2|s_1)\cdots P(s_k|s_1, s_2, \cdots, s_{k-1}) \\ &= P(s_1)\prod_{k=2}^{K} P(s_k|s_{k-1}). \end{aligned} \tag{4}$$

In standard HMM models, the causal relationship within a single process and each HMM state causes an effect on the next state. However, in many real-world scenarios there may be more than one process interacting with each other.

The variants of HMM exist to model such inter–process causal relationships (or couplings) can be divided into two major groups based on whether the processes are dependent or independent. In case of *independent processes*, the coupling is at the output level only and is called Factorial HMM (FHMM, Fig. 2(a)). A number of HMM variants exist for modeling *dependent processes*, based on their degree of dependence: Linked HMM (LHMM, Fig. 2(b)), Hidden Markov Decision Tree (HMDT, Fig. 2(c)) and Coupled HMM (CHMM, Fig. 2(d)). In general, for a graph of N_G nodes, the joint distribution is given by [26]

$$P(S) = \prod_{n=1}^{N_G} P(s_n|s_{lk}), \tag{5}$$

where s_{lk} denotes the set of states in a graph that have a direct link toward s_n. In Factorial HMM, signals are factored as component processes which are modeled independently. The full posterior of the C-chain FHMM can be computed as

$$P(S^C|O) = \frac{1}{P(O)} \prod_{c}^{C} \left(P(s_1^c)P(o_1^c|s_1^c) \prod_{k=2}^{K} P(s_k^c|s_{k-1}^c)P(o_{k-1}^c|s_{k-1}^c) \right) \tag{6}$$

In LHMM the current state of a process is dependent on its previous state and the current state of the neighboring process. The full posterior of the LHMM can be computed as

$$P(S^C|O) = \frac{1}{P(O)} \prod_{c}^{C} \left(P(s_1^c)P(o_1^c|s_1^c) \times \right.$$
$$\left. \times \prod_{k=2}^{K} \left(P(o_k^c|s_k^c)P(s_k^c|s_{k-1}^c) \prod_{d}^{C} P(s_k^c|s_{k-1}^d) \right) \right) \tag{7}$$

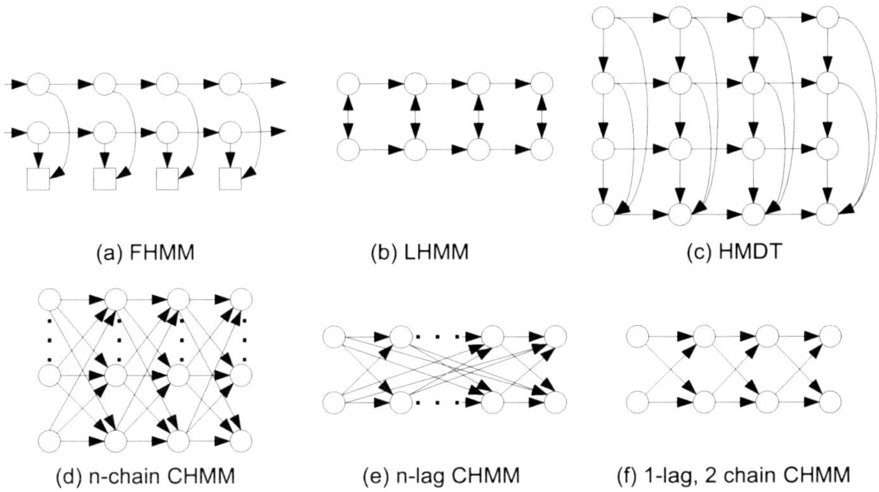

Fig. 2. Graphical representation of HMM variants. (a) FHMM, (b) LHMM, (c) HMDT, (d) n-chain CHMM, (e) n-lag CHMM and (f) 1-lag, 2 chain CHMM. FHMM models independent processes while the rest model dependent processes (Key: Circles: state variables; squares: output variables).

Contrary to LHMM, in HMDT the current state of any process is dependent on the current state of all the interaction processes as well as the previous state of the process. The full posterior in this case can be computed as

$$P(S^C|O) = \frac{1}{P(O)} \prod_{k=2}^{K} \Big(P(s_k^1|s_{k-1}^1) P(s_k^2|s_{k-1}^2, s_k^1) \cdots \\ \cdots P(s_k^C|s_{k-1}^C, s_k^1, \cdots, s_k^{C-1}) \Big) \prod_{c}^{C} P(o_1^c|s_1^c) \quad (8)$$

CHMMs are an example of the full coupling and the process state does not only depend upon their previous state, but also on the previous state of all the interacting processes. The full posterior of the C-chain CHMM $P(S^C|O)$ can be computed as

$$P(S^C|O) = \frac{1}{P(O)} \prod_{c}^{C} \left(P(s_1^c) P(o_1^c|s_1^c) \prod_{k=2}^{K} \left(P(o_k^c|s_k^c) \prod_{e}^{C} P(s_k^c|s_{k-1}^e) \right) \right), \quad (9)$$

where $P(s_t^c|s_{t-1}^e)$ is the state transition probability at time k for object c to state s_k^c given the state s_{k-1}^e of the interacting object e at time $k-1$. In CHMMs, each discrete hidden node is coupled also with all the previous nodes of all the C chains. Such a model is called n-lag C-chain model.

To reduce the complexity while keeping the interaction modeling, the 1-lag model is considered, the posterior of which, $P(S^C|O)$, is given in Eq. (9). In the case of only two objects p and q, only a 2-chain CHMM is required, for which the posterior can be further simplified to

$$P(S|O) = \frac{P(s_1^p)P(o_1^p|s_1^p)P(s_1^q)P(o_1^q|s_1^q)}{P(O)} \times$$
$$\times \prod_{k=2}^{K} \Big(P(s_k^p|s_{k-1}^p)P(s_k^q|s_{k-1}^q)P(s_k^p|s_{k-1}^q) \times$$
$$\times P(s_k^q|s_{k-1}^p)P(o_k^p|s_k^p)P(o_k^q|s_k^q) \Big), \qquad (10)$$

where s_k^p is the state of chain p and s_k^q is the state of chain q at time k. The problem of interaction event modeling involves analyzing jointly the states of multiple objects to model their influence on each other. This implies the need of accounting for the influence of the current state of the interacting processes while estimating their next. The problem of estimating the most probable next state of the interaction is similar to the single process case (see Eq. (1)), except that the posterior $P(S|O)$ is estimated as in Eq. (10) to take into account the coupling.

4.2 Interaction among Dynamic and Static Objects

In the case of one object being static and another dynamic, it is enough to estimate the state of one object only. This implies generating the optimal state sequence observed by one of the object with respect to the other. To find the single most probable state sequence, S_k, a quantity $\delta_{i,k}$ needs to be defined as

$$\delta_{i,k} = \max_{s_1,\cdots,s_{k-1}} P(s_1,\cdots,s_{k-1},s_k = s_i, o_1 \cdots o_k|\lambda), \qquad (11)$$

where $s_1,\cdots,s_{k-1}$ are the most probable states and $o_1 \cdots o_k$ are the sequence of observations from time 1 to k.

$$\delta_{i,k} = \max_{1 \le t \le k-1} P(s_t = s_i|O,\lambda). \qquad (12)$$

According to the Markovian assumption, the conditional probability distribution of future states depends on the current state only and not on past states, hence using the Forward Viterbi we have

$$\delta_{j,k+1} = \max_{1 \le i \le N} (\delta_{i,k} a_{ij}) b_{j,o_l}. \quad 1 \le l \le k \qquad (13)$$

Finally, the most likely hidden state sequence S_K up to time $k+1$ can be computed as

$$s_{k+1} = \arg\max_{1 \le i \le N} (\delta_{i,k} a_{ij}). \qquad (14)$$

This simple Hidden Markov Model is unable to completely model certain events due to the duration distribution of the observation sequence for a certain state.

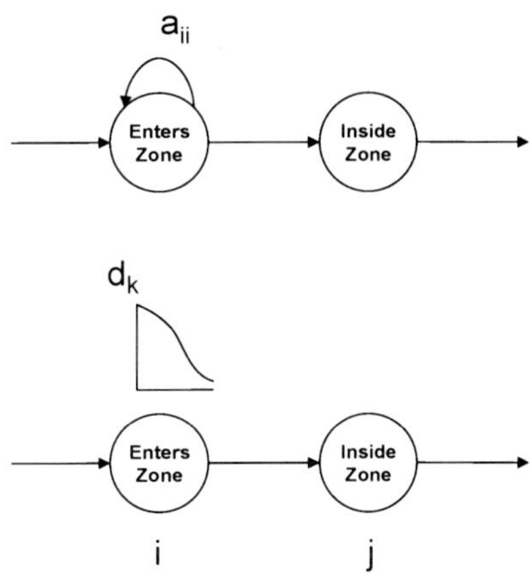

Fig. 3. Examples of self-transition modeling for an Hidden Markov Model: (top) self-transition probability (a_{ii}); (bottom) self-transition replaced with a state occupancy duration *pdf*.

The Markovian assumption constrains the state occupancy distribution to be exponentially distributed [46]. Therefore the estimation of the most likely path S_K is problematic, because a state with high self-transition probability a_{ii} can cause the algorithm to stay in this state for a longer interval. To avoid such self-transitions, Hidden Semi-Markov Models (HSMM) [47] are used to enable the explicit modeling of duration probability distribution d_k. The duration probability distribution is the probability of staying at least for a duration τ in the state s_j, with $1 \leq \tau \leq D_j$ (Fig. 3). To compute the most likely state sequence S_K using the durational distribution, the joint probability need to be maximized by re-writing Eq. (12) as

$$\delta_{i,k} = \max_{1 \leq t \leq k-1} P(s_t = s_i, O|\lambda). \tag{15}$$

Using the forward Viterbi algorithm the Eq. (13) can be solved as

$$\delta_{j,k+1} = \max_{\substack{1 \leq j \leq N \\ 1 \leq \tau \leq D_j}} \left(\max_{1 \leq i \leq N} (\delta_{i,k} a_{ij}) d_\tau b_{k,o_l} \right), \tag{16}$$

with $1 \leq l \leq k$. Given λ the duration probability distribution d_k, the best state sequence can be computed using Eq. (16) by performing the HMM decoding using the Viterbi algorithm. The state transition probabilities a_{ij} can be defined empirically or, if there is sufficient training data, calculated using the Baum-Welch algorithm [48]. In order to use the Viterbi algorithm with duration probability distribution d_k and the observation sequence need to be modeled.

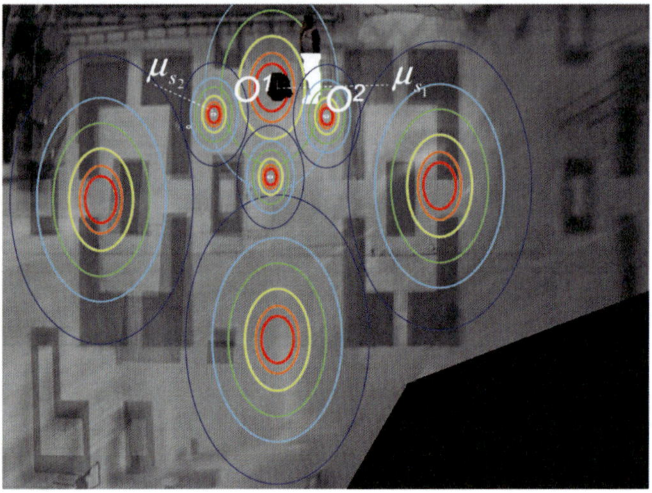

Fig. 4. Multivariate object-centric distribution model. The distribution of the states is placed on the line joining the centroid of the objects.

Duration probability distribution

The duration probability distribution d_k can be modeled using different parametric duration distributions. Here we show a comparison between two distributions, namely the *half-normal distribution* and the *triangular distribution*, which are well adapted to the problem at hand. The half-normal distribution, d_k, can be expressed as

$$d_k = \frac{1}{\sigma}\sqrt{\frac{2}{\pi}}\exp\left(-\frac{1}{2}\left(\frac{k-k_e}{\sigma}\right)^2\right), \tag{17}$$

where σ is the standard deviation, computed as $3\sigma = \tau$, and k_e is the mean and is the time when the object transits into the state and $k_e \leq k \leq K + k_e$. The triangular distribution, d_k, can be expressed as

$$d_k = \frac{2(\tau + k_e - k)}{\tau^2}. \tag{18}$$

In case of events with high self-transitions, a uniform distribution can be used that implicitly converts a HSMM to a HMM. The selection of the appropriate distribution, for the specific event or activity, can be done using the Chi-square test.

Object-centric and scene-centric models

The estimation of the emission probabilities b_{jk} depends on the possible states of the object. These states can either be associated with the static object with

(a) (b)

Fig. 5. Multivariate scene-centric distribution model. (a) Sequence AP-11 C4; (b) Sequence BE-19 C1.

respect to other interacting objects in the scene or can be associated with the dynamic object with respect to the static object with which it is interacting. Here two models for b_{jk} are discussed, namely a scene-centric and an object-centric model. In both these models the b_{jk} can be modeled using any appropriate distribution such as uniform, normal, Poisson or mixture of multiple distributions based on the application. For e.g. for each jth state, using a multi-variate Gaussian distribution $\mathcal{N}(\mu_j, \Sigma_j)$, with mean μ_j and covariance Σ_j, each emission probability b_{jk} can be defined as

$$b_{jk} = \frac{1}{\sqrt{(2\pi)^n |\Sigma_j|}} \exp\left(-\frac{1}{2}(O_k^r - \mu_j)^T \Sigma_j^{-1} (O_k^r - \mu_j) \right), \tag{19}$$

where $|\Sigma_j|$ is the determinant of the covariance matrix Σ_j and n is the number of dimensions in the state. In the *scene-centric* approach, the values for μ_j and Σ_j are set based on the contextual information specific to the task at hand (Fig. 5).

In the *object-centric* approach, the values of the elements in Σ_j depend on the state to be modeled, whereas the value of μ_j is assigned dynamically. This is the key point of the *object-centric* modeling. The value of μ_j of the first state is set as the centroid of the reference object O_k^{ref} (Fig. 4). The remaining state distributions are then placed around O_k^{ref} to estimate the possible state of O_k^{ref} with respect to the objects O_k^r. The μ_j of the other states are positioned on the line passing through the centroid of the two objects (O_t^{ref} and O_t^r) at a distance that is a function of the variances of the states to be detected. The rationale for using Gaussian functions instead of hard boundaries and fixed thresholds is to increase the flexibility of the algorithm in order to detect several different events in different scenarios.

Algorithm 1. Event Detection

$S = \{s_1, \cdots, s_{N_s}\}$: events (states that an object can acquire)
a_{ij} : state transition probabilities between state i to j
μ_j : mean for each state j; Σ_j : covariance matrix for each state j
$O^r_{1:K}$: observation for object r upto time K ; *count* : counter

for $k = 1$ to K **do**
  for $j = 1$ to N_T **do**
    Compute b^r_{jk} :
    $b^r_{jk} = \frac{1}{\sqrt{(2\pi)^n |\Sigma_j|}} \exp\left(-\frac{1}{2}(O^r_k - \mu_j)^T \Sigma_j^{-1} (O^r_k - \mu_j)\right)$
  end for
  $count = count + 1$
  if $count = n$ **then**
    Initialize initial state s^r_0
    if $s_0 = -1$ **then**

$$s^r_0 = \zeta(\max_{j=1\ldots l} b^r_{jk})$$

      where ζ returns s_j corresponding to b^r_{jk}
    end if
    Apply Forward Viterbi Algorithm:

$$\delta_k = \max_i(\delta^r_{k-1} a_{ij}) b^r_{jk}$$

$$\delta_{j,k+1} = \max_{\substack{1 \le j \le N \\ 1 \le \tau \le D_j}} \left(\max_{1 \le i \le N} (\delta_{i,k} a_{ij}) d_\tau b_{k,O_l} \right), \quad 1 \le l \le k$$

$$s^r_K = \arg\max_i(\delta^r_{k-1} a_{ij})$$

$$s^r_0 = s^r_k$$

  end if
end for

Moreover, as the behaviors of objects in real scenarios are generally characterized by fuzzy boundaries between different states, a progressive transition from one state to another is preferred to a fixed threshold-based transition [49].

The estimation of the emission probabilities b_{jk} completes the computation of the HMM parameters. These parameters are now used to compute the most likely state sequence S_K for each object r by applying the *Forward Viterbi algorithm* every K observations. The last state s_k of the state sequence is then used as the initial state s_0 for next computation. The event detection algorithm using the *forward Viterbi algorithm* for HSMM is summarized in Algorithm 1.

4.3 Example-1: Object-Centric Model

The *object-centric* HMM model is used for detecting abandoned baggage scenarios where the states are *attended*, *unattended* and *abandoned*. Figure 6 shows

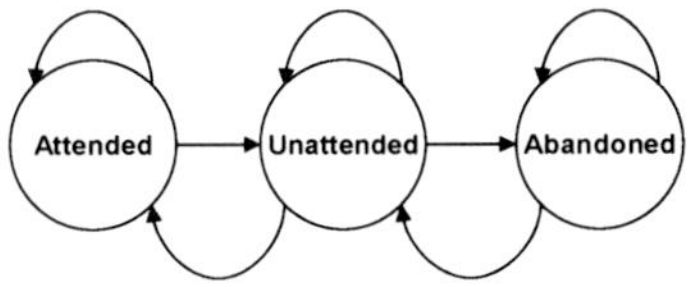

Fig. 6. HMM model for baggage detection on the PETS and CAVIAR datasets. Each state represents an event. The initial state is selected as the state with the maximum emission probability b_{jk} at time 0.

the HMM model used to detect these states. The object is considered *attended* when the owner is within 2m distance of the baggage. The warning of baggage being *unattended* is raised when the object mover further then 3m of the baggage. The baggage is considered *abandoned* when it is unattended for more then 30sec. These events are detected on real surveillance scenarios from the PETS, the CAVIAR datasets.

Here each object at each time k is represented by a 4D state-space consisting of position (x, y) and size (w, h). $n = 4$, $\mu_j = \{\mu_x, \mu_y, \mu_w, \mu_h\}$ and $\Sigma_j = diag[\sigma_x^2, \sigma_y^2, \sigma_w^2, \sigma_h^2]$. For the PETS sequences, the baggages are detected based on their size and aspect ratio (ranging between 1 and 1.8). For the *attended baggage* (s_1) event, $\sigma_x = \sqrt{2 * 36}$ and $\sigma_y = \sqrt{2 * 96}$ respectively, whereas for the *unattended baggage* (s_2) and the *abandoned baggage* (s_3) events the values are $\sigma_x = \sqrt{36/2}$ and $\sigma_y = \sqrt{96/2}$. These values are based on the calculation that, for this scenario, 1m in world-coordinates corresponds in the ground plane to 36 pixels along the x-axis and to 96 pixels along the y-axis. A baggage is considered *unattended* when its related object (the *owner*) is 2m away. A baggage is considered *abandoned* when its related object is $3m$ away, for at least 30 seconds. For the CAVIAR sequences, the baggages are detected in a similar fashion and the parameters of the events are defined as follows. For the *attended baggage* (s_1) event, $\sigma_x = \sqrt{2 * 36}$ and $\sigma_y = \sqrt{2 * 36}$ respectively, whereas for *unattended baggage* (s_2) the values are $\sigma_x = \sqrt{36}$ and $\sigma_y = \sqrt{48}$ and for *abandoned baggage* (s_3) the values are $\sigma_x = \sqrt{24}$ and $\sigma_y = \sqrt{24}$.

Evaluation metrics

To evaluate the event detection results three measures namely the *accuracy*, the *precision* and the *sensitivity* of the event detector are estimated. Let FP be the number of false positive detections, TP the number of true positive detections, and FN the number of false negative detections. Moreover, let GT be the frame number corresponding to an event in the ground truth and AD the frame number identified by the event detector for the same event. The *accuracy* gives an indication of the frame-level performance of the algorithm, and is defined as $\gamma = \left(1 - \frac{|GT-AD|}{NF}\right) \times 100$, with NF representing the minimum duration of an

Fig. 7. Sample event detection results for the PETS 2006 dataset. (First row): Sequence S1-T1-C3, frames 1955, 2754 and 2790. (Second row): Sequence S5-T1-G3, frames 2020, 2833 and 2890. The evaluation of the event detection accuracy is discussed in the text.

event. The *precision* is defined as $TP/(TP + FP)$ and the *sensitivity* is defined as $TP/(TP + FN)$.

Analysis of results

Figure 7 shows sample event detection results on the sequences S1-T1-C3 and S5-T1-G3 of the PETS 2006 dataset. The images show the detection of the object around which the model is built (the bag) and the subsequent sequence of events, namely a *warning* (unattended baggage) and an *alarm* (abandoned baggage). To evaluate the results the accuracy of the detection is computed: for the sequence S1-T1-C3, the accuracy for the *warning* event is 90.5% and for the *alarm* event is 92.9%; for the sequence S3, the accuracy is 100% for both events; for the sequence S5, the accuracy is 88.8% and 83.02% for *warning* and *alarm*, respectively, and for the sequence S6 the accuracy is 98.5% and 95.5%. Both the precision and sensitivity scores for PETS dataset are unitary as the object-centric approach selects events associated with detected objects only and the baggage was detected.

Figure 8 shows sample event detection results on the two sequences CL1 and CL2 of the CAVIAR dataset. Figure 8(a) shows the detection of the *abandoned baggage* events. Figure 8(b) shows that the person has left the baggage at the end of the stairs moving toward the kiosk machine and hence the *attended* and then *unattended baggage* events are generated. In accordance with the ground truth available for the CAVIAR dataset, the accuracy of the detection of the activities related to the baggage is computed rather than the accuracy for the alarm and

(a) (b)

Fig. 8. Example of left baggage detection on the CAVIAR dataset using the object-centric event modeling. (a) *Abandoned* baggage event in sequence CL1 (frame 1334); (b) *unattended* baggage event in sequence CL2 (frame 721).

the warning events. For the sequence CL1, the event initialization accuracy is 95% and the event termination accuracy is 94.66%. The event initialization accuracy for the sequence CL2 is 97.33% and the termination accuracy is 95.60%. Improvements in the object detection accuracy will help in further enhancing the event detection accuracy as problems such as blob merging are one of the reasons for not having even higher values. Similarly to the PETS dataset, the precision and sensitivity scores for the CAVIAR dataset are unitary as all events are detected.

4.4 Example-2: Scene-Centric Model

The *scene-centric* HMM model is used for monitoring activities such as *enter zone*, *inside zone*, *exit zone*, *change zone*, *opens*, *closes*, *go up stairs*, *go down stairs*, *empty area*, and *stopped object*. Figure 6 shows the HMM model used to detect these states. The activities are detected on ETISEO datasets. The same evaluation metrics is used here as in this case of object-centric model example. Table 1 shows the performance comparison between three algorithms: (i) HMM [17], (ii) HSMM using triangular distribution (HSMM-TRI) [18] and (iii) HSMM using half normal distribution (HSMM-HN) [18]. The comparison was done on the ETISEO sequences AP-11 (C1 and C4) and BE-19 (C1) using the CREDS protocol [50] which provides a weighted sum of the true positive, false positive and false negative detections [50]. It is possible to notice that the duration modeling in HMM improves the results and that the modeling using the triangular distribution outperformed by 3.75% the half-normal distribution. In summary, the HSMM model with triangular distribution performed at 96%, the HSMM model with half-normal distribution obtained a score of 92.5% and the HMM scored 85%. Figure 10 shows detection results on the ETISEO dataset for the *enter zone, inside zone, stopped* and *empty area* events. In this case event detection is performed on the image plane. The green rectangle drawn on the tarmac is the zone considered for triggering the events *enter zone, inside zone*

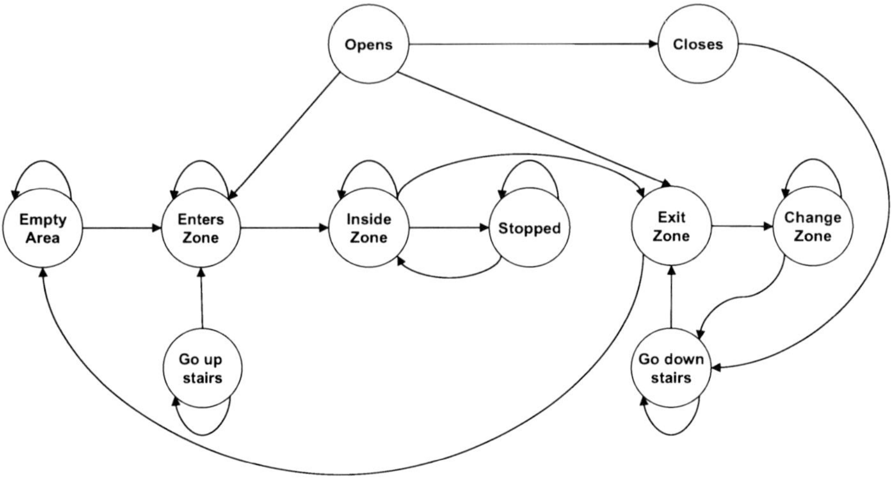

Fig. 9. Scene-centric HMM model for activity monitoring on the ETISEO dataset. Each state represents an event. The initial state is selected as the state with the maximum emission probability b_{jt} at time t_0.

Table 1. Performance comparison between HSMM algorithm with half-normal and triangular distribution for state occupancy duration and event detection using HMM without state duration modeling.

	HMM	**HSMM-TRI**	**HSMM-HN**
AP	0.882	0.980	0.956
BE	0.790	0.966	0.980
Total	0.847	0.965	0.975

and *empty area*. The *stopped* event is detected anywhere in the scene. Table 2 shows the accuracy for the detected events in all ETISEO sequences.

4.5 Interactions among Dynamic Objects

When multiple interacting objects are dynamic, the state sequence for such objects need to be estimated. Furthermore, since their states are dependent on the states of other interacting objects it is important to model the coupling between them. Here we first consider a few examples of interaction events where for simplicity only two objects are considered. Methods for modeling these interactions will be discussed in the next sub section. Let the objects O^p and O^q be at positions (x^p, y^p) and (x^q, y^q) at time k_0 with a certain velocity on a random path. Let at a certain time $k_1 (> k_0)$, the two objects start approaching a meeting point (x_m, y_m) such that at time $k_1 + n_r$, with $r \in \{p, q\}$, they arrive at the meeting point either together $(n^p = n^q)$ or one after another $(n^p \neq n^q)$. After staying at the meeting point for a duration K, the two objects start following a

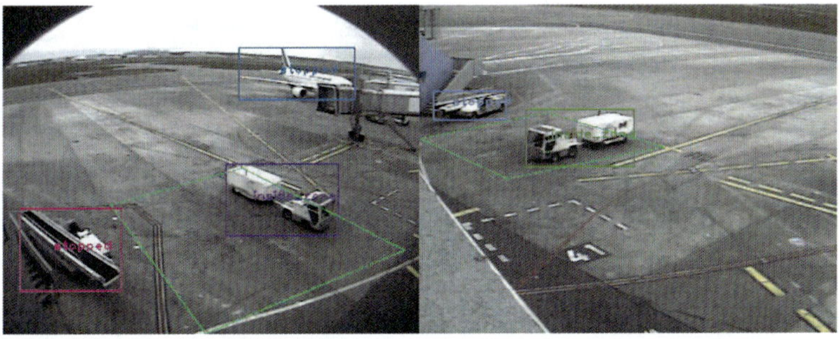

(a) (b)

Fig. 10. Sample tracking and event detection results for the ETISEO dataset using the scene-centric event modeling. The detected events are *stopped* and *inside zone*. (a): ETI-VS2-AP-11, Camera 4, frame 760. (b): ETI-VS2-AP-11, Camera 7, frame 760.

common or separate random path. The goal is to model the interaction between these objects by considering their joint states.

Let there exist a full coupling between the target states modeled as 1-lag, 2-chain Coupled Hidden Markov model (Fig. 2(f)). Let O^p and O^q represent the set of observations; S^p, S^q be the set of states for the two chains and $P(s^i_{k+1}|s^j_k)$, $i, j \in \{p, q\}$ be the transition probabilities among these states. The observation O_k is the feature vector computed using the current and the previous positions, of the interacting objects, obtained from their trajectories $O^r_{k_0:k}$ up to time k for object $r \in p, q$.

Let us define the five types of interactions, namely *follow-reach-go together*, *approach-meet-go separately*, *approach-meet-go together*, *change direction approach-meet-go separately* and *change direction-approach-meet-go together*. Figure 11 shows the schematic diagram of these interactions.

In the *follow-reach-go together* interaction (E1), two objects O^p and O^q follow a random path. At time k, O^q starts approaching the current position of O^p. At $k + n_1$, O^q approaches O^p, i.e. $O^q_{k+n_1}(x, y) = O^p_k(x, y)$. O^q then starts following O^p's path with higher speed and approaches O^p after n_2 time stamps at time $k + n_1 + n_2$. From time $k + n_1 + n_2$ both objects move together (Fig. 11(a)).

In the *approach-meet-go separately* interaction (E2), two objects O^p and O^q starts with moving on a random path. At time k, both objects start approaching a meeting point and wait for each other. After arriving, both objects first wait and then go separately on different paths. The *approach-meet-go together* interaction (E3) is similar to E2, with the difference that after meeting the two objects move together (Fig. 11(b)).

In the *change direction-approach-meet-go separately* interaction (E4), objects O^p and O^q starts moving on a random path as in case of E2. At time k, O^q starts approaching O^p with an increased speed and changing direction

Table 2. Event detection precision and sensitivity for 5 test sequences of the ETISEO dataset

	Start frame			End frame		
	GT	AD	Accuracy	GT	AD	Accuracy
AP-11-C4 (Precision: 1.00, Sensitivity: 0.56)						
empty area	1	12	98.53	689	664	96.67
enters zone	675	664	98.53	720	728	98.93
inside zone	690	731	94.53	804	803	99.87
stopped	1	2	99.87	804	803	99.87
stopped	1	3	99.73	804	803	99.87
All			**98.24**			**99.04**
AP-11-C7 (Precision: 1.00, Sensitivity: 0.50)						
empty area	1	187	75.20	689	653	95.20
enters zone	675	658	97.73	720	695	96.67
inside zone	690	696	99.20	804	803	99.87
stopped	1	2	99.87	804	803	99.87
All			**93.00**			**97.90**
BE-19-C1 (Precision: 0.65, Sensitivity: 0.65)						
closes	335	371	95.20	453	450	99.60
opens	258	250	98.93	320	300	97.33
opens	366	395	96.13	400	407	99.07
stopped	270	283	98.27	1025	1024	99.87
All			**97.13**			**98.97**
BE-19-C4 (Precision: 0.87, Sensitivity: 0.35)						
inside zone	185	180	99.33	245	338	87.60
opens	77	101	96.80	150	180	96.00
opens	737	717	97.33	780	776	99.47
stopped	170	206	95.20	950	1048	86.93
All			**97.17**			**92.50**
RD-06-C7 (Precision: 1.00, Sensitivity: 0.25)						
stopped	570	559	98.53	710	743	95.60
All			**98.53**			**95.60**

continuously to reach O^p. At $k+n$, O^q approaches O^p, i.e., $O^q_{k+n}(x,y) = O^p_{k+n}(x,y)$. After arriving, both objects first wait and then move separately on some random path (Fig. 11(c)).

Finally, the *change direction-approach-meet-go together* interaction (E5) is similar to E4, with the difference that after meeting the two objects move together. The corresponding sample trajectories are shown in Fig. 12.

Interaction features

The interactions could be identified by analyzing if the two objects coexist at a certain time in the same region. However, this would make the approach

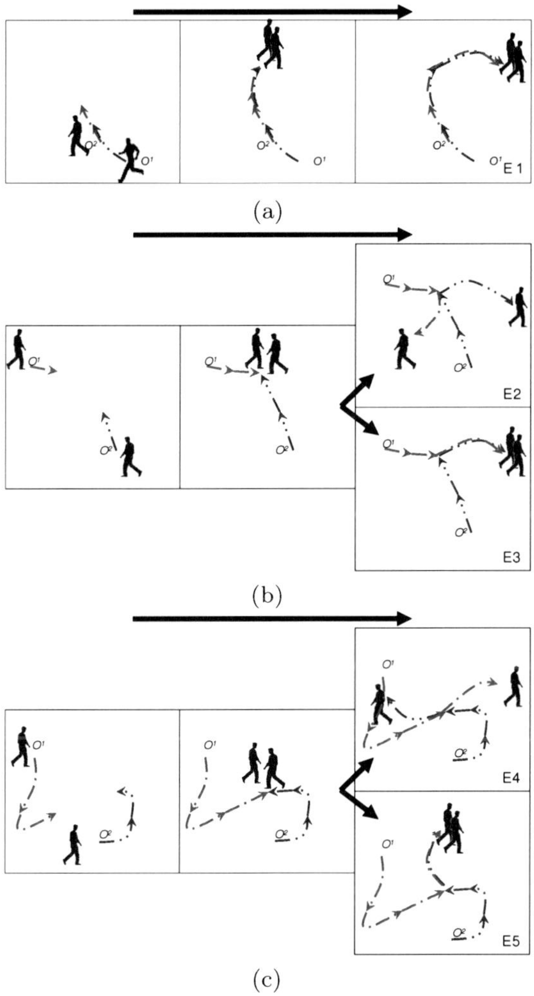

Fig. 11. Schematic diagram showing different stages of five interactions (green single dotted line represents object O^1 and blue double dotted line represents object O^2). (a) follow-reach-go together (E1), (b) approach-meet-go separately (E2) and approach-meet-go together (E3), (c) change direction-approach-meet-go separately (E4) and change direction-approach-meet-go together (E5).

heavily dependent on contextual information (e.g., the presence of a kiosk). The five interactions E1 to E5 can be divided into the following sub-events: *following*, *approaching*, *meeting*, *going together* and *walking/going separately*. These sub-events have similar values for features (such as speed, distance from scene boundary, and direction), to other non-interacting objects. To identify such interactions, the combination of relative features among objects offers a better representation of the states.

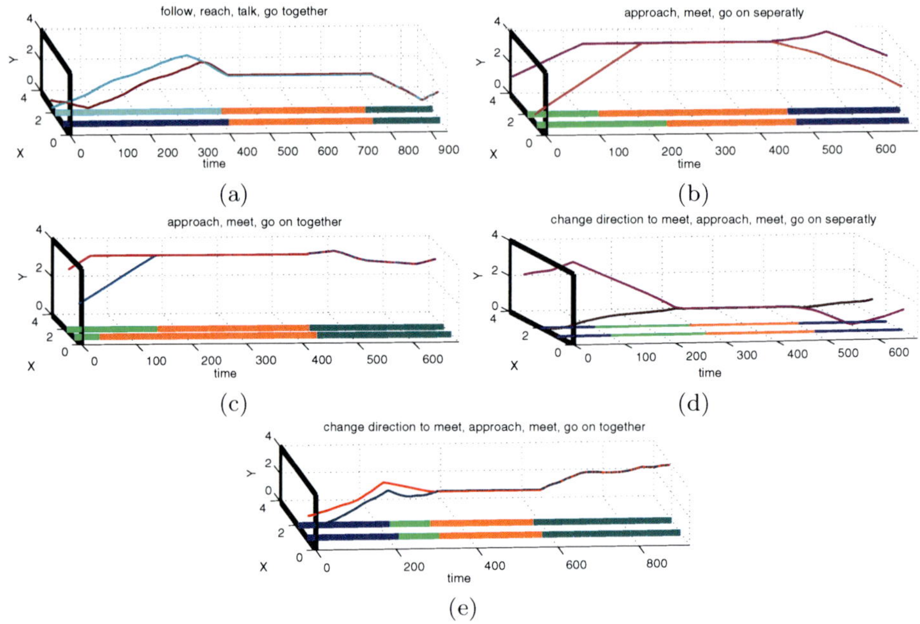

Fig. 12. Sample trajectories for each interaction (E1 to E5) between 2 objects. (a) follow-reach-go together (E1), (b) approach-meet-go separately, (E2) approach-meet-go together (E3), (c) change direction-approach-meet-go separately (E4) and change direction-approach-meet-go together (E5). The horizontal colored lines along the time axis shows the change in states (dark blue: walking/going separately; green: approaching; brown: meeting and waiting to meet; teal: going together; cyan: follow).

For example in both the *following* and the *approaching* states the objects will be getting closer to each other and therefore the relative distance will be a continuously decreasing (Fig. 13(c)). However, the relative direction will have distinguishing properties as in the case of *following*, and the two objects will have the same direction. Similarly, in the case of *meeting* and *going together*, the relative distance will be approximately zero; and the magnitude of the speed will be the differentiating factor (Fig. 13(a)). The relative direction together with the relative distance helps in differentiating between *going together* and *going separately* (Fig. 13(b) and (c)), as two objects going in the same direction with same speed can be far from each other. The relative distance helps in solving such ambiguities as its value will be high in such situations. The clustering is therefore performed on a 5-dimensional feature space $\mathbf{f}$, represented as

$$\mathbf{f}_k^{(p,q)} = \left(\nu_k^{(p)},\ \nu_k^{(q)},\ \phi_k^{(p,q)},\ \varrho_k^{(p,q)},\ \dot{\varrho}_k^{(p,q)} \right), \tag{20}$$

where $\nu_k^{(p)}$ and $\nu_k^{(q)}$ are the magnitudes of speed; i.e.,

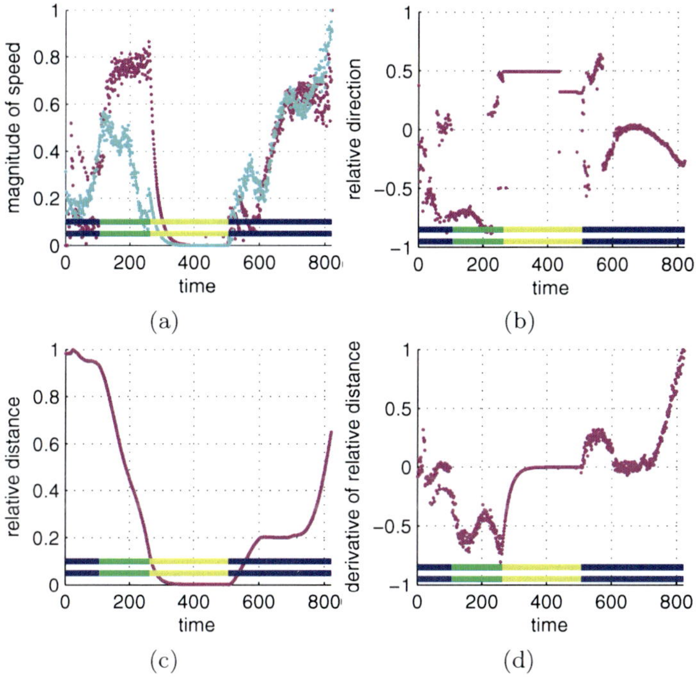

Fig. 13. Sample features for a 2-target interaction. The colored bar at the bottom of each graph is the ground truth of the objects individual states for interaction E3 (dark blue: walking/going separately; green: approaching; brown: meeting and waiting to meet). (a) the magnitude of speed for both objects, (b) the angle between the two targets, (c) relative distance between the objects, and (d) derivative of the relative distance between the objects.

$$\nu_k^{(p)} = \sqrt{\nu_{x_k^p}^2 + \nu_{y_k^p}^2}, \tag{21}$$

with ν_{x^pk} and ν_{y^pk} representing the horizontal and vertical components of the velocity. Likewise $\nu_k^{(q)}$, $\phi^{(p,q)}$ is the relative direction, defined as

$$\phi_k^{(p,q)} = \arctan\left(\frac{y_k^p - y_{k-1}^p}{x_k^p - x_{k-1}^p}\right) - \arctan\left(\frac{y_k^q - y_{k-1}^q}{x_k^q - x_{k-1}^q}\right), \tag{22}$$

with $\varrho_k^{(p,q)}$ being the relative distance, given as

$$\varrho_k^{(p,q)} = \sqrt{(x_k^p - x_k^q)^2 + (y_k^p - y_k^q)^2}, \tag{23}$$

and $\dot{\varrho}^{(p,q)}$ is the derivative of the relative distance, computed as

$$\dot{\varrho}_k^{(p,q)} = \frac{\varrho_k^{(p,q)} - \varrho_{k-1}^{(p,q)}}{\Delta k}. \tag{24}$$

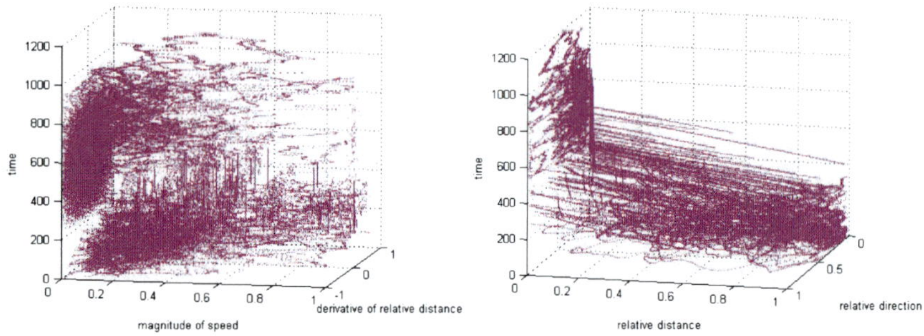

Fig. 14. Sample extracted features from 50 trajectories for the interaction E3. (a) Magnitude of speed vs. relative distance over time (3 clusters). (b) Relative direction vs. derivative of relative distance over time (2 clusters).

Note that the relative direction and distance are computed based on the current position of two interacting objects and not from a fixed reference point. Figure 14 shows the projection of the features computed on 50 trajectory pairs (E3). For visualization, 2 features per graph only are presented here. It can be seen that these features form certain clusters in the space.

Interaction event sequence estimation

Interaction event recognition (IER) can be performed with a trained CHMM. The states are discrete random variables (both the initial state and the state transition will result in one of the possible states). A distribution suited to model this discrete random variable is a multinomial distribution, as each trial results in exactly one out of a fixed finite number of possible outcomes. Hence both state transition and initial state probabilities of each chain can be chosen to be multinomial distributions. On the other hand, the emitting symbols are continuous random variables modeled using a continuous distribution. This distribution can be chosen to be a multivariate Gaussian, representing the projection of the trajectory onto the 5-dimensional feature space $f : X_k^p \times X_k^q \rightarrow \Re^5$, $\forall\ p, q$. The CHMM can be trained using the Expectation Maximization (EM) algorithm [51] using either the Maximum Likelihood (ML) [40] or Maximum a posterior (MAP) approach [22]. The MAP approach [22] is proven to perform better as it incorporates the prior distribution over the quantities to be estimated for improved optimization.

The recognition of the interactions is performed by applying the CHMM Viterbi decoding [52] using the trained model parameters. The posterior $P(O, S|\lambda)$, in Viterbi decoding, is calculated using Eq. (10). The decoding strategy is preferred here over the evaluation, as it does not require the event templates to be recognized but allow generating a sequence of activities performed by the objects. The steps for IER are summarized in Algorithm 2.

Algorithm 2. Interaction Event Recognition

λ : model parameters vector containing state transition probabilities, initial state probabilities and observation model parameters
μ_{S^i} : mean for each state i;
Σ_{S^i} : covariance for each state i;
μ^i_{GMM} : mean of ith cluster of GMM;
Σ^i_{GMM} : covariance of ith cluster of GMM;
N_S : total number of states;
η : threshold of log-posterior;
$O^r_{1:k}$: trajectory for target r at time t ; *count* : k;
f : feature vector;

Compute:

$$\mathbf{f}_k^{(p,q)} = \left(\nu_k^{(p)},\ \nu_k^{(q)},\ \phi_k^{(p,q)},\ \varrho_k^{(p,q)},\ \dot{\varrho}_k^{(p,q)}\right),$$

Train CHMM:
Initialize CHMM using GMM clustering on **f**
for $i = 1$ to N_S **do**
  $\mu_{S^i} = \mu^i_{GMM}$
  $\Sigma_{S^i} = \Sigma^i_{GMM}$
end for
Apply EM to train CHMM by maximizing $\mathcal{Q}(\lambda)$
repeat $\kappa = \kappa + 1$
  E-step compute: $\mathcal{Q}(\lambda; \lambda_\kappa)$
  M-step :

$$\lambda_{i+1} = \arg\max_{\lambda} \mathcal{Q}(\lambda; \lambda_\kappa)$$

until $\mathcal{Q}(\lambda_{\kappa+1}; \lambda_\kappa) - \mathcal{Q}(\lambda_\kappa; \lambda_{\kappa-1}) \leq \eta$
$\hat{\lambda} = \lambda_{\kappa+1}$
Apply CHMM Viterbi Algorithm using $\hat{\lambda}$ to estimate the state sequence.

Table 3 shows the comparison of the two CHMM's, CHMM-ML and CHMM-MAP, trained on the selected feature set on synthetic data. The synthetic dataset consists of 100 trajectory pairs for each event and of approximately 750 data points per trajectory and is equally divided into training and test sets. CHMM-MAP achieves both the better mean μ accuracy and lower standard deviation σ as compared to CHMM-ML. Some examples of interaction events detected using CHMM-MAP on real datasets are shown in Fig. 15. The real data consists of

Table 3. Accuracy comparison on test data for interaction detection using a Baseline method, a DBN, an HMM, a CHMM-ML and a CHMM-MAP

		E1	E2	E3	E4	E5
CHMM-ML	μ	0.8563	0.7881	0.5419	0.7976	0.7527
	σ	0.0798	0.0717	0.1861	0.0574	0.0851
CHMM-MAP	μ	0.8665	0.8688	0.7650	0.8376	0.8049
	σ	0.0682	0.0575	0.1175	0.0557	0.0804

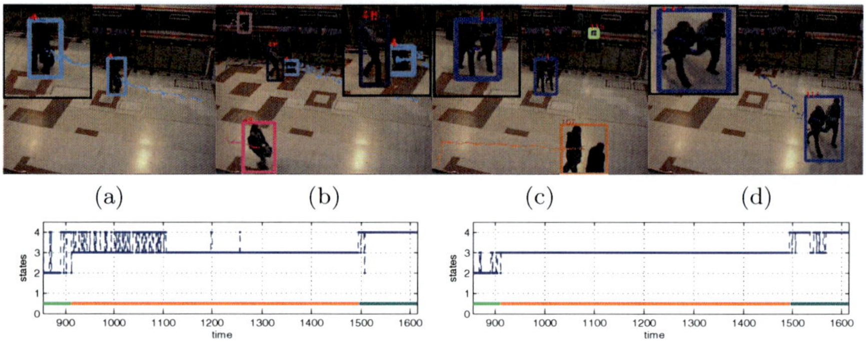

(a) (b) (c) (d)

(e)

Fig. 15. Sample detection, tracking and interaction event recognition results on a real surveillance scenario. (a-d) Frames 805, 902, 1535 and 1600 of the sequence S2-T3-C3 from the PETS2006 dataset, showing an approach-meet-go together interaction (E3). (j) CHMM-MAP generated sequence of interactions and the ground truth (light green: approaching; brown: meeting and waiting to meet; dark green: going together).

surveillance scenarios. The surveillance video is from the train station scenario S2-T3-C3 of the PETS2006 dataset and consists of 2551 frames (25 fps, resolution of 720 × 576 pixels). The CHMM trained on synthetic data can be applied to detect events on trajectories from real data, without retraining, by projecting them on the same space as the one used for generating synthetic data. This is important for practical reasons as it is difficult and time-consuming to obtain enough training samples from real data.

In case real data contains large number of objects and although in such cases the complexity of CHMM is combinatorial, however, the computational cost can be reduced through gating. Gating can be applied such that it eliminates objects that do not lie within 3σ of the Gaussian windows centered on the mid-point of the base of the bounding box of each object. The improvement in performance depends upon the covariance of the Gaussian window, which depends upon the average size and speed of the objects. In these results it is set empirically to 100 and 125 pixels along the horizontal and vertical directions. The evaluation of the IER via state estimation using Viterbi decoding is shown in Table 3. The accuracy is computed as $\wp = \frac{\zeta(GT_k \cap AR_k)}{N_{GT}}$, $k = k_0, \cdots, k_0 + N_{GT}$ where AR_k is the automatic recognition result, N_{GT} is the duration of the event in the ground truth, ζ gives the number of elements in the interaction set and k_0 is the starting time of the event. GT_k is the ground truth and is a discrete variable containing the state of the interacting objects at each time instance during the event span $k_0, \cdots, k_0 + N_{GT}$. The accuracy measure computes the frame level overlap of the estimated state sequence with the ground truth.

In S2-T3-C3 of the PETS2006, two persons meet and then leave the scene together (Fig. 15(a-d)) resulting in the detection of interaction sequence E3. A sample frame of the sequence, with ground truth, is shown in Fig. 15. The

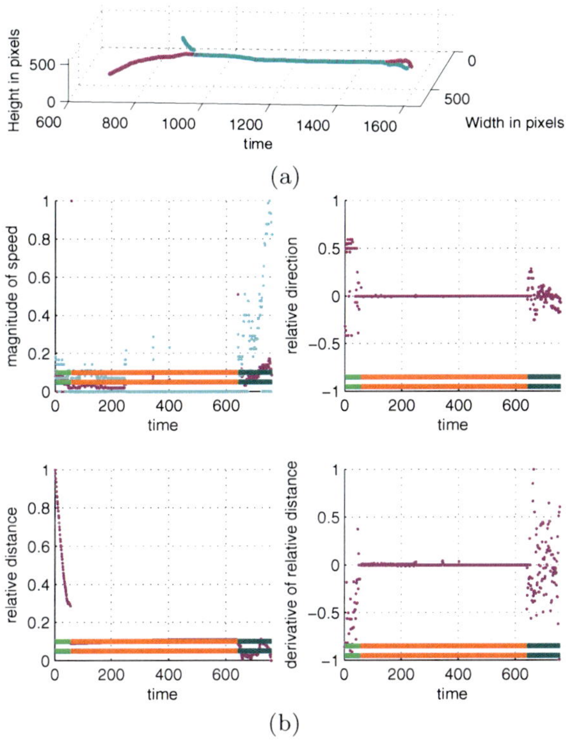

Fig. 16. Sample interacting trajectories and their features for real data. (a) Trajectories of two persons in the scene S2-T3-C3 of the PETS2006 data, showing interaction E3. (b) Normalized features for the trajectories of two persons.

accuracy for this recognition is 97.71%. It can be seen that the relative distance (Fig. 16(d)) between the two objects remain relatively high compared to the synthetic data as objects are in the close field of the camera. In such cases, even when the objects are close to each other, there exists a distance between the mid-point of the bases of their bounding boxes, which is much larger than in the case of point objects. This problem could be solved by defining a normalizing function based on the objects state space (position and size), as opposed to the target positions only.

5 Summary

This chapter has discussed the use of detection and tracking information for analyzing the interaction between objects. We showed that the activity detection problem can be modeled using time-series analysis and that graphical methods, and in particular Hidden Markov Models and their variants, are well suited for this analysis with spatio-temporal independence. Two cases of interaction event detection were considered. The first involved the interaction of dynamic objects with static objects. In this case the state sequences of one object only need to be estimated. We considered *object-centric* and *scene-centric* models to detect such interactions. In the *object-centric* model, the state of the temporarily static object is estimated with respect to the position of the dynamic objects. In the *scene-centric* model, the collective state of the dynamic objects is estimated with respect to static objects. For improving the event detection accuracy, explicit duration modeling can be performed using Hidden-Semi Markov Models to avoid prolonged self-transitions. The detection of interactions among multiple dynamic objects requires the state estimation for each object. To this end, the full coupling between the multiple object states was modeled using Coupled HMMs on relative features among the objects under analysis.

References

1. Andrade, E.L., Blunsden, S., Fisher, R.B.: Modelling crowd scenes for event detection. In: Proc. of IEEE Conf. on Pattern Recognition, Hong Kong, CN (2006)
2. Weinland, D., Ronfard, R., Boyer, E.: Free viewpoint action recognition using motion history volumes. Elsevier Journal of Computer Vision and Image Understanding 104 (2006)
3. Fathi, A., Mori, G.: Action recognition by learning mid-level motion features. In: Proc. of IEEE Int. Conf. on Computer Vision and Pattern Recognition, Anchorage, AK, USA (2008)
4. Stauffer, C., Grimson, W.: Learning patterns of activity using real-time tracking. IEEE Trans. on Pattern Analysis and Machine Intelligence 22, 747–757 (2000)
5. Taj, M., Maggio, E., Cavallaro, A.: Multi-feature graph-based object tracking. In: Stiefelhagen, R., Garofolo, J.S. (eds.) CLEAR 2006. LNCS, vol. 4122, pp. 190–199. Springer, Heidelberg (2007)
6. Taj, M., Maggio, E., Cavallaro, A.: Objective evaluation of pedestrian and vehicle tracking on the CLEAR surveillance dataset. In: Stiefelhagen, R., Bowers, R., Fiscus, J.G. (eds.) RT 2007 and CLEAR 2007. LNCS, vol. 4625, pp. 160–173. Springer, Heidelberg (2008)
7. Cavallaro, A., Ebrahimi, T.: Interaction between high-level and low-level image analysis for semantic video object extraction. EURASIP Journal on Applied Signal Processing 6, 786–797 (2004)
8. Wu, B., Nevatia, R.: Detection of multiple, partially occluded humans in a single image by bayesian combination of edgelet part detectors. In: Proc. of IEEE Int. Conf. on Computer Vision, Washington, DC, USA, pp. 90–97 (2005)
9. Viola, P., Jones, M., Snow, D.: Detecting pedestrians using patterns of motion and appearance. In: Proc. of Int. Conf. on Computer Vision Systems, Nice, FR (2003)
10. Shafique, K., Shah, M.: A noniterative greedy algorithm for multiframe point correspondence. IEEE Trans. on Pattern Analysis and Machine Intelligence 27, 51–65 (2005)
11. Yilmaz, A., Javed, O., Shah, M.: Object tracking: A survey. ACM Computing Surveys (CSUR) 38, 1–45 (2006)
12. Maggio, E., Smeraldi, F., Cavallaro, A.: Adaptive multifeature tracking in a particle filtering framework. IEEE Trans. on Circuits System and Video Technology 17, 1348–1359 (2007)

13. Maggio, E., Piccardo, E., Regazzoni, C., Cavallaro, A.: Particle PHD filter for multi-target visual tracking. In: Proc. of IEEE Int. Conf. on Acoustics, Speech, and Signal Processing, Honolulu, HI, USA (2007)
14. Karlsson, S., Taj, M., Cavallaro, A.: Detection and tracking of humans and faces. EURASIP Journal on Image and Video Processing, 1–9 (2008)
15. Zhou, H., Taj, M., Cavallaro, A.: Target detection and tracking with heterogeneous sensors. IEEE Journal of Selected Topics In Signal Processing 2 (2008)
16. Taj, M., Cavallaro, A.: Multi-camera track-before-detect. In: Proc. of ACM/IEEE Int. Conf. on Distributed Smart Cameras, Como, IT (2009)
17. Taj, M., Cavallaro, A.: Multi-camera scene analysis using an object-centric continuous distribution hidden Markov model. In: Proc. of IEEE Int. Conf. on Image Processing, San Antonio, TX, USA (2007)
18. Taj, M., Cavallaro, A.: Object and scene-centric activity detection using state occupancy duration modeling. In: Proc. of IEEE Int. Conf. on Advanced Video and Signal Based Surveillance, Santa Fe, NM, USA (2008)
19. Laptev, I., Marszalek, M., Schmid, C., Rozenfeld, B.: Learning realistic human actions from movies. In: Proc. of IEEE Int. Conf. on Computer Vision and Pattern Recognition, Anchorage, AK, USA (2008)
20. Liu, J., Ali, S., Shah, M.: Recognizing human actions using multiple features. In: Proc. of IEEE Int. Conf. on Computer Vision and Pattern Recognition, Anchorage, AK, USA (2008)
21. Velipasalar, S., Brown, L., Hampapur, A.: Specifying, interpreting and detecting high-level, spatio-temporal composite events in single and multi-camera systems. In: Proc. of IEEE Int. Conf. on Computer Vision and Pattern Recognition, NY, USA (2006)
22. Rezek, I., Gibbs, M., Roberts, S.J.: Maximum a posteriori estimation of coupled hidden Markov models. Journal of VLSI Signal Processing Systems 32, 55–66 (2002)
23. Mahmood, T.S., Vasilescu, A., Sethi, S.: Recognizing action events from multiple view points. In: Proc. of IEEE Workshop on Detection and Recognition of Events in Video, Madison, WI, USA (2001)
24. Ghanem, N., DeMenthon, D., Doermann, D., Davis, L.: Representation and recognition of events in surveillance video using Petri nets. In: Proc. of IEEE Int. Conf. on Computer Vision and Pattern Recognition, Washington, DC, USA (2004)
25. Wang, Y.: The variable-length hidden Markov model and its applications on sequential data mining. Technical report, Tsinghua University, Beijing, CN (2006), http://learn.tsinghua.edu.cn:8080/2001315444/VLHMM/icdm-techreport.pdf (last accessed: June 9, 2008)
26. Bishop, C.M.: Pattern Recognition and Machine Learning. Springer, Heidelberg (2006)
27. Medioni, G.G., Cohen, I., Bremond, F., Hongeng, S., Nevatia, R.: Event detection and analysis from video streams. IEEE Trans. on Pattern Analysis and Machine Intelligence 23, 873–889 (2001)
28. Rabiner, L.R.: A Tutorial on Hidden Markov Models and Selected Applications in Speech Recognition. Kaufmann, San Mateo (1990)
29. Natarajan, P., Nevatia, R.: Coupled hidden semi Markov models for activity recognition. In: IEEE Int. Workshop on Motion and Video Computing, Austin, TX, USA (2007)
30. i LIDS Team, Imagery library for intelligent detection systems (i-lids); a standard for testing video based detection systems. In: Proc. of IEEE Int. Carnahan Conf. on Security Technology, pp. 75–80 (2006)

31. Cher, D.: ETISEO Metrics Definition. Silogic, Toulouse Cedex 1, FR (2006), https://www-sop.inria.fr/orion/ETISEO/iso_album/eti-metrics_definition-v2.pdf (last accessed: June 30, 2009)
32. Ferryman, J.: Performance evaluation of tracking and surveillance. In: Conj. with IEEE Int. Conf. on Computer Vision and Pattern Recognition (2006), http://www.cvg.rdg.ac.uk/PETS2006/data.html (last accessed: June 30, 2009)
33. Fisher, R.: Caviar: Context aware vision using image-based active recognition (2001-2005), http://homepages.inf.ed.ac.uk/rbf/CAVIAR/caviar.htm (last accessed: June 30, 2009)
34. Zotkin, D., Duraiswami, R., Davis, L.: Multimodal 3-D tracking and event detection via the particle filter. In: Proc. of IEEE Workshop on Detection and Recognition of Events in Video, Vancouver, CA (2001)
35. Andrade, E.L., Blunsden, S., Fisher, R.B.: Detection of emergency events in crowded scenes. In: IEE Int. Symp. on Imaging for Crime Detection and Prevention, London, UK (2006)
36. Wu, G., Wu, Y., Jiao, L., Wang, Y., Chang, E.Y.: Multi-camera spatio-temporal fusion and biased sequence-data learning for security surveillance. In: Proc. of ACM Int. Conf. on Multimedia, NY, USA (2003)
37. Ke, Y., Sukthankar, R., Hebert, M.: Efficient visual event detection using volumetric features. In: Proc. of IEEE Int. Conf. on Computer Vision, Beijing, CN (2005)
38. Morris, R.J., Hogg, D.C.: Statistical models of object interaction. Int. Journal on Computer Vision 37, 209–215 (2000)
39. Brand, M.: Coupled hidden Markov models for modeling interacting processes. MIT media lab perceptual computing / learning and common sense technical report 405, Massachusetts Institute of Technology (1997), http://citeseer.ist.psu.edu/7422.html (last accessed: December 30, 2008)
40. Oliver, N., Rosario, B., Pentland, A.: A bayesian computer vision system for modeling human interactions. IEEE Trans. on Pattern Analysis and Machine Intelligence 22, 831–843 (2000)
41. Chartrand, G.: Introductory Graph Theory. In: Directed Graphs as Mathematical Models, ch. 1, pp. 16–19. Dover Publications, New York (1985)
42. Murphy, K.: Dynamic Bayesian networks: Representation, inference and learning. PhD thesis, Department of Computer Science, UC Berkeley (2002)
43. Zhang, L., Samaras, D., Klein, N.A., Volkow, N., Goldstein, R.: Modeling neuronal interactivity using dynamic bayesian networks. In: Weiss, Y., Schölkopf, B., Platt, J. (eds.) Advances in Neural Information Processing Systems, vol. 18, pp. 1593–1600. MIT Press, Cambridge (2006)
44. Brand, M., Kettnaker, V.: Discovery and segmentation of activities in video. IEEE Trans. on Pattern Analysis and Machine Intelligence 22, 844–851 (2000)
45. Galata, A., Cohn, A., Magee, D., Hogg, D.: Modeling interaction using learnt qualitative spatio-temporal relations and variable length Markov models. In: Proc. of European Conf. on Artificial Intelligence, Lyon, FR (2002)
46. Marhasev, E., Hadad, M., Kaminka, G.A.: Non-stationary hidden semi Markov models in activity recognition. In: Proc. of the AAAI Workshop on Modeling Others from Observations, Boston, MA, USA (2006)
47. Russell, M., Moore, R.: Explicit modelling of state occupancy in hidden Markov models for automatic speech recognition. In: Proc. of IEEE Int. Conf. on Acoustics, Speech, and Signal Processing, Tampa, FL, USA (1985)

48. Burshtein, D.: Robust parametric modeling of durations in hidden Markov models. IEEE Trans. on Speech and Audio Processing 4, 240–242 (1996)
49. Auvinet, E., Grossmann, E., Rougier, C., Dahmane, M., Meunier, J.: Left-luggage detection using homographies and simple heuristics. In: Joint IEEE Int. Workshop on Visual Surveillance and Performance Evaluation of Tracking and Surveillance, NY, USA (2006)
50. RATP FR, Call for Real-Time Event Detection Solutions (CREDS) for Enhanced Security and Safety in Public Transportation (2005)
51. McLachlan, G., Krishnan, T.: The EM Algorithm and Extensions, vol. 2. John Wiley & Sons, New York (1996)
52. Rabiner, L.R.: A tutorial on hidden Markov models and selected applications in speech recognition. Proc. of IEEE, 267–296 (1990)

Robust Audio Visual Biometric Person Authentication with Liveness Verification

Girija Chetty

Faculty of Information Sciences and Engineering,
University of Canberra, Australia
girija.chetty@canberra.edu.au

Abstract. In this paper we propose liveness verification for enhancing the robustness of audio-visual biometric person authentication systems. Liveness verification ensures that biometric cues are acquired from a live person who is actually present at the time of capture for authenticating the identity. The proposed liveness checking technique based on cross-modal association models involves hybrid fusion of acoustic and visual speech correlation features, which measure the degree of synchrony between the lips and the voice extracted from speaking face video sequences. Performance evaluation in terms of DET (Detector Error Tradeoff) curves and EERs (Equal Error Rates) on publicly available audiovisual speech databases show a significant improvement in robustness of system against different types of simulated replay attacks.

1 Introduction

Most of the commercial biometric identity authentication systems currently deployed are based on modeling the identity of a person based on unimodal information, i.e. face, voice, or fingerprint features. Also, many current interactive civilian remote human computer interaction applications are based on speech based voice features, which achieve significantly lower performance for operating environments with low signal-to-noise ratios (SNR). For a long time, use of acoustic information alone has been a great success for several automatic speech processing applications such as automatic speech transcription or speaker authentication, while face identification systems based visual information alone from faces also proved to be of equally successful. However, in adverse operating environments, performance of either of these systems could be suboptimal. Use of both visual and audio information can lead to better robustness, as they can provide complementary secondary clues that can help in the analysis of the primary biometric signals (Potamianos et al (2004)). The joint analysis of acoustic and visual speech can improve the robustness of automatic speech recognition systems (Liu et al(2002), Gurbuz et al (2002).

There have been several systems proposed on use of joint face-voice information for improving the performance of identity authentication systems. However, most of these state-of-the-art authentication approaches are based on independently processing the voice and face information and then fusing the scores – the score fusion (Chibelushi et al (2002), Pan et al (2000), Chaudari et. al.(2003)). A major weakness of these systems is that they do not take into account fraudulent replay attack scenarios into consideration, leaving them vulnerable to spoofing by recording the voice of

H.T. Sencar et al. (Eds.): Intel. Multimedia Analysis for Security Appli., SCI 282, pp. 59–78.
springerlink.com

the target in advance and replaying it in front of the microphone, or simply placing a still picture of the target's face in front of the camera. This problem can be addressed with liveness verification, which ensures that biometric cues are acquired from a live person who is actually present at the time of capture for authenticating the identity. With the diffusion of Internet based authentication systems for day-to-day civilian scenarios at a astronomical pace (Chetty and Wagner (2008)), it is high time to think about the vulnerability of traditional biometric authentication approaches and consider inclusion of liveness checks. Though there is some work in finger print based liveness checking techniques (Goecke and Millar (2003), Molhom et al (2002)), there is hardly any work in liveness checks based on user-friendly biometric identifiers (face and voice), which enjoy more acceptability for civilian Internet based applications requiring person identity authentication.

A significant progress however, has been made in independent processing of face only or voice only based authentication approaches (Chibelushi et al (2002), Pan et al (2000), Chaudari et. al.(2003)), in which until now, inherent coupling between jointly occurring primary biometric identifiers were not taken into consideration. Some preliminary approaches such as the ones described in (Chetty and Wagner (2008), Goecke and Millar (2003)), address liveness checking problem by using the traditional acoustic and visual speech features for testing liveness. Both these approaches, neither considered an inherent coupling between speech and orafacial articulators (lips, jaw and chin) during speech production, nor used a solid pattern recognition based evaluation framework for the validating the performance of co-inertia features.

In this Chapter we propose a novel approach for extraction of audio-visual correlation features based on cross-modal association models, and formulate a hybrid fusion framework for modelling liveness information in the identity authentication approach. Further, we develop a sound evaluation approach based on Bayesian framework for assessing the vulnerability of system at different levels of replay attack complexity. The rest of the Chapter is organized as follows. Section 2 describes the motivation for using the proposed approach, and the details the cross-modal association models are described in Section 3. Section 4 describes the hybrid fusion approach for combining the correlation features with loosely couple and mutually independent face-speech components. The data corpora used and the experimental setup for evaluation of the proposed features are described in Section 5. The experimental results, evaluating proposed correlation features and hybrid fusion technique is discussed in Section 6. Finally, Section 7 summarises the conclusions drawn from this work and plans for further research.

2 Motivation for Cross Modal Association Models

The motivation to use cross-modal association models is based on the following two observations: The first observation is in relation to any video event, for example a speaking face video, where the content usually consists of the co-occurring audio and the visual elements. Both the elements carry their contribution to the highest level semantics, and the presence of one has usually a "priming" effect on the other: when hearing a dog barking we expect the image of a dog, seeing a talking face we expect the presence of her voice, images of a waterfall usually bring the sound of running water etc. A series of psychological experiments on the cross-modal influences (Molhom et al (2002), MacDonald and McGurk(1978)) have proved the importance of synergistic fusion of the multiple modalities in the human perception system. A typical example of this kind is the well-known McGurk effect (MacDonald and McGurk(1978)). Several independent studies by cognitive psychologists suggest that the type of multi-sensory interaction between acoustic and orafacial articulators occurring in the McGurk effect involves both the early and late stages of integration processing (MacDonald and McGurk(1978)). It is likely that a human brain uses a hybrid form of fusion that depends on the availability and quality of different sensory cues.

Yet, in audiovisual speech and speaker verification systems, the analysis is usually performed separately on different modalities, and the results are brought together using different fusion methods. However, in this process of separation of modalities, we lose valuable cross-modal information about the whole event or the object we are trying to analyse and detect. There is an inherent association between the two modalities and the analysis should take advantage of the synchronised appearance of the relationship between the audio and the visual signal. The second observation relates to different types of fusion techniques used for joint processing of audiovisual speech signals. The late-fusion strategy, which comprises decision or the score fusion, is effective especially in case the contributing modalities are uncorrelated and thus the resulting partial decisions are statistically independent. Feature level fusion techniques, on the other hand, can be favoured (only) if a couple of modalities are highly correlated. However, jointly occurring face and voice dynamics in speaking face video sequences, is neither highly correlated (mutually dependent) nor loosely correlated nor totally independent (mutually independent). A complex and nonlinear spatiotemporal coupling consisting of highly coupled, loosely coupled and mutually independent components may exist between co-occurring acoustic and visual speech signals in speaking face video sequences (Jiang et al(2002), Yehia et al (1999)). The compelling and extensive findings by authors in Jiang et al(2002), validate such complex relationship between external face movements, tongue movements, and speech acoustics when tested for consonant vowel (CV) syllables and sentences spoken by male and female talkers with different visual intelligibility ratings. They proved that there is a higher correlation between speech and lip motion for C/a/ syllables than for C/i/ and C/u/ syllables. Further, the degree of correlation differs across different places of articulation, where lingual places have higher correlation than bilabial and glottal places. Also, mutual coupling can vary from talker to talker; depending on the gender of the talker, vowel context, place of articulation, voicing, and manner of articulation and the size of the face. Their findings also suggest that male speakers show higher correlations than female speakers. Further, the authors in Yehia et al (1999), also validate the complex, spatiotemporal and non-linear nature of the coupling between the vocal-tract and the facial articulators during speech production, governed by human physiology and language-specific phonetics. They also state that most likely connection between the tongue and the face is indirectly by way of the jaw. Other than the biomechanical coupling, another source of coupling is the control strategy between the tongue and cheeks. For example, when the vocal tract is shortened the tongue does not get retracted.

Due to such a complex nonlinear spatiotemporal coupling between speech and lip motion, this could be an ideal candidate for detecting and verifying liveness, and modelling the speaking faces by capturing this information can make the biometric authentication systems less vulnerable to spoof and fraudulent replay attacks, as it would be almost impossible to spoof a system which can accurately distinguish the artificially manufactured or synthesized speaking face video sequences from the live video sequences. Next section briefly describes the proposed cross modal association models based on cross-modal association models.

3 Cross-modal Association Models

In this section we describe the details of extracting audio-visual features based on cross-modal association models, which capture the nonlinear correlation components between the audio and lip modalities during speech production. This section is organised as follows: The details of proposed audio-visual correlation features based on different cross modal association techniques: Latent Semantic Analysis (LSA) technique, Cross-modal Factor Analysis (CFA) and Canonical Correlation Analysis (CCA) technique is described next.

3.1 Latent Semantic Analysis (LSA)

Latent semantic analysis (LSA) is used as a powerful tool in text information retrieval to discover underlying semantic relationships between different textual units e.g. keywords and paragraphs (Li et al(2003), Li et al(2001)). It is possible to detect the semantic correlation between visual faces and their associated speech based on the LSA technique. The method consists of three major steps: the construction of a joint multimodal feature space, the normalization, the singular value decomposition (SVD), and the semantic association measurement.

Given n visual features and m audio features at each of the t video frames, the joint feature space can be expressed as:

$$X = [V_1, \ldots, V_i, \ldots, V_n, A_1, \ldots, A_i, \ldots A_m] \tag{1}$$

where

$$V_i = (v_i(1), v_i(2), \ldots, v_i(t))^T \tag{2}$$

and

$$A_i = (a_i(1), a_i(2), \ldots, a_i(t))^T \tag{3}$$

Various visual and audio features can have quite different variations. Normalization of each feature in the joint space according to its maximum elements (or certain other statistical measurements) is thus needed and can be expressed as:

$$\hat{X}_{ij} = \frac{X_{kl}}{\max_{k,l}(abs(X_{kl}))} \quad \forall k,l \tag{4}$$

After normalisation, all elements in the normalised matrix $\hat{X}$ have values between – 1 and 1. SVD can then be performed as follows:

$$\hat{X} = S \cdot V \cdot D^T \tag{5}$$

where S and D are matrices composed of left and right singular vectors and V is the diagonal matrix of singular values in descending order.

Keeping only the first k singular vectors in S and D, we can derive an optimal approximation of with reduced feature dimensions, where the semantic correlation information between visual and audio features is mostly preserved. Traditional Pearson correlation or mutual information calculation (Li et al (2003), Hershey and Movellan(1999), Fisher et al(2000)) can then be used to effectively identify and measure semantic associations between different modalities. Experiments in Li et al(2003), have shown the effectiveness of LSA and its advantages over the direct use of traditional correlation calculation.

The above optimization of $\hat{X}$ in the least square sense can be expressed as

$$\hat{X} \cong \tilde{X} = \tilde{S} \,.\, \tilde{V} \,.\, \tilde{D} \tag{6}$$

where $\tilde{S}, \tilde{V}, and\ \tilde{D}$ consist of the first *k* vectors in *S*, *V*, and *D*, respectively.

The selection of an appropriate value for *k* is still an open issue in the literature. In general, *k* has to be large enough to keep most of the semantic structures. Eq. (6) is not applicable for applications using off-line training since the optimization has to be performed on the fly directly based on the input data. However, due to the orthogonal property of singular vectors, we can rewrite Eq. (6) in a new form as follows

$$\hat{X} \cong \tilde{X} = \tilde{S} \cdot V \cdot \tilde{D}^T \tag{7}$$

Now we only need the $\tilde{D}$ matrix in the calculation, which can be trained in advance using ground truth data. This derived new form is important for those applications that need off-line trained SVD results.

3.2 Cross Modal Factor Analysis (CMA)

LSA does not distinguish features from different modalities in the joint space. The optimal solution based on the overall distribution, which LSA models, may not best represent the semantic relationships between the features of different modalities, since distribution patterns among features from the same modality will also greatly impact the results of the LSA. A solution to the above problem is to treat the features from different modalities as two separate subsets and focus only on the semantic patterns between these two subsets. Under the linear correlation model, the problem now is to find the optimal transformations that can best represent or identify the coupled patterns between the features of the two different subsets. We adopt the following optimization criterion to obtain the optimal transformations:

Given two mean-centred matrices X and Y, which consist of row-by-row coupled samples from two subsets of features, we want orthogonal transformation matrices A and B that can minimise the expression:

$$\|XA - YB\|_F^2 \tag{8}$$

where

$$A^T A = I \text{ and } B^T B = I .$$

$\|M\|_F$ denotes the Frobenius norm of the matrix M and can be expressed as:

$$\|M\|_F = \left(\sum_i \sum_j |m_{ij}|^2 \right)^{1/2} \tag{9}$$

In other words, *A* and *B* define two orthogonal transformation spaces where coupled data in *X* and *Y* can be projected as close to each other as possible. Since we have

$$\|XA - YB\|_F^2 = trace((XA - XB).(YA - YB)^T) \tag{10}$$

$$= trace(XAA^T X^T + YBB^T Y^T - XAB^T Y^T - YBA^T X^T)$$

$$= trace((XX^T) + trace(YY^T) - 2 \cdot trace(XAB^T Y^T))$$

where the trace of a matrix is defined to be the sum of the diagonal elements. We can easily see from above that matrices *A* and *B* which maximise *trace* (XAB^TY^T) will minimise Eq. (10). It can be shown (Li et al(2003)), that such matrices are given by

$$\begin{cases} A = S_{xy} \\ B = D_{xy} \end{cases} \quad \text{where } X^T Y = S_{xy} \cdot V_{xy} \cdot D_{xy} . \tag{11}$$

With the optimal transformation matrices *A* and *B*, we can calculate the transformed version of *X* and *Y* as follows:

$$\begin{cases} \tilde{X} = X \cdot A \\ \tilde{Y} = Y \cdot B \end{cases} \tag{12}$$

Corresponding vectors in $\tilde{X}$ and $\tilde{Y}$ are thus optimised to represent the coupled relationships between the two feature subsets without being affected by distribution patterns within each subset. Traditional Pearson correlation or mutual information calculation (Li et al (2003), Hershey and Movellan(1999), Fisher et al(2000)) can then be performed on the first and most important *k* corresponding vectors in $\tilde{X}$ and $\tilde{Y}$, which similar to those in LSA preserve the principal coupled patterns in much lower dimensions.

In addition to feature dimension reduction, feature selection capability is another advantage of CFA. The weights in *A* and *B* automatically reflect the significance of individual features, clearly demonstrating the great feature selection capability of CFA, which makes it a promising tool for different multimedia applications including audiovisual speaker identity verification.

3.3 Canonical Correlation Analysis (CCA)

Following the development of the previous section, we can adopt a different optimization criterion: Instead of minimizing the projected distance, we attempt to find transformation matrices A and B that maximise the correlation between X_A and Y_B. This can be described more specifically using the following mathematical formulations:

Given two mean centered matrices *X* and *Y* as defined in the previous section, we seek matrices A and B such that

$$correlation(XA, XB) = correlation(\tilde{X}, \tilde{Y}) = diag(\lambda_1, \cdots \lambda_i, \cdots, \lambda_l) \quad (13)$$

where $\tilde{X} = Y \cdot B$, and $1 \geq \lambda_1 \geq, \cdots, \lambda_i, \cdots, \geq \lambda_l \geq 0$. λ_i represents the largest possible correlation between the ith translated features in $\tilde{X}$ and $\tilde{Y}$. A statistical method called canonical correlation analysis (Lai and Fyfe(1998), Tabanick and Fidell(1996)] can solve the above problem with additional norm and orthogonal constraints on translated features:

$$E\{\tilde{X}^T \cdot \tilde{X}\} = I \text{ and } E\{\tilde{Y} \cdot \tilde{Y}\} = I \quad (14)$$

In CCA, A and B are calculated as follows:

$$A = \sum\nolimits_{xx}^{-1/2} . S_K \quad , \quad B = \sum\nolimits_{yy}^{-1/2} . D_K \quad (15)$$

where

$$\sum\nolimits_{xx} = E\{X^T X\}, \sum\nolimits_{yy} = E\{Y^T Y\}, \sum\nolimits_{xy} = E\{X^T Y\}$$

and

$$L = \sum\nolimits_{xx}^{-1/2} \sum\nolimits_{xy} \sum\nolimits_{yy}^{-1/2} = S_K . V_K . D_K^T \quad (16)$$

The CCA is described in further details in Hotelling (1936) and Hardoon et al. (2004). The optimization criteria used for all three cross modal associations CFA, CCA and LSA exhibit a high degree of noise tolerance. Hence the correlation features extracted perform better as compared to normal correlation analysis against noisy environmental conditions.

4 Hybrid Audiovisual Fusion

In this Section, we describe the fusion approach used for combing the extracted audio-lip correlated components with mutually independent audio and visual speech features.

4.1 Feature Fusion of Correlated Components

The algorithm for fusion of audiovisual feature extracted using the cross modal association (CMA) models (a common term being used here to represent LSA, CFA or CCA analysis methods) can be described as follows:

Let f_A and f_L represent the audio MFCC and lip-region eigenlip features respectively. A and B represent the CMA transformation matrices (LSA, CFA or CMA matrices). One can apply CMA to find two new feature sets $f_A^{'} = A^T f_A$ and $f_L^{'} = B^T f_L$ such that the between-class cross-modal association coefficient matrix of $f_A^{'}$ and $f_L^{'}$ is diagonal with maximised diagonal terms. However, maximised diagonal terms do not necessarily mean that all the diagonal terms exhibit strong cross modal association. Hence, one can pick the maximally correlated components that are above a certain correlation threshold θ_k. Let us denote the projection vector that corresponds to the diagonal terms larger than the threshold θ_k by $\tilde{w}_A$. Then the corresponding projections of f_A and f_L are given as:

$$\tilde{f}_A = \tilde{w}_A^T . f_A \text{ and } \tilde{f}_L = \tilde{w}_L^T . f_L \tag{10}$$

Here $\tilde{f}_A$ and $\tilde{f}_L$ are the correlated components that are embedded in f_A and f_L. By performing feature fusion of correlated audio and lip components, we obtained the CMA optimised feature fused audio-lip feature vector:

$$\tilde{f}_{AL}^{LSA} = \left[\tilde{f}_A^{LSA} \quad \tilde{f}_L^{LSA}\right] \tag{11}$$

$$\tilde{f}_{AL}^{CFA} = \left[\tilde{f}_A^{CFA} \quad \tilde{f}_L^{CFA}\right] \tag{12}$$

$$\tilde{f}_{AL}^{CCA} = \left[\tilde{f}_A^{CCA} \quad \tilde{f}_L^{CCA}\right] \tag{13}$$

4.2 Late Fusion of Mutually Independent Components

In the Bayesian framework, late fusion can be performed using the product rule assuming statistically independent modalities, and various methods have been proposed in the literature as alternatives to the product rule such as max rule, min rule and the reliability-based weighted summation rule (Nefian et al(2002), Movellan and Mineiro(1997)). In fact, the most generic way of computing the joint scores can be expressed as a weighted summation

$$\rho(\lambda_r) = \sum_{n=1}^{N} w_n \log P(f_n \mid \lambda_r) \quad for\ r = 1,2,\ldots,R \tag{14}$$

where $\rho_n(\lambda_r)$ is the logarithm of the class-conditional probability, $P(f_n|\lambda_r)$, for the nth modality f_n given class λ_r, and w_n denotes the weighting coefficient for modality n, such that $\Sigma_n w_n = 1$. Then the fusion problem reduces to a problem of finding the optimal weight coefficients. Note that when $w_n = \frac{1}{N}$ $\forall n$, Eqn. 14 is equivalent to the product rule. Since the w_n values can be regarded as the reliability values of the classifiers, this combination method is also referred to as RWS (Reliability Weighted Summation) rule (Jain et al(2005), Nefian et al(2002)).The statistical and the numerical range of these likelihood scores vary from one classifier to another. Using sigmoid and variance normalization as described in (Jain et al(2005)), the likelihood scores can be normalised to be within the (0, 1) interval before the fusion process.

The hybrid audiovisual fusion vector in this Chapter was obtained by late fusion of feature fused correlated components ($\tilde{f}_{AL}^{LSA}, \tilde{f}_{AL}^{CFA}, \tilde{f}_{AL}^{CCA}$) with uncorrelated and mutually independent implicit lip texture features, and audio features with weights selected using the an automatic weight adaptation rule and is described in the next Section.

4.3 Automatic Weight Adaptation

For the RWS rule, the fusion weights are chosen empirically, whereas for the automatic weight adaptation, a mapping needs to be developed between modality reliability estimate and the modality weightings. The late fusion scores can be fused via sum rule or product rule. Both methods were evaluated for empirically chosen weights, and it was found that the results achieved for both were similar. However, sum rule for fusion has been shown to be more robust to classifier errors in literature (Jain et al(2005), Sanderson(2008)), and should perform better when the fusion weights are automatically, rather than empirically determined. Hence the results for additive fusion only, are presented here. Prior to late fusion, all scores were normalised to fall into the range of [0,1], using min-max normalisation.

$$\begin{aligned} P(S_i|x_A, x_V) &= \alpha P(S_i|x_A) + \beta P(S_i|x_v) \\ P(S_i|x_A, x_V) &= P(S_i|x_A)^{\alpha} \times P(S_i|x_v)^{\beta} \end{aligned} \tag{15}$$

where

$$\alpha = \begin{Bmatrix} 0 & c \le 1 \\ 1+c & -1 < c < 0 \\ 1 & c \ge 0 \end{Bmatrix} \quad and \quad \beta = \begin{Bmatrix} 1 & c \le 0 \\ 1-c & 0 < c < 1 \\ 0 & c \ge 1 \end{Bmatrix} \tag{16}$$

where, x_A and x_V refer to the audio test utterance and visual test sequence/image respectively.

To carry out automatic fusion, that adapts to varying acoustic SNR conditions, a single parameter c, the *fusion parameter*, was used to define the weightings; the audio weight α and the visual weight β, i.e., both α and β dependent on c. Higher values of c (>0) place more emphasis on the audio module whereas lower values (<0) place more emphasis on the visual module. For $c \geq 1$, $\alpha = 1$ and $\beta = 0$, hence the audiovisual fused decision is based entirely on the audio likelihood score, whereas, for $c \leq -1$, $\alpha = 0$ and $\beta = 1$, the decision is based entirely on the visual score. So in order to account for varying acoustic conditions, only c has to be adapted.

The reliability measure was the audio log-likelihood score $\rho_n(\lambda_r)$. As the audio SNR decreases, the absolute value of this reliability measure decreases, and becomes closer to the threshold for client likelihoods. Under clean test conditions, this reliability measure increases in absolute value because the client model yields a more distinct score. So, a mapping between ρ and c can automatically vary α and β and hence place more or less emphasis on the audio scores. To determine the mapping function $c(\rho)$, the values of c which provided for optimum fusion, c_{opt}, were found by exhaustive search for the N tests at each SNR levels. This was done by varying c from -1 to $+1$, in steps of 0.01, in order to find out which c value yielded the best performance. The corresponding average reliability measures were calculated, ρ_{mean}, across the N test utterances at each SNR level.

$$c(\rho) = c_{os} + \frac{h}{\exp[d.(\rho + \rho_{os})]} \tag{17}$$

A sigmoid function was employed to provide a mapping between the c_{opt} and the ρ_{mean} values, where c_{os} and ρ_{os} represent the offsets of the fusion parameter and reliability estimate respectively; h captures the range of the fusion parameter; and d determines the steepness of the sigmoid curve. The sigmoidal parameters were determined empirically to give the best performance. Once the parameters have been determined, automatic fusion can be carried out. For each set of N test scores, the ρ value was calculated and mapped to c, using $c = c(\rho)$, and hence, α and β can be determined. This fusion approach is similar to that used in (Sanderson(2008)) to perform speech recognition. The method can also be considered to be a secondary classifier, where the measured ρ value arising from the primary audio classifier is *classified* to a suitable c value; also, the secondary classifier is *trained* by determining the parameters of the sigmoid mapping.

The described method can be employed to combine any two modules. It can also be adapted to include a third module. We assume here that only the audio signal is degraded when testing, and that the video signal is of fixed quality. The third module we use here is an audio-lip correlation module, which involves a cross modal transformation of feature fused audio-lip features based on CCA, CFA or LSA cross modal analysis as described in Section 3.

An overview of the fusion method described is given in Figure 1. It can be seen that the reliability measure, ρ, depends only on the audio module scores. Following

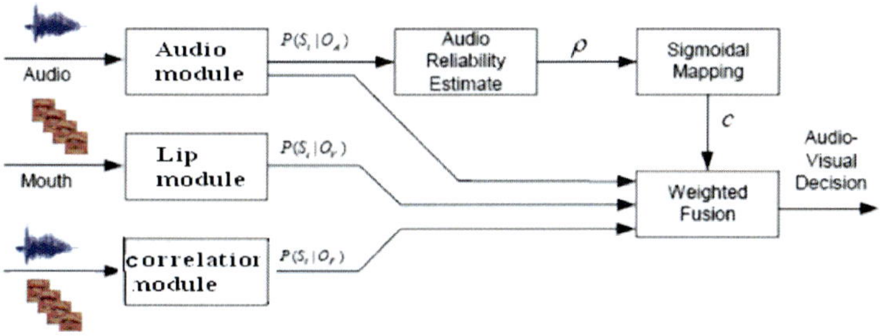

Fig. 1. System Overview of Hybrid Fusion Method

the sigmoidal mapping of ρ, the fusion parameter c is passed into the fusion module along with the three scores arising from the three modules; fusion takes place to give the audiovisual decision.

5 Data Corpora and Experimental Setup

A experimental evaluation of proposed correlation features based on cross-modal association models and their subsequent hybrid usion was carried out with two different audio-visual speaking face video corpora VidTIMIT (Sanderson(2008)) and (DaFEx (Battocchi et al (2004), Mana et al (2006)). Figure 2 show some images from the two corpora. The details of the two corpora are given in VidTIMIT (Sanderson(2008), DaFEx (Battocchi et al (2004), Mana et al (2006)).

The pattern recognition experiments with the data from the two corpora and the correlation features extracted from the data involved two phases, the training phase and the testing phase. In the training phase a 10-mixture Gaussian mixture model λ of a client's audiovisual feature vectors was built, reflecting the probability densities for the combined phonemes and visemes (lip shapes) in the audiovisual feature space. In the testing phase, the clients' live test recordings were first evaluated against the client's model λ by determining the log likelihoods $log\ p(X|\lambda)$ of the time sequences X of audiovisual feature vectors under the usual assumption of statistical independence of successive feature vectors.

For testing replay attacks, we used a two level testing, a different approach from traditional impostor attacks testing used in identity verification experiments. Here the impostor attack is a surreptitious replay of previously recorded data and such an attack can be simulated by synthetic data. Two different types of replay attacks with increasing level of sophistication and complexity were simulated: the "*static*" replay attacks and the "*dynamic*" replay attacks.

For testing "*static*" replay attacks, a number of "fake" or synthetic recordings were constructed by combining the sequence of audio feature vectors from each test utterance with ONE visual feature vector chosen from the sequence of visual feature vectors and keeping that visual feature vector constant throughout the utterance. Such a synthetic sequence represents an attack on the authentication system, carried out by replaying an audio recording of a client's utterance while presenting a still photograph

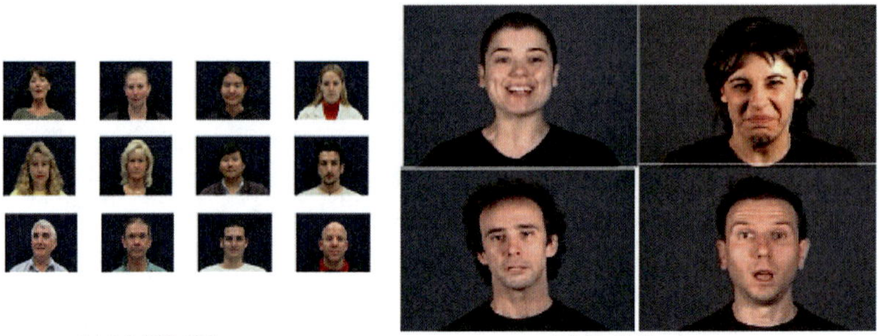

(a) VidTIMIT corpus (b) DaFeX corpus

Fig. 2. Sample Images from VidTIMIT and DaFeX corpus

to the camera. Four such fake audiovisual sequences were constructed from different still frames of each client test recording. Log-likelihoods *log p(X'|λ)* were computed for the fake sequences *X'* of audiovisual feature vectors against the client model λ. In order to obtain suitable thresholds to distinguish live recordings from fake recordings, detection error trade-off (DET) curves and equal error rates (EER) were determined.

For testing "*dynamic*" replay attacks, an efficient photo-realistic audio-driven facial animation technique with near-perfect lip-synching of the audio and several image key-frames of the speaking face video sequence was done to create a artificial speaking character for each person (Chetty and Wagner(2008), Sanderson(2008).

In Bayesian framework, the liveness verification task can be essentially considered as a two class decision task, distinguishing the test data as a genuine client or an impostor. The impostor here is a fraudulent replay of client specific biometric data. For such a two-class decision task, the system can make two types of errors. The first type of error is a False Acceptance Error (FA), where an impostor (fraudulent replay attacker) is accepted. The second error is a False Rejection (FR), where a true claimant (genuine client) is rejected. Thus, the performance can be measured in terms of False Acceptance Rate (FAR) and False Reject Rate (FRR), as defined as (Eqn. 18) :

$$FAR\ \% = \frac{I_A}{I_T} \times 100\ \% \qquad FRR\ \% = \frac{C_R}{C_T} \times 100\ \% \tag{18}$$

where I_A is the number of impostors classified as true claimants, I_T is the total number of impostor classification tests, C_R is the number of true claimants classified as impostors, and C_T is the total number of true claimant classification tests. The implications of this is minimizing the FAR increases the FRR and vice versa, since the errors are related. The trade-off between FAR and FRR is adjusted using the threshold θ, an experimentally determined speaker-independent global threshold from the training/enrolment data. The trade-off between FAR and FRR can be graphically represented by a Receiver Operating Characteristics (ROC) plot or a Detection Error Trade-off (DET) plot. The ROC plot is on a linear scale, while the DET plot is on a normal-deviate logarithmic scale. For DET plot, the FRR is plotted as a function of FAR. To quantify the performance into a single number, the Equal Error Rate (EER) is often used. Here the system is configured with a threshold, set to an operating point when FAR % = FRR %.

It must be noted that the threshold θ can also be adjusted to obtain a desired performance on test data (data unseen by the system up to this point). Such a threshold is known as the aposteriori threshold. However, if the threshold is fixed before finding the performance, the threshold is known as the apriori threshold. The apriori threshold can be found via experimental means using training/enrolment or evaluation data, data which has also been unseen by the system up to this point, but is separate from test data.

Practically, the a priori threshold is more realistic. However, it is often difficult to find a reliable apriori threshold. The test section of a database is often divided into two sets: evaluation data and test data. If the evaluation data is not representative of the test data, then the apriori threshold will achieve significantly different results on evaluation and test data. Moreover, such a database division reduces the number of verification tests, thus decreasing the statistical significance of the results. For these reasons, many researchers prefer to use the aposteriori and interpret the performance obtained as the expected performance.

Different subsets of data from the VidTIMIT and DaFeX were used. The gender-specific universal background models (UBMs) were developed using the training data from two sessions, Session 1 and Session 2, of the VidTIMIT corpus, and for testing Session 3 was used. Due to the type of data available (test session sentences differ from training session sentences), only text-independent speaker verification experiments could be performed with VidTIMIT. This gave 1536 (2*8*24*4) seconds of training data for the male UBM and 576 (2*8*19*4) seconds of training data for the female UBM. The GMM topology with 10 Gaussian mixtures was used for all the experiments. The number of Gaussian mixtures was determined empirically to give the best performance. For the DaFeX database, similar gender-specific universal background models (UBMs) were obtained using training data from the text-dependent subsets corresponding to neutral expression. Ten sessions of the male and female speaking face data from these subsets were used for training and 5 sessions for testing.

For all the experiments, the global threshold was set using test data. For the male only subset of the VidTIMIT database, there were 48 client trials (24 male speakers x 2 test utterances in Session 3) and 1104 impostor trials (24 male speakers x 2 test utterances in Session 3 x 23 impostors/client), and for the female VidTIMIT subset, there were 38 client trials (19 male speakers x 2 test utterances in Session 3) and 684 impostor trials (19 male speakers x 2 test utterances in Session 3 x 18 impostors/client). For the male only subset for DaFeX database, there were 25 client trials (5 male speakers x 5 test utterances in each subset) and 100 impostor trials (5 male speakers x 5 test utterances x 4 impostors/client), and for the female DaFeX subset, there were similar numbers of the client and impostor trials as in the male subset as we used 5 male and 5 female speakers from different subsets.

Different sets of experiments were conducted to evaluate the performance of the proposed correlation features based on cross modal association models (LSA, CCA and CMA), and their subsequent fusion in terms of DET curves and equal error rates (EER). Next Section discusses the results from different experiments.

6 Experimental Results

Figure 3 plots the maximised diagonal terms of the between class correlation coefficient matrix after the LSA, CCA and CFA analysis of audio MFCC and lip-texture

Table 1. Results for correlation features based in CMA models: EERs at varying correlation coefficient threshold values (θ) with the corresponding projection dimension (dim)

	EER(%) at (θ, dim)						
Θ	0.0	0.1	0.2	0.3	0.4	0.5	0.6
Dim	40	15	12	10	8	6	4
$\tilde{f}_{AL}^{CFA}$	6.8	4.7	5.3	5.0	4.7	7.4	10.3
$\tilde{f}_{AL}^{CCA}$	7.5	5.18	5.84	5.5	5.18	8.16	11.36
$\tilde{f}_{AL}^{LSA}$	11.7	8.09	9.12	8.6	8.09	12.74	17.74

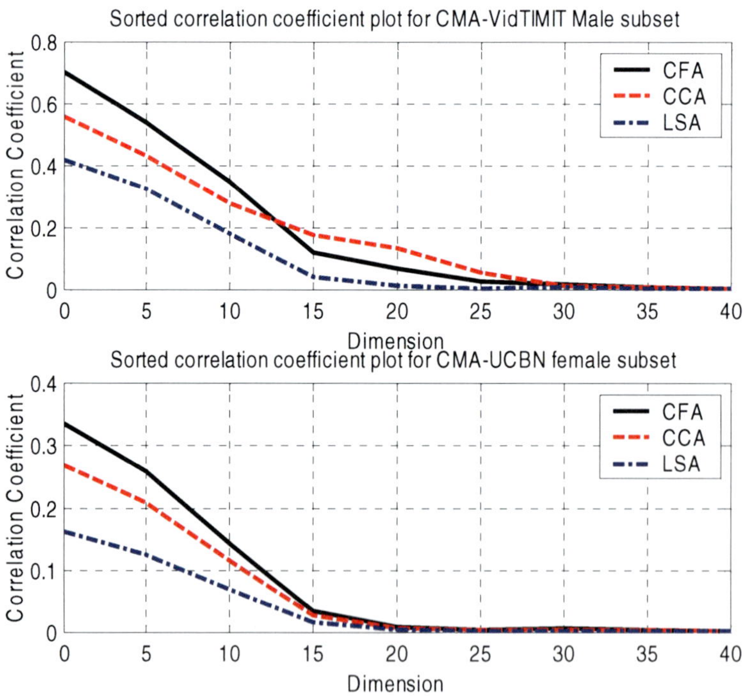

Fig. 3. Sorted correlation coefficient plot for audio and lip texture cross modal analysis

(f^{N}_{eigLip}) features. Results for the CFA analysis technique for the VidTIMIT male subset are only discussed here. As can be observed from Figure 3, the maximum correlation coefficient is around 0.7 and 15 correlation coefficients out of 40 are higher than 0.1.

Table 1 presents the EER performance of the feature fusion of correlated audio-lip fusion features (cross modal features) for varying correlation coefficient threshold θ. Note that, when all the 40 transformed coefficients are used, the EER performance is 6.8%. The EER performance is observed to have a minimum around 4.7% for threshold values from 0.1 to 0.4. The optimal threshold that minimises the EER performance and the feature dimension is found to be 0.4.

As can seen in Table 2 and Figure 4, for static replay attack scenarios (from the last four rows in Table 2), the nonlinear correlation components between acoustic and orafacial articulators during speech production is more efficiently captured by hybrid fusion scheme involving late fusion of audio f_{mfcc} features, f_{eigLip} lip features, and

Table 2. EER performance for static replay attack scenario with late fusion of correlated components with mutually independent components: (+) represents RWS rule for late fusion, (-) represents feature level fusion)

	VidTIMIT male subset			DaFeX male subset		
Modality	CFA EER (%)	CCA EER (%)	LSA EER (%)	CFA EER (%)	CCA EER (%)	LSA EER (%)
f_{mfcc}	4.88	4.88	4.88	5. 7	5. 7	5. 7
f_{eigLip}	6.2	6.2	6.2	7.64	7.64	7.64
$f_{mfcc-eigLip}$	7.87	7.87	7.87	9.63	9.63	9.63
$\tilde{f}_{mfcc-eigLip}$	3.78	2.3	2.76	4.15	2.89	3.14
$f_{mfcc} + f_{mfcc-eigLip}$	2.97	2.97	2.97	3.01	3.01	3.01
$f_{mfcc} + \tilde{f}_{mfcc-eigLip}$	0.56	**0.31**	0.42	0.58	0.38	0.57
$f_{mfcc} + f_{eigLip} + f_{mfcc-eigLip}$	6.68	6.68	6.68	7.75	7.75	7.75
$f_{mfcc} + f_{eigLip} + \tilde{f}_{mfcc-eigLip}$	0.92	0.72	0.81	0.85	0.78	0.83

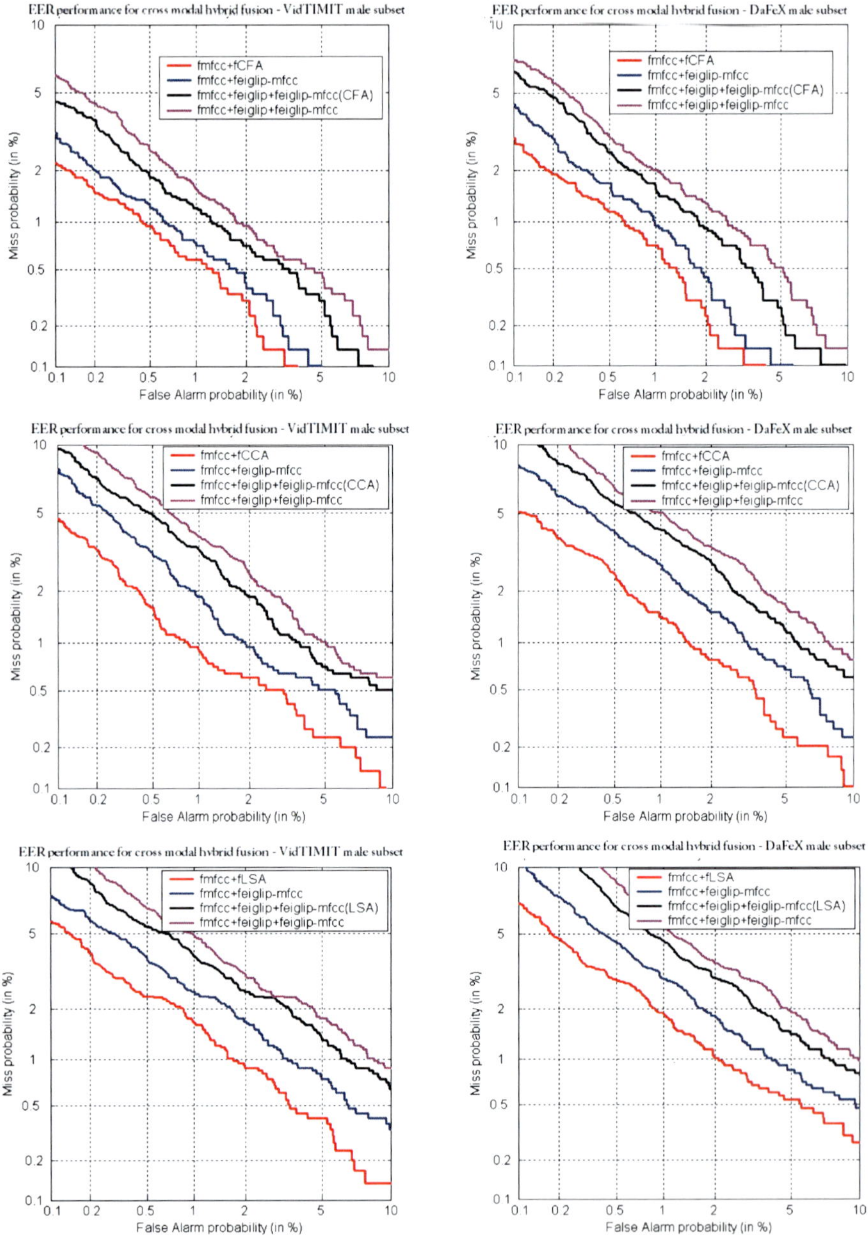

Fig. 4. DET curves for hybrid fusion of correlated audio-lip features and mutually independent audio-lip features for static replay attack scenario

Table 3. EER performance for dynamic replay attack scenario with late fusion of correlated components with mutually independent components

	VidTIMIT male subset			DaFeX male subset		
Modality	CFA EER (%)	CCA EER (%)	LSA EER (%)	CFA EER (%)	CCA EER (%)	LSA EER (%)
f_{eigLip}	36.58	36. 58	36. 58	37.51	37. 51	37. 51
$f_{mfcc-eigLip}$	27.68	27.68	27.68	28.88	28.88	28.88
$\tilde{f}_{mfcc-eigLip}$	24.48	22.36	23.78	26.43	24.67	25.89
$f_{mfcc} + f_{mfcc-eigLip}$	22.45	22.45	22.45	23.67	23.67	23.67
$f_{mfcc} + \tilde{f}_{mfcc-eigLip}$	17.89	16.44	19.48	18.46	17.43	20.11
$f_{mfcc} + f_{eigLip} + f_{mfcc-eigLip}$	21.67	21.67	21.67	25.42	25.42	25.42
$f_{mfcc} + f_{eigLip} + \tilde{f}_{mfcc-eigLip}$	14.23	10.06	12.27	16.68	12.36	13.88

feature-level fusion of correlated audio-lip $\tilde{f}_{mfcc-eigLip}$ features). This could be due to modelling of identity specific mutually independent, loosely coupled and closed coupled audio-visual speech components with this approach, resulting in an enhancement in overall performance.

Though all correlation features performed well, the CCA features appear to be the best performer for static attack scenario, with an EER of 0.31%. This was the case for all the subsets of data shown in Table 2. Also, the EERs for hybrid fusion experiments with $\tilde{f}_{mfcc-eigLip}$ correlated audio lip features performed better as compared to ordinary feature fusion of $f_{mfcc-eigLip}$ features. EERs of 0.31% and 0.72% were achieved for $f_{mfcc} + \tilde{f}_{mfcc-eigLip}$ and $f_{mfcc} + f_{eigLip} + \tilde{f}_{mfcc-eigLip}$ features, the hybrid fusion types involving CMA optimised correlated features, as compared to an EER of 2.97% for $f_{mfcc} + f_{mfcc-eigLip}$ features and 6.68% for $f_{mfcc} + f_{eigLip} + f_{mfcc-eigLip}$ features, which are hybrid fusion types based on ordinary feature fusion of uncorrelated audio-lip features. This shows that correlation features based on proposed cross-modal association models can extract the intrinsic nonlinear temporal correlations between audio-lip features and could be more useful for checking liveness.

The EER table in Table 3 shows the evaluation of hybrid fusion of correlated audio-lip features based on cross modal analysis (CFA, CCA and LSA) for dynamic replay attack scenario. As can be seen, the CMA optimized correlation features perform better as compared to uncorrelated audio-lip features for complex dynamic attacks. Further, for the VidTIMIT male subset, it was possible to achieve the best EER of 10.06% for $f_{mfcc} + f_{eigLip} + \tilde{f}_{mfcc-eigLip}$ features, a hybrid fusion type involving feature fusion of correlated audio-lip features based on CCA analysis.

7 Conclusion

In this Chapter, we have proposed liveness verification for enhancing the robustness of biometric person authentication systems against impostor attacks involving fraudulent replay of client data. Several correlation features based on novel cross-modal association models have been proposed as an effective countermeasure against such attacks. These new correlation measures model the nonlinear acoustic-labial temporal correlations for the speaking faces during speech production, and can enhance the system robustness against replay attacks.

Further, a systematic evaluation methodology was developed, involving increasing level of difficulty in attacking the system – moderate and simple static replay attacks, and, sophisticated and complex dynamic replay attacks, allowing a better assessment of system vulnerability against attacks of increasing complexity and sophistication. For both static and dynamic replay attacks, the EER results were very promising for the proposed correlation features, and their hybrid fusion with loosely coupled (feature-fusion) and mutually independent (late fusion) components, as compared to fusion of uncorrelated features. This suggests that it is possible to perform liveness verification in authentication paradigm. and thwart replay attacks on the system. Further, this study shows that, it is difficult to beat the system, if underlying modelling approach involves efficient feature extraction and feature selection techniques that can capture intrinsic biomechanical properties accurately.

However, though the EER performance appeared to be very promising for static replay attack scenarios (EER of 0.31 % for CCA features), the deterioration in performance for more sophisticated - dynamic replay attack scenario (EER of 10.06 % for CCA features), suggests that, there is an urgent need to investigate more robust feature extraction, feature selection, and classifier approaches, as well as sophisticated replay attack modelling techniques. Further research will focus on these two aspects.

References

1. Battocchi, A., Pianesi, F.: DaFEx: Un Database di Espressioni Facciali Dinamiche. In: Proceedings of the SLI-GSCP Workshop, Padova, Italy (2004)
2. Chaudhari, U.V., Ramaswamy, G.N., Potamianos, G., Neti, C.: Information Fusion and Decision Cascading for Audio-Visual Speaker Recognition Based on Time- Varying Stream Reliability Prediction. In: IEEE International Conference on Multimedia Expo., Baltimore, USA, vol. III, pp. 9–12 (2003)
3. Chibelushi, C.C., Deravi, F., Mason, J.: A Review of Speech-Based Bimodal Recognition. IEEE Transactions on Multimedia 4(1), 23–37 (2002)

4. Chetty, G., Wagner, M.: Robust face-voice based speaker identity verification using multi-level fusion. Image and Vision Computing 26(9), 1249–1260 (2008)
5. Fisher III, J.W., Darrell, T., Freeman, W.T., Viola, P.: Learning joint statistical models for audio-visual fusion and segregation. In: Advances in Neural Information Processing Systems (NIPS), pp. 772–778 (2000)
6. Potamianos, G., Neti, C., Luettin, J., Matthews, I.: Audio-Visual Automatic Speech Recognition: An Overview. Issues in Visual and Audio-Visual Speech Processing (2004)
7. Goecke, R., Millar, J.B.: Statistical Analysis of the Relationship between Audio and Video Speech Parameters for Australian English. In: Schwartz, J.L., Berthommier, F., Cathiard, M.A., Sodoyer, D. (eds.) Proceedings of the ISCA Tutorial and Research Workshop on Auditory-Visual Speech Processing AVSP 2003, St. Jorioz, France, pp. 133–138 (2003)
8. Gurbuz, S., Tufekci, Z., Patterson, T., Gowdy, J.N.: Multi-Stream Product Modal Audio-Visual Integration Strategy for Robust Adaptive Speech Recognition. In: Proc. IEEE International Conference on Acoustics, Speech and Signal Processing, Orlando (2002)
9. Hershey, J., Movellan, J.: Using audio-visual synchrony to locate sounds. In: Proc. Advances in Neural Information Processing Systems (NIPS), pp. 813–819 (1999)
10. Hotelling, H.: Relations between two sets of variates. Biometrika 28, 321–377 (1936)
11. Hardoon, D.R., Szedmak, S., Shawe-Taylor, J.: Canonical Correlation Analysis: An Overview with Application to Learning Methods. Neural Computation 16(12), 2639–2664 (2004)
12. Jain, A., Nandakumar, K., Ross, A.: Score Normalization in Multimodal Biometric Systems. Pattern Recognition (2005)
13. Jiang, J., Alwan, A., Keating, P.A., Auer Jr., E.T., Bernstein, L.E.: On the Relationship between Face Movements, Tongue Movements, and Speech Acoustics. EURASIP Journal on Applied Signal Processing 11, 1174–1188 (2002)
14. Lai, P.L., Fyfe, C.: Canonical correlation analysis using artificial neural networks. In: Proc. European Symposium on Artificial Neural Networks, ESANN (1998)
15. Li, M., Li, D., Dimitrova, N., Sethi, I.K.: Audio-visual talking face detection. In: Proc. International Conference on Multimedia and Expo. (ICME), Baltimore, MD, pp. 473–476 (2003)
16. Li, D., Wei, G., Sethi, I.K., Dimitrova, N.: Person Identification in TV programs. Journal on Electronic Imaging 10(4), 930–938 (2001)
17. Liu, X., Liang, L., Zhaa, Y., Pi, X., Nefian, A.V.: Audio-Visual Continuous Speech Recognition using a Coupled Hidden Markov Model. In: Proc. International Conference on Spoken Language Processing (2002)
18. MacDonald, J., McGurk, H.: Visual influences on speech perception process. Perception and Psychophysics 24, 253–257 (1978)
19. Mana, N., Cosi, P., Tisato, G., Cavicchio, F., Magno, E., Pianesi, F.: An Italian Database of Emotional Speech and Facial Expressions. In: Proceedings of Workshop on Emotion: Corpora for Research on Emotion and Affect, in association with 5th International Conference on Language, Resources and Evaluation (LREC 2006), Genoa (2006)
20. Molholm, S., et al.: Multisensory Auditory-visual Interactions During Early Sensory Processing in Humans: a high-density electrical mapping study. Cognitive Brain Research 14, 115–128 (2002)
21. Movellan, J., Mineiro, P.: Bayesian robustification for audio visual fusion. In: Jordan, M.I., Kearns, M.J., Solla, S.A. (eds.) Proceedings of the Conference on Advances in Neural information Processing Systems, Denver, Colorado, United States, vol. 10, pp. 742–748. MIT Press, Cambridge (1997)

22. Nefian, V., Liang, L.H., Pi, X., Liu, X., Murphy, K.: Dynamic Bayesian Networks for Audio-visual Speech Recognition. EURASIP Journal on Applied Signal Processing, 1274–1288 (2002)
23. Pan, H., Liang, Z., Huang, T.: A New Approach to Integrate Audio and Visual Features of Speech. In: Proc. IEEE International Conference on Multimedia and Expo., pp. 1093–1096 (2000)
24. Potamianos, G., Neti, C., Luettin, J., Matthews, I.: Audio-Visual Automatic Speech Recognition: An Overview. Issues in Visual and Audio-Visual Speech Processing (2004)
25. Sanderson, C.: Biometric Person Recognition: Face, Speech and Fusion. VDM-Verlag (2008) ISBN 978-3-639-02769-3
26. Tabachnick, B., Fidell, L.S.: Using multivariate statistics. Allyn and Bacon Press, Boston (1996)
27. Yehia, H.C., Kuratate, T., Vatikiotis-Bateson, E.: Using speech acoustics to drive facial motion. In: Proc. the 14th International Congress of Phonetic Sciences, San Francisco, Calif, USA, pp. 631–634 (1999)

Multiple Traits for People Identification

Maria De Marsico[1], Riccardo Distasi[2], Michele Nappi[2], and Daniel Riccio[2]

[1] DI — Dipartimento di Informatica, Università di Roma "La Sapienza", Italy
demarsico@di.uniroma1.it
[2] DMI — Dipartimento di Matematica e Informatica, Università di Salerno, Fisciano (SA), Italy
{ricdis,mnappi,driccio}@unisa.it

Summary. Present biometric systems mostly rely on a single physical or behavioral feature for either identification or verification. However, day to day use of single biometries in massive or uncontrolled scenarios still has several shortcomings. These can be due to complex or unstable hardware settings, to changing environmental conditions or even to immature software procedures: some classification problems are intrinsically hard to solve. Possible spoofing of single biometric features is an additional issue. Last but not least, some features may occasionally lack the requisite of universality. As a consequence, biometric systems based on a single feature often have poor reliability, especially in applications where high security is needed.

Multimodal systems, i.e., systems that concurrently exploit multiple features, are a possible way to achieve improved effectiveness and reliability. There are several issues that must be addressed when designing such a system, including the choice of the set of biometric features, the normalization method, the integration schema and the fusion process, and the use of a measure of reliability for each subsystem on a single response basis. This chapter describes the state of the art regarding such issues and sketches some suggestions for future work.

1 Introduction

The aim of securing access to specific resources has been a major issue for a long time. An increasing number of government agencies and private companies are investing significant resources to improve security systems or implement reliable remote authentication. The circumstances that require identity checks through fast and effective automatic systems include applications that have already been refined and improved, such as controlling restricted access areas or locating dangerous individuals, and more recent additions, such as electronic transactions. Since ancient times, the solution adopted has been oriented towards the knowledge or the ownership of immaterial or material codes (e.g. the secret password to access a sect, or the pentacle tattooed over Pythagorean hands). From this point of view, traditional automatic identification systems can be either knowledge-based, relying on information retained by the user such as a username and password pair, or token-based, relying on the ownership of specific objects such as smart cards. Both approaches require preliminary delivery

H.T. Sencar et al. (Eds.): Intel. Multimedia Analysis for Security Appli., SCI 282, pp. 79–95.
springerlink.com

of immaterial or material items, which creates security problems in itself: information might be forgotten or wormed out of the owner, while physical objects might be lost or stolen. On the other hand, biometrics allows the recognition of individual physical or behavioral features. Persons are required to pass through equipment catching their image or some other physical element like their fingerprints; the number of subjects undergoing the acquisition phase varies with the kind of application, as does the number of matches actually performed in working situations.

There are two basic operation modes: verification and identification. Verification entails a one-to-one comparison. It is mostly used when positive recognition is performed, as in access control. In this case, the aim is to prevent multiple people from using the same identity. Only authorized subjects will have to undergo the acquisition phase, and a single identity will be matched to validate a person's claim. On the contrary, identification entails a one-to-many comparison, as when identifying dangerous or missing people. A recognition failure implies that the checked person is not among those we are looking for. In this case, features for "interesting" people will be collected, and all such identities will be matched to ascertain an individual's identity. The main use of identification mode is within negative recognition, where the goal is to find out if a person is who he/she denies to be in order to prevent a single person from using multiple identities. Identification can also be used for positive recognition, when users are not required to claim their identity. Further basic concepts and definitions about biometric recognition can be found in several published works, for example [17].

It is worth underlining that, although promising, the biometric approach in most cases is still limited to controlled experimental settings. The "academic" scenario, where aware and consenting users stay still in selected poses, contrasts with the "everyday" scenario, where the users are unaware, freely moving, and often not consenting. Much research is still needed to exploit all the potentialities of biometric recognition in everyday applications. Moreover, the acquisition of certain images (e.g., fingerprints or iris scans) requires cooperation from the users. Therefore, the collection of different biometric data, e.g. face, ear, and fingerprints, at nearly the same time for many people freely moving in space in real-time is impossible at present—unless the system directs the subjects. This implies, for example, setting up an equipped track where the users are driven as quickly as possible through a cascade of acquisition devices. This might seem too demanding in terms of required time and involvement. However, in most situations it is realistic to expect cooperation: on one hand, a user concerned with security in a public space or when entering home, though not happy, will be willing to undergo a reliable checking procedure; on the other hand, a malicious user will probably try to avoid attracting attention by complaining, hoping in a negative recognition failure.

A useful biometry must meet the requirements of universality, distinctiveness, permanence and collectability of related features. Fingerprints are the best-known biometric method. More biometries include iris scanning, hand conformation, features from face, ears or voice or handwriting. At present, DNA and iris are the most trustworthy biometric features [17]. The spatial patterns in the human iris make up texture information which is highly distinctive of an individual, due to differences in the development of anatomical body structures [8]. In particular, biomedical evidence suggests that the iris is as distinct as are the patterns of retinal blood vessels [3]; but an iris image can be more easily obtained than a retina image. For further references from clinical observation and developmental biology, see [43]. The other side of the coin is that techniques related to iris and DNA have not yet gained widespread acceptance despite their reliability, so their diffusion is limited. Furthermore, DNA tests require a time-consuming procedure, so they are not viable for on-line systems.

Today's systems mostly rely on a single biometry. The main drawback is that they are vulnerable to intentional attacks and not very robust to acquisition errors (possibly due to bad hardware, such as a dirty sensor), and actual distortions of the biometric feature (e.g. a voice altered by a cold or a dimly lit face). Another issue that has to be considered is the possible non-universality of the biometric feature. For example: voice recognition cannot be performed on deaf-mute subjects, and there are classes of users where fingerprints have little or no use, such as some heavy manual workers or patients assuming drugs that can blur or delete fingerprints [33].

Due to the above problems, the present trend of research is toward multimodal systems, i.e., systems that concurrently exploit more physical or behavioral features. Such systems are in turn not free from problems, but they do have better robustness and flexibility. There are specific issues must be addressed when designing a multimodal system; they will be listed in the next section.

This chapter is divided into seven major sections. Section 2 outlines active research areas in the field of multimodal biometric systems, while Section 3 describes a particular set of biometries that can be used to implement single subsystems to be combined in a multimodal system. Section 4 discusses multimodal architectures, while Sections 5 and 6 introduce normalization functions and reliability measures for recognition results. Section 7 sketches some possibilities for setting up experiments to validate a multimodal system. Finally, Section 8 provides a few concluding remarks.

2 Active Research Areas

A multimodal system would effectively solve most of the problems related to a single kind of feature, as flaws of an individual system can be compensated by the availability of several alternative biometries. Five main aspects must be taken into account during the design of a multimodal architecture:

1. The biometry set. Biometries that are efficient but not sufficiently effective when used alone, i.e., face and ear, can be combined with other biometries such as fingerprints, which yield a high recognition rate, albeit with slower response times. This would improve processing times without losing too much precision, so to provide applications able to address both massive and security-critical operational settings. Of course, speed and reliability are

somewhat conflicting aims: the actual processing modalities and parameters should be chosen to meet the desired tradeoff point.

2. The normalization method. Each system may return results using different dimensionalities and scales [17]. When the results are merged, it is important that they are comparable, so that they can be considered as produced by the same system. This is particularly important when trivial voting procedures are used for merging classification scores, for example when the highest score wins.
3. The integration schema. Present systems follow three possible design choices: parallel, serial or hierarchic [34]. In parallel mode, information from multiple subsystems is used simultaneously; in serial mode, the output from one biometry is used as the input for the next one; in hierarchical mode, the different subsystems are combined into a tree-structured design.
4. A reliability measure. Each subsystem should return a measure of how much its single result can be trusted. As a matter of fact, it often happens that not all subsystems are equally reliable, so the responses from the single biometries deserve different confidence. This is important while fusing the single results; however, this aspect is often neglected. Moreover, it is also important to compare the reliability of different responses from the same subsystem. It may happen that not all of them deserve the same confidence: for example, a temporary hardware fault may affect a single acquisition, but this should not' compromise the overall system reliability.
5. The fusion process. The integration of information by different biometries is possible in three different moments: feature extraction, matching, and decision [34]. The sooner a fusion is performed, the higher the amount of extracted information that can be saved. Performing fusion in the matching module seems to be a profitable choice, where a "weighted" integration strategy can be employed. As for the fusion rule, one could trivially decide to accept the identity with the highest returned score, or to rely on a linear combination of the subsystem responses, or even to apply a more complex fusion schema.

3 The Set of Biometries

In this chapter we will sketch some issues related to face-, ear- and fingerprint-based biometric systems. They may be used to provide a fast but less accurate analysis in a first recognition step (face and ear), possibly followed by a validation or refinement of results with a slower but more accurate biometry (fingerprints). Fingerprints were first mentioned by Purkyne in 1823 [32]. The FBI has been using them since 1924, and they are by now legally acknowledged as a reliable biometric technique. As a consequence of such early acceptance, there is a vast corpus of related research and a variety of effective algorithms [24]. Many of them are based on the computationally expensive extraction of features, called minutiae, from the fingerprint ridge. Related recognition techniques are slow, so that fingerprints cannot be used in on-line systems and are better suited to identity verification than to identification. Research has then turned towards other physical features allowing faster and less invasive recognition techniques.

Many researchers have focused on face recognition [45]. The face is the most natural way for a human to recognize a known person. Moreover, people are especially used to be photographed, and this makes face-based recognition methods sufficiently accepted. Even if some people complain about violations of privacy [18], there have been several attempts to take theory into everyday practice [10, 37]. Face recognition, however, raises a number of nontrivial issues that reduce the reliability of related methods, which are still far from the recognition rates usually provided by fingerprints. Face features appear to be too sensitive to a number of intra-class variations, such as expression, lighting, pose, occlusions and age [45]. Different solutions have been attempted: as an example, graph algorithms have been proposed to solve pose and lighting problems [44], while neural networks deal with the problem of occlusions [22]. However, the fact that no system based on face biometry reached the same large scale diffusion as fingerprints suggests that facial features are not yet able to provide an adequate level of confidence by themselves.

A further alternative is ear recognition. In-depth medical investigations has shown that the ear complies with the basic principle of distinctiveness, and is a biometry in every respect. Though counterintuitive, this thesis has been especially supported by Iannarelli [15], who concluded that ear structure is sufficiently rich to distinguish two subjects even if very similar. Ear-based techniques offer further advantages when compared to face-based ones. The limited surface allows a faster processing, while the lack of expressive variations reduces intra-class variations [7, 42]. However, even the ear is not completely reliable. When the ear is not perfectly frontal with respect to the acquisition device, small variations can significantly modify the image. Moreover, occlusion due to earrings or hair might be much more serious than occlusion on a face, since the usable surface area is limited. Differently from fingerprints and face, investigation of this biometric feature is still in its infancy, and only few algorithms are available—notably, Burge and Burger's work [6] and techniques based on force fields [14].

4 Multimodal Architectures

A typical biometric system includes four main components: a sensor, a feature extraction module, a matching-score module and a decision-making module. The combination of single biometries and the fusion of several modalities can be performed within any such module.

Multimodal biometric systems have further attracted the scientific community's attention to fusion/combination strategies for multiple classifiers. In a multibiometric scenario, each classifier generally holds its own representation of the (possibly different) input patterns. It is often shown that there is no overall winning combining rule and that both bad classifiers and bad feature sets may contain valuable information. In order to make the most of it, it is necessary to devise an effective combination method. Fusion of biometric information and related recognition results is possible at different stages:

- Feature extraction. Features extracted from different biometrical traits are combined into a unique feature vector. However, such a procedure makes sense only if the modalities or the features extracted are sufficiently independent. Fusion at this level is expected to provide better accuracy, as all original information is available for further processing. However, when combining modules available off-the-shelf to build a combined system, it is quite difficult to perform this kind of fusion, because access to this stage is generally not granted. This fusion mode might make little sense when using multiple classifiers for a single biometry, if each classifier requires a specific representation.
- Score level. Each biometric system produces scores that describe the similarities, or distances, between the biometric features acquired and their templates. Score level fusion combines such scores. In general, each system can use different algorithms and metrics, so the results to be fused might span different ranges of values. For this reason, a preliminary normalization is required: the scores have to belong to a common domain before the combination. For some normalization techniques, it is also desirable to know the possible extremes of each range, in order to obtain better normalization results—that is, values well contained in the prescribed range and better reflecting the original distributions.
- Decision level. This fusion mode consists in combining the decisions taken by each biometric system to obtain the final one. Different strategies are possible, ranging from simple ones such as majority voting, to more complex combination schemes.

It is argued that the sooner the fusion, the better the results and the less the lost information [35]. On the other hand, there are legitimate reason for delaying the fusion process. In particular, the sooner fusion is performed, the harder it is to manage the obtained data. As an example, score level fusion allows each classifier to operate asynchronously, which is not possible if the fusion is performed at the feature level. Moreover, additional modalities can be easily incorporated by simply adding a suitable module and expanding the fusion algorithms. Compared with decision level fusion, the detailed matching scores contain more information than just the final system decisions, so it might be argued that the combination of matching scores can lead to better performance. Furthermore, in the event of an unexpected result it is easier to find the reason why it happened.

Not all systems actually employ multiple modalities, but all of them exchange data at some point. As an example, the upper part of Fig. 1 shows the operation of the N-Cross Testing Protocol [2]. In this protocol, the different subsystems individually perform a first identification phase, after which they transmit their ordered lists of identities to one another. Such lists are used in a subsequent verification phase, which is in turn followed by a final voting procedure. The lower part of Fig. 1 depicts a more traditional hierarchical protocol, where more efficient but less reliable subsystems perform a first reduction of the search space, before passing their result to more reliable biometries. Systems that use data fusion combine different data to enhance the recognition process and increase both performance and robustness. In summary, they can operate in different settings [17, 35]:

- Single biometry, multiple sensors. The same biometry is acquired by different sensors. For instance, the face might be acquired by both a standard camera and a Near InfraRed (NIR) camera, or two cameras may be used to acquire the face to get both 2D and 3D representations.
- Multiple biometries. Different biometries of a same person are acquired. This approach is the one that might actually use different modalities, e.g. images for the face and the fingerprints, sounds for the voice, and video streams for the gait. A different sensor is typically employed for each biometry.
- Single biometry, multiple units. Different units of the same biometry are acquired, e.g. two different fingers or the two irises.
- Single biometry, multiple snapshots. More instances of the same biometric unit are acquired by the same sensor, e.g. multiple samples of the voice, or multiple images of the face with possibly slightly different conditions.
- Single biometry, multiple representations and classifiers. The same biometry is acquired once by a single sensor and then different approaches to features extraction and matching are combined. This typically implies the use of different classifiers.

From an architectural point of view, the subsystems might operate according to three main modes:

- Parallel mode. All the subsystems work simultaneously. The final response can be computed as soon as all subsystems have reached the point of response fusion (see below). The recognition time will therefore be conditioned by the slowest subsystem.
- Serial mode. The different subsystems work one after the other, each one passing its output (e.g. a subset containing the individuals that are most similar to the probe) to the next, until the final response is reached—possibly before even activating the last module. Typically, fast biometries are used to narrow down the search space before running modules for slower but more accurate ones. While this is very attractive in theory, in practice it may happen that the right answer is prematurely rejected by a poorly accurate system. It is therefore necessary to exercise some care when choosing the strategies for reducing the search space (e.g. similarity thresholds).
- Hierarchical mode. The different subsystems work according to a hierarchical design, modelled as a tree structure, where subsets of subsystems work in parallel before passing data to the following ones. This kind of architecture can merge the advantages of the above ones, while limiting their flaws.

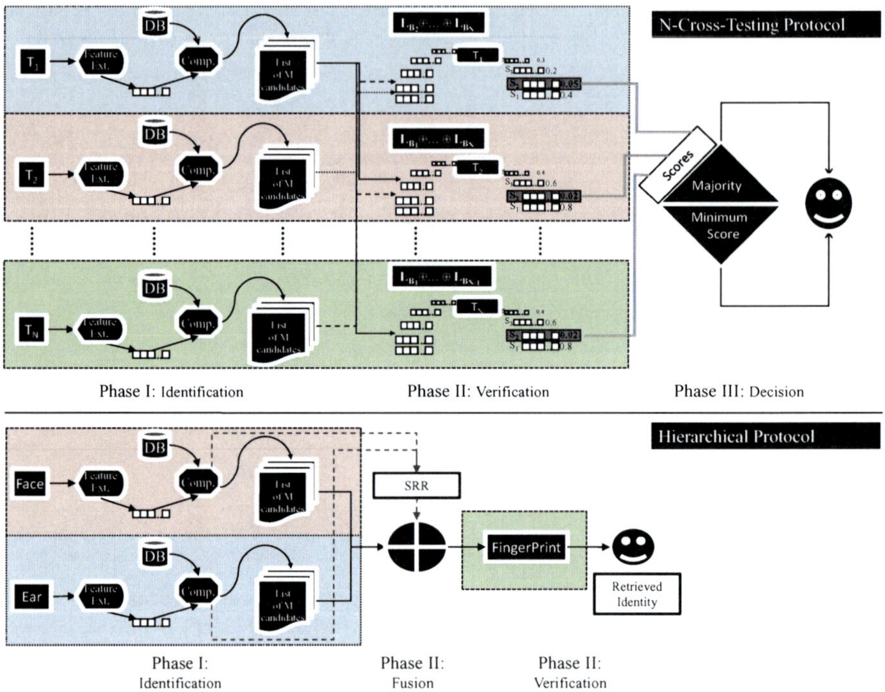

Fig. 1. Graphical representation of the N-Cross Testing Protocol and the Hierarchical Protocol

5 Data Normalization and Fusion

Normalization is important for data fusion. In general, the single subsystems produce output lists containing numeric values in some range, e.g. $[0, \infty)$. Such scores come from measuring different features, using different procedures and different scales, so they must be normalized before being combined. Several solutions have been proposed [16]; however, each solution has issues that should be considered.

The min-max normalization technique performs a "shifting" of the minimum and maximum values in the interval between 0 and 1. Given a set of matching scores $\{x_k\}, k = 1, 2, \ldots, N$, the corresponding normalized values are given by

$$x'_k = \frac{x_k - \min}{\max - \min} \,. \tag{1}$$

Such rule assumes that the minimum and maximum ever generated by a matching module are known.

The Z-score technique is the most widespread and uses the arithmetic mean μ and standard deviation σ of the scores returned by a single subsystem. The normalized values are given by

$$x'_k = \frac{x_k - \mu}{\sigma} . \tag{2}$$

The main drawback is that Z-score does not guarantee a common interval for values from different subsystems.

The median/MAD uses the median and the MAD (median of absolute value). The normalized value is given by

$$x'_k = \frac{x_k - \text{median}}{\text{MAD}} , \tag{3}$$

where MAD = median($|x_k$ − median $|$). As compared with Z-score, median/MAD is quite robust and more effective when values have a Gaussian distribution; otherwise median and MAD are not significant, and the normalization technique neither preserves the original value distribution nor transforms the values in a common interval.

The hyperbolic tangent function tanh guarantees projection in the open interval $(0, 1)$. However, despite robustness, normalized data tend to concentrate around the centre of such interval. The normalized values are defined as:

$$x'_k = \frac{1}{2}\left(\tanh\left(0.01\frac{x_k - \mu}{\sigma}\right) + 1\right) . \tag{4}$$

There is one more family of functions often used for score normalization: sigmoid functions. They have the following form.

$$x'_k = \frac{1}{1 + ce^{-kx}} . \tag{5}$$

The codomain of normalized values is the open interval $(0, 1)$. However, this function introduces excessive distortion at the extremes, and its shape depends on the parameters c and k, whose optimal values in turn strongly depend on the domain of x. Sanderson and Paliwal in [36] proposed an improvement by considering the mean and standard deviation of the original data. However, the authors assumed a Gaussian distribution for the input, which is not always the case, and did not consider the distortion at the extremes.

We compared the amount of distortion introduced by the normalization methods described above on the two test functions

$$f_1(x) = 2(\cos x + 1)$$
$$f_2(x) = 2\log(x + 1)(\cos x + 1)$$

in the interval $[0, 2\pi]$. Normalization is trivial when minimum and maximum values are known; we pretended to miss the maximum for both test functions and chose the average in its place, to place further stress on the normalization functions. Figure 2 shows how they act upon their input: values normalized by min-max, Z-scores and median/MAD do not necessarily fall within $[0, 1]$; Z-scores and median/MAD also introduce a sharp distortion in the original distribution. The tanh and sigmoidal functions guarantee a limited codomain $(0, 1)$, but the former excessively concentrates values around the interval centre, while the latter introduces significant distortion.

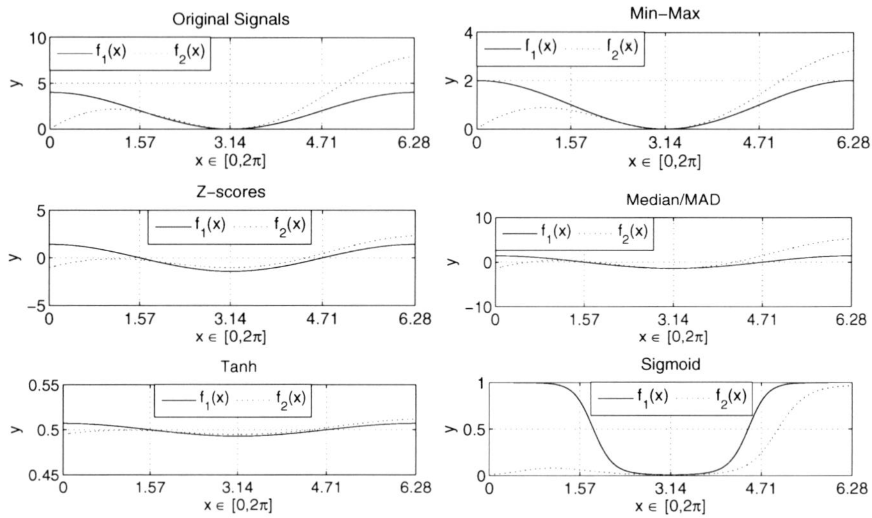

Fig. 2. Comparison of the distortion introduced by the different normalization functions

6 System Response Reliability

Classifier fusion can be considered one of the most significant advances in pattern classification in recent years [20], but only few studies investigated the optimization of fusion performance. Some more investigated points relating to multibiometric systems are data normalization [16, 38] discussed above, investigation of efficient fusion rules [34, 39], and design of specific architectures using the single biometries in a more sophisticated way [1]. However, subsystems might not be all equally reliable, due to the possibly different accuracy of their procedures, and not all responses by a single subsystem might be equally reliable, due to the possibly different quality of input from time to time. An unreliable response may not necessarily cause a reject. Rather, it may represent a valid reason to perform a further check. The definition of a measure for response reliability is then a crucial step when fusing the single results.

A possible approach for reliability assessment relies on input data quality. However, it is hard to devise a metric estimating the confidence in an identification result given the quality of a biometric acquisition. Such a metric must also depend on the classification method. As an example, using occlusion to estimate input quality would be appropriate in weighting the response of a system based on principal component analysis, yet almost needless with a method robust to occlusion. A further fault of absolute metrics is that they do not consider the relation between the datasets used for enrollment and for testing. If subjects are enrolled/verified in non optimal yet constant conditions, it is likely that the reliability will be adequate. For example, face images acquired by a constant light from one fixed side, both for enrollment and for testing, will probably be matched satisfactorily.

The quality of input data is considered by Kryszczuk et al. [21]. Bayesian networks are used to evaluate the probability Pr(MR|E) of a correct verification decision by a classifier MR given the available evidence E coming from several sources, which include a vector of signal-domain quality measures built considering the most suitable characteristics of each biometry and of each classification method.

The alternative of existing performance measures such as the recognition rate RR is not viable, because such measures only provide a global a-posteriori estimate of the ability to recognize a subject; they do not say anything about the reliability of a single testing operation. Relying on the single value of the distance/similarity between the biometric keys of the probe and those of the first retrieved subject makes even less sense: this distance alone does not account for individual distances between the first retrieved subject and the rest of the gallery. In other words, the results must be evaluated differently based on how similar gallery subjects are.

Some solutions use confidence measures based on margins. Each margin measures the amount of "risk" associated with a single subsystem response after observing its scores. Margins can be applied to any subsystem, whatever the nature of its input. Poh and Bengio introduced a confidence margin based on False Acceptance Rate (FAR) and False Rejection Rate (FRR) of a biometric system [31], showing how it behaves better than those by Freund and by Vapnik [11, 40], discussed in the framework of boosting and statistical learning theory, respectively. These two confidence margins, however, can only be calculated supposing that the target output (in this case: the impostor/genuine label) is known in advance, and therefore they are only significant in training. On the other hand, since Poh and Bengio's margin relies on an estimate of the actual distribution of genuine/impostor subjects' scores (FAR, FRR), a large number of responses is marked as reliable. Although the error rate of the global system is reduced, this modus operandi might not be appropriate for applications requiring a very high security level. Kryszczuk et al. also define a margin similar to Poh and Bengio's, defined in terms of Correct Acceptance (CA) and Correct Rejection (CR) accuracies at a given acceptance threshold [21]. It should be noted that the frequentist approach to reliability can be considered valid only by assuming that the scores of the testing set originate from a similar distribution as the scores from the development set.

How to integrate a reliability index into a fusion protocol? One possible way is to have each biometric module in a multimodal system produce a reliability measure for each response. Then, the fusion module has to implement an integration policy able to weight the single responses based on the corresponding reliability. Different choices exist for integration, all equally sound, yet possibly leading to very different results.

A simple fusion rule could be implemented based on a majority criterion, but there are more sophisticated methods. Veeramachaneni et al. published a study about decision level fusion in a binary hypotheses-testing problem, where each one of several different classifiers makes its own binary decision regarding the class of the observation [41]. The authors use a particle swarm optimization algorithm [19] that simultaneously selects the best decision threshold for each classifier and the best rule for fusion, in order to optimize FAR and FRR after fusion [28]. The number of explored rules grows exponentially with the number of subsystems, and if the performance of one system changes, the fusion rule must be determined again. The interesting outcome is that, for fusion with two classifiers, AND and OR rules constitute a very important set (out of the 16 possible ones), and that, as the number of classifiers increases, the optimal fusion rules are constructed from these two, giving better results than rules based on averaged sums. This suggests that, despite the variety of fusion policies, AND and OR do represent a suitable choice.

As an example, let us consider a system composed by N subsystems $T_k, k = 1 \ldots, N$, each one using its own gallery G_k. All N galleries have the form $G_k = \{g_1, \ldots, g_{|G_k|}\}$ and in general contain different numbers of acquisitions for the same number of identities. Each subsystem T_k sorts the acquisitions in its gallery into a list $L(1, \ldots, |G_k|)$ according to their distance $d_k(p, g_i)$ from the probe image p, and computes a numeric estimate RR_k of its response reliability. The N galleries might have different sizes, as each biometry might require a different number of acquisitions for each identity. A consistent data fusion process should enforce the constraint $\sum_k \mathrm{RR}_k = 1$ by normalizing the single indexes:

$$w_k = \frac{\mathrm{RR}_k}{\sum_{i=1}^{N} \mathrm{RR}_i} \tag{6}$$

For each subsystem T_k, a threshold is estimated, and it is used to accept the subsystem's responses as reliable according to integration policies such as the OR, AND, and SUM fusion rules presented in Table 1.

It can be observed that the OR and AND integration policies can follow a twofold schema, to better integrate reliability estimate within an identification

Table 1. Description of fusion rules

Rule	Description
OR	The combined response is valid only if at least one subsystem response reliability is above the corresponding threshold; the system returns the first identity from the list of the subsystem that has the higher reliability above the corresponding threshold.
AND	The combined response is valid only if all subsystem response reliabilities are above the corresponding thresholds; the system returns the identity with the minimum weighted sum of distances from the probe, where weights are the reliability degrees of the different subsystems.
SUM	The combined response is valid only if the sum of the reliability measures is equal or above the sum of the corresponding thresholds; the returned identity is chosen as in the AND policy.

problem. To determine global system reliability, a decision fusion under binary hypotheses is performed first (classes are reliable/not reliable), using OR or AND operator. Afterwards, a different score fusion is used to determine the returned identity, depending on the operator used previously. The SUM policy also follows a similar schema, but the decision fusion does not involve OR and AND operators.

7 Testing Benchmarks

System performance can be analyzed using Recognition Rate (RR) and Equal Error Rate (EER). These two measures provide a sufficiently complete system analysis, as they have complementary properties. RR is mainly used to evaluate the ability to retrieve the correct subject of $1 : m$ identification systems. On the other hand, measures such as EER (as well as FAR/FRR from which it is derived) evaluate the ability to recognize the right subject, but also to refuse impostors with a high certainty degree, which makes them more suited to assess $1 : 1$ verification systems.

The use of EER for the evaluation of identification systems might be criticized in favor of other measures such as the Cumulative Match Characteristic (CMC) curve. The percentage of times that one of the top n numerically ranked similarity scores is the correct match for all individuals in a probe set is referred to as the "rank n score" or Cumulative Match Score (CMS). A plot of n versus CMS is called a Cumulative Match Characteristic (CMC) curve. In other words, CMC expresses the probability that a given user appears in different sized candidate lists. The faster the CMC curve approaches 1, indicating that the user always appears in the candidate list of specified size, the better the matching algorithm. However, it can be shown that there is a precise relation between CMC and FAR/FRR, given that a $1 : 1$ matcher is used for sorting identification scores [5]. In such a case, CMC can be computed from FAR and FRR, so that the latter can be used for an identification system as well.

Further popular performance metrics are Receiver Operating Characteristic (ROC) and Detection Error Trade-off (DET) curves. A ROC curve plots, parametrically as a function of the acceptance decision threshold, the rate of false positives (i.e., impostors incorrectly accepted) on the x-axis, against the corresponding rate of true positives (i.e., genuine attempts correctly accepted) on the y-axis. It is threshold independent, allowing performance comparison of different systems under similar conditions, or of a single system under differing conditions. A DET curve is a modified ROC curve, which is often preferred in the case of biometric systems. A DET curve plots error rates on both axes, giving uniform treatment to both types of error. The graph can be drawn using a logarithmic scale, which makes it easier to distinguish different well-performing systems.

The goal of a recognition system is the correct classification of future test samples through the discrimination abilities acquired while working on likely dissimilar training samples. The generalization ability of a classifier refers to its performance in classifying test patterns which were not used during the training stage. For this reason, in order for the error estimate to be reliably predictive of future classification performance, training and test set should be sufficiently large and independent. This has often been a problem for early research on multibiometric systems, due to the lack of significatively large corpora of combined biometric data. Some examples of multimodal databases are M2VTS [23], XM2VTS (extended M2VTS) [26], and BANCA [4].

Because of the shortage of useful biometric data available for research, a number of projects were started with the explicit aim of producing such research corpora. Meanwhile, a number of researchers went on using so-called *chimeric* users, i.e., users virtually produced by combining images from different unimodal databases. Very popular databases used to assess recognition performance are FERET [29] and AR-Faces [25] for faces, and Notre-Dame [27] for the ear. The assignment of different biometric data to the same user must be consistent throughout the experiments. It is worth making a digression about the use of chimeric users, since there is an open debate about the equivalence between chimeric and true users in authentication experiments [13, 30].

Poh and Bengio use the XM2VTS multimodal database [26], containing 200 clients and 95 casual impostors, where each client is characterized by a set of different biometric traits (true identity match). The authors also build 1000 samples of random identity match (chimeric). In brief, they check if the half total error rate (HTER)—i.e., the mean of FAR and FRR—of true identity match falls within the HTER range of the 1000 samples of random identity match. Only about two thirds of the data indicate that chimeric users can be used. However, in the other cases the computed differences are very small. Moreover, a sample of size 1000 might be too small to reach a conclusion when compared to the possible 200! combinations. A further objection comes from the work by Dorizzi, Garcia-Salicetti and Allano: a bias in the experiments might be due to particular distributions of the chosen training and test sets. In [9], the authors examine 100 different splits for the true multimodal BIOMET database [12], and for each of the 1000 virtual databases built from it, FAR and FRR are computed for each value of the decision, obtaining a detection error trade-off (DET) curve. They then combine the 100 splits for each database into an average DET curve, which produces one real and 1000 virtual curves. They also compute the average of the virtual DET values and compare the behavior of the real curve with those of the virtual ones. The real curve is always inside the band generated by the 1000 virtual ones. Even more interestingly, for values of the threshold close to the EER, the real curve is close to the average virtual one. This suggests that experimental evaluations, using EER as a performance measure, would not be significantly affected by the use of chimeric users. This is a possible argument for adopting EER as a performance measure.

8 Conclusions

Recent research in the biometric field has been very intense, and many physical and behavioral features have been studied. While some of them are more reliable than others, none is free from limitations. Therefore, the present trend is to combine more biometries into one system. We examined three biometries: face, ear, and fingerprints. We then discussed multimodal architectures, from the points of view of composition schemes and fusion strategies. We also sketched some further related issues, such as data normalization and reliability indexes that a system can associate to its own responses. Each of the considered issues makes an individual research area by itself. However, we are convinced that better achievements can be obtained by addressing all such issues within a unified framework, as each variation in one processing step can modify the overall system performance.

References

1. Aaraj, N., Ravi, S., Raghunathan, A., Jha, N.K.: Architectures for efficient face authentication in embedded systems. In: DATE 2006: Proc. Conf. Design, Automation and Test in Europe, pp. 1–6. European Design and Automation Association, Leuven (2006)
2. Abate, A., Nappi, M., Riccio, D., De Marsico, M.: Face, ear and fingerprint: Designing multibiometric architectures. In: 14th Int. Conf. on Image Analysis and Processing (ICIAP 2007), Modena, Italy, September 2007, pp. 437–442 (2007) ISBN 978-0-7695-2877-9
3. Adler, F.: Physiology of the Eye: Clinical Applications. C.V. Mosby Company (1965)
4. Bailly-Baillière, E., Bengio, S., Bimbot, F., Hamouz, M., Kittler, J., Mariethoz, J., Matas, J., Messer, K., Popovici, V., Porée, F., Ruiz, B., Thiran, J.: The BANCA database and evaluation protocol. In: Kittler, J., Nixon, M.S. (eds.) AVBPA 2003. LNCS, vol. 2688, pp. 625–638. Springer, Heidelberg (2003)
5. Bolle, R.M., Connell, J.H., Pankanti, S., Ratha, N.K., Senior, A.W.: The relation between the ROC curve and the CMC. In: AUTOID 2005: Proc. 4th IEEE Workshop on Automatic Identification Advanced Technologies, pp. 15–20. IEEE Computer Society, Washington (2005)
6. Burge, M., Burger, W.: Ear biometrics in computer vision. In: Int. Conf. on Pattern Recognition (ICPR 2000), vol. 2, pp. 822–826 (2000)
7. Chang, K., Bowyer, K.W., Sarkar, S., Victor, B.: Comparison and combination of ear and face images in appearance-based biometrics. IEEE T. Pattern Anal. 25(9), 1160–1165 (2003)
8. Davson, H.: Physiology of The Eye, 5th edn. Pergamon, New York (1990)
9. Dorizzi, B., Garcia-Salicetti, S., Allano, L.: Multimodality in biosecure: Evaluation on real vs. virtual subjects. In: 2006 IEEE Int. Conf. on Acoustic, Speech, and Signal Processing (ICASSP 2006), vol. 5, pp. 1089–1092 (2006)
10. FaceIt, `http://www.l1id.com/pages/71-facial-screening` (21/09/2009)
11. Freund, Y., Schapire, R.: A short introduction to boosting. J. Japan. Soc. for Artif. Intel. 14(5), 771–780 (1999)
12. Garcia-Salicetti, S., Beumier, C., Chollet, G., Dorizzi, B., Jardins, J.L.L., Lunter, J., Ni, Y., Petrovska-Delacrétaz, D.: BIOMET: A multimodal person authentication database including face, voice, fingerprint, hand and signature modalities. In: Kittler, J., Nixon, M.S. (eds.) AVBPA 2003. LNCS, vol. 2688, pp. 845–853. Springer, Heidelberg (2003)

13. Garcia-Salicetti, S., Mellakh, M., Allano, L., Dorizzi, B.: A generic protocol for multibiometric systems evaluation on virtual and real subjects. In: Kanade, T., Jain, A., Ratha, N.K. (eds.) AVBPA 2005. LNCS, vol. 3546, pp. 494–502. Springer, Heidelberg (2005)
14. Hurley, D.J., Nixon, M.S., Carter, J.N.: Force field energy functionals for image feature extraction. Image and Vision Comput. 20, 311–317 (2002)
15. Iannarelli, A.: Ear Identification. Paramont Publishing, Fremont (1989)
16. Jain, A., Nandakumar, K., Ross, A.: Score normalization in multimodal biometric systems. Pattern Recog. 38(12), 2270–2285 (2005)
17. Jain, A.K., Ross, A., Prabhakar, S.: An introduction to biometric recognition. IEEE T. Circ. Syst. Vid. 14, 4–20 (2004)
18. Johnson, M.L.: Biometrics and the threat to civil liberties. Computer 37(4), 90–92 (2004)
19. Kennedy, J., Eberhart, R.: Particle swarm optimization. In: Proc. IEEE Int. Conf. on Neural Networks, pp. 1942–1945 (1995)
20. Kittler, J., Hatef, M., Duin, R.P., Matas, J.: On combining classifiers. IEEE T. Pattern Anal. 20(3), 226–239 (1998)
21. Kryszczuk, K., Richiardi, J., Prodanov, P., Drygajlo, A.: Reliability-based decision fusion in multimodal biometric verification systems. EURASIP Journal on Advances in Signal Processing, Article ID 86572, 9 pages (2007), doi: 10.1155/2007/86572
22. Kurita, T., Pic, M., Takahashi, T.: Recognition and detection of occluded faces by a neural network classifier with recursive data reconstruction. In: IEEE Conf. Advanced Video and Signal Based Surveillance, p. 53. IEEE Computer Society, Los Alamitos (2003)
23. M2VTS: Multi-modal verification for teleservices and security, `http://www.tele.ucl.ac.be/PROJECTS/M2VTS/m2fdb.html` (21/09/2009)
24. Maltoni, D., Maio, D., Jain, A.K., Prabhakar, S.: Hand Book of Fingerprint Recognition. Springer, Heidelberg (2003)
25. Martínez, A.M.: Recognizing imprecisely localized, partially occluded, and expression variant faces from a single sample per class. IEEE T. Pattern Anal. 24(6), 748–763 (2002)
26. Matas, J., Hamouz, M., Jonsson, K., Kittler, J., Li, Y., Kotropoulos, C., Tefas, A., Pitas, I., Tan, T., Yan, H., Smeraldi, F., Begun, J., Capdevielle, N., Gerstner, W., Ben-Yacoub, S., Abdeljaoued, Y., Mayoraz, E.: Comparison of face verification results on the XM2VTS database. In: Proc. 15th Int. Conf. on Pattern Recognition, vol. 4, pp. 858–863 (2000)
27. Notre Dame Ear Database, `http://www.nd.edu/~cvrl/CVRL/Data_Sets.html` (21/09/2009)
28. Osadciw, L., Veeramachaneni, K.: Sensor network management through fitness function design in multi-objective optimization. In: Proc. Asilomar Conf. on Signals, Systems, and Computers (November 2007)
29. Phillips, P.J., Huang, J., Rauss, P.: The FERET database and evaluation procedure for face recognition algorithms. Image and Vision Comput. 16, 295–306 (1998)
30. Poh, N., Bengio, S.: Can chimeric persons be used in multimodal biometric authentication experiments? In: Renals, S., Bengio, S. (eds.) MLMI 2005. LNCS, vol. 3869, pp. 87–100. Springer, Heidelberg (2006)
31. Poh, N., Bengio, S.: Improving fusion with margin-derived confidence in biometric authentication tasks. In: Kanade, T., Jain, A., Ratha, N.K. (eds.) AVBPA 2005. LNCS, vol. 3546, pp. 474–483. Springer, Heidelberg (2005)

32. Purkyne, J.E.: Commentatio de examine physiologico organi visus et systematis cutanei. Habilit. dissert. f. d. Professur d. Physiol. and k. Universität zu Breslau. Also in Op. Omn. 1, 163–194 (1823)
33. Reuters: Cancer patient held at airport for missing fingerprint, http://www.reuters.com/article/oddlyEnoughNews/idUSTRE54Q42P20090527 (May 27, 2009)
34. Ross, A., Jain, A.: Information fusion in biometrics. Pattern Recogn. Lett. 24(13), 2115–2125 (2003)
35. Ross, A., Jain, A.K.: Multimodal biometrics: An overview. In: Proc. 12th European Signal Processing Conf., pp. 1221–1224 (2004)
36. Sanderson, C., Paliwal, K.K.: Noise compensation in a person verification system using face and multiple speech features. Pattern Recog. 36(10), 293–302 (2003)
37. SecurIDent biometric face recognition products, http://www.cryptometrics.com/products_securident.php (21/09/2009)
38. Snelick, R., Indovina, M., Yen, J., Mink, A.: Multimodal biometrics: issues in design and testing. In: Oviatt, S.L., Darrell, T., Maybury, M.T., Wahlster, W. (eds.) Proc. of the 5th Int. Conf. on Multimodal Interfaces (ICMI 2003), November 5-7, pp. 68–72. ACM, New York (2003)
39. Toh, K.A., Yau, W.Y.: Combination of hyperbolic functions for multimodal biometrics data fusion. IEEE T. Syst. Man Cy. B 34(2), 1196–1209 (2004)
40. Vapnik, V.N.: Statistical Learning Theory. Wiley-Interscience, Hoboken (1998)
41. Veeramachaneni, K., Osadciw, L., Varshney, P.: An adaptive multimodal biometric management algorithm. IEEE T. Syst. Man Cy. C 35(3), 344–356 (2005)
42. Victor, B., Bowyer, K., Sarkar, S.: An evaluation of face and ear biometrics. In: Proc. 16th Int. Conf. on Pattern Recognition (ICPR 2002), vol. 1, pp. 429–432. IEEE Computer Society, Washington (2002)
43. Wildes, R.: Iris recognition: An emerging biometric technology. Proc. IEEE 85(9), 1348–1363 (1997)
44. Wiskott, L., Fellous, J.M., Krüger, N., von der Malsburg, C.: Face recognition by elastic bunch graph matching. IEEE T. Pattern Anal. 19(7), 775–779 (1999)
45. Zhao, W., Chellappa, R., Phillips, P.J., Rosenfeld, A.: Face recognition: A literature survey. ACM Comput. Surv. 35(4), 399–458 (2003)

Intelligent Multimedia Analysis for Emerging Biometrics

Georgios Goudelis, Anastasios Tefas, and Ioannis Pitas

Dept. of Informatics, Aristotle University of Thessaloniki, Greece
goudelis@aiia.csd.auth.gr, tefas@aiia.csd.auth.gr, pitas@aiia.csd.auth.gr

Summary. Various anthropometric studies have been conducted in the last decade in order to investigate how different physiological or behavioral human characteristics can be used as identity evidence to prove the individuality of each person. Some of these characteristics are: face, eyes, ears, teeth, fingers, hands, feet, veins, voice, signature, typing style and gait. Since the first biometric security systems appeared in the market, an increasing demand for novel techniques that will cover all different scenarios, has been observed. Every new method appears to outmatch some of its competitors but, at the same time, presents disadvantages compared to others. However, there is still no method that consists a single panacea to all different scenarios and demands for security. This is the reason for which researchers are on a continuous effort for more efficient and generic biometric modalities that can be used in various applications. In this chapter, emerging biometric modalities that appeared in the last years in order to improve the performance of biometric recognition systems, are presented. The presented methods are divided in two major categories, intrusive and non-intrusive ones, according to the level of user nuisance that each system sets off.

1 Introduction

Biometric recognition is a well-known research area that aims to provide more efficient solutions to the ever-growing growing human need for security. Biometrics refers to methods that can be used for uniquely recognizing humans based upon one or more intrinsic physical or behavioral characteristics. In information technology, in particular, biometrics are used as a tool for efficient and reliable identity management and access control. Biometrics are also used in surveillance applications in order to identify individuals. The last decade, a large number of novel biometric modalities, such as palm, gait and veins, as well as new methods for well-known biometric modalities, such as face, voice and finger have been proposed in the literature.

Probably the most interesting problem to researchers working on the specific field, is the problem of automatic recognition/verification for security applications. On the one hand, person recognition refers to the problem of recognizing the identity of a test person (using one or more of its biometric characteristics) by selecting the most similar (best match) or the N most similar persons from a

H.T. Sencar et al. (Eds.): Intel. Multimedia Analysis for Security Appli., SCI 282, pp. 97–125.
springerlink.com

given database [13], [11]. In many recognition systems, the final decision is taken by a human expert. On the other hand, the automatic rejection or acceptance of an individual's claim for a specific identity is referred as verification. More specifically, a verification system must take the right decision when a person claims an identity that is included in the internal database. Several studies focus on the verification problem as it appears to concern many researchers working on different modalities [12]. Biometric verification is mostly used for access control whereas biometric recognition is used for surveillance applications and forensics. Although recognition and verification are different in principle, the majority of the proposed biometric technologies in the literature can be easily used both for recognition and verification applications.

By many researchers working on biometrics, face is considered as one of the most attractive modalities for recognition/verification, especially due to the "nature" of the body part that it concerns (face is usually exposed and an image of it can be easily obtained). For this reason it has received significant attention [13], [5], [15]. Automatic recognition of subjects based on their faces attracts the interest of researchers coming from different fields such as pattern recognition, neural networks, computer vision, computer graphics and psychology. As mentioned above, there is a large number of studies and proposed methods for a variety of applications. However, face recognition/verification and its efficiency, depends on many factors such as illumination, general environmental and recording conditions and for this reason it still remains an open issue [16], [17].

Another modality, that has been under extensive research, is voice [81], [18]. Voice recognition analyzes the audio signal produced by speech and uses the features extracted from it to perform identification. Noise cancellation techniques and high quality microphones can positively affect the efficiency while other conditions like background and channel noise, variable and inferior microphones and telephones , extreme hoarseness, fatigue or vocal stress can dramatically affect the result[19].

Probably the first biometric used for many years as a proof of a person's identity and consequently the most widely known is fingerprints. It remains the most commonly used forensic evidence worldwide and thus engrosses many researchers. The technology is mostly based on analysis of two-dimensional data maps obtained from different types of sensors. When a finger touches or rolls onto a surface, the elastic skin deforms. The quantity and direction of the pressure applied by the user, the skin conditions and the projection of an irregular 3D object (the finger) onto a 2D flat plane introduce distortions, noise and inconsistencies in the captured fingerprint image. These problems result in inconsistent, irreproducible and non-uniform contacts during each acquisition. Although fingerprint techniques are easy to implement, their performance is easily affected by factors as dryness, dirt or ageing [21], [22].

As biometrics spread, many review articles have been published reporting the advantages and disadvantages of each of the well-known methods. However, it is important to note that even though current recognition systems have reached a certain level of maturity, their success is limited by the conditions imposed by many real applications [5]. Besides effectiveness, the availability and the affordability of biometric technologies appear to be important requirements for biometric systems.

There is no doubt that security has been always one of the main concerns of the human species. In our days though, the need for security seems to be essential. Since there is still no method able to guarantee ultimate efficiency while the variety of possible applications is so wide, new biometric modalities emerge trying to provide solutions to the existent complicated problems. In this chapter, emerging technologies on biometrics are reviewed. There is a variety of categories that biometric systems could fall into (e.g. image or non-image based, 2D or 3D based etc). We could also classify them according to principal attributes that make a biometric system practical, like universality, uniqueness, permanence, collectability and acceptability. However, for simplicity we shall separate them to intrusive and non-intrusive, according to the level of nuisance that each system sets off to the user. The separation of the modalities to intrusive and non-intrusive ones, is indeed a difficult demarcation problem which may cause different arguments. However in this chapter, a modality is characterized as intrusive only if the cooperation of the subject is indispensable during biometric testing and as non-intrusive if it is not.

It should also be mentioned that the aim of this chapter is not to provide analytical investigation and comparisons of the already commonly known modalities such as iris, fingerprint, face, voice, retina and signature, but only to focus on emerging biometric technologies. Another thing that should be noted, is that the authors of this chapter do not allege that the emerging modalities should have better performance than any of the well known modalities. All of the presented methods have just emerged and it is obvious that time is required until these methods are thoroughly evaluated.

2 Non Intrusive Systems

The number of non-intrusive biometric systems so far is very small. Most of the developed techniques require the imminent cooperation of the person that is to be recognized. Although this voluntary cooperation is accordant with the idea of protection of the personal data, intrusive methods cannot be applied in all circumstances.

The most known method in this category is face recognition. Face recognition can be applied, in most cases, without the subject's cooperation. It usually requires only a camera pointing to the subject's direction (a face detector is used to locate the face). However, there are many face recognition algorithms that require certain position of the person in front of the camera, as well as absence of facial expressions.

Over the last twenty years, numerous algorithms have been proposed for face recognition. The latest achievements on non-intrusive biometrics present new technologies that promise to change the way of thinking in this direction.

Thermogram, lip recognition, smile identification and hyperspectral analysis seem to be the most important and promising techniques.

2.1 Thermogram

Recognition based only on the visual spectrum has shown deficiencies in performing well under uncontrolled operating conditions. Face recognition accuracy degrades quickly when the lighting conditions are deficient or when face is not uniformly illuminated. Light reaching the sensor after it has been reflected from human faces may varies as it depends on the skin color which may also vary from person to person. Thermal infrared (IR) sensors measure heat energy emitted, not reflected, from the objects. Hence thermal imaging has an important advantage in face recognition in low illumination conditions or even in total darkness, where visual face recognition techniques fail. Researchers have found that the facial heat distribution patterns obtained by infrared sensors are unique to every person. The means used for measuring the heat coming from the body is called thermography. Equinox, is an example of a database containing facial thermal images [23]. Equinox database is a collection of face imagery, in the following modalities: coregistered broadband-visible/longwave infrared (8-12 microns), midwave infrared (3-5 microns), shortwave infrared (0.9-1.7 microns). A sample taken from the database is shown in Figure 1.

Researchers in [91] examine the performance of Long-Wave Infrared (LWIR) imagery tested on different illuminations. For the evaluation, two face recognition algorithms are applied (eigenfaces [25] and Arena [26]) on LWIR and visual images respectively. For the experiments, prerecorded infrared videos of 91 subjects were used. The best performance of Arena algorithm on LWIR imagery reported to be 99% while the worst was 97%. Cases with facial expressions and glasses, proved to affect dramatically the results, as in such cases the performance is reduced. The experiment was repeated for eigenfaces applied on the same datasets used for Arena, giving 96% and 87% respectively. In a newer study, researchers in [24] indicated that thermal images perform better or similar to visible light images in many cases using Monte Carlo analysis of performance. Recognition performance increases further, when algorithms applied to visible and thermal infrared images are fused.

In [27], the authors propose a two stage method based on infrared images and statistical modelling of visible images. Bessel modelling is applied exclusively to the facial region by pipelining a classification algorithm to produce a unique solution. Matching is performed using a Bayesian classifier. The experiments indicated that segmentation of the facial regions, results in a better classification performance.

Eyeglasses is one of the major problems in thermal image based approaches as they are opaque to IR radiation. Attempting to reduce the error caused, researchers in [28], fuse visual and thermal imagery exploiting the fact that visible-light based recognition is less sensitive to the presence of eyeglasses. The fusion performed was pixel-based and operated in the wavelet domain. To decide how to combine thermal with visible information, genetic algorithms were employed.

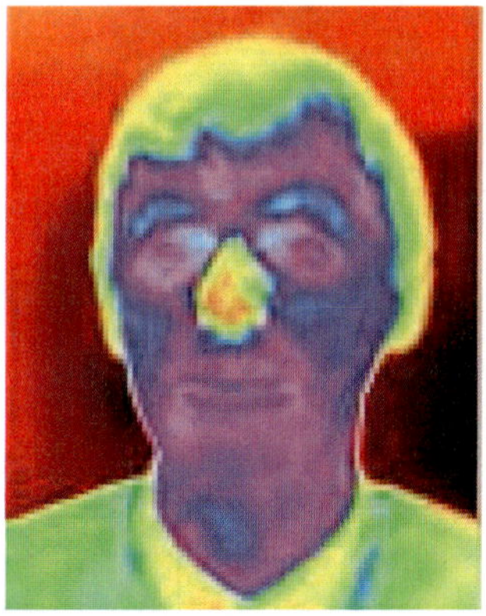

Fig. 1. Sample image from Equinox database [23].

Although the approach managed to improve the performance presented by the single thermal IR methods, it was not able to fully eliminate illumination effects present in the visible (not IR) images. The experimental results though, according to the authors, show substantial improvements in the overall recognition performance.

The most recent advance in thermal IR is outlined in [87] where the novelty of the approach is the use of characteristic and time-invariant physiological information to construct the feature space. The motivation behind this effort is to concentrate on the permanency of innate characteristics that are under the skin. The researchers claim that although thermal facial maps shift over time, the contrast between the superficial vasculature and surrounding tissue remains invariant. This physiological feature has permanence and is very difficult to be altered as it is found under the skin. Therefore, it gives a potent advantage to any face recognition method that may use it. The method uses a novel Bayesian segmentation algorithm to separate the facial tissue from the background. Additionally, it extracts the vascular contour network from the surface of the skin by using white top hat segmentation preceded by anisotropic diffusion. Thermal Minutia Points (TMPs) are localized in order to create a feature vector. Finally, recognition is performed by matching TMP-based feature vectors. Tests with 500 thermal face images from 50 subjects show an Equal Error Rate of $\simeq 6\%$.

The great advantage of thermograms as opposed to visible light images for face recognition, is their independence of ambient illumination which makes them ideal for covert surveillance. However, researchers working on this modality still need to solve several challenging problems. Thermal signatures are subject to changes according to body temperature caused by physical exercise or ambient temperature. As mentioned above, eyeglasses may result in loss of useful information around the eyes in thermal face images since glass material blocks a large portion of thermal energy. Finally, the cost of thermal cameras is much higher compared to the cost for a conventional camera which may explain the lack of new papers on the specific modality the very last years.

2.2 Hyperspectral Images

Hyperspectral cameras may provide data beyond the normal visible range which can include useful and discriminant information not obtained by any other imaging methods [29], [30]. According to authors of the above papers, subsurface tissue structure is significantly different from person to person. Features obtained from hyperspectral images are relatively invariant to illumination, although they cannot be obtained in complete darkness. The experimental results of the above works, show that spectral properties of human tissue are not considerably affected by face orientation and expressions. This observation is quite important since it reveals a great advantage of hyperspectral information for face recognition, as it can be used over a large range of poses and expressions.

In this specific work, the database used consists of near-infrared hyperspectral images of 200 subjects. The face recognition algorithm proposed, exploits the spectral measurements for multiple facial tissue types. A CCD camera equipped with a liquid crystal tunable filter, collected the hyperspectral images provided 31 bands over the near-infrared (0.7μm-1.0μm) as shown in Figure 2. Each face image was represented using spectral reflectance vectors that are extracted from small facial regions (e.g. forehead, lefft/right cheek, hair, lips). Evaluation was performed by the squared Mahalanobis distance. The above technique proved to be robust against facial expressions and facial pose presence. Authors report significant improvements over current face recognition systems on rotated faces. Moreover, they argue that by modelling the spectral reflectance changes due to the face orientation changes, performance can be further improved.

As an extension of their previous work researchers in [88], present results on recognizing 200 human subjects under unknown outdoor illumination in hyperspectral face images. For each subject, several near IR (NIR) images with

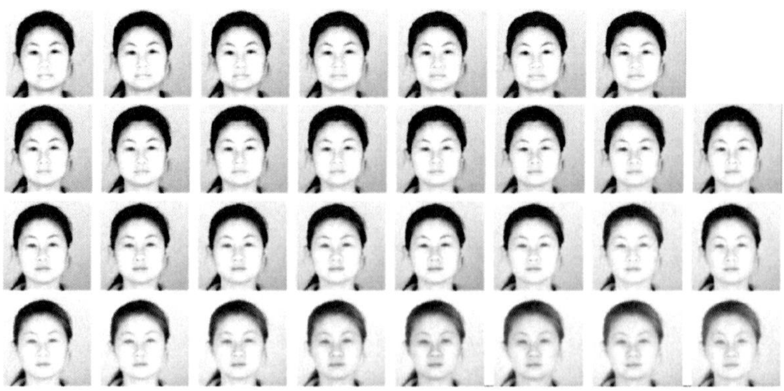

Fig. 2. Thirty-one bands of hyperspectral images of one subject [30] (©2003 IEEE) .

different facial expressions and face orientations were acquired on different days under various natural illumination conditions. A set of 7258 global spectral irradiance functions were used to synthesize reflected radiance images of each subject. A low-dimensional linear model for each tissue type for each subject was used to model illumination variation in radiance images. Authors advocate their system claiming that the algorithm provides accurate recognition performance for front-view probes, with or without facial expression changes. They also add that the results are promising for face recognition under unknown outdoor illumination and with various face orientations.

Another biometric solution that uses NIR imaging hardware, algorithms, and system design, has been proposed in [90]. The system is presented as another solution to illumination variation problems in face recognition modalities. An illumination invariant face representation is obtained by extracting local binary pattern (LBP) features of NIR images to compensate for the monotonic transform, thus deriving an illumination invariant face representation. Using statistical learning algorithms the most discriminative features are extracted from a large pool of invariant LBP features and construct a highly accurate face matching engine. For the dimensionality reduction and classification, LBP+LDA and LBP+AdaBoost methods have been developed. For the experiments 10000 face images of about 1000 people, all Chinese, were used for training the system. The testing dataset contained 3,237 images from a total of 35 persons and the accuracy reported by authors was 94.4%.

In [92] the novelty compared to other NIR systems, is the use of constant illumination for face recognition. Authors argue that active NIR illumination provides a constant invisible illumination condition and facilitates the automatic eye detection by introducing bright pupils. The results provided indicate that the actively illuminated faces show better separability for all classifiers than faces under varying ambient illumination. More specifically, radial basis function (RBF), Adaboost and support vector machines (SVM) classifiers where applied on 2360 face images from 295 subjects where SVM, achieved the best results with 0% error rate.

Another study that examines the effectiveness of NIR images for face recognition [93], ascribes the success of the system presented, firstly to NIR images that facilitate the classification process and secondly, to the learning based methods with local features, proposed in the paper. Evaluation of the system on 1470 persons indicated an equal error rate of 0.3%.

The main advantage of the hyperspectral image based techniques as already mentioned, is that they overcome problems due to illumination. As illumination is almost eliminated, the use of NIR images is also supposed to provide advantages over other methods on rotated faces, expressions and robustness over time. However, this specific modality does not seem yet to be suitable for uncooperative user applications such as face recognition in video surveillance [90]. Although many methods present impressive performance on both indoor and outdoor conditions, hyperspectral technology so far, is mainly suggested for indoor biometric applications.

2.3 Smile Recognition

Researchers in [31], proposed a system that was equipped with a high speed camera with strong zoom lens. This system was able to produce smile maps that could be used as features for identification, under the claim that they are unique for each person. The system examines the way the skin around the mouth moves over in the video frames, by tracking the tiny wrinkles in the skin and analyzing their motion. That is, the system probes the characteristic pattern of muscle deformations beneath the skin of the face. The extracted data is used to produce a motion vector containing the information about the deformation of the facial region. The deformation is insusceptible to the influence of facial make-up or smile size and is only controlled by the pattern of muscles under the skin. According to the authors, the feature map can be created, even if the smile movement is not full, as system's sensitivity is high. They also argue that smile maps can be produced without the cooperation of the person tested and this is the reason why they describe the technique as "invisible". An alternative application of the proposed method could be its use for medical reasons indicating distinctive asymmetries in movement of facial muscles caused by nerve disorders.

The method has been tested so far on samples of 4 lab members while smiling. Further experiments on a larger dataset of 30 persons have been announced. No results have been published though.

2.4 Lip Deformation Recognition

In [32], the shape similarity of lips when vowels are uttered is examined. More specifically, a person recognition method is proposed based on a mathematical morphology analysis of the lip area using three different structuring elements. The proposed structuring elements are the square, the vertical and the horizontal line and they are used for deriving a pattern spectrum of the lip images. The shape vector is compared with the reference vector to recognize an individual from its lip shape as shown in Figure 3.

Experiments on eight Japanese persons (classified with 100% accuracy by their lips) indicated that the shape vector obtained, contains enough information for recognition. Beyond doubt, the dataset used for the experimental procedure is very small and authors state clearly that results may be biased. As authors

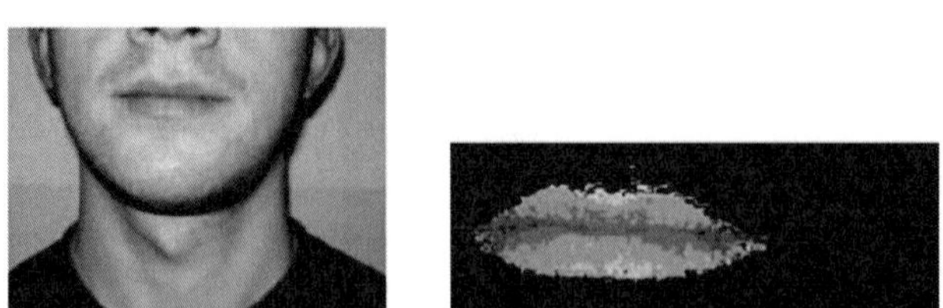

Fig. 3. Example of lower face and extracted corresponding lips area in [85].

state, classification accuracy should be improved by a more sophisticated system. Proposed improvements refer to consideration of other structuring elements (for instance, rectangle, ellipse or asymmetric shape). A larger database is strongly required in order to evaluate better the efficiency of this method.

Another approach considers lips' shape and color features in order to determine human identity [85]. More specifically, the method calculates color features of the masked out lips and merges them with shape features of the binarized lips. Color statistics and moments as well as a set of standard geometrical parameters and the moments of Hu and Zernike are used for feature extraction. The feature vector that finally describes the lips area, consists of a selection of the most discriminant information of: Hu and Zernike moments, central moments, standard geometrical parameters, statistical color features in RGB, YUV and HSV color spaces. Experiments on a database of 38 subjects showed that the method was able to recognize successfully 76% of the test set.

Although the results are promising for such an emerging technology, it is obvious that further improvement is strongly required for a stand alone application. It is also mentioned that reliable lip detection, acquired using surveillance cameras, consists a major drawback of the system.

2.5 Gait

Gait is one of the modalities that has been under research for more than a decade [8], [9]. However, the biometric research community still pays attention on gait, since on one hand it seems to have great potentials while on the other hand, there is still no commercial product available based on gait so far. Many psychological studies have shown that a person can be recognized by the way it walks. Thus, many research papers have proposed gait as suitable modality for biometric applications. Here, the most representative methods will be presented.

Boyd and Little in [7] define gait to be "the coordinated cyclic combination of movements that result in human locomotion". The movements are coordinated in the sense that they must occur with a specific temporal pattern for the gait to occur. The set of movements that consist a full gait cycle repeat in every cycle. The periodicity of these movements as well as the coordinated and cyclic motion of gait makes it a unique modality. The basic techniques used in gait and motion analysis systems are: background substraction, silhouette detection, optical flow and motion energy/ history images. There is a variety of methods that are used for gait recognition and according to [7], they can be categorized by the data type that they use: shape, joint trajectory, self similarity, and pixels.

An a example of a system using joint trajectories is given in [3]. The method extracts a hip joint trajectory from a sequence of images. The trajectory for the hip closest to the camera is acquired. Subsequently, recognition is performed based on the Fourier components of the trajectory. The method was tested on a dataset of 10 individuals (ten repetitions each). The classification analysis showed that the phase-weighted Fourier magnitude offered a classification rate of (100%) while for just the Fourier magnitude the classification rate was (80%).

A self similarity based method in [2], suggests that the image self-similarity plots of a moving person encode much information about gait motion patterns providing discriminant information for gait sequences that is useful for gait recognition. Researchers construct a self-similarity image from the image sequence, in which pixel intensities indicate the extent to which two images in the sequence are alike, i.e., pixel (i, j) in the self-similarity image indicates the similarity of the images at times t_i and t_j.

A system based on pixel oscillation [1], demonstrates how the frequency of the gait and the timing of the component movements, determine the frequency and phase of the pixel oscillations. More specifically, authors demonstrated that an array of phase-locked loops (PLL), one per pixel, can synchronize internal oscillators to the frequency and phase of pixel oscillations. This synchronization process inherently performs frequency entrainment and phase locking. Boyd [1] uses a phasor, a complex number that represents a rotating vector, to represent the magnitude and phase of the oscillations at each pixel. Thus, once the PLL synchronization occurs, one can construct a complex image of phasors in which each pixel indicates the extent to which there are oscillations and the relative timing of the oscillations.

A more recent method [4], is based on the anthropometric proportions of human limps and the characteristics of the gait task. The system uses a single camera, does not require camera calibration and works with a wide range of directions of walking. The proposed gait analysis is based on two consecutive steps: a motion estimation method which extracts the limbs orientations with respect to the image reference system and a view-point independent gait reconstruction algorithm that normalizes and corrects the limbs inclinations in the lateral reference system. For the experiments 200 video sequences of 3 subjects, viewed at 6 different camera inclinations have been used. The authors report that the specific method has been proved to be viewpoint invariant thus, more efficient than others for real scenarios. However, discriminability reduces as the camera moves to capture a subject walking in frontal view.

Another recent study in gait identification [6], examines the effects of covariation on the recognition process. Authors show how these factors can separately affect the walking pattern. Further, they assess the contribution and discriminatory significance of the gait dynamics used for recognition. On a database of 440 samples, a recognition rate of 73.4% was achieved using a k-nearest neighbor (KNN) classifier. Authors argue, that the results confirm that person identification using dynamic gait features is still achievable with better recognition rate.

Gait presents advantages compared to other biometric modalities such as iris or fingerprints. Its main advantage is that it is effective from a distance or where only low resolution images/ video is available (e.g CCTV cameras). However, there are many factors that can negatively influence the accuracy of a gait recognition system. The speed at which someone walks or runs has little effect on the biometric, but wearing a trench coat can mask the feet, and using flip-flops can also affect the results. With respect to gait security, studies also indicated that gait biometric is robust against minimal effort impersonation attacks. However, impostors who know their closest person in the database or the gender of the users in the database can be a threat to a gait authentication system. Although gait is a subject of research for many years, it is still not suggested as a stand alone application and it is usually proposed for multi-modal biometrics where it is supposed to increase the overall performance of the system.

3 Intrusive Systems

The majority of the recently developed biometric systems belong in the category of intrusive systems. These systems require the cooperation of the subject to be recognized. The level of intrusiveness is determined by the level of cooperation that they demand from the subject. Many well-known biometric modalities like iris, retina, fingerprints, keystroke dynamics and dental analysis are representative examples. In the following, the latest developments within this class of biometric recognition/verification techniques are presented.

3.1 Finger-vein Patterns

A method based on finger vein patterns is proposed in [36], [37]. Finger vein patterns found to be distinctive between twins and even between a person's left and right hand. For this reason the specific biological information has judged to be ideal for person authentication. The way vein maps are produced is illustrated in Figure 4. Authors believe, that the images obtained are very rich in information content. Irregular shading produced by various thicknesses of the finger bones and muscles, contribute to the final feature map extracted by an infrared image of a finger. The technology uses near-infrared light to detect vein vessel patterns. Thus, fluctuations in brightness due to variations in the light power or the thickness of the finger occur. The extraction of the patterns from an original image (line-tracking operations) with randomly chosen start points is repeatedly carried out. Two commonly known methods for line-shaped patterns, structural matching [38] and template matching [39], are used for matching. The method was tested on 678 subjects and the evaluation results showed an equal error rate (EER) of 0.145%. The proposed scheme proved to be robust against brightness fluctuations compared to conventional feature extraction techniques.

The authors in [82] based in their previous papers ([36], [37]), proposed a scheme based on finger vein patterns for biometric identification utilizing biological information. They proposed a scheme for extracting global finger vein patterns by iteratively tracking local lines from various positions to robustly extract finger vein patterns from such fuzzy images. The proposed method extracts the centerlines of the veins from the unclear image by calculating the curvature of the cross-sectional profile of the image. To obtain the vein pattern spreading in an entire image, all the profiles in a direction are analyzed. All the profiles in four directions are also analyzed in order to obtain the vein pattern spreading

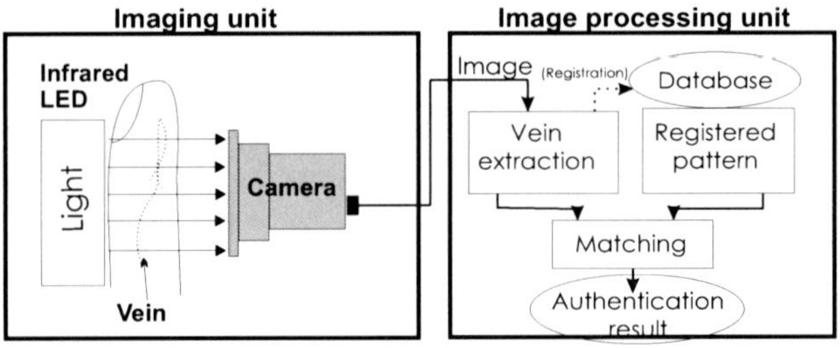

Fig. 4. Flow of finger-vein person identification [82].

in all directions. Matching, was performed using a commonly known method for line-shaped patterns (template matching) proposed in authors' previous works.

The proposed scheme appears to be robust against brightness fluctuations, compared with the conventional feature extraction schemes. The method was tested on 678 subjects and the evaluation results showed an equal error rate (EER) of 0.0009%.

The researchers report that cold weather affects the result as mismatch ratio appears to be higher in cold conditions as the veins become less visible. It is clear that for the stability of the performance of such a technology, more efficient capture devices are needed while a more sophisticated algorithm which will be robust against these fluctuations should be designed. There are also other directions that authors consider to drive their research to. Rotation of fingers degrades identification accuracy. So, redesigning their application in such a way that would make the user to place his finger in the same position every time, would improve the results. An advantage of the method is that by its nature, can be easily combined with other biometrics, such as fingerprints and finger/hand geometry.

A commercial product based on finger vein imaging called "SecuaVeinAttestor", has already been released. Characteristics and full specification of this product can be found in [40].

3.2 Skin Spectroscopy

The measure of the intensity of light taken as a function of wavelength of energy, is called spectrum (hyperspectral is part of spectral imaging class and more specifically is a set of contiguous bands usually received from one sensor). Human skin contains unique spectral properties. Skin is a tissue formed in a layered structure and has complex interaction with light. The way multiple layers and biochemical substances (collagen fibres hair follicles etc.) are set in the skin, differ from person to person. Cell size and density within the skin layers, as well as in the chemical makeup of these layers, also vary from person to person. The penetration depth of the light emitted to the skin, depends on one hand on the wavelength of light and on the other hand depends on the natural color of the skin. The reflectance spectrum of skin provides useful information about

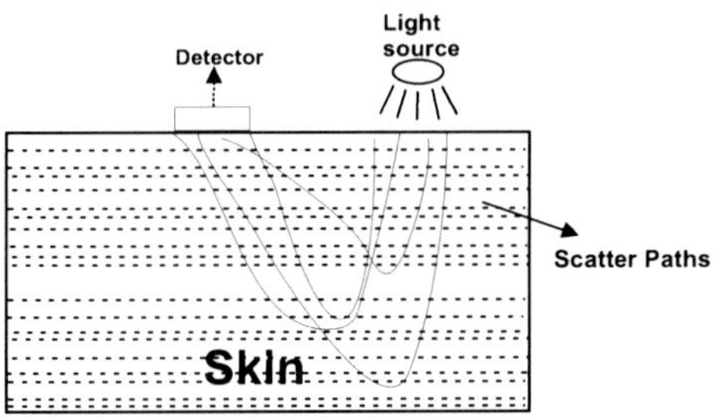

Fig. 5. Illustration showing light undergoing optical scatter as it passes through skin, resulting in a portion of light that is diffusely reflected. Different wavelengths provide different information about the skin [42].

the location and concentration of various substances that appear and is highly person dependant. Thus, spectroscopic measurements can be successfully used as a biometric.

A new commercial product based on these properties is presented in [42]. This product claims to recognize the individuality of each person's skin by measuring the optical effects they produce. The developed sensor illuminates a small (0.4 inch diameter) patch of skin at multiple wavelengths ("colors") of visible and near infrared light. The light that is diffusely reflected back, after being scattered in the skin, is then measured for each of the wavelengths (Figure 5). As light changes passing through the skin, is analyzed and processed to extract a characteristic optical pattern. Afterwards, this pattern is compared to the database of patterns stored in the device in order to authenticate an individual.

The optical differences caused by changes to properties of the human skin, also provide the advantage of sensitive and easy aliveness detection. The optical signal received by non-human tissue or synthetic materials has different optical properties than the human skin. The sensor used to perform these measurements comprises several LEDs that send light at different wavelenghts into the skin, and photodiodes that read the scattered light, which is analyzed to perform the authentication. They all are embedded in an alumina ceramic housing shown in Figure 6 [43].

The sensing system has been designed to fulfill the demanding requirements of incorporating a biometric sensor in a personal portable electronic device such as a cellular telephone, laptop or PDA.

Inventors of the solid-state spectral biometric sensor [43] conducted experiments for a 4 months period, in several sessions, on 113 volunteers (11,000 samples total). 59 males and 54 females of different nationalities participated in the experiments. The age of the participants varied from 20 to 71 years. The equal error rate (EER) obtained for single-try data was 2.7%. However, researchers found that familiarization of the user with the system, increases remarkably the overall performance. After each person successfully used the sensor 20 times the overall EER obtained was decreased to 1.7%.

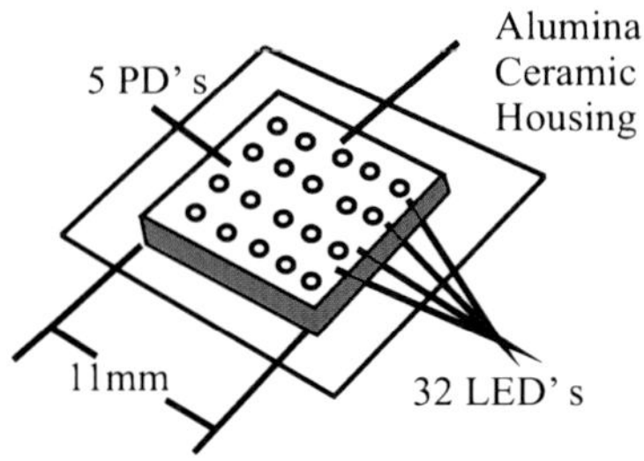

Fig. 6. Solid-state biometric sensor [43].

The spectroscopic approach offers a great advantage over other conventional technologies. Since skin is such a complex organ, it cannot be copied or replaced by synthetic materials, offering in the same time liveness detection. Such an approach that examines spectroscopy as liveness detection solution for biometric systems, is presented in [80].

3.3 Thermal Palm Recognition

Palm as a biometric modality is used for more than a decade [84]. Person verification using the thermal images of palm-dorsa vein patterns (instead of conventional visual images) captured by an infrared camera, is presented in [33] (Figure 7). The region of interest on the thermal images, is defined by capturing automatically two of the finger webs as the datum points. Extraction of the feature points of the vein patterns (FPVPs) is achieved by a watershed transform modified according to the properties of thermal images. The watershed transform calculates the locations of region basin minima (or maxima) [34]. The region maximum method in this case, is used to extract the FPVPs. The region maximum method selects pixels with a high gray value within a region as candidates for watershed points. Two extra restrictions have been added. The first restriction is, that the pixel with a high regional maximum value is also the central point of the region. The other is that its gray value must be larger than the mean of the pixel value inside the region.

Multiple multi-resolution filters (MRFs) are used to create multi-resolution representations of images with FPVPs. MRFs extract the dominant points by filtering miscellaneous features for each FPVP. Three different MRFs, retain the properties of multiple features of the FPVPs at the next level resolution. The first MRF is used to construct multi-scale feature point images (FPIs) and is called moment filter. The second is called mean filter and computes the means of the x and y coordinates as representation for the next level resolution, while the third is called count filter and counts the N feature points inside local square windows for a representation of the next level resolution. In order the former to be integrated, an inter-to-intra personal variation ratio (weights) is proposed, while a positive Boolean function is used to integrate the latter.

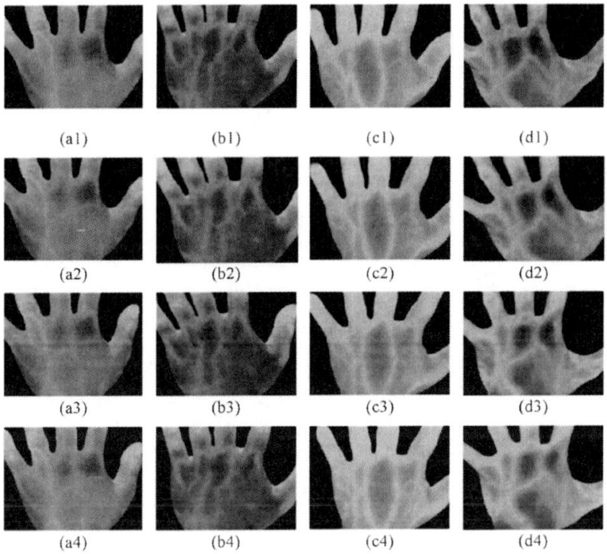

Fig. 7. Thermal images captured from four different palm-dorsa: (a1-a4), (b1-b4), (c1-c4), and (d1-d4) [33] (©2004 IEEE).

Experiments on the specific technology indicated a decent performance (false rejection rate: 2.3% and false acceptance rate: 2.3%)[33]. However, the method is still immature and a number of problems still remain unsolved. The degree of venous engorgement, ambient temperature, the condition of the vein walls and the contiguity of the vein to the surface, are some of the factors that have undesirable influence to the recognition performance. Environmental temperature may dramatically affect the results as it leads to unstable distribution patterns. Hence, it is obvious that a stand alone product based only on the vein-pattern features in palm-dorsum thermal images, could be easily proved unreliable.

3.4 Nail ID

Researchers in [41] suggested a truly novel modality capable of identifying a person from his nailbed. More specifically, the specific commercial product reads the information hidden in the finger nail (nail and nailbed are illustrated in Figure 8). The nailbed is the vascular epidermis upon which most of the fingernail or toenail rests that has a longitudinally-ridged surface often visible through the nail. Blue thin lines observed under nail after small nail injuries is actually blood from a damaged blood vessel from inside the nailbed. The epidermal network beneath the nail is mimicked on the outer-surface of the nail. Rotating one's fingernail under a light reveals parallel lines spaced at intervals. As the nail is produced by the root, it streams down along the nailbed, which adds material to the undersurface of the nail making it thicker. It is important for normal nail growth that the nailbed is smooth.

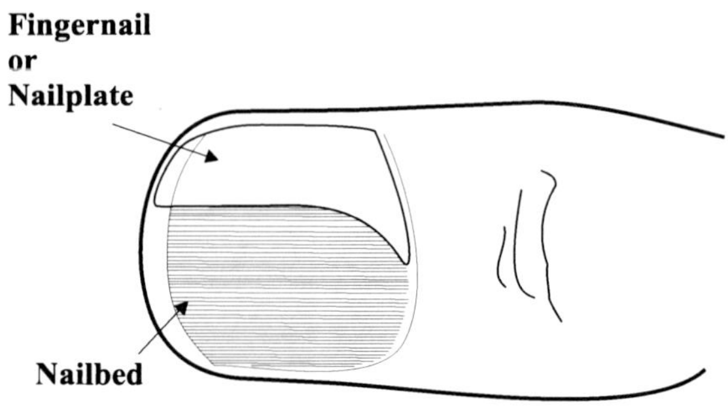

(a)

Fig. 8. Schematic representation of nail [41].

Keratin microfibrils within the nailbed are located at the interface of the nailbed and the nailplate, or fingernail. A broadband interferometer technique detects polarized phase changes in back-scattered light introduced through the nailplate and into the birefringent cell layer. By measuring the phase of the maximum amplitude polarized optical signal, one can reconstruct the nailbed dimensions using a pattern recognition algorithm on the interferometric data. According to inventors, a numerical string (one-dimensional map) much like a "barcode" unique to each person, is generated during the identification process. The use of the specific "barcode" results in an inexpensive commercial product.

The nailbed presents a number of advantages compared to other relevant modalities such as fingerprints and hand geometry. Since nailbed is beneath the nailplate, is extremely difficult to be copied or duplicated. The company that produces this system claims that the system is able to read the hidden barcode even in the case where the user is wearing surgical gloves. However, since this is a commercial product, there are no published results showing the exact performance of the system.

3.5 Ear Prints

Biometrics based on ear shape is still a subject of discussion. Indeed there is still an open discussion between researchers whether ear is unique enough to constitute a proof of individuality. However, it is an interesting topic and still attracts researchers' attention. In [44] authors argue that ear biometrics is a "viable and promising new passive approach to automated human identification". According to a possible scenario, a mugger listens at a door that he is about to break, leaves a print crated by oils and waxes on the ear. This print can be easily obtained by techniques similar to the ones used when lifting fingerprints. An attempt to correlate such an ear print to a person, was part of a research project with the collaboration of several European institutes. The study was presented in [45].

Earprints, photographs and video are some of the ways ear data can be acquired. However, all methods require the ear to be pressed against a firmed transparent material such as a glass. A sample of ear print geometry is shown in Figure 9. The polar axis shown in the figure, is a common tangent to inner edge of the impression of the (onset of the) *crus* of *helix* and the tip of *tragus*. The ear print geometry is based on the following metrical characteristics ('cues'):

- (A) Intersection of the 290ø line from *tragus* tip O with the median line of the *anthelix* impression
- (B) tangent point on the tip of the *antitragus* of a perpendicular from the polar axis
- (C) tangent point of tip of polar axis with the median line of the (onset of the) *crus* of *helix* impression
- (D) intersection point of the line extending OA with the median line of the outer *helix* impression
- (E) intersection point of the 345ø line from *tragus* tip O with the median line of the upper *helix* impression
- (O) tangent point of polar axis with the tip of *tragus*

Some researchers report advantages of the ear in comparison to face based approaches [46]. Working on ear and face recognition, using standard principal component analysis, they found recognition performance to be almost equal between face an ear. For the experiments they involved 197 subjects. Face and ear images were taken under the same conditions and the same image acquisition session. The dataset was tested under pose and lighting variation. The results indicated that the performance is not significantly different between the two biometric modalities. The published results report a recognition rate of 70.5% and 71.6% with 29.5% and 28.6% false recognition rate for the face and the ear respectively.

A number of other papers [47] suggest ear based solutions but the performance is not sufficient yet. The main disadvantage of all ear based applications is that the subject has to turn its head perpendicularly to the camera axis or, even worse, to stick its ear in a board in order the ear sample to be taken. Therefore, such a security system would be considerably intrusive.

3.6 Mouse Dynamics

It is known that most of the currently available biometric technologies typically require special and often expensive equipment that hinders their widespread use. An advantageous solution is based on mouse dynamics [50].

It employs a similar idea to keystroke dynamics. Keystroke dynamics is a common and widely known technique since the beginning of the past decade [51]. The keystroke dynamics method measures two distinct variables: "dwell time", which is the amount of time one holds down a particular key and the

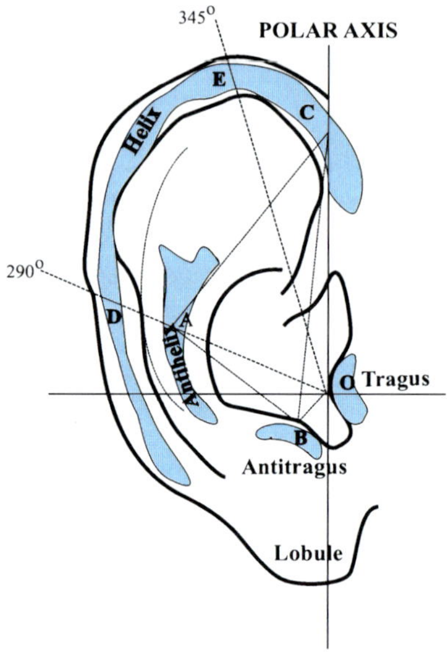

Fig. 9. Reference points for metrical characteristics ('cues') of an earprint [49].

"flight time", which is the amount of time it takes a person to search and press the next appropriate key.

According to the researchers, the proposed method uses state of the art pattern recognition algorithms combined with artificial intelligence to provide a biometric layer over traditional password based security. The system learns an optimum set of mouse-movement characteristics unique to the user's mouse-written signature and uses them to authenticate later signatures. It can also learn over time to include changes of the user's mouse signature characteristics. The main idea of this method is illustrated in Figure 10. First the user's mouse dynamics data are collected through an application that monitors the mouse movement for the specified duration. Certain signature characteristics are extracted in the mouse dynamics patterns, such as double-clicking speed, movement velocity and acceleration per direction.

In order to increase the performance of the system, researchers combined the conventional keystroke dynamics method with mouse dynamics. This way, a user must pass two distinct tests to gain access to restricted content. The first examines the typing style of the password and the second the dynamics of the mouse based signature. The additional level of security can vary according to application needs. In trials with 41 participants, a false acceptance rate of 4.4% and a false rejection rate of 1% where obtained respectively. In these trials, it was assumed that the password was known, whereas in reality it would not be.

In [79], the behavior characteristics from the captured data is modelled using artificial neural networks. A graphical based application involving general mouse movement, silence, drag and drop behavior, point and click behavior, is

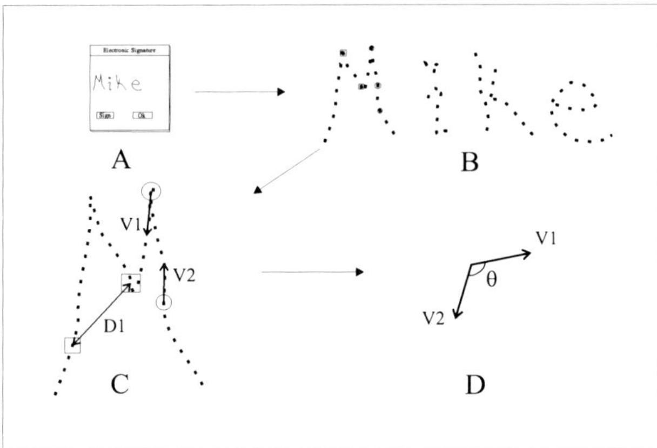

Fig. 10. Main idea of the mouse dynamics recognition system [50].

used to measure several attributes with respect to the users exertion. The authors develop a mouse dynamic signature (MDS) for each user using a variety of machine learning techniques. The data collected for the experiments comprises of 22 participants and was used in an off-line approach to evaluate their detection system. The subjects were separated into two categories (clients and impostors) and the features obtained were used to train a neural network that performs the classification. The EER obtained for this study was 2.46%. According to the authors, this approach could also be applied for continuous user authentication.

A number of advantages are given for this method: The system builds on already familiar user skills, like mouse movements and users can reliably reproduce complex mouse based signatures. The system based on neural networks, can learn over time to incorporate changes of the users typing and mouse signature characteristics. The specific modality, is mostly proposed as an on-line biometric verification solution. On-line banking, shopping, or accessing web based e-mail could be a few of the possible applications. In addition, the technique can be used to validate computer-controlled access to rooms or buildings, confirming identity at security checkpoints without expensive special equipment.

3.7 Electrocardiogram (ECG)

Electrocardiogram (ECG) is the usual technique for the investigation of heart diseases. The rate and the regularity of heartbeats as well as the size and the position of the chambers, the presence of any damage to the heart, are all measured by ECG. Recent studies, indicated that ECG is unique to each person [52], [53], [54]. In [55] ECG with quantifiable metrics is proposed for biometric security purposes. Data filters have been designed based on the observed noise signals. The filtered data were used for the digital extraction of the fiducial points. Stable features computed on the fiducial points characterize the individuality of a

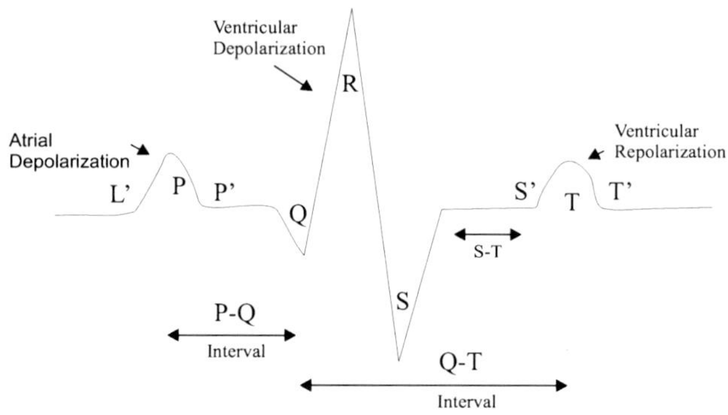

Fig. 11. ECG data for one subject across three sessions [77].

subject according to its ECG. The fiducial points are marks on the ECG signal used for measurements. The locations of the fiducial positions, noted by an apostrophe ('), are illustrated in Figure 11. Physically, the L' and P' fiducials indicate the start and end of the atrial depolarization. The corresponding S' and T' positions indicate the start and end of ventricular repolarization. The extracted features are based upon cardiac physiology and have fixed positions relative to the heartbeat.

Tests performed using the specific cardiac signal, show that results are not dependent on the location of the sensor and they are not dependent on the psychologic situation of the subject. 29 subjects participated in the experiments (males and females) between 22 and 48 years old. Every session contained a set of recordings during 7 two-minute tasks. The tests were designed in such a way that different states of anxiety to be stimulated. The hardware used for the collection of the required data was using high temporal resolution of 1000Hz unlike conventional ECG data. The measures were taken from neck and chest at the same time. For the two different measure points, a heartbeat classification of 82% and 72% was achieved respectively, while the person classification was found to be 100%.

Further data collection is needed in order to obtain more features and prove stability over time and scalability of the features on large population. At the moment, the technique is highly intrusive as it requires placement of several electrodes on subject's body while the enrolment procedure is time consuming. An evaluation on how easy an ECG biometric system can be fooled by the morphology of the electrocardiogram can be found in [76].

3.8 Cognitive Biometrics

A recent work in [73] presents an alternative biometric based on cognitive abilities. The study was motivated by independent studies in cognitive neuroscience and psychiatry, reporting that the generation of random rhythms or numbers is a demanding cognitive task that carries enough information to discriminate between different clinical populations. The authors of the paper, believe that if a person generates random numbers (e.g via a keyboard) there is a cognitive load implied. This is due to the close interaction between short-term memory and internalized decision making mechanisms. Another related task, is the generation of the random tapping rhythms. Finger tapping, for instance, requires sensorimotor interaction and specific cortical networks. Interestingly, it has been demonstrated that everyone has his own eigen-rhythms regulating spontaneous finger tapping.

While many modalities have been suggested, according to authors, this is the first approach where human-generated time-series of random latencies are tested as biometric. For the generation of RTI signal a simple approach was used. A computer based exercise requires from the user to press the space key by the index finger as irregularly as possible, until an "end" message is displayed on the screen. For this purpose a square 4x4 cm, appears and disappears at random frequency while synchronized beep sounds are emitted. Authors state, that the exact reproduction of the given frequency is not the objective of this particular task. The specific example is indicative of the sort of time series one has to create.

When realizations from different subjects were compared, the dynamics indicated a prominent idiosyncratic character. To experimentally verify that it is feasible to identify a person from Random Time-Intervals (RTI) signals and systemize such comparisons, researchers established an appropriate similarity measure. Incorporating this measure in a support vector machine (SVM) based verification system, an equal error rate of $\simeq 5\%$ was achieved. For the experiments a database of 40 person was used. Although the above approach shows potentials for future research, at the moment has a main drawback for commercial use. The enrolment procedure takes almost two minutes and requires full user cooperation. Further investigation of the extracted features as well as more efficient ways for their extraction is highly required.

3.9 Eye Movement

The characteristics of the eye movement (the reaction of the eye to visual stimulation) is examined as a biometric solution in [65]. The user interface of the system, requires the subject to follow a point on a computer's monitor. All the necessary information is related to the eye movement performed during the test. To collect this information, an accurate eye tracker based on infrared reflection is used. The dataset collected consists of probes. The main challenge of this method was the extraction of a set of features suitable for identification. Each probe was collected after an eight seconds stimulus of the eye and consisted of 2048 single measurements. Each measurement consists of six values which describe the position of the point on the screen and the position of the points the right and the left eye are following respectively. Cepstrum coefficients [66] were used as discriminant features.

Every subject in the database (9 in total) was enrolled more than 30 times. 270 probes were used for training the system and 30 for testing. The average false acceptance rate for the proposed method performing person identification, found to be 2% while a high average false rejection rate of 25% was obtained.

The continuous movement of the eye for biometric purposes is also suggested in [96]. The authors, have conducted a case study to investigate the potential of the eye-tracking signal as biometric. They argue that the distance between the eyes proved to be the most discriminant feature (90% identification rate). The best dynamic feature was received from the delta pupil size which corresponds to the variation of the pupil size in time (60% identification success). The information obtained by measuring the size of the pupil itself proved to be weak giving 40% identification. Combination of different features does not seem to offer any considerable improvement. For the experiments 12 subjects participated with normal or corrected to normal vision.

In order to investigate whether there were any individual differences in eye-movement dynamics, researchers had to remove the static properties from the signal. By taking the time averages for each subject, the authors created a static user template. As long-term statistics, these were expected to carry the information about the physiological properties of the subject's eyes. By considering as feature vector the time signal, the dynamic user templates were formed. In summary, eye movement seems to provide discriminatory information. Considering that both the training and test signals had a duration of 1 second, the recognition accuracy of 40-90% can be considered, according to authors, as high, especially taking into account the low sampling rate (50 Hz).

Eye movement as a biometric combines both behavioral and physiological characteristics compared to others like fingerprints or face which use exclusively physiological characteristics. Thus, aliveness detection is embedded in the system providing a major advantage to it. On the other hand, this technique is highly depended on the physical situation of the subject. It requires conscious effort on behalf of the subject and it would fail in cases such as drunkenness.

3.10 Dental Biometrics

Dental radiographs provide dentists with useful information concerning the condition of teeth, their roots, jaw placement and the overall composition of the facial bones. The radiographs acquired after a person's death are called post-mortem (PM) and the radiographs acquired while the person is alive are called antemortem (AM). Such images, are commonly used for human identification in crime investigations. An automated method for identification based on information provided by digital or digitized radiographs is proposed in [67]. The paper proposes an automatic technique for matching dental radiographs for identification in forensics. Feature extraction is achieved using anisotropic diffusion to enhance the images and a Gaussian mixture model to segment any dental work that may appear. In case of dental work presence, an area-based metric is used for matching it.

Matching consists of three steps. At first, a shape registration method aligns the teeth contours and computes the between distance. In the presence of dental work, matching is performed using an area-based metric. The two matching

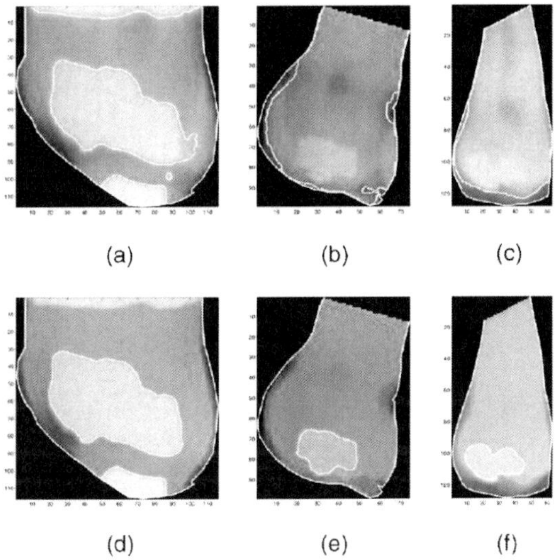

Fig. 12. The original images and the extracted contours of dental work (a), (b) and (c). The corresponding smoothed images and the extracted contours (d), (e), and (f) [67] (©2005 IEEE).

distances are then combined using posterior probabilities. Afterwards, the correspondence established for PM and AM images is used to compute the similarity between them. Finally, the distances calculated between individuals, are used for identification. Some examples of extracted teeth shapes are presented in Figure 12. For the identification task, 11 PM subjects were matched against 25 AM subjects. The results produced show an accuracy of 72%, 91% and 100% according to the number of top retrievals that have been used (1, 4 and 7, respectively).

In a more recent work, researchers use exclusively dental work (DW) to perform identification [95]. The technique is based on three main processing steps: a) the dental radiographs are pre-processed to segment the dental work. The information obtained contains details about the DW (position on both jaws, size, distance between vicinal DW), b) All the collected features form a "dental code" (DC), c) This code is finally used for matching with the corresponding DC in the database.

To segment the DW from the rest of the denture, the authors use active contours (snakes). A separate snake is used for each of the DW found. The process is hastened by computing all the DWs from a binary mask. Matching is performed using Levenshtein distance. 46 subjects participated in the evaluation of the method (68 radiographs collected). The equal error rate obtained for the proposed method performing identification using the above dataset, was 11%.

Although experimental results show that dental based approaches are promising, there is still a number of challenges to overcome according to the authors [67]. First of all, for both techniques, the experiments should run on a larger database.

Shape extraction is a problem for dental radiographs. For subjects with missing teeth, other features for identification must be explored. The method, as it is presented, examines the identification of individuals in the forensic domain but it can be also applied to living persons. However, a radiographic test procedure would be extremely intrusive and undesirable due to X-ray radiation hazards to human health. Another image acquisition device not based on radio-activity should be applied. Such a devise is not available nowadays.

4 Conclusion

A large number of human physiology characteristics such as iris, retina, fingerprints, face, palm, and personal behavior characteristics such as smile or gait, and different imaging methods have been suggested as biometric recognition/verification solutions. None of these modalities though, is able to provide acceptable performance in order to satisfy every market demand so far. The need for more robust and reliable security systems is continuously increasing, leading to the discovery of new human personal characteristics, able to provide sufficient individuality and more efficient biometric methods.

Recent studies propose advanced techniques and sophisticated electronic devices while various human features have been suggested. Thermogram, hyperspectral images, finger veins, ECG, mouth, teeth, are some of the proposed modalities. All emerging techniques provide a number of advantages, compared to others, according to specific market requirements. However, most of the systems have only been tested on very small datasets, require expensive equipment of high technology while others, require time consuming enrolment procedures. Mulimodality and intrusiveness are only two of the issues that must be extensively investigated in the years to come.

The methods presented here have been separated for simplicity, into two main categories (intrusive and non-intrusive) according to the level of vexation that they induce to the user. Both categories contain techniques that have rather high accuracy rates and are based on very promising ideas. Although, as mentioned, none of them can guarantee satisfying performance, the intrusive methods seem to have the precedence in terms of performance. However, the impressive number and the variety of the proposed developments clearly indicate the thirsty market demands, as well as the impressive forthcoming biometric achievements that the society will encounter.

References

1. Boyd, J.E.: Video phase-locked loops in gait recognition. In: International Conference on Computer Vision, Vancouver, BC, pp. 696–703 (2001)
2. Abdelkader, B., Cutler, R., Nanda, H., Davis, L.: Eigengait: motion-based recognition of people using image self-similarity. In: Audio- and Video-Based Biometric Person Authentication, Halmstad, Sweden (2001)

3. Cunado, D., Nixon, M.S., Carter, J.N.: Automatic extraction and description of human gait models for recognition purposes. Computer Vision and Image Understanding 90, 1–41 (2003)
4. Goffredo, M., Seely, R.D., Carter, J.N., Nixon, M.S.: Markerless View Independent Gait Analysis with Self-camera Calibration. In: IEEE International Conference on Automatic Face and Gesture Recognition, Amsterdam, The Netherlands, pp. 17–19 (2008)
5. Zhao, W., Chellappa, R., Phillips, P., Rosenfeld, A.: Face recognition: A literature survey. ACM Comput. Surv. Dec. 335(4), 399–468 (2003)
6. Bouchrika, I., Nixon, M.: Exploratory Factor Analysis of Gait Recognition. In: 8th IEEE International Conference on Automatic Face and Gesture Recognition, Amsterdam, The Netherlands (2008)
7. Boyd, J.E., Little, J.J.: Biometric Gait Recognition. In: Tistarelli, M., Bigun, J., Grosso, E. (eds.) Advanced Studies in Biometrics. LNCS, vol. 3161, pp. 19–42. Springer, Heidelberg (2005)
8. Nixon, M.S., Tan, T.N., Chellappa, R.: Human Identification based on Gait. Springer, New York (2005)
9. Nixon, M.S., Carter, J.N.: Automatic Recognition by Gait. Proceedings of the IEEE 94(11), 2013–2024 (2006)
10. Jain, A.K., Flynn, P., Ross, A.A.: Handbook of Biometrics. Springer, New York (2007)
11. Kotropoulos, C., Tefas, A., Pitas, I.: Frontal face authentication using discriminating grids with morphological feature vectors. IEEE Transactions on Multimedia 2, 14–26 (2000)
12. Wayman, J.L., Jain, A.K., Maltoni, D., Maio, D.: Biometric Systems: Technology, Design and Performance Evaluation. Springer, New York (2004)
13. Jain, A.K., Li, S.Z.: Handbook of Face Recognition. Springer, New York (2005)
14. Zhao, W., Chellappa, R., Rosenfeld, A., Phillips, P.: Face recognition: A literature survey, UMD CfAR Technical Report CAR-TR-948 (2000)
15. Tefas, A., Kotropoulos, C., Pitas, I.: Using support vector machines to enhance the performance of elastic graph matching for frontal face authentication. IEEE Transactions on Pattern Analysis and Machine Intelligence 23, 735–746 (2001)
16. Messer, K., Kittler, J., Sadeghi, M., Marcel, S., Marcel, C., Bengio, S., Cardinaux, F., Sanderson, C., Czyz, J., Vandendorpe, L., Srisuk, S., Petrou, M., Kurutach, W., Kadyrov, A., Paredes, R., Kepenekci, B., Tek, F., Akar, G., Deravi, F., Mavity, N.: Face verification competition on the xm2vts database. In: Kittler, J., Nixon, M.S. (eds.) AVBPA 2003. LNCS, vol. 2688, pp. 964–974. Springer, Heidelberg (2003)
17. Messer, K., Kittler, J., Sadeghi, M., Hamouz, M., Kostin, A., Cardinaux, F., Marcel, S., Bengio, S., Sanderson, C., Poh, N., Rodriguez, Y., Czyz, J., Vandendorpe, L., McCool, C., Lowther, S., Sridharan, S., Chandran, V., Vidal, E., Bai, L., Shen, L., Wang, Y., Yueh-Hsuan, C., Hsien-Chang, L., Yi-Ping, H., Heinrichs, A., Muller, M., Tewes, A., von der Malsburg, C., Wurtz, R., Wang, Z., Xue, F., Ma, Y., Yang, Q., Fang, C., Ding, X., Lucey, S., Goss, R., Schneiderman, H.: Face authentication test on the banca database. In: International Conference in Pattern Recognition (ICPR 2004), Cambridge, United Kingdom, pp. 523–532 (2004)
18. Wang, Y., Acero, L.D.: Spoken language understanding. IEEE Signal Processing Magazine 22, 16–31 (2005)
19. Markowitz, J.A.: Voice biometrics. Communications of the Association Computing Machinery (ACM) 43, 66–73 (2000)
20. Faundez-Zanuy, E., Monte-Moreno, M.: State-of-the-art in speaker recognition. IEEE Aerospace and Electronic Systems Magazine 20, 7–12 (2005)

21. Blackburn, T., Butavicius, M., Graves, I., Hemming, D., Ivancevic, V., Johnson, R., Kaine, A., McLindin, B., Meaney, K., Smith, B., Sunde, J.: Biometrics technology review. Australian department of Defense science and Technology Organisation (2002)
22. Cappelli, R., Maio, D., Maltoni, D., Wayman, J., Jain, A.: Performance evaluation of fingerprint verification systems. IEEE Transactions on Pattern Analysis Machine Intelligence 28, 3–18 (2006)
23. Equinox: Face database, `http://equinoxsensors.com/products/HID.html`
24. Socolinsky, D.A., Selinger, A., Neuheisel, J.D.: Face recognition with visible and thermal infrared imagery. Computer Vision and Image Understanding 91, 72–114 (2003)
25. Turk, M., Pentland, A.P.: Eigenfaces for recognition. Journal of Cognitive Neuroscience 3, 71–86 (1991)
26. Sim, T., Sukthankar, R., Mullin, M., Baluja, S.: Memory-based face recognition for visitor identification. In: FG 2000: Proceedings of the Fourth IEEE International Conference on Automatic Face and Gesture Recognition 2000, p. 214. IEEE Computer Society, Washington (2000)
27. Buddharaju, P., Pavlidis, I., Kakadiaris, I.: Face recognition in the thermal infrared spectrum. In: Computer Vision and Pattern Recognition Workshop (CVPRW 2004), vol. 8, p. 133. IEEE Computer Society, Washington (2004)
28. Gyaourova, A., Bebis, G., Pavlidis, I.: Fusion of infrared and visible images for face recognition. In: Pajdla, T., Matas, J(G.) (eds.) ECCV 2004. LNCS, vol. 3024, pp. 456–468. Springer, Heidelberg (2004)
29. Zhihong Pan, M.P., Healey, G., Tromberg, B.: Face recognition in hyperspectral images. In: IEEE Computer Society Conference on Computer Vision and Pattern Recognition (CVPR 2003), Madison, WI, USA, June 2003, pp. 334–339 (2003)
30. Zhihong Pan, M.P., Healey, G., Prasad, M., Tromberg, B.: Face recognition in hyperspectral images. IEEE Transactions on Pattern Analysis and Machine Intelligence 25, 1552–1560 (2003)
31. Guan, E., Rafailovich-Sokolov, S., Afriat, I., Rafailovich, M., Clark, R.: Analysis of the facial motion using digital image speckle correlation. In: Mechanical Properties of Bio-Inspired and Biological Materials, V. MRS fall Meeting (2004)
32. Makoto Omata, T.H., Hangai, S.: Lip recognition using morphological pattern spectrum. In: Bigun, J., Smeraldi, F. (eds.) AVBPA 2001. LNCS, vol. 2091, pp. 108–114. Springer, Heidelberg (2001)
33. Lin, C., Fan, K.: Biometric verification using thermal images of palm-dorsa vein patterns. IEEE Transactions on Circuits Systems and Video Technologie 14, 199–213 (2004)
34. Beucher, S.: The watershed transformation applied to image segmentation. In: Conference on Signal and Image Processing in Microscopy and Microanalysis, September 1991, pp. 299–314 (1991)
35. Ferrer, M., Travieso, C., Alonso, J.: Using hand knuckle texture for biometric identification. In: 9th Annual Inernational Carnahan Conferennce on Security Technology (CCST 2005), pp. 74–78 (2005)
36. Miura, N., Nagasaka, A., Miyatake, T.: Feature extraction of finger vein patterns based on iterative line tracking and its application to personal identification. Systems and Computers in Japan 35, 61–71 (2004)
37. Miura, N., Nagasaka, A., Miyatake, T.: Feature extraction of finger-vein patterns based on repeated line tracking and its application to personal identification. Machine Vision Applications 15, 194–203 (2004)

38. Zhang, W., Wang, Y.: Core-based structure matching algorithm of fingerprint verification. In: 16th International Conference on Pattern Recognition (ICPR 2002), vol. 1, p. 10070. IEEE Computer Society, Washington (2002)
39. Jain, A.K., Duin, R.P.W., Mao, J.: Statistical pattern recognition: A review. IEEE Transactions on Pattern Analysis and Machine Intelligence 22, 4–37 (2000)
40. Hitachi, `http://www.hitachi-hec.co.jp/virsecur/secuavein/vein01.htm`
41. Nail ID. BIOPTid the human barcode 2009, Biometrics systems division (2009), `http://www.humanbarcode.com`
42. Lumidigm 2009 (2009), `http://www.lumidigm.com/index.html`
43. Rowe, R.K., Corcoran, S.P., Nixon, K.: Biometric identity determination using skin spectroscopy. Lumidigm, Inc., 800 Bradbury SE, Suite 213, Albuquerque, NM, USA 87106 (2004), `http://www.lumidigm.com`
44. Burge, M., Burger, W.: Ear biometrics in computer vision. In: International Conference in Pattern Recognition (ICPR), pp. 2822–2826 (2000)
45. Meijerman, L., Sholl, S., De Conti, F., Giacon, M., van der Lugt, C., Drusini, A., Vanezis, P., Maat, G.: Exploratory study on classification and individualisation of earprints. Forensic Science International 140, 91–99 (2003)
46. Chang, K.I., Bowyer, K.W., Sarkar, S., Victor, B.: Comparison and combination of ear and face images in appearance-based biometrics. IEEE Transactions on Pattern Analysis and Machine Intelligence 25, 1160–1165 (2003)
47. Moreno, B., Sanchez, A., Velez, J.: On the use of outer ear images for personal identification in security applications. In: IEEE 33rd Annual International Carnahan Conference on Security Technology, pp. 469–476 (1999)
48. Forensic Evidence, `http://forensic-evidence.com/site/id/idearnews.html`
49. Maat, G.: Ear print project-brief report on the pilot study. In: Barge's Anthropologica, Leiden University Medical Centre (1999)
50. McOwan, P., Everitt, R.: Artificial intelligence to increase security of online shopping and banking, Queen Mary, University of London, `http://www.qmw.ac.uk/poffice/nr270803.shtml`
51. Bleha, S., Slivinsky, C., Hussien, B.: Computer-access security systems using keystroke dynamics. IEEE Transactions on Pattern Analysis and Machine Intelligence, 1217–1222 (1990)
52. Biel, L., Pettersson, O., Philipson, L., Wide, P.: ECG analysis: A new approach in human identification. IEEE Transactions on Instrumentation and Measurement 50, 808–812 (2001)
53. Hoekema, R., Uijen, G., van Oosterom, A.: Geometrical aspect of the interindividual variability of multilead ECG recordings. IEEE Transactions on Biomedical Eng. 48, 551–559 (2001)
54. Irvine, J., Wiederhold, B., Gavshon, L., Israel, S., McGehee, S., Meyer, R., Wiederhold, M.: Heart rate variability: a new biometric for human identification. In: International Conference on Artificial Intelligence (IC-AI 2001), Las Vegas, Nevada, pp. 1106–1111 (2001)
55. Israel, S.A., et al.: ECG to identify individuals. Pattern Recognition 38, 133–142 (2005)
56. Marcel, S., del Millán J.R.: A new method to identify individuals using signals from the brain. In: 4th International Conference on Information Communication and Signal Processing (ICICS), Singapore, pp. 15–18 (2003)
57. Poulos, V.C.M., Rangoussi, M., Evangelou, A.: Person identification based on parametric processing on the EEG. In: 6th International Conference on Electronics, Circuits and systems (ICECS), vol. 1, pp. 283–286

58. Marcel, S., del Millán J.R.: Person authentication using brainwaves (EEG) and maximum a posteriori model adaptation. IEEE Transactions on Pattern Analysis and Machine Intelligence Special Issue on Biometrics 29(4), 743–752 (2007)
59. Millán, J.: Brain-computer interfaces. In: Arbib, M.A. (ed.) The Handbook of Brain Theory and Neural Networks, 2nd edn. The MIT Press, Cambridge (2002)
60. Palaniappan, R., Mandic, D.P.: EEG Based Biometric Framework for Automatic Identity Verification. Journal of VLSI Signal Processing Systems 49(2), 243–250 (2007)
61. Palaniappan, R., Mandic, D.P.: Biometrics from brain electrical activity: A machine learning approach. IEEE Transactions on Pattern Analysis and Machine Intelligence 29, 738–742 (2007)
62. Hoppe, U., Weiss, S., Stewart, R.W., Eysholdt, U.: An automatic sequential recognition method for cortical auditory evoked potentials. IEEE Transactions on Biomedical Engineering 48(2), 154–164 (2001)
63. Dietl, H., Weiss, S.: Cochlear hearing loss detection system based on transient evoked otoacoustic emissions. In: IEEE EMBSS Postgraduate Conference, Southampton (2004)
64. Dietl, H., Weiss, S.: Parameterisation of transient evoked otoacoustic emissions. In: Biosignal International EURASIP Conference (2004)
65. Kasprowski, P., Ober, J.: Eye movements in biometrics. In: Maltoni, D., Jain, A.K. (eds.) BioAW 2004. LNCS, vol. 3087, pp. 248–258. Springer, Heidelberg (2004)
66. Rabiner, L.R., Schafer, R.W.: Digital Processing of Speech Signals. Prentice Hall, Englewood Cliffs (1978)
67. Chen, H., Jain, A.K.: Dental biometrics: Alignment and matching of dental radiographs. IEEE Transactions on Pattern Analysis and Machine Intelligence 27, 1319–1326 (2005)
68. International Biometric Group, `http://www.biometricgroup.com/index.html`
69. Meeker-O'Connell, A.: How evidence works, How stuff works, `http://www.howstuffworks.com/dna-evidence.htm`
70. Pouthas, F., Gentil, C., Cote, D., Bockelmann, U.: DNA detection on transistor arrays following mutation-specific enzymatic amplification. Applied Physics Letters 84, 1594–1596 (2004)
71. Pouthas, F., Gentil, C., Cote, D., Zeck, G., Straub, B., Bockelmann, U.: Spatially resolved electronic detection of biopolymers. Physical Review E 70(3), 031906(1-8) (2004)
72. Yen, R.C.: DNA typing and prospects for biometrics. In: Biometric identification Seminar Forensic, Department of Defence, Biometric Management Office. Miltreck Systems Inc. USA (2004)
73. Laskaris, N., Zafeiriou, S., Garefa, L.: Use of random time-intervals (RTIs) generation for biometric verification. Pattern Recognition 42(11), 2787–2796 (2009)
74. Swabey, M.A., Chambers, P., Lutman, M.E., White, N.M., Chad, J.E., Brown, A.D., Beeby, S.: The biometric potential of transient otoacoustic emissions. International Journal of Biometrics 1(3), 349–364 (2009)
75. Sun, S.: Multitask learning for EEG-based biometrics. In: 19th International Conference on Pattern Recognition, ICPR 2008, pp. 1–4 (2008)
76. Tsao, Y.T., Shen, T.W., Ko, T.F., Lin, T.H.: The Morphology of the Electrocardiogram for Eevaluating ECG Biometrics. In: 9th International Conference on e-Health Networking, Application and Services, pp. 233–235 (2007)
77. Chan, A.D.C., Hamdy, M.M., Badre, A., Badee, V.: Person Identification using Electrocardiograms. In: Canadian Conference on Electrical and Computer Engineering, CCECE 2006, pp. 1–4 (2006)

78. Revett, K., Jahankhani, H., Magalhaes, S.T., Santos, H.M.D.: A Survey of User Authentication Based on Mouse Dynamics. Communications in Computer and Information Science 12, 210–219 (2008)
79. Ahmed, A.E., Traore, I.: A New Biometric Technology Based on Mouse Dynamics. Dynamics. IEEE Transactions on Dependable and Secure Computing 4(3), 165–179 (2007)
80. Davar, P.: Spectroscopically Enhanced Method and System for Multi-Factor Biometric Authentication. IEICE - Transactions on Information and Systems E91-D(5), 1369–1379 (2008)
81. Campbell, J.P.: Speaker Recognition: A Tutorial. Proceedings of the IEEE 85(9), 1437–1462 (1997)
82. Miura, N., Nagasaka, A., Miyatake, T.: Extraction of Finger-Vein Patterns Using Maximum Curvature Points in Image Profiles. The Institute of Electronics, Information and Communication Engineers 90(8), 1185–1194 (2007)
83. Ravikanth, C., Kumar, A.: Biometric Authentication using Finger-Back Surface. In: IEEE Computer Society Conference on Computer Vision and Pattern Recognition (CVPR), Los Alamitos, CA, USA, pp. 1–6 (2007)
84. Kong, A., Zhang, D., Kamel, M.: A Survey of Palmprint Recognition. Pattern Recognition 42(7), 1408–1418 (2009)
85. Choras, M.: Human Lips Recognition. Computer Recognition Systems 2, 45, 838–843 (2008)
86. Pavlidis, I., Tsiamyrtzis, P., Manohar, C., Buddharaju, P.: Biometrics: Face Recognition in Thermal Infrared. In: Biomedical Engineering Handbook. CRC Press, Boca Raton (2006)
87. Buddharaju, P., Pavlidis, I.T., Tsiamyrtzis, P., Bazakos, M.: Physiology-Based Face Recognition in the Thermal Infrared Spectrum. IEEE Transactions on Pattern Analysis and Machine Intelligence 29(4), 613–626 (2007)
88. Zhihong, P., Healey, G., Tromberg, B.: Hyperspectral face recognition under unknown illumination. Optical Engineering 46(7), 077201–077209 (2007)
89. Robila, S.A.: Toward hyperspectral face recognition. In: Image Processing: Algorithms and Systems VI, vol. 6812, pp. 68120X–6812X9 (2008)
90. Chu, R., Liao, S., Zhang, L.: Illumination invariant face recognition using near-infrared images. IEEE Transactions on Pattern Analysis and Machine Intelligence 29, 627–639 (2007)
91. Socolinsky, D., Wolff, L., Neuheisel, J., Eveland, C.: Illumination invariant face recognition using thermal infrared imagery. In: Computer Vision and Pattern Recognition Conference (CVPR), Kauai, HI (December 2001)
92. Zou, X., Kittler, J., Messer, K.: Face Recognition Using Active Near-IR Illumination. In: British Machine Vision Conference (2005)
93. Li, S.Z., Chu, R.F., Ao, M., Zhang, L., He, R.: Highly Accurate and Fast Face Recognition Using Near Infrared Images. In: International Conference on Biometrics, International Association for Pattern Recognition (IAPR), pp. 151–158 (2006)
94. Li, S.Z., Zhang, L., Liao, S., Zhu, X., Chu, R., Ao, M., He, R.: A Near-infrared Image Based Face Recognition System. In: Proceedings of the 7th International Conference on Automatic Face and Gesture Recognition, pp. 455–460 (2006)
95. Hofer, M., Maranara, A.N.: Dental Biometrics: Human Identification Based On Dental Work Information. In: IEEE XX Brazilian Symposium on Computer Graphics and Image Processing, pp. 281–286 (2007)
96. Bednarik, R., Kinnuenen, T., Mihaila, A., Franti, P.: Eye-Movements as a Biometric. In: Kalviainen, H., Parkkinen, J., Kaarna, A. (eds.) SCIA 2005. LNCS, vol. 3540, pp. 780–789. Springer, Heidelberg (2005)

Overview of Audio Forensics

Robert C. Maher

Montana State University
Electrical & Computer Engineering Department
Bozeman, MT 59717-3780 USA
rob.maher@montana.edu

Abstract. Audio forensics applies the tools and techniques of audio engineering and digital signal processing to study audio data as part of a legal proceeding or an official investigation of some kind. This chapter summarizes the principal audio forensic tasks, including authentication, enhancement, and interpretation. The chapter explains the relevant procedural and historical background, presents several examples of audio forensic applications, and reviews several important areas for future research and development.

1 Introduction

The field of *audio forensics* involves the scientific interpretation of audio recordings that are obtained from a formal civil investigation or a criminal legal proceeding. Audio forensic evidence is often obtained deliberately from an acoustical recording system such as a cockpit voice recorder, an automated call center recording, or a surveillance tape acquired in the course of a criminal investigation by a law enforcement agency. In other cases the evidence may be collected inadvertently, such as a soundtrack extracted from an electronic news gathering rig. In any case, the audio evidence must be evaluated to determine its authenticity, the likelihood that its contents can be enhanced and interpreted, and its relevance to the goals of the investigation [25].

1.1 Types of Audio Forensic Investigations

Authenticity

An investigation in which audio material is presented for forensic examination may have several needs and goals. One of the common requirements is to determine the *authenticity* of the recording. The audio forensic examiner seeks to verify that the recording was produced under controlled circumstances, was maintained in a documented chain-of-custody, and was not inadvertently or deliberately altered prior to examination.

Enhancement

Forensic audio examinations often involve recordings that were made surreptitiously or under circumstances that did not permit ideal microphone placement or optimized signal-to-noise ratio. Therefore, the quality of the audio may be compromised by

H.T. Sencar et al. (Eds.): Intel. Multimedia Analysis for Security Appli., SCI 282, pp. 127–144.
springerlink.com

additive noise, distortion, poor equalization, or excessive reverberation. Among the most frequent enhancement tasks involve noise reduction of recorded speech to improve intelligibility so that an accurate written transcript can be prepared.

Interpretation

Following authentication and enhancement, the audio material for forensic examination ultimately must be evaluated and interpreted to discover its relevance and importance to the investigation. In the case of a speech recording, this often includes preparation of a transcript, identification of the talkers, interpretation of any background sounds that might uniquely identify the circumstances of the conversation, and so forth. Other types of recordings, such as audio evidence obtained from accident or crime scenes, require specialized analysis to document all tell-tale sounds and timing relationships within the recording.

2 History and Examples of Audio Forensics Investigations

Forensic audio examination traces its roots to the 1950s, with the advent of live recording systems for use outside of the recording studio. In the United States, the Federal Bureau of Investigation (FBI) has developed expertise since the early 1960s in audio forensics for the purposes of speech intelligibility enhancement and authentication of recordings [18].

2.1 Audio Forensics and the Law

The seminal legal case in the United States that dealt directly with recorded conversations is the 1958 ruling in *United States v. McKeever* (169 F.Supp. 426, 430, S.D.N.Y. 1958). The judge in the McKeever case was asked, for the first time, to determine the legal admissibility of a tape recorded conversation involving the defendant. The judge ultimately allowed in court the use of a written transcript of the recorded conversation [25].

The McKeever ruling is particularly important because the judge cited seven specific requirements necessary for a recording to be accepted in court, and these requirements are now assumed by most state and federal courts in the United States.

Table 1. Seven Tenets of Audio Authenticity (the McKeever case).

(1) That the recording device was capable of taking the conversation now offered in evidence.
(2) That the operator of the device was competent to operate the device.
(3) That the recording is authentic and correct.
(4) That changes, additions or deletions have not been made in the recording.
(5) That the recording has been preserved in a manner that is shown to the court.
(6) That the speakers are identified.
(7) That the conversation elicited was made voluntarily and in good faith, without any kind of inducement.

In summary, the seven tenets require that audio forensic material for use in court must be obtained and preserved in a documented manner, be unaltered, and contain recognizable talkers and other sound sources—or have witnesses to the recording who can verify its veracity.

2.2 The Watergate Tapes

The Watergate scandal of the mid-1970s had many ramifications for the legal system. The revelation in 1973 by White House aid Andrew Butterfield that U.S. President Richard M. Nixon had installed an audio recording system in the White House and in the Executive Office Building resulted in an order by Judge John J. Sirica of the U.S. District Court for the District of Columbia that the recordings be turned over to the court for transcription. During the ensuing investigation in 1974 it was discovered that the recording of a White House conversation between President Nixon and Chief of Staff H.R. Haldeman recorded in the Executive Office Building in 1972 contained an unexplained 18 ½ minute segment where a buzzing sound completely obscured the speech presumably contained on the tape. The question for the court was whether the gap was caused by a malfunction at the time of the recording, or by some subsequent accidental or deliberate action that destroyed that portion of the recorded conversation.

Chief Judge Sirica appointed a group of technical experts to comprise a special Advisory Panel on White House Tapes to devise and implement a complete physical analysis of the tape itself, the magnetic signals on it, the electrical and acoustical signals generated by playback of the tape, and the properties of the recording equipment used to produce the magnetic signals on the tape. After performing a comprehensive set of tests, the Panel concluded that the 18 ½ minute gap was caused by multiple overlapping passes on the tape by the magnetic erase head of a specific model of tape recorder that differed from the device that produced the original recording. The Panel's tests clearly showed the characteristic magnetic patterns on the tape caused by the recording and the erase heads of the available recording devices [1].

The work by the Advisory Panel on White House Tapes was highly influential in the field of audio forensics. The Panel's methodology is now widely accepted as the model for judging the authenticity of audio recordings. The five steps are summarized in Table 2.

2.3 Other High Profile Cases

Several other highly publicized forensic audio cases have helped shape the techniques and reputation of the field. Acoustic evidence and reconstructions have been used in

Table 2. Advisory Panel on White House Tapes procedure.

(1) physically observe the entire length of the tape (or other data storage medium)
(2) document the total length and mechanical integrity of the storage medium
(3) verify that the recording is continuous with no unexplained stop/start sequences or erasures
(4) perform critical listening of the entire tape
(5) use non-destructive signal processing as needed for intelligibility enhancement

the ongoing investigations surrounding the 1961 assassination of President John F. Kennedy in Dallas [29]. Acoustic evidence has also been discussed in connection with the assassination of presidential candidate Sen. Robert Kennedy in 1968 in Los Angeles. Other important applications of forensic audio include interpretation of conversations and background sounds from cockpit voice recorder data following a commercial aircraft accident [12], and authenticity assessment and enhancement of recordings purportedly made by terrorists [32].

3 Qualifications of Audio Examiners and Expert Witnesses

Audio forensic examiners are often called upon to render their opinions regarding the authenticity of a recording, the source of the recorded sounds, and the identity of the talkers in the recording. In the United States there are a variety of standards for admitting the testimony of topical experts that vary from state to state and between state and federal jurisdictions. The standards for expert testimony often cite the 1923 Frye case (*Frye v. United States*, 54 App. D.C. 46, 293F.1013, DC Ct App 1923), the Daubert case (*Daubert v. Merrell Dow Pharmaceuticals*, 509 U.S. 579 1993), or the more recent *Kumho Tire Co. v. Carmichael* (526 U.S. 137 1999). In general, the expertise standards require that the examiner use methods and develop findings in a manner that is generally accepted by the professional scientific and engineering communities [2, 3, 33].

A forensic audio examiner and expert witness will be the most effective when the examiner can demonstrate a pertinent sequence of formal professional education and training, relevant experience in the audio engineering field, and evidence of ongoing education and professional practice. The list of professional accomplishments should ordinarily include a complete listing of the examiner's prior forensic audio investigations, then number of prior appearances as an audio expert in legal proceedings, a list of formal, peer-reviewed publications authored by the examiner, and evidence of membership in appropriate technical organizations, such as the Audio Engineering Society and the American College of Forensic Examiners. Audio examiners and expert witnesses must also have experience working with attorneys and other legal professionals, as well as the ability to explain the often complicated and arcane principles of audio forensics to judges and juries in layman's language appropriate for the nontechnical triers of fact. [19].

4 Initiation of an Investigation

An audio forensic investigation typically commences with a request for assistance by an investigative body, a legal representative, or a law enforcement agency. The request may involve determination of authenticity, enhancement of speech intelligibility, identification of talkers, interpretation of sounds in the recording, or some combination of these tasks.

4.1 Basic Equipment and Laboratory Setup

A contemporary audio forensic laboratory needs a variety of hardware and software to support authenticity evaluations, signal enhancement, and audio interpretation. The basic complement of equipment can include [18, 19]:

- Acoustically isolated and quiet laboratory (e.g., ambient < 25dBA SPL).
- Analog and digital playback equipment for common storage media, such as analog compact and mini cassette, Minidisc, CD/DVD, flash memory (CF and SD), etc.
- Provision for accommodating non-standard and proprietary media playback.
- Computer-based audio acquisition/playback systems with low-noise A/D and D/A.
- Audio editing, spectral analysis, and display software.
- Reliable and spectrally-flat headphones and amplifier.

Authentication investigations may require specialized techniques and equipment, such as magnetic development of record and erase signatures on analog tape [4, 18, 30].

4.2 Handling of Audio Evidence

Unlike ordinary audio studio work, audio forensic examinations must maintain proper procedures and documentation for handling evidence [2, 3, 33].

When the examiner receives an audio forensic assignment, the accompanying information should include all of the relevant circumstances and documentation regarding the evidentiary recording. If the audio material was recorded using a proprietary format or non-standard device or medium, the proper device and instructions must be provided with the recording.

Upon receipt, the examiner needs to document the physical condition of the evidence, noting any damage, markings, serial numbers, lot numbers, format indications, presence of erase-prevention tabs, and other characteristics of the material. The examiner labels the evidence with a permanent marker to show the date of receipt and the examiner's initials [2].

Following the physical observations, the examiner carefully produces a high-quality digital recording of the audio material, either by a direct digital transfer, if possible, or via a low-noise analog-to-digital conversion in the case of analog source material. This digital recording serves as a laboratory back-up copy of the evidence, and as the starting point for non-destructive signal enhancement and interpretation.

The examiner also performs a critical listening session of the entire recording, noting the general characteristics of the audio material, and carefully listening for any apparent alterations or irregularities. The examiner must listen specifically for any audible discontinuities or subtle changes in background sounds that could indicate edits or splices.

5 Methodology for Interpreting Authenticity

Determining the authenticity of audio evidence requires several types of observations, generally following the McKeever standard and the procedures of the Advisory Panel on White House Tapes. The examiner needs to perform visual, physical, electrical, and acoustical tests that include [2, 3, 18, 19, 33]:

(1) Review the documented history of the evidence: the circumstances of the recording, stated content, and the subsequent chain of custody.
(2) Verify that the recording device was operating properly and was capable of producing the type and format of the tape or data.
(3) Determine that the recording medium is intact, unaltered, and bears identifying marks, lot numbers, or similar markings consistent with the documented time frame of the recording.
(4) Perform critical listening of the entire audio recording.
(5) Verify that the recording is continuous with no unexplained stop/start sequences or erasures. Look and listen for any changes, additions or deletions in the recording that are unaccounted for in the documentation.
(6) Use short-time spectral analysis software and other signal processing procedures to identify any irregularities.

The necessary examination steps may vary from project to project, but the general requirements are [25]:

Physical inspection
The examiner documents the condition and properties of the audio recording medium. In the case of analog or digital tape, the examiner verifies the length and condition of the tape, the condition of reels and housing, any manufacturing serial numbers or batch numbers, and the magnetic configuration on the tape (number of tracks, mono or stereo, etc.). The tape itself is examined for any physical damage or tape splices.

Critical listening
The examiner carefully listens to the entire recording and notes any apparent alterations or irregularities. Any audible evidence of edits, splices, or audible discontinuities in background sounds, buzzes, tones, etc., are noted.

Magnetic signature and waveform observations
If the evidence is a physical audio tape, the magnetic signals can be examined using magnetic development techniques, and compared to reference signatures of recordings obtained from the same recording device. The distinctive magnetic patterns, or signatures, caused by the record and erase heads during transitions from stop to record, record to pause, and punch-in overdub recording are examined for consistency with the properties of a continuous, unaltered recording.

The playback signal from the storage medium is also observed using a spectrographic analyzer or software package to look for tell-tale signal evidence of a discontinuous or otherwise altered recording. For example, an audio spectrogram may reveal discontinuities in the recorded material, as depicted in Fig. 1. In this example, a section of the original audio recording was abruptly edited, resulting in a broadband event indicated by the arrows in the figure. Note that a more careful smoothing of the edit point could reduce the likelihood of detection.

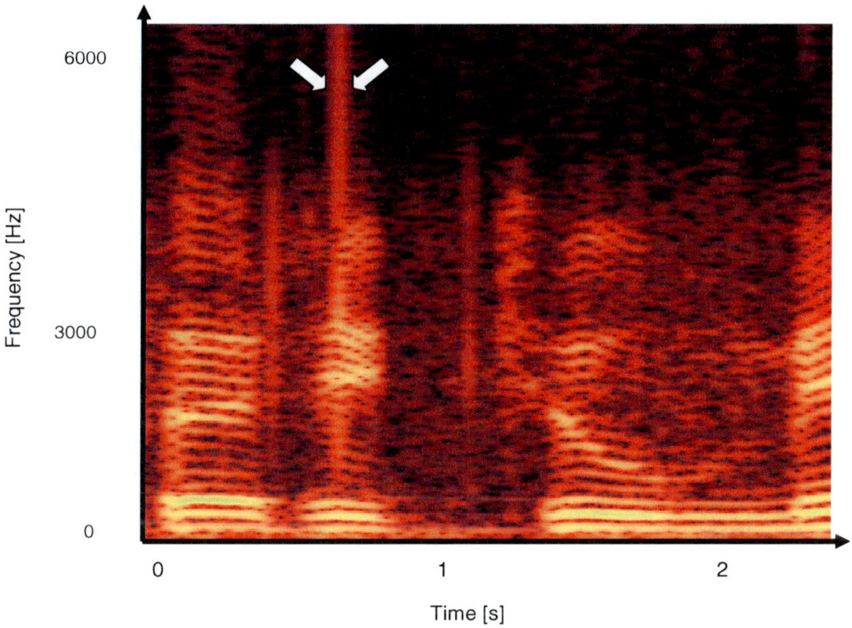

Fig. 1. Spectrogram of a digital audio recording of speech, with indications of a possible edit at the point in time indicated by the arrows. An authentic audio file has no evidence of edits or alterations, so a suspicious spectral signature requires additional investigation by the forensic examiner.

Another example indicating questionable authenticity of a recording is shown in Fig. 2. In this example a word has been inserted into a recording, and the insertion is easily detected by an abrupt change in the background noise (broadband speckle) visible in the spectrogram.

Report preparation
Finally, the examiner prepares a report describing the assessment procedure and the examiner's evaluation of whether the tape is believed to be authentic, a copy, or altered in any manner.

5.1 Planning for Authenticity Verification

In cases where a recording is to be made deliberately for subsequent authentication, several steps can be taken to aid in the authentication process. For example, a surveillance recording should always be made in one continuous recording operation, with no start/stop sequences or pauses. The start of the recording should be audibly marked with a spoken statement giving all of the relevant information surrounding the recording process: date, time, location, identity of participants, model and serial number of the recording device, the type and position of the microphone, and so forth [25].

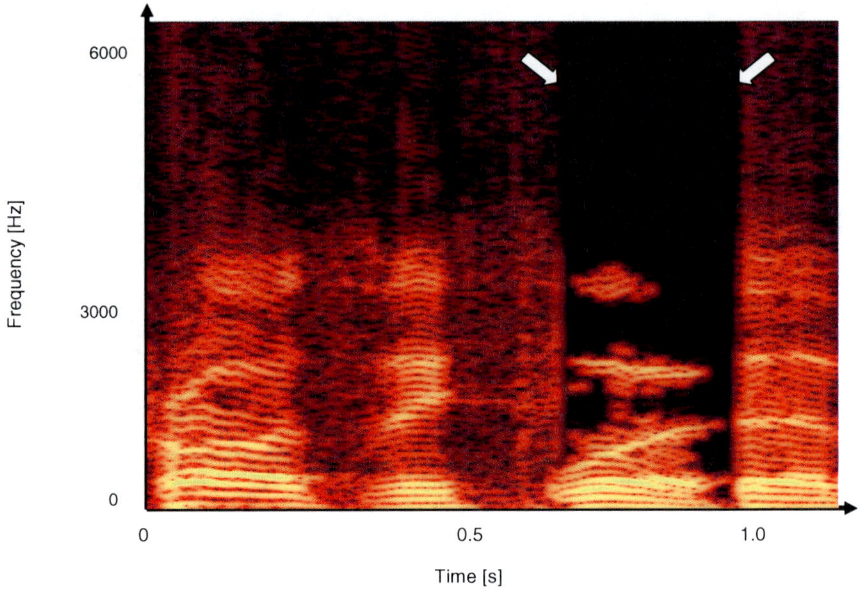

Fig. 2. Spectrogram of a speech recording indicating a likely alteration in the form of an insertion (indicated between the arrows), showing an abrupt change in the character of the background noise.

A forensic recording may also be made easier to authenticate by deliberately including uniquely identifiable background sounds during the recording process, such as a radio broadcast or the natural sounds found in the recording venue. As depicted in Fig. 2, such aleatoric background sounds are virtually impossible to leave unchanged if the foreground sounds have been edited [18].

A recently developed procedure to assist in authentication uses the residual pickup of electrical power line magnetic fields by the audio recording device. The electrical network frequency (ENF) can sometimes be detected by analyzing the AC power network signal in the audio band. Because the power network operates synchronously in a large geographic area of the power grid, the ENF, nominally 60 Hz in the United States and 50 Hz in many other parts of the world, is not precisely constant but varies up to +/- 0.5 Hz from time to time in an unpredictable fashion due to small mismatches between the electrical system load and system generation. Thus, a comparison of the measured ENF extracted from an audio recording with a database of known ENF measurements from the electrical grid may be able to show whether the audio recording was made at the reported time and place [9, 10, 13, 15, 16].

6 Methodology for Audio Enhancement

Forensic audio recordings are often made in nonideal surroundings with nonoptimal microphone placement. Thus, forensic recordings typically suffer from noise, distortion, interfering sounds, and other examples of signal degradation.

A forensic audio examiner may be called upon to perform non-destructive signal processing that might allow a listener to produce a more accurate transcript of a recorded conversation, a higher degree of confidence in assessing the identity of a particular participant, less aural fatigue when listening to an annoyingly noisy recording, or possibly improving the audibility of subtle background sounds that are meaningful to the investigation.

6.1 General Enhancement Steps

When assigned an audio recording for enhancement, the examiner must determine the purpose of the investigation and select an appropriate processing strategy. The examiner first listens to the entire recording in order to determine the scope of the problem and the candidate techniques for enhancement. In some cases, such as speech transcription, the examiner may determine that the highest intelligibility will be obtained by working with the original, unprocessed recording.

A common request is to perform broadband noise reduction on a forensic audio recording [5, 14, 20, 26, 28, 35]. The noise reduction process is applied to a digital copy of the original forensic recording so that several different enhancement procedures can be used and compared without damaging the original audio evidence.

Enhancement methods

Audio forensic enhancement is accomplished with processes operating in both the time domain (noise gates and automatic gain controls) and in the frequency domain (frequency-selective filters).

Automatic gain control

Time-domain enhancement usually involves gain adjustments to normalize the amplitude envelope of the recorded audio signal. One common technique is to apply automatic gain control, or *gain compression/expansion*, to try to keep the sound level relatively constant during playback: portions of the recording attributable only to noise are made quieter, low-amplitude signal passages are amplified, and loud passages are attenuated or left alone.

One traditional approach is to apply a *noise gate* or *squelch* process on the noisy signal. The noise gate is either an electronic device designed for the purpose, or it can be implemented as a software "plug-in" for processing with a computer. The noise gate compares the short-time level of its input signal with a pre-determined level threshold. If the signal level is below the threshold, the gate closes and no signal is let through, while if the signal level is above the threshold, the gate opens and allows the signal to pass. The examiner must adjust the threshold so that the gate passes the desired speech or other audio content, but turns off the noisy background sound that occurs between words and sentences, or during pauses in the conversation. A noise gate can help the listener understand a signal that is perceived to be less noisy because the background sound is gated off during pauses in the conversation. However, the simple noise gate cannot do anything to selectively reduce the noise and boost the signal when both are present simultaneously and the gate is open [25].

More advanced noise gate systems and software use digital signal processing techniques to perform gating separately in different frequency bands. This allows the examiner to tailor the gating effect to the particular types of noise and hiss present in the recording.

Frequency-selective filters

In some cases the quality of a forensic recording can be improved by selectively attenuating tonal components in the spectrum, such as power-related hum and buzz signals. The use of a multi-band audio equalizer can also be helpful in reducing out-of-band noise while still retaining the frequency band of interest, such as the speech frequency range from 200 Hz to 5kHz.

Spectral subtraction

Spectral subtraction refers to a digital signal processing technique in which an estimate of the short-term noise spectrum is determined, and the estimate is then subtracted from the spectrum of short frames of the noisy input signal. The spectrum following the subtraction is used to reconstruct the noise-reduced frame of the output signal, and the process continues for subsequent frames to create the entire output signal via an overlap-add procedure [5, 26].

The effectiveness of spectral subtraction hinges on the reliability of the noise spectrum estimate. The estimate is usually obtained from an input signal frame that is known to contain only the background noise, such as a pause between sentences in a recorded conversation. It is therefore desirable to update the noise spectrum estimate on a regular basis in the recording so that changes in the background noise spectrum can be accommodated.

More sophisticated noise reduction methods combine the time-domain level detection and the frequency-domain spectral subtraction concepts. Additional signal models and rules are utilized to help separate signal components that are most likely to be part of the desired signal from components that are likely to be additive noise [14, 21, 27, 28].

It is important to note that forensic audio enhancement requires careful experimentation, experience, and patience to produce useful results. The procedures are highly subjective and rely upon the training and skill of the examiner.

An example of noise reduction for forensic audio enhancement is shown in Figures 3 and 4. Figure 3(a) shows the time domain waveform of a section of noisy speech, and the corresponding spectrogram is shown in Figure 3(b). A spectral noise reduction process [21] results in the time domain waveform and spectrogram of Figure 4 (a) and (b), respectively. Note that the apparent noise level has been reduced by the enhancement processing.

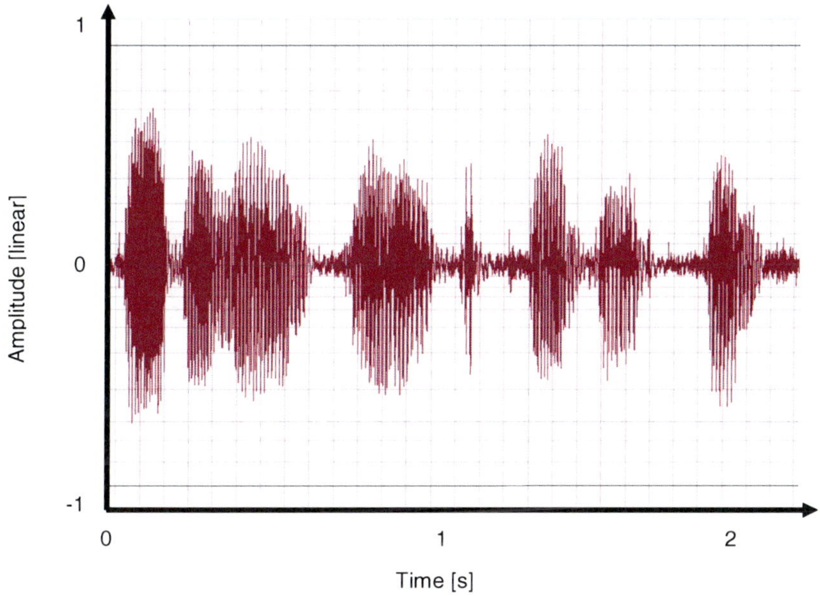

Fig. 3(a). Example of forensic audio enhancement: original time waveform of a noisy speech recording.

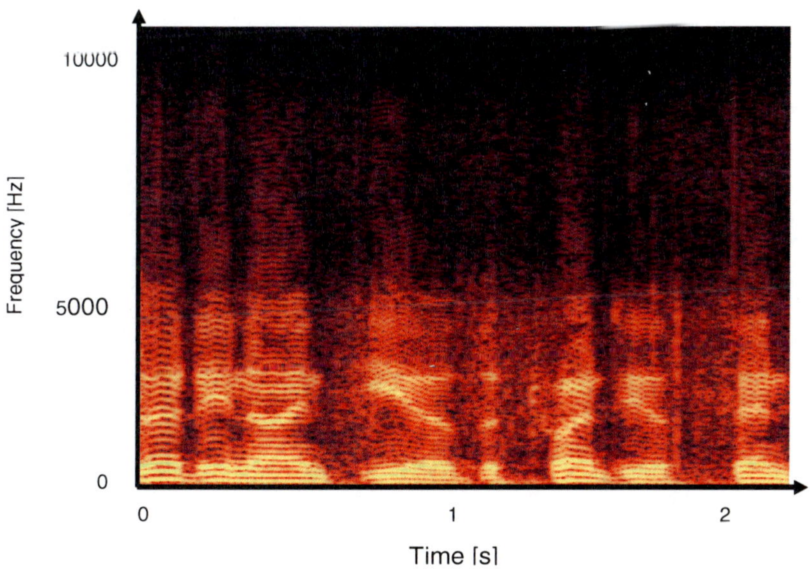

Fig. 3(b). Example of forensic audio enhancement: original spectrogram of the noisy speech recording of Fig. 3(a).

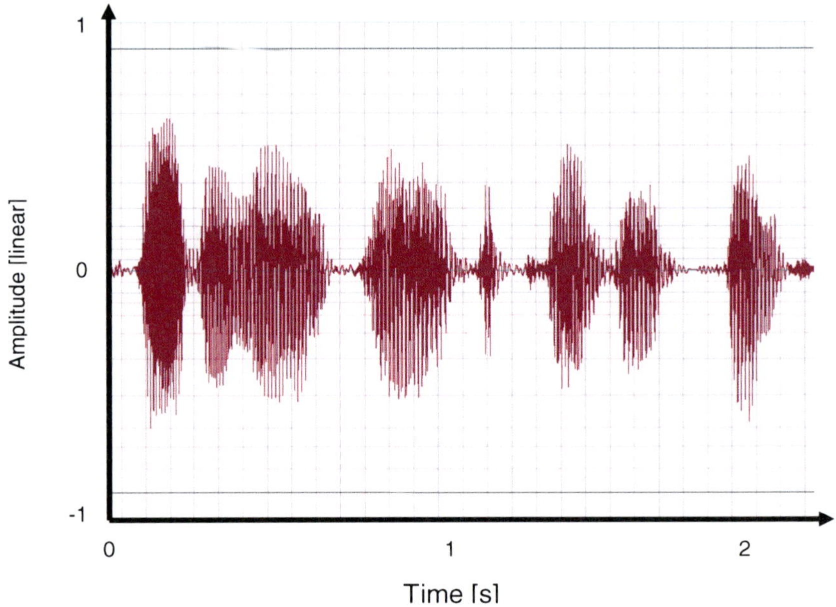

Fig. 4(a). Time waveform of signal shown in Fig. 3(a) following spectral noise reduction process.

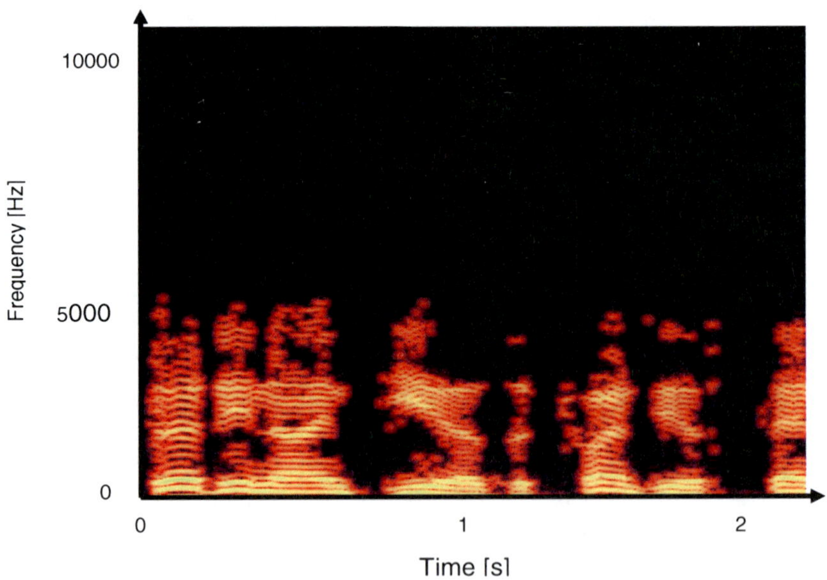

Fig. 4(b). Enhanced spectrogram of the noisy signal from Fig. 3(b) following the spectral noise reduction process.

7 Audio Forensic Interpretation Examples

As noted above, audio forensics projects may involve authentication, enhancement, and interpretation. Here are several examples of the interpretation phase of audio forensics projects.

7.1 Gunshot Acoustical Analysis

Modern crime scenes may involve audio recordings of gunshots, typically from news gathering crews or from tapes of emergency center telephone calls. The characteristics and timing of the gunshots can help the authorities reconstruct the sequence of events at the crime scene, and in some cases determine the orientation of the gun barrel and the type of firearm.

The acoustical characteristics of a gunshot include the boom of the muzzle blast, the arrival of sound reflected from the ground and other nearby surfaces, and possibly some evidence of an acoustic shock wave if the bullet is traveling at supersonic speed in the general direction of the microphone. If the microphone is close to the firearm, it is also possible that the tell-tale sounds of the weapon's mechanical action can be detected in the recording [11, 22, 23, 24].

Typical firearms use rapid combustion of gunpowder to accelerate the bullet out of the barrel, and the expanding gases emanating from the muzzle create the acoustic

Fig. 5. Gunshot sounds can be very distinctive, but most forensic recordings also contain acoustic reflections and reverberation from the ground and other obstacles surrounding the firearm and the microphone.

muzzle blast. The high acoustic intensity of the muzzle blast generally drives the microphone and downstream electronics into clipping, and so the precise details of the acoustic signature are usually obscured. The peak sound pressure levels at the muzzle can exceed 150 dB re 20 μPa. The extremely rapid acoustic pressure rise times are generally also distorted by the recording system, especially for recordings obtained via telephone.

If the microphone is located at a great enough distance that the electronics do not become overloaded, the recording typically will contain significant multi-path arrivals of sound reflections and reverberation.

If the bullet travels at supersonic speed, the acoustic evidence may include a shock wave signature as the bullet travels through the air [11, 24]. The shock wave itself propagates at the speed of sound outward from the bullet's path, expanding as a cone trailing the bullet. The shock wave cone has an inner angle θ_M that is related to the speed of the bullet by the formula $\theta_M = \arcsin(c/V)$, where c is the speed of sound and V is the speed of the bullet. This means that if V is much greater than the speed of sound, a very narrow shock wave cone is produced, resulting in the shock wave propagating nearly perpendicular to the bullet's trajectory. For example, the speed of sound at room temperature is approximately 343 m/s, so a rifle bullet traveling at 914 m/s produces a shock wave angle of $\theta_M = {\sim}22°$. If sufficient information is available regarding the geometry of the crime scene, the speed and trajectory of the bullet, etc., the forensic audio examiner may be able to verify several parameters of the shooting scenario [22, 23].

7.2 Aural-spectrographic Voice Identification

Recorded conversations obtained from legal wiretaps and authorized surveillance operations often include the speech of individuals who either were not physically present at the recording location or whose identity cannot be corroborated by eyewitnesses. For example, a suspect in a criminal or civil investigation may deny being the individual who uttered the recorded words in a telephone conversation. The question for the audio forensic examiner is whether the recorded words can be attributed to the suspect, or if the recorded voice can be excluded as being the suspect.

Some audio forensic examiners specialize in the *aural-spectrographic method* of voice identification. The examiner compares the recorded speech of the unknown (or disputed) talker with one or more examples of known speech (called *exemplars*) uttered by the suspect. The trained examiner critically listens to the unknown speech and to the known speech, and also visually compares the spectrograms of the speech samples [6, 7, 8, 17, 31].

The examiner performs a recording session with the suspect to create the exemplar recordings. The phrases used for exemplars are selected to match as closely as possible the timing and emphasis of selected phrases of the talker in the unknown recording. The examiner instructs the suspect to repeat the exemplar phrases several times in order to get a good comparison with the unknown talker.

The examiner then uses aural and visual observations to form an *opinion* about the likelihood that the exemplars match or do not match the unknown recording. The examiner's report provides one of the following conclusions:

1. Positive identification (the exemplar recordings positively match the unknown speech)
2. Probable identification
3. No decision
4. Probable elimination
5. Positive elimination

Despite the widespread use of the aural-spectrographic method for forensic voice identification, there remains some dispute about the reliability and statistical error rate for this type of subjective analysis [7, 31]. There is considerable interest in replacing the subjective experience of the examiner with a possibly more objective analysis by a computerized automatic speaker recognition system, but as of now there are no court cases in the United States in which computer-based transcription and recognition evidence has been admitted.

7.3 Aircraft Accident Investigations

Commercial passenger aircraft accident investigations increasingly rely upon information recovered from the onboard flight data recorder system and the cockpit voice recorder (CVR) system. The transcript of cockpit conversations can help accident investigators determine the circumstances leading up to an aircraft accident. The cockpit voice recording can also be used to detect audible warning and alert signals, mechanical noises from the air frame, and the sound from the aircraft's engines.

The flight data recorder stores a plethora of digital parameters from the engines, avionics, flight control surfaces, and other sensors, while the CVR contains several channels of audio signals from the radio communications with flight controllers on the ground, as well as an acoustic recording from a microphone located in the cockpit itself. The CVR system is generally designed to record two hours of audio in a loop, providing 120 minutes of cockpit sounds leading up to the crash before being over-recorded. [12].

In one significant case involving audio forensic investigation using CVR data, examiners from the U.S. National Transportation Safety Board used a careful analysis of audio CVR material from the September, 1994, crash near Pittsburgh of USAir Flight 427 (Boeing 737 aircraft), to understand the behavior of the aircraft's engines and the timing, reactions, and effort of the pilot and first officer during the incident. Among other details, the investigation included experiments to determine the ability of the cockpit microphone to pick up sound through structure-borne vibration [12].

A 1997 investigation of the CVR data from a Beechcraft 1900C commuter aircraft accident that occurred in 1992 used signal characteristics from both the cabin microphone and from an unused CVR channel to study the theory that an in-flight engine separation was preceded by evidence of propeller whirl flutter attributable to a cracked truss in the engine mount [34].

8 Areas for Future Research

There are many current and emerging research issues for audio forensic examination. Among the key challenges are verifying the authenticity of digital audio data, improving speech intelligibility in the presence of noise and distortion, and automated methods for speech transcription and speaker recognition.

8.1 Authenticating Digital Data

An important issue for digital data is the possibility that a digital recording has been copied, edited or otherwise modified using a computer, then the modified data has been written to a new file on a different recording medium. Even if the original recording was encrypted or encoded with a digital watermark, it is conceivable that a clandestine decoding, editing, and re-encoding sequence could be perpetrated by a determined individual. The evidence of such an alteration would have to be determined from an examination of the audio signal itself, since the low-level data transport and storage signatures would only reveal a continuously recorded file. Although crude digital editing can be revealed using conventional techniques (see Fig. 1 and Fig. 2), more sophisticated manipulation will require new methods for assessment and evaluation.

The electrical network frequency (ENF) concept mentioned previously is among the emerging secondary techniques for authenticity assessment. It appears that the most productive areas for future authenticity research in surveillance applications will incorporate end-to-end encoding, embedded special signals, and a carefully documented methodology to maintain integrity in the chain-of-custody.

8.2 Speech Intelligibility Enhancement

Single-ended noise reduction of digital recordings has been investigated for many years in the telecommunications and broadcasting fields, as well as in the audio forensic field. The fundamental challenge when reducing noise in forensic applications is to ensure that the intended quality enhancement does not inadvertently degrade the speech inflections, nuances, and essential intelligibility needed for interpretation. In a legal proceeding the Court will need to be convinced that the enhancement procedure has not altered the nature and content of the recorded conversation. For example, subtle phoneme differences may result from spectral threshold methods, causing a phrase such as "*I saw him kick off the mat*" to be interpreted as "*I saw him pick up the bat.*" Both the prosecutor and the defendant in a court case will reasonably expect that the enhanced recording properly reflects the actual conversation, so there remains a need for research into the most reliable and effective enhancement techniques that can be explained and demonstrated to the Court.

8.3 Automated Speech and Speaker Recognition

At present, courts of law exclusively rely upon human experts to transcribe dialog and to assess the likelihood that the speech of a particular individual is present in a forensic audio recording. A typical situation occurs when a police detective believes that a criminal suspect has uttered the words in a recorded telephone conversation, but the suspect denies that it is his voice in the recording. The forensic examiner can provide an opinion based on a review of the aural-spectrographic evidence, but the reliability and objective standards of such a subjective examination can be disputed. Thus, new techniques that can be demonstrated with known performance and reliability statistics would be particularly valuable to supplement human listeners, transcribers, and the subjective aural-spectrographic methodology.

9 Conclusion

The field of audio forensics requires expertise in a variety of audio, acoustics, and signal processing fields. The increasing availability of low-cost digital recorders and other means for obtaining speech and audio data indicates that there will be future demand for audio forensic techniques and services. The importance of employing data handling procedures that meet the requirements for admissibility in legal proceedings will remain a key attribute of audio forensic investigations.

References

[1] Advisory Panel on White House Tapes, The executive office building tape of June 20, 1972: report on a technical investigation. United States District Court for the District of Columbia (1974)

[2] Audio Engineering Society, AES27-1996: AES recommended practice for forensic purposes – Managing recorded audio materials intended for examination (1996)

[3] Audio Engineering Society, AES43-2000: AES standard for forensic purposes – Criteria for the authentication of analog audio tape recordings (2000)

[4] Begault, D.R., Brustad, B.M., Stanley, A.M.: Tape analysis and authentication using multi-track recorders. In: Proc. Audio Eng. Soc. 26th Conf. Audio Forensics in the Digital Age, Denver, CO (2005)

[5] Boll, S.: Suppression of acoustic noise in speech using spectral subtraction. IEEE Trans. Acoust. Speech and Signal Processing ASSP-29, 113–120 (1979)

[6] Bolt, R.H., Cooper, F.S., David, E.E., Denes, P.B., Pickett, J.M., Stevens, K.N.: Identification of a speaker by speech spectrograms. Science 166, 338–342 (1969)

[7] Bolt, R.H., Cooper, F.S., David, E.E., Denes, P.B., Pickett, J.M., Stevens, K.N.: Speaker identification by speech spectrograms: a scientist's view of its reliability for legal purposes. J. Acoust. Soc. Am. 47, 597–612 (1970)

[8] Bolt, R.H., Cooper, F.S., Green, D.M., Hamlet, S.L., McKnight, J.G., Pickett, J.M., Tosi, O.I., Underwood, B.D.: On the theory and practice of voice identification. Nat. Acad. Sci. (1979)

[9] Brixen, E.B.: Techniques for the authentication of digital audio recordings. In: Proc. Au dio Eng. Soc. 122nd Conv. Paper 7014 (2007)

[10] Brixen, E.B.: ENF—quantification of the magnetic field. In: Proc. Audio Eng. Soc. 33rd Conf. Audio Forensics—Theory and Practice, Denver, CO (2008)

[11] Brustad, B.M., Freytag, J.C.: A survey of audio forensic gunshot investigations. In: Proc. Audio Eng. Soc. 26th Conf. Audio Forensics in the Digital Age, Denver, CO (2005)

[12] Byrne, G.: Flight 427: anatomy of an air disaster. Springer, New York (2002)

[13] Cooper, A.J.: The electric network frequency (ENF) as an aid to authenticating forensic digital audio recordings – an automated approach. In: Proc. Audio Eng. Soc. 33rd Conf. Audio Forensics—Theory and Practice, Denver, CO (2008)

[14] Godsill, S., Rayner, S.P., Cappé, O.: Digital audio restoration. In: Kahrs, M., Brandenburg, K. (eds.) Applications of Digital Signal Processing to Audio and Acoustics. Kluwer Academic Publishers, Dordrecht (1998)

[15] Grigoras, C.: Digital audio recording analysis: the electric network frequency (ENF) criterion. Int. J. Speech Language and the Law 12, 63–76 (2005)
[16] Grigoras, C.: Application of ENF analysis method in authentication of digital audio and video recordings. In: Proc. Audio Eng. Soc. 123rd Conv. Paper 1273 (2007)
[17] Koenig, B.E.: Spectrographic voice identification: a forensic survey. J. Acoust. Soc. Am. 79, 2088–2091 (1986)
[18] Koenig, B.E.: Authentication of forensic audio recordings. J. Audio Eng. Soc. 38, 3–33 (1990)
[19] Koenig, B.E., Lacey, D.S., Killion, S.A.: Forensic enhancement of digital audio recordings. J. Audio Eng. Soc. 55, 252–371 (2007)
[20] Lim, J.S., Oppenheim, A.V.: Enhancement and bandwidth compression of noisy speech. Proc. IEEE 67, 1586–1604 (1979)
[21] Maher, R.C.: Audio enhancement using nonlinear time-frequency filtering. In: Proc. Audio Eng. Soc. 26th Conf. Audio Forensics in the Digital Age, Denver, CO (2005)
[22] Maher, R.C.: Modeling and signal processing of acoustic gunshot recordings. In: Proc. IEEE Sig. Proc. Soc. 12th DSP Workshop, Jackson, WY (2006)
[23] Maher, R.C.: Acoustical characterization of gunshots. In: Proc. IEEE SAFE 2007: Workshop on Signal Processing Applications for Public Security and Forensics, Washington, DC (2007)
[24] Maher, R.C., Shaw, S.R.: Deciphering gunshot recordings. In: Proc. Audio Eng. Soc. 33rd Conf. Audio Forensics—Theory and Practice, Denver, CO (2008)
[25] Maher, R.C.: Audio forensic examination: authenticity, enhancement, and interpretation. IEEE Sig. Proc. Mag. 26, 84–94 (2009)
[26] McAulay, R., Malpass, M.: Speech enhancement using a soft-decision noise suppression filter. IEEE Trans. Acoust. Speech and Signal Processing ASSP-28, 137–145 (1980)
[27] Moorer, J., Berger, M.: Linear-phase bandsplitting: theory and applications. J. Audio Eng. Soc. 34, 143–152 (1986)
[28] Musialik, C., Hatje, U.: Frequency-domain processors for efficient removal of noise and unwanted audio events. In: Proc. Audio Eng. Soc. 26th Conf. Audio Forensics in the Digital Age, Denver, CO (2005)
[29] National Academy of Sciences, Report of the Committee on Ballistic Acoustics. National Academy Press, Washington (1982)
[30] Owen, T.: Forensic audio and video—theory and applications. J. Audio Eng. Soc. 36, 34–40 (1988)
[31] Poza, F., Begault, D.R.: Voice identification and elimination using aural-spectrographic protocols. In: Proc. Audio Eng. Soc. 26th Conf. Audio Forensics in the Digital Age, Denver, CO (2005)
[32] Sachs, J.S.: Graphing the voice of terror. Popular Science (2003), `http://www.popsci.com/scitech/article/2003-02/graphing-voice-terror` (Cited August 7, 2009)
[33] Scientific Working Group on Digital Evidence, SWGDE best practices for forensic audio, Version 1.0 (2008), `http://www.swgde.org/documents/swgde2008/SWGDEBestPracticesforForensicAudioV1.0.pdf` (Cited August 7, 2009)
[34] Stearman, R.O., Schulze, G.H., Rohre, S.M.: Aircraft damage detection from acoustic and noise impressed signals found by a cockpit voice recorder. In: Proc. Nat. Conf. on Noise Control Eng., vol. 1, pp. 513–518 (1997)
[35] Tsoukalas, D.E., Mourjopoulos, J.N., Kokkinakis, G.: Speech enhancement based on audible noise suppression. IEEE Trans. Speech Audio Processing 5, 479–514 (1997)

Printer and Scanner Forensics: Models and Methods

Pei-Ju Chiang, Nitin Khanna, Aravind K. Mikkilineni,
Maria V. Ortiz Segovia, Jan P. Allebach,
George T.C. Chiu, and Edward J. Delp

Purdue University, West Lafayette, Indiana USA
{pchiang,khannan,amikkili,mortz,allebach,gchiu,ace}@ecn.purdue.edu

1 Introduction

Contrary to popular opinion, the use of paper in our society will not disappear any time during the foreseeable future. In fact, the use of paper continues to grow rather than decline. It is certainly true that as individuals, we may be printing less than we used to. And the role of paper has been transformed from the archival record of a document to a convenient and aesthetically appealing graphical user interface. The use of paper is now intimately linked to the electronic systems that capture, process, transmit, generate, and reproduce textual and graphical content. Paper can be thought of as an interface between humans and the digital world. If this interface is not secure, the entire system becomes vulnerable to attack and abuse. Although paper is read by humans in the same way that it has been for millennia, and has had the same fundamental form and composition for almost that long as well, the technologies for printing and scanning documents and capturing their content have evolved tremendously, especially during the last twenty years. This has moved the capability to generate printed documents from the hands of a select few to anyone with access to low-cost scanners, printers, and personal computers. It has greatly broadened the opportunities for abuse of trust through the generation of fallacious documents and tampering with existing documents, including the embedding of messages in these documents.

In today's digital world, securing different forms of content is very important in terms of protecting copyright and verifying authenticity. Foreign trade losses due to piracy of books between 1995 and 2001 are estimated to be between $633 million and $695 million per year. In addition, printed material is a direct accessory to many criminal activities such as forgery or alteration of documents used for purposes of identity or recording of transactions. Printed material may also be used in the course of conducting illicit activities. Examples include instruction manuals, team rosters, meeting notes, and correspondence. In both cases, the ability to identify the device or type of device used to print or scan the material in question would provide a valuable aid for law enforcement and intelligence agencies.

H.T. Sencar et al. (Eds.): Intel. Multimedia Analysis for Security Appli., SCI 282, pp. 145–187.
springerlink.com

Similarly, digital images and scanned documents generated by various sources are widely used in a number of applications ranging from medical imaging and law enforcement to banking and daily consumer use. The increasing functionality of image editing software allows even an amateur to easily manipulate images. In some cases a digitally scanned image can meet the threshold definition requirements of a "legal duplicate" if the document can be properly authenticated [2]. Forensic tools that help establish the origin, authenticity, and the chain of custody of such digital images are essential to a forensic examiner. These tools can prove to be vital whenever questions of digital image integrity are raised. Therefore, a reliable and objective way to examine digital image authenticity is needed.

A basic imaging pipeline commonly encountered in an office or household environment is composed of a printer, flatbed scanner, and a computer. A typical usage of this pipeline, in the case of duplicating a document, involves first scanning the document, processing and modifying it using editing tools, and finally printing the modified version. Notice that whereas the scanner converts a hard copy document into a digital one, the printer accomplishes the opposite. Within this imaging pipeline, there exist several stages during which the document can be altered or corrupted either intentionally or accidentally.

Signal processing plays a key role in the characterization of scanners and printers [17, 38, 39, 42, 45, 59]. Development of appropriate signal processing tools allows the ability to prevent or discourage unauthorized use of scanned or printed materials. Two strategies can be used to achieve these goals. The first strategy is passive and applies to both printers and scanners. It involves characterizing a device through the use of intrinsic features present in the scanned image or printed document that are distinctive for the particular device, model, or manufacturer's products. This is referred to as the *intrinsic signature*. The intrinsic signature requires an understanding and modeling of the device mechanism, and the development of analysis tools for the detection of the signature in a scanned or printed document with arbitrary content.

The second strategy is active and has, to our knowledge, only been applied to printers. It involves embedding an *extrinsic signature* in a printed page either by modifying the document to be printed before it is sent to the printer, or by modifying process parameters in the printer mechanism to encode identifying information such as the printer serial number and date of printing. The extrinsic signature also requires an understanding and modeling of the device mechanism, and the development of modulation techniques for the embedding of the signature in a printed document with arbitrary content. It also requires the development of image analysis tools for detecting the presence of the extrinsic signature and demodulating and decoding its content. Several scanning and printing technologies currently dominate the non-commercial market. Scanners differ based on the type of sensor and mechanism that they use to image a document. Printers differ based on the mechanism and marking technology used to form the image on the printed page. The details of these technologies are discussed in later sections.

2 Scanners

In a conventional desktop scanner, a hard copy document is placed face-down on a glass window, a bright light illuminates the page, and the printed patterns are reflected back to a 1-D array of photosensitive elements. The scanning head is attached to a belt and carries the light source and the photosensitive element array in one direction underneath the glass until the selected area of the page is covered. A series of mirrors and lenses direct the reflected light onto the sensor array that converts it into electrical signals. To complete the process, electrical signals produced by the sensor array are digitized by an analog-to-digital converter (ADC) and are sent to the host computer. The operational diagram for a typical scanner is shown in Figure 1.

2.1 Scanner Architecture

Figure 2 shows the architecture of a typical flatbed scanner. The document is placed on the scanner bed and the acquisition process starts. The lamp used to illuminate the document is either a cold cathode fluorescent lamp (CCFL), xenon lamp, or light emitting diodes (LEDs), while older scanners may use a standard fluorescent lamp. Using a belt and a stepper motor, the scan-head slowly translates linearly to capture the image. Velocity fluctuations in the constant speed portion of the scan-head's motion may lead to color registration errors in the

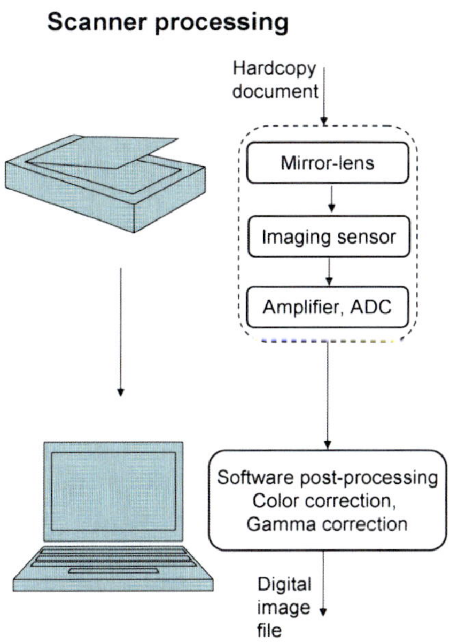

Fig. 1. Block diagram of a typical scanning process.

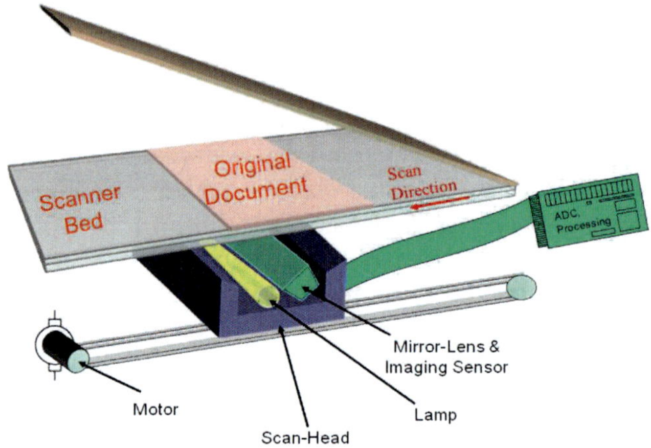

Fig. 2. Flatbed scanner architecture.

scanned document [53]. The scan-head includes a set of lenses, mirrors, filters, and the imaging sensor. Most desktop scanners use charge-coupled device (CCD) imaging sensors. Other scanners use complementary metal-oxide semiconductor (CMOS) imaging sensors, contact image sensors (CIS), or photomultiplier tube (PMTs). The native resolution of the scanner is specified in terms of the scan direction and cross-scan direction resolutions. The number of elements in the linear CCD sensor determines the cross-scan direction optical resolution. The step size of the motor controlling the scan-head and the sensor data retrieval time determines the scan-direction resolution.

The process of manufacturing imaging sensors introduces various defects that create noise in the pixel values [30, 32]. Sensor noise, which is of interest for use in forensic characterization, can be described in three forms. The first type of noise is caused by array defects. These include point defects, hot point defects, dead pixels, pixel traps, column defects, and cluster defects. These defects cause pixel values in the image to deviate greatly from their neighbors. For example, dead pixels show up as black in the image and hot point defects show up as very bright pixels in the image, regardless of image content. The second type of noise is pattern noise, which refers to any spatial pattern that does not change significantly from image to image. Pattern noise is caused by dark current and photo-response non-uniformity (PRNU). Dark currents are stray currents from the sensor substrate into the individual pixels. This varies from pixel to pixel and the variation is known as fixed pattern noise (FPN). FPN is caused by differences in detector size, doping density, and foreign matter trapped during fabrication. PRNU is the variation in pixel responsivity that is present when the device is illuminated. This noise is caused by variations between pixels such as detector size, spectral response, thickness in coatings, and other imperfections created during the manufacturing process. The third type of noise consists of random noise components which vary from frame to frame. This random noise is inevitable and cannot be removed by calibration. However, its statistical characteristics may give some clues about the source scanning device. The first type of noise leads to large deviations in pixel values and is easily corrected in most of the devices available in market. The second type of noise does not lead to large variations in pixel values; and algorithms such as flat-fielding used to correct it are difficult to implement inside the imaging device.

2.2 Scanner Signatures

Recent advances in source scanner identification have been based upon techniques for source camera identification using sensor noise [15, 43, 48]. These source camera identification techniques typically involve a training phase, in which reference patterns for individual cameras are estimated and stored in a database. The source of an image of unknown origin is then determined by estimating the similarity of its noise to a set of known reference patterns, measured using a suitable distance measure. A direct extension of the sensor noise based source camera identification algorithm can be used for the source scanner identification problem [26]. The source camera identification schemes use 2-dimensional reference patterns. In case of source scanner identification, using 1-dimensional reference patterns gives slightly higher classification accuracies compared to using 2-dimensional reference patterns. However, experiments on images scanned at the scanners' native resolution as well as at lower resolutions have shown lower classification accuracies than those achieved in similar experiments for source camera identification [26].

Another class of intrinsic signatures consists of statistical features of the sensor noise estimates obtained through the use of denoising filters, high-frequency wavelet coefficients, and neighborhood prediction errors [27]. Experiments using such statistical features from images scanned at low resolution from seven different scanners show an average classification accuracy of over 90%.

A set consisting of statistical features of the sensor noise, exploiting the way in which the sensor is moved within the scanner mechanism, proves to be a much better intrinsic signature for source scanner identification [37, 41, 44]. This scheme uses an estimate of the fixed "row-pattern" of the noise in the scanned image. A robust intrinsic signature can be obtained in this manner since the same sensor elements are used to scan each row of the image. Using a set of statistical features derived from the row-pattern allows source scanner identification even from images scanned at low resolutions, or that have undergone post-processing such as JPEG compression and contrast stretching. The rest of this section presents details of the features used in [41, 44] and some key results.

A scanned image is generated by translating a linear sensor array along the length of the scanning bed. Therefore, each row of the resulting digital image is generated by the same set of sensor pixels. Thus, for scanned images, the average of all the rows of sensor noise will give an estimate of the fixed "row-pattern". Averaging will reduce the random component while at the same time enhancing the fixed component of the noise. Various statistical features are obtained from the "row-pattern". A similar set of features is estimated in the column direction to be compared with the "row-pattern" statistics.

Let I denote the input image of size $M \times N$ pixels (M rows and N columns) and I_{noise} be the noise corresponding to the image. Let $I_{denoised}$ be the result of applying a denoising filter to I. Then as in [47], an estimate of image noise (I_{noise}) can be obtained by subtracting the denoised image ($I_{denoised}$) from the original image (I). Since the three color channels of a scanned image are independently generated, the same set of features are extracted from each color channel to get the complete feature vector. Let $\widetilde{I}^r_{noise}$ and $\widetilde{I}^c_{noise}$ denote the average of all the rows and columns of the noise (I_{noise}) respectively (Equations 1 and 2). Let $\rho_{row}(i)$ denote the correlation value between the average of all the rows ($\widetilde{I}^r_{noise}$) and the ith row of the noise (I_{noise}). Similarly, $\rho_{col}(j)$ denotes the value of correlation between the average of all the columns ($\widetilde{I}^c_{noise}$) and j^{th} column of the noise (I_{noise}). The correlation between two vectors $\mathbf{X}$ and $\mathbf{Y}$ of length N are estimated using Equation 3.

$$\widetilde{I}^r_{noise}(j) = \frac{1}{M}\sum_{i=1}^{M} I_{noise}(i,j); \quad 1 \le j \le N \tag{1}$$

$$\widetilde{I}^c_{noise}(i) = \frac{1}{N}\sum_{j=1}^{N} I_{noise}(i,j); \quad 1 \le i \le M \tag{2}$$

$$Corr(\mathbf{X},\mathbf{Y}) = \frac{\sum_{i=1}^{N}\left((\mathbf{X}(i)-\overline{X})(\mathbf{Y}(i)-\overline{Y})\right)}{\left(\sum_{i=1}^{N}(\mathbf{X}(i)-\overline{X})^2\right)^{0.5}\left(\sum_{i=1}^{N}(\mathbf{Y}(i)-\overline{Y})^2\right)^{0.5}} \tag{3}$$

The periodicity between rows of the fixed component of the sensor noise of a scanned image leads to ρ_{row} having larger values than ρ_{col}. The statistical properties of ρ_{row}, ρ_{col}, $\widetilde{I}^r_{noise}$, and $\widetilde{I}^c_{noise}$ capture the essential properties of the image that are useful for discriminating between different scanners. As an example, for a low-quality scanner having a large amount of random noise, such as that due to fluctuations in lighting conditions, values of ρ_{row} will be comparatively small and close to the values of ρ_{col}. On the other hand, a high quality scanner is not expected to have as much random noise and thus the values of ρ_{row} will be comparatively much larger than the values of ρ_{col}. Furthermore, for a low-quality scanner, $\widetilde{I}^r_{noise}$ and $\widetilde{I}^c_{noise}$ will have much higher energy than corresponding values for a high quality scanner.

The mean, standard deviation, skewness, and kurtosis of ρ_{row} and ρ_{col} are the first eight features extracted from each color channel of the input image. The standard deviation, skewness, and kurtosis of $\widetilde{I}^r_{noise}$ and $\widetilde{I}^c_{noise}$ correspond to features 9 through 14. The last feature for every channel is given by Equation 4 which is representative of the relative difference in periodicity along the row and column directions of the sensor noise. Since we expect ρ_{row} to be large for high quality scanners and small for low-quality scanners, f_{15} will have a high positive value for all but very low-quality scanners or for images which have undergone post-processing operations such as very heavy down-sampling or JPEG compression which have a large impact on the sensor noise.

$$f_{15} = \left(1 - \frac{\frac{1}{N}\sum_{j=1}^{N} \rho_{col}(j)}{\frac{1}{M}\sum_{i=1}^{M} \rho_{row}(i)}\right) * 100 \quad (4)$$

By extracting these 15 features from each of the three color channels, a 45 dimensional feature vector is obtained for each scanned image. To capture the three color channels, some scanners use three different linear sensors while others use a single imaging sensor in coordination with a tri-color light source. To capture this difference among scanners of different make and models, six additional features are used. These features are obtained by taking mutual correlations of $\widetilde{I}^{r}_{noise}$ from different color channels (same for $\widetilde{I}^{c}_{noise}$). Hence, in total each scanned image has a 51 dimensional feature vector associated with it.

Using a denoising filter bank comprising of four different denoising algorithms [44]: LPA-ICI (Local polynomial approximation - intersection of confidence intervals) denoising scheme [23], median filtering (size 3×3), and Wiener adaptive image denoising for neighborhood sizes 3×3 and 5×5, gives much higher accuracy and robustness than using only a single filter [41]. Using a set of denoising algorithms helps to better capture different types of sensor noise [27]. These denoising algorithms are chosen based on the performance of the complete filter bank in scanner identification. Initial experiments on different linear filtering algorithms such as those using an averaging filter and a Gaussian filter demonstrated that the linear filtering algorithms are not as effective in scanner identification as those used in [44]. Each denoising algorithm is independently applied to each color band in the image. The features extracted from individual blocks of the filter bank are concatenated to create the final feature vector for each scanned image. Since, for each of the four denoising filters, there are 15 features per color channel (total $15 \times 3 = 45$ features for a color image) and six cross-channel correlations. There are 51 features from each of the four denoising filters. Hence, each scanned image has a 204 ($51 \times 4 = 204$) dimensional feature vector associated with it. To reduce the dimensionality of the feature vectors, linear discriminant analysis (LDA) [22] is used and a ten-dimensional feature vector is obtained for each image. Each component of the ten-dimensional feature vector is then a linear combination of the original 204 features. Finally a SVM classifier is used to classify these ten-dimensional feature vectors. This procedure [44] for source scanner identification using statistical features of the sensor noise is summarized in Figure 3. The `LIBSVM` package [13, 31] is used in this study. All the features are scaled to the range $[-1, 1]$ before using the SVM classifier. The mapping is decided by the values of the features in the training set and the same mapping is applied to the features in the testing set. A radial basis function (RBF) is chosen as the kernel function and a grid search is performed to select the best parameters for the kernel. SVM training and testing steps are repeated multiple times using a random selection of images for the training testing sets to generate the final confusion matrices. In all the experiments

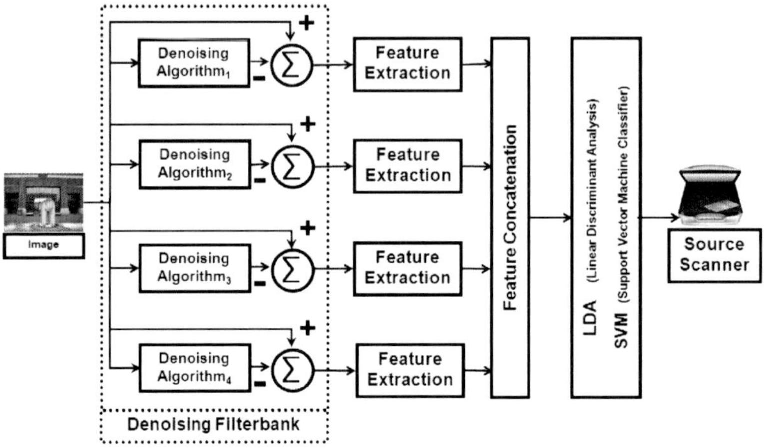

Fig. 3. Block diagram of statistical-feature-based scanner identification method.

Table 1. Scanners used for our experiments.

	Make/Model	Sensor	Native Resolution
S_1	Epson Perfection 4490 Photo	CCD	4800 dpi
S_2	HP ScanJet 6300c-1	CCD	1200 dpi
S_3	HP ScanJet 6300c-2	CCD	1200 dpi
S_4	HP ScanJet 8250	CCD	4800 dpi
S_5	Mustek 1200 III EP	CCD	1200 dpi
S_6	Visioneer OneTouch 7300	CIS	1200 dpi
S_7	Canon LiDE 25	CIS	1200 dpi
S_8	Canon LiDE 70	CIS	1200 dpi
S_9	OpticSlim 2420	CIS	1200 dpi
S_{10}	Visioneer OneTouch 7100	CCD	1200 dpi
S_{11}	Mustek ScanExpress A3	CCD	600 dpi

on source scanner identification, 80 randomly chosen images from each source class are used for training the classifier and the remaining images are used for testing the classifier.

This intrinsic signature based on statistical features of the pattern noise gives high classification accuracies over a wide range of scanning scenarios and post-processing conditions. Figure 4 shows a scatter plot of the first two features obtained after application of LDA on the 204-dimensional feature vectors corresponding to the uncompressed TIFF images from six scanner classes. Out of the ten scanner classes (Table 1), these six scanner classes S_1, $(S_2 + S_3)$, S_4, S_5, S_{10}, and S_{11} have the largest separation in this two-dimensional feature space. This scatter plot gives an indication of the high accuracy of these intrinsic

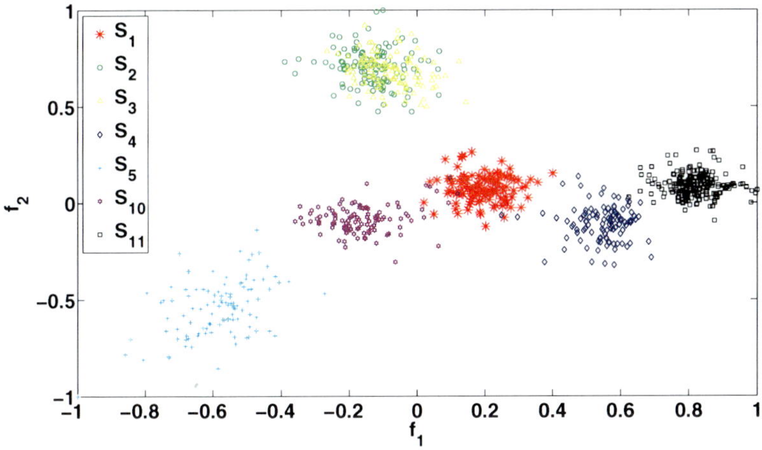

Fig. 4. Scatter plot of first two features obtained after applying LDA on statistical features of sensor noise for six classes.

signatures [44] since even in the two dimensional feature space, six scanner classes can be easily separated. In this two-dimensional feature space the features for images from S_2 and S_3 are non-separable. However, together they form one cluster which is separate from all other classes. This is due to S_2 and S_3 both being of the same make and model. Degradation in the characteristics of sensor noise due to heavy down-sampling prevents successful separation of images scanned from the two scanners of exactly the same make and model.

Table 1 lists the set of scanners used in our experiments. One-hundred images were obtained from each scanner at a low resolution of 200 dpi. Statistical-feature-based classification gives average classification accuracies of 99.9% and 92.3% for uncompressed and compressed images, respectively. Confusion matrices for these experiments are shown in Tables 2 and 3. Note that scanners S_2 and S_3 of the same make and model are treated as one class. These signatures can successfully differentiate between scanners of same make and model for images scanned at the native resolution of the scanner. For images scanned at a non-native resolution, these signatures assign images from scanners with same make and model to a single class. Further experiments also show that these signatures are robust to post-processing operations such as image sharpening and contrast stretching [44].

The intrinsic signatures successfully used for source scanner identification [41, 44] are also capable of performing image source classification [36, 40]. Table 4 shows the sources of different classes of digital images used in experiments on image source classification. 1000 photorealistic computer generated (PRCG) images, 1050 digital camera generated (CG) images, and 1100 scanner generated (SG) images (scanned at 200 dpi) were used for the results reported here. Using the same features and classifiers as those described for source scanner

Table 2. Confusion matrix for scanner identification using a general classifier. Set of images consists of 200 dpi scanned TIFF images.

		Predicted									
		S_1	S_2+S_3	S_4	S_5	S_6	S_7	S_8	S_9	S_{10}	S_{11}
Actual	S_1	100	0	0	0	0	0	0	0	0	0
	S_2+S_3	0	100	0	0	0	0	0	0	0	0
	S_4	0	0	100	0	0	0	0	0	0	0
	S_5	0	0	0	100	0	0	0	0	0	0
	S_6	0	0	0	0	100	0	0	0	0	0
	S_7	0	0	0	0	0	100	0	0	0	0
	S_8	0	0	0	0	0	0	100	0	0	0
	S_9	0	0	0	0	0	0	0	100	0	0
	S_{10}	0.3	0	0	0	0	0	0.3	0	99.4	0
	S_{11}	0	0	0	0	0	0	0	0	0	100

Table 3. Confusion matrix for scanner identification using a general classifier. Set of images consists of 200 dpi scanned images, JPEG compressed at several quality factors (Q=90, 80, 70).

		Predicted									
		S_1	S_2+S_3	S_4	S_5	S_6	S_7	S_8	S_9	S_{10}	S_{11}
Actual	S_1	89.4	5.4	0.1	0.6	0	1.5	1.3	0.7	0.6	0.3
	S_2+S_3	0.6	96.9	0.0	0.1	0.0	0.2	1.1	0.9	0.1	0.1
	S_4	1.0	1.0	97.4	0.1	0	0.1	0	0.1	0.3	0
	S_5	2.3	0.8	0.1	92.9	1.0	0.7	0.8	0.5	1.0	0
	S_6	0.0	0.0	0	0.9	97.2	0.5	0.2	0.7	0.4	0
	S_7	1.2	1.4	0.0	0.4	1.5	85.2	8.2	1.5	0.3	0.2
	S_8	0.5	6.5	0.1	0.3	0.0	5.6	85.5	1.1	0.4	0.0
	S_9	2.5	1.9	0.1	0.7	0.7	1.5	2.7	89.5	0.4	0
	S_{10}	3.6	1.0	0.5	1.9	0.9	0.5	1.7	0.8	89.2	0
	S_{11}	0.0	0	0	0.0	0.1	0.3	0	0	0	99.5

Table 4. Image sources used for evaluation of image source classification method.

Image Class	Image Source
Digital Camera CG	Canon PowerShot SD200, Nikon Coolpix 4100, Nikon Coolpix 7600
Computer Generated PRCG	www.3dlinks.com, www.irtc.org, www.raph.com, www.digitalrepose.com, www.maxon.net, www.realsoft.com
Flatbed Scanners SG	Epson Perfection 4490 Photo, HP ScanJet 6300c-1, HP ScanJet 6300c-2, HP ScanJet 8250, Mustek 1200 III EP, Visioneer OneTouch 7300, Canon LiDe 25, Canon Lide 70, OpticSlim 2420, Visioneer OneTouch 7100, Mustek ScanExpress A3

Table 5. Confusion matrix for image source classification using statistical features of JPEG compressed images (Q=70).

		Predicted		
		Camera	PRCG	Scanner
Actual	Camera	91.9	2.7	5.4
	PRCG	6.1	87.8	6.2
	Scanner	1	1.1	97.8

identification, an average classification accuracy of 92.5% is obtained when all the images are saved as JPEG (Q=70) before feature extraction, for classifier training and testing. The corresponding confusion matrix is shown in Table 5.

3 Electrophotographic (Laser) Printers

Signal processing plays a key role in the characterization of scanners and printers. Development of appropriate signal processing tools allows the ability to prevent or discourage unauthorized use of scanned or printed materials. We use our knowledge of the printer mechanism and the results of the printer characterization to determine the printer process parameters that can be modulated to encode the desired identifying information. The modulation of these parameters will require modification of the printer mechanism.

The goal of the printing operation is to transform the document into dots on a piece of paper. Figure 5 shows two different approaches to printing, namely *host-based printing* and *controller-based printing*. Host-based printing refers to the case in which the printer driver uses the host computer to do everything required to produce the printer-ready halftone page image. The printer driver interprets the Page Description Language (PDL) file to create a display list. Then the objects in the display list are rendered. The output of this interpretation stage is an 8-bit CMYK or RGB continuous-tone document image. This continuous-tone document image is subsequently modified by color space conversion, gamut mapping, and finally halftoning. When these steps are completed, the printer driver sends the halftone data to the printer to produce a hardcopy document.

In a controller-based printer, the document is sent to the printer in the form of a high-level description language. In this case, the process of conversion from the high-level description language to the halftoned image described for host-based printing is performed by the printer controller instead of the printer driver.

3.1 Electrophotographic Printer Architecture

Electrophotography (EP), also referred to as xerography, is the underlying marking technology for laser printers and office copiers. There are six steps in a typical EP process: charging, exposure, developing, transferring, fusing, and cleaning. Figure 6 shows the architecture of a typical laser printer. Typically, an organic

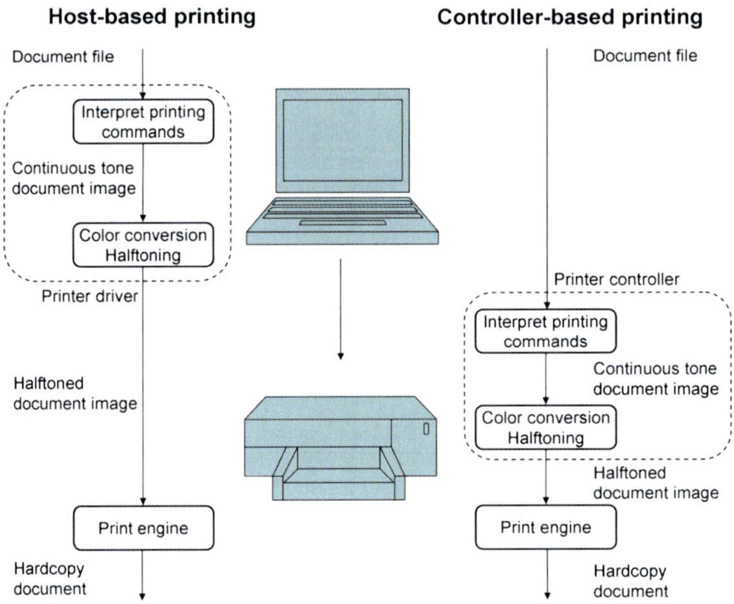

Fig. 5. Block diagrams of operation for typical printer.

photoconductive (OPC) drum, also called a photoreceptor, rotating at a constant angular velocity is electro-statically charged through a charger roller. A latent image is then exposed onto the OPC drum by scanning a pulsed laser beam with a rapidly spinning polygon mirror and discharging specific locations on the OPC surface. A toner image is developed by electrostatically adhering toner particles to the areas of the OPC surface that were discharged by the laser beam. The developed image is then transferred electrostatically onto the output media (paper) through a charged transfer roller. The toner image on the paper is then fused to the paper through heat and pressure by the fuser. To prepare to print the next page, the surface of the OPC drum is cleaned to remove any residual toner by a cleaning blade that scrapes across the OPC surface.

In the exposure stage, specific locations of the OPC drum surface are discharged by a laser beam. The laser beam intensity (LBI) profile is commonly modeled as the 2D-Gaussian function [69] in Equation 5, where $I_o(t)$ is the laser beam power, α is the standard deviation in the process direction y, and β is the standard deviation in the scan direction x. In turn, the laser beam power $I_o(t)$ can be expressed by Equation 6. The rising and falling times t_r and t_f respectively, depend mostly on the laser beam mechanism, the polygon mirror velocity, the printer resolution, and the OPC angular velocity.

$$I(x, y, t) = I_o(t)e^{-\left(\frac{y^2}{2\alpha^2} + \frac{x^2}{2\beta^2}\right)} \quad (5)$$

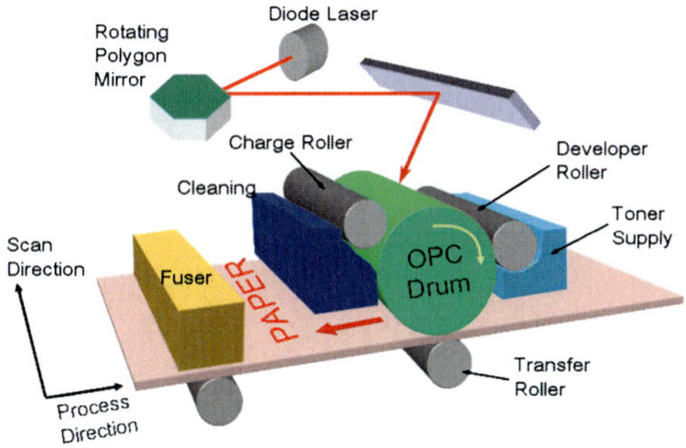

Fig. 6. Architecture of a typical electrophotographic printer.

$$I_o(t) = \begin{cases} 0, & t < 0, \\ I_{max}\left(1 - e^{-\frac{t}{t_r}}\right), & 0 < t < t_{off}, \\ I_{max}\left(1 - e^{-\frac{t_{off}}{t_r}}\right) e^{-\frac{t - t_{off}}{t_f}}, & t > t_{off} \end{cases} \tag{6}$$

The exposure due to the pixel (m, n) at the position (x, y) of the OPC surface is linked to the LBI by Equation 7. Since the exposure is assumed to be additive, the total exposure at position (x, y) is simply the sum of the exposure values due to all the neighboring pixels that fall on that position, as shown in Equation 8.

$$E_{mn}(x, y) = \int I(x, y, t)dt \tag{7}$$

$$E(x, y) = \sum_{mn} E_{mn}(x, y) \tag{8}$$

After the latent image is drawn on the OPC drum, the charge of the toner powder is set to be the opposite of the charge of the latent image [62]. The vicinity of toner particles to the latent image generates electric fields that attract the toner to the OPC surface. Electrostatic phenomena determine the amount of toner that adheres to the OPC drum. The process by which toner particles are attracted to the latent image is governed by the Coulomb force $F = QE$, where E is the electric field on the latent image and Q is the charge of the toner powder.

Development systems are commonly divided in two categories [62]: dual-component systems, and mono-component systems. In a dual-component system the toner travels on carrier beads from the toner reservoir to the OPC drum. The latent image should overcome the adhesion force between carrier and toner to develop the image on the OPC surface. On the other hand, the mono-component system utilizes only a developer roller to transfer the toner powder to the OPC surface, as seen in Figure 7.

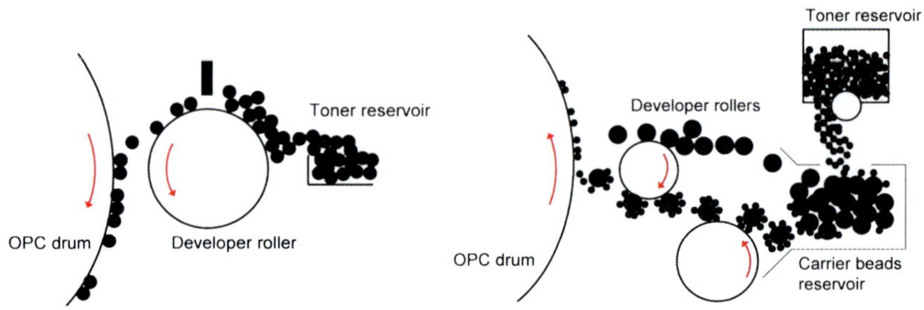

Fig. 7. Mono-component development system (left) and dual-component development system (right).

The ratio of the developed toner mass per unit area M/A is usually understood as the quantity of toner that is placed on the latent image. Ideally, the development on the OPC stops when the toner charge per unit area σ_t is the same as the OPC charge per unit area σ_p, i.e. when the charges are neutralized as described by Equation 9. The charge neutralization idea is a concept rarely observed in practice because of the phenomena complexity and the number of variables that impact the development process. In spite of these difficulties, it has been observed in experiments that the toner-charge-to-mass ratio $\frac{Q}{M}$ is proportional to the toner concentration C_t on the OPC surface [61]. This relationship is shown in Equation 10, where V is the electrostatic potential on the OPC surface, ϵ_o is the free space permittivity, K_s is the OPC dielectric constant, and d_s is the OPC thickness.

$$\sigma_p = \sigma_t = \frac{Q}{A} = \frac{M}{A}\frac{Q}{M}. \tag{9}$$

$$C_t^{-1} = \frac{M}{A} = \frac{V\epsilon_o K_s}{(Q/M)d_s}. \tag{10}$$

Many efforts have been made to reduce the number of variables to model toner development. A probabilistic model of toner distribution on the latent image is presented in [54]. The placement of toner powder on the OPC surface is analyzed as being both uniformly and non-uniformly distributed. Parameters such as the size of the toner particles, the spread of the laser beam energy, and the number of toner particles likely to adhere to the paper per unit length were determined through comparison of simulated printed dots to microscopic samples of actual dots. In [73], an empirical model is proposed based on physical observations. An electrostatic voltmeter and a laser power modulator were utilized to predict the OPC surface potential and calculate the electrical field between the OPC and the development system. With the value of the OPC voltage, the density of developed toner can be estimated from Equation 10. Also, measurements of the reflectance of printed samples can be plotted against their theoretical exposure values to observe the relationship between exposure and developed toner. This approach is taken by [35], in which the exposure to absorptance relationship is found through the use of constant exposure halftone patches.

In general, the development process is considered to be the sum of a non-linear function ϕ of the exposure $E(x, y)$ and noise $n(x, y)$, as seen in Equation 11. The function ϕ depends on the exposure of the OPC drum, and accounts for complex electrostatic phenomena.

$$W(x, y) = \phi(E(x, y)) + n(x, y). \tag{11}$$

So far we have described the printing pipeline and printer architecture. Whereas the printing pipeline helps us visualize the stage of the printing process in which the signature occurs, the printer mechanism analysis enables a more formal approach by providing the mathematical basis to describe the signature extraction and embedding processes. In the following two sections we will illustrate different methods that have made use of this knowledge to develop forensic applications that involve intrinsic and extrinsic signatures.

3.2 Intrinsic Signatures for Electrophotographic Printers

Inherent artifacts generated by EP printers due to their physical components such as the gear mechanism, polygon mirror imperfections, and optical photoconductor (OPC) angular velocity can be used as an intrinsic signature of the device. Laser printers can be characterized using intrinsic signatures such as banding [4]. Banding is an artifact caused by fluctuations of the OPC angular velocity and errors in the gear transmission mechanism. It appears as non-uniform light and dark lines perpendicular to the process direction. This is the direction in which the paper moves through the printer. Different printers have different sets of banding frequencies depending upon brand and model.

Several techniques can be used to intrinsically characterize EP printers including measures of image sharpness, toner fusing characteristics, dot gain, and asymmetry of toner distribution [55]. In addition, an optical effect due to the toner particles lying on top of the paper is characteristic of EP printers [71]. A surface profiling apparatus can display how the printed areas extend above the surface of the paper.

In the following sections we will describe several approaches to intrinsic characterization of EP printers. These approaches are illustrative of how signal processing techniques can be applied toward characterization of EP printers.

Banding-Based Signatures

Banding-based identification uses frequency domain analysis of a one-dimensional projected signal of large mid-tone regions of the document, typically occurring in printed images. Fourier analysis of the signal yields the banding frequencies. The method is detailed in Figure 8. Here, the 193 cycles/inch peak corresponds to the tooth-to-tooth error of the large gear, and the 24 cycle/inch peak corresponds to the eccentricity error associated with the small gear. With tooth-to-tooth error,

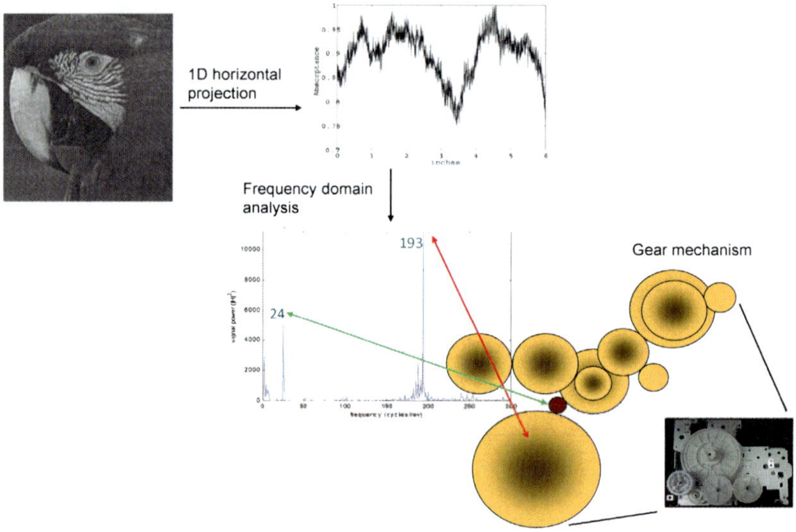

Fig. 8. Intrinsic signature of a laser printer based on banding. A 1-D signal is obtained from the horizontal projection of a printed image. Frequency analysis of the projected signal provides a set of banding frequencies that are characteristic of the gear train mechanism of the laser printer that produced the image.

each cycle of the banding noise corresponds to the angular separation between succeeding gear teeth. With eccentricity error, each cycle of the banding noise corresponds to a complete revolution of the gear. Since one revolution of the small gear corresponds to the meshing with a large number of teeth of the large gear, the noise associated with the large gear will be at a much higher frequency than the noise associated with the small gear. This conclusion is based on a precision measurement of the individual gear components of an actual printer unit, as well as an analysis of the manner in which the fluctuations propagate through the gear train to the OPC drum [14].

One technique for identifying a printer is based on principle components analysis (PCA) of extracted banding signals. An n-dimensional vector X can be represented by the summation of n linearly independent vectors

$$X = \sum_{i=1}^{n} y_i \phi_i = \Phi Y, \tag{12}$$

where y_i is the i-th principal component and the ϕ_i's are the basis vectors obtained from the eigenvectors of the covariance matrix of X. Using only $m < n$ basis vectors ϕ_i's, the vector X can be approximated as

$$\hat{X}(m) = \sum_{i=1}^{m} y_i \phi_i + \sum_{j=m+1}^{n} b_j \phi_j, \tag{13}$$

where the b_i's are constants. The coefficients b_i and the vectors ϕ_i are to be determined so that X can be best approximated. If the first m y_i's are calculated, the resulting error is

$$\Delta X(m) = X - \hat{X}(m) = \sum_{i=m+1}^{n} (y_i - b_i)\phi_i. \tag{14}$$

The set of m eigenvectors of the covariance matrix of X, which correspond to the m largest eigenvalues, minimizes the error over all choices of m orthonormal basis vectors. The expansion of a random vector over the eigenvectors of the covariance matrix is also called the discrete version of the Karhunen-Loeve expansion [25].

A test page consisting of the letter "I" in 10 point, 12 point, and 14 point size in Arial font was generated. Each test page has 40-100 letters. From each letter, a one dimensional projected signal is extracted by simply summing the pixels in the horizontal direction. The projected signals are mean subtracted and normalized. This step is performed to remove variability due to long term trends, such as cartridge depletion and printer wear, and other factors which are not stable intrinsic features. The projected signals from different printers are concatenated into a large data matrix. The Canonical Variates method is applied to this data matrix to get the principal components [5].

The PCA using five different printer models is shown in Figures 9 and 10. Figure 9 illustrates the process of obtaining 1-D projections of individual characters from the scanned test document, and performing PCA on the resulting projections. Each projection in this case has 168 samples. The high dimensional data is represented only by the first two principal components. The classes (different printers) are well separated. A sixth printer is added as a test printer. The sixth printer is an HP LaserJet 4050 and the projections from this printer overlap with those of the other LaserJet 4050. The projections from the LaserJet 1000 and LaserJet 1200 overlap because of the similarities in their banding characteristics. It should be noted that the Samsung ML-1450 and the Okipage 14e show well-separated classes.

Texture-Based Signatures

The PCA method described in the previous section uses 1-D projections of text characters which capture the banding signal present in the printer. Another way to look at the banding signal is as a texture within the printed region of each character. In this case, texture features estimated from individual text characters, can be used to capture the intrinsic signature. Graylevel co-occurrence features [19, 29], as well as two pixel based features [49], are used to describe the texture.

Graylevel co-occurrence texture features assume that the texture information in an image is contained in the overall spatial relationships among the pixels in the image [29]. The features are obtained from the Graylevel Co-occurrence Matrix (GLCM) which is an estimate of the second order probability density function of the pixels in the image. The features are then statistics of the GLCM.

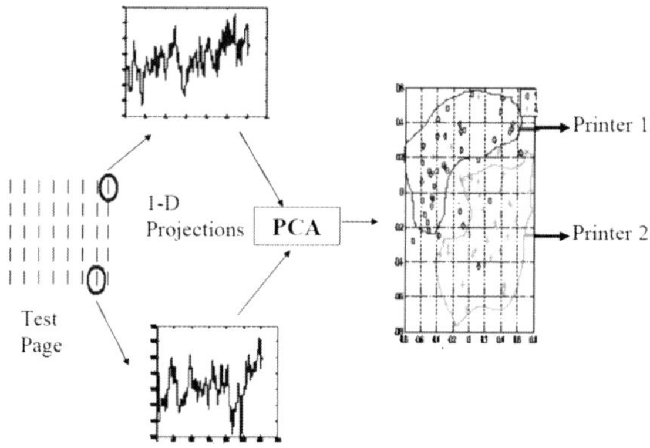

Fig. 9. Principle components analysis using 1-D projected signal.

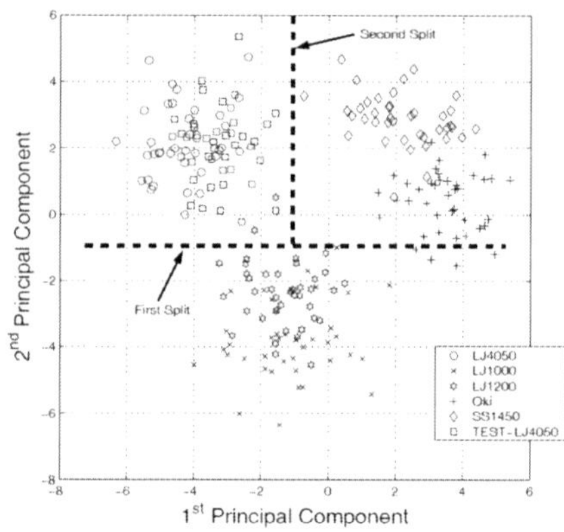

Fig. 10. Representation of the projected signals by the first two principle components.

The GLCM, defined in Equation 15, has entries $C(n, m)$ which are equal to the number of occurrences of pixels with graylevels n and m respectively with a separation of (dr, dc) pixels. The support of the region of interest is denoted by the set S, and consists of all the pixels contained within a given text character. If the GLCM is normalized with respect to $|S|$, its entries then represent the probability of occurrence of pixel pairs with graylevels n and m with separation (dr, dc). Twenty-two features such as the mean, variance, and entropy of C are obtained and together form a feature vector [50, 51].

$$C(n,m) = \sum_{(i,j),(i+dr,j+dc)\epsilon S} 1_{\{I(i,j)=n, I(i+dr,j+dc)=m\}} \tag{15}$$

It has been shown that high classification accuracy can be achieved among 10 printers by using support vector machine (SVM) classification [12, 20, 33] with

Table 6. Confusion matrix for GLCM based printer identification using 22 features with SVM classifier. (blank indicates -0-)

train \ test	lj5m	lj6mp	lj1000	lj1200	E320	ml1430	ml1450	hl1440	1250w	14e
lj5m	296	2		1		1				
lj6m	1	256	6		17			15	5	
lj1000	2	2	284	12						
lj1200	7	2	2	289						
E320					300					
ml1430	1					299				
ml1450							300			
hl1440		28				5	2	259	6	
1250w								3	292	5
14e								17	67	216

GLCM based texture features [51]. Table 6 shows the confusion matrix for classification between 10 printers. 600 letter "e"s were printed on each of the 10 printers and GLCM features were extracted from each "e". 300 of the resulting feature vectors from each printer were used to train the SVM classifier, and the other 300 were used as a test set. The entries in Table 6 indicate the number of those 300 feature vectors in the test set that were originally from the printer indicated by the row, and classified by the printer indicated by the column. For example, 296 "e"s were correctly classified as being printed by the "lj5m" printer, while 2 were incorrectly classified as being printed by the "lj6mp" printer printer, 1 by the "lj1200" printer, and 1 by the "ml1430" printer. A majority vote across these individual classifications would result in correctly identifying the "lj5m" printer.

In a typical forensic printer identification scenario, it may not be possible to obtain known documents from a set of printers with characteristics matching that of the document in question. Paper type, font size, font type, and even the age of the consumables in the printer may vary. We have shown that texture based features are consistent across several variables listed in Table 7 [49].

Four cases are explored using 10 different printers. In each case the training set consists of 500 "e"s and the test set consists of 300 "e"s. The first case considered is where the printer identification system is trained using data of font size fs_{train} and tested using data of font size fs_{test} with all other variables held constant (ft =Times; pt =PT01). It is assumed that printing the training and test data immediately after one another holds age constant.

In the second case the system is trained using data of font type ft_{train} and tested using data of font type ft_{test} with all other variables held constant (fs =12pt; pt =PT01).

In the third case the system is trained using data of paper type pt_{train} and tested using data of paper type pt_{test} with all other variables held constant (fs =12pt; ft =Times).

Table 7. Four variables considered for forensic identification experiment.

Category	Sub-Types
Font Size (fs)	08 point 10 point 12 point 14 point 16 point
Font Type (ft)	Arial Courier Garamond Impact Times
Paper Type (pt)	PT01: 20lb., 84brt PT02: 28lb., 97brt PT03: 32lb., 100% cotton
Age (consumables)	-

Finally we consider the case where the system is trained on old data and tested on new data. Testing and training data sets printed 5 months apart are used. Ten sub-cases are considered by testing and training using data from the sets $\{fs_x$,Times,PT01$\}$ and $\{$12pt,ft_x,PT01$\}$. This is representative of a forensic scenario where the printing device that created a suspect document needs to be identified given only the document in question and newly generated test and training data from the printer.

The results for Case 1 are shown in Table 8. The rows of the table correspond to the value of fs_{train} and the columns correspond to the value of fs_{test}. Each entry contains two values. The first value is the percent correct classification of the system (i.e. the percentage of printers that were correctly identified after majority vote by training on font size fs_{test} and testing on font size fs_{train}). The second value, enclosed in parentheses, is the percent correct classification of the individual feature vectors immediately after the SVM and before the majority vote (i.e. if the percentage in parenthesis is $x\%$ and the test page from each printer contained 300 "e"s, then there are a total of 3000 "e"s in the test set across all 10 printers, $x\%$ of which were correctly classified). The table shows that when the font sizes of the training and testing data are within 2 points of each other, at least 9 out of 10 printers are correctly classified.

The results for Case 2 are shown in Table 9. These results show that the current feature set is font dependent. If $ft_{train} = ft_{test}$, then 9 out of 10 printers can be classified correctly. At most 7 out of 10 printers are classified correctly if $ft_{train} \neq ft_{test}$. Even though the font size is 12pt for each font type, the height of the "e" in each instance is different. It is possible that this implicit font size difference partly causes the low classification rates for different font types. The Times "e" and Courier "e" are approximately the same height; and

Table 8. Percent correct classification for varying font size. (% after SVM)

$fs_{train} \setminus fs_{test}$	8pt	10pt	12pt	14pt	16pt
8pt	100 (87.6)	90 (82.9)	80 (61.0)	50 (43.0)	40 (35.1)
10pt	100 (78.3)	100 (95.3)	90 (72.9)	70 (56.3)	50 (47.9)
12pt	80 (58.3)	90 (73.3)	100 (93.0)	100 (84.1)	80 (66.0)
14pt	50 (43.6)	70 (62.7)	100 (88.9)	90 (89.7)	90 (81.2)
16pt	40 (37.6)	50 (48.1)	80 (74.4)	90 (84.2)	90 (89.5)

Table 9. Percent correct classification for varying font type. (% after SVM)

$ft_{train} \setminus ft_{test}$	Arial	Courier	Garamond	Impact	Times
Arial	90 (84.1)	40 (35.0)	40 (26.0)	20 (17.8)	40 (34.7)
Courier	20 (23.0)	90 (86.8)	50 (43.8)	0 (2.6)	50 (49.3)
Garamond	10 (12.4)	40 (43.2)	90 (82.3)	10 (11.9)	20 (27.8)
Impact	10 (16.8)	10 (10.4)	10 (11.4)	90 (82.9)	10 (17.9)
Times	20 (30.1)	70 (57.0)	40 (33.0)	10 (6.6)	90 (84.0)

the classification rate for training on Times and testing on Courier is shown to be 70%.

The results for different paper types, Case 3, are shown in Table 10. 100% correct classification is achieved if both the training and testing sets use the same paper type. If paper type PT01 or PT02 is used for training, and PT01 or PT02 for testing, then at least 9 out of 10 printers are classified correctly. The same is not true with paper type PT03. Paper types PT01 and PT02 are both visually similar except that PT02 appears slightly smoother and brighter. PT03 has a visually rougher texture than the other two paper types. The features used for classification may be affected by the paper texture as well as textures from the printer itself.

Table 11 shows the results for the Case 4, training with new data and testing with old data. At least 7 out of 10 printers are correctly identified in each sub-case. The individual SVM classifications show that in each of these sub-cases the lj1200 was classified as an lj1000. This behavior was observed in previous

Table 10. Percent correct classification for varying font size. (% after SVM)

$pt_{train} \setminus pt_{test}$	PT01	PT02	PT03
PT01	100 (93.0)	90 (83.3)	60 (47.2)
PT02	90 (75.2)	100 (93.2)	40 (32.4)
PT03	50 (40.4)	30 (28.1)	100 (93.0)

Table 11. Percent correct classification for varying age. (training and testing data generated 5 months apart)

$fs_{train} \setminus fs_{test}$	08pt	10pt	12pt	14pt	16pt
PT01	90 (66.0)	90 (76.3)	90 (72.3)	80 (66.9)	80 (67.8)
Train Test	Arial	Courier	Garamond	Impact	Times
PT01	70 (64.6)	70 (62.6)	80 (67.3)	80 (67.0)	80 (58.5)

work and attributed to the fact that the two printers appear to have the same or similar print engine.

3.3 Extrinsic Signatures for Electrophotographic Printers

Printers are complex electromechanical devices. Imperfections in the printer mechanism such as imperfect gear meshing or motor speed fluctuations are always present. These imperfections directly affect the printed output. For example, as discussed earlier, fluctuations in the optical photoconductor drum angular velocity will cause the inter-scan-line spacing to vary which creates banding in the printed page. The effect on the printed output can be directly correlated with the mechanical properties of the printer. It is because of this direct correlation that these features can be used as an intrinsic signature of the printer. The intrinsic signature can be used to determine the device that created a document, and in some circumstances also provide a level of tamper-proofing for the document. However, in many security applications there is a need to embed additional content into the printed form of the document that cannot be achieved with intrinsic signatures. Examples include a secure hash of the document, serial number of the printer, or date and time of the printing.

Various methods exist for embedding additional content into a document which fall under two basic categories. The first contains methods that embed security information into the document before it is sent to the printer. These methods are typically designed such that the embedded security features are able to survive the printing process. The second category contains those methods which embed the security features at the printer mechanism level. Performing the embedding in the printer allows a broader domain for marking the print and makes attacking the security features more difficult.

Embedding at the Document Level

Early techniques for embedding information into a printed document relied on modification of the electronic document in such a way that the embedded information survived the printing process. Many of these techniques are extensions of digital watermarking methods widely used to secure digital images. Once the electronic document is marked and printed, the embedded information can be extracted through analysis of a scanned image of the document.

One of the earliest methods for securing printed text involves the shifting of elements in a text document [10]. In this approach, methods are described for providing copyright protection for text documents by embedding information specific to the recipient or source in order to deter illegal distribution. The methods developed to encode this information into each page involve the shifting of textual elements in the document by amounts imperceptible to the human viewer to encode individual bits of data. These textual elements can be lines, words, or individual characters. Line shifting can be implemented by shifting every other line slightly up or down, approximately 1/600th inch, to encode a one or zero. To detect the shifts in a document, no prior information about the original is necessary since the information is embedded by shifting every other line. The baseline locations can be estimated in the scanned document; and by measuring the relative distances between the baselines of adjacent text lines, an estimate of the embedded data can be found. This type of encoding is robust to scan-print attacks and photocopy generation loss, although some variability in the detection arises due to errors in the scan process such as rotation. Another detection method for line shifting, more robust to imaging errors such as scan rotation, is the use of the relative distance between centroids of adjacent lines of text.

Word and character coding allow a higher data density, but are not as robust as line coding due to the fact that each shift is encoded in a smaller portion of the printed page. Also, most word processors will vary the spacing between adjacent words and characters, so estimation of shifts using the differential method will not work unless the original document is also available for comparison. These shifts could instead be used as a fragile watermark to detect alterations to a document.

The shifting method deals with encoding information in text; but documents may also contain halftone images. Halftoning is the process of converting a continuous-tone image into an image having only a finite number of levels, typically two for printed images. Numerous methods have been developed for watermarking halftone images. Most involve modifying the halftone patterns used when printing the image. The three basic methods are the use of multiple dither matrices, conjugate halftone screens, angular variation of the halftone screen [24], or angular variation of the halftone dots themselves [11]. In the first method, the dither matrix changes from tile to tile and can be used to encode information. Detection involves estimating the statistical properties of the halftone patterns and their variation across the image. The second method involves conjugate halftone screens, two screens are used to form two halftone images and the data are embedded through the correlations between these two screens. The third basic method involves encoding the data in the angular orientation of the screen in each halftone cell. In this case, each angular orientation can represent multiple bits depending on the number of angles at which the halftone screen dot can be generated and detected.

Another method of data hiding using halftone patterns relies on images that are printed opposite each other on the front and back of the page using conjugate halftone patterns such that when the page is held up to a light source a fake watermark will appear [60]. This technique requires a high degree of control over the registration of each side of the document to make sure the halftone patterns line up. A similar technique known as visual watermarking can be used for one-sided printing using a transparent mark for verification.

Watermarking of images in documents can also be performed by using continuous-tone image watermarking techniques [8]. These methods first embed a watermark into the continuous-tone image. The marked image is then printed at a high resolution to create the document. To detect the watermark, the document is scanned and transformed back into a continuous-tone image, after which an appropriate method for detecting the watermark is used. The type of watermark embedded has to be one that can survive the print-scan process.

Detectability of the embedded information and the printed image quality can both be improved by considering the printer's halftoning process while embedding [34]. This approach uses a modified version of direct binary search (DBS) halftoning in which each iteration jointly optimizes both the watermark detectability and perceptual image quality metrics of the halftone image. This method was shown to be robust against many common image processing operations such as JPEG compression and histogram equalization when compared to prior methods which do not take into account the printing process.

Embedding at the Printer Mechanism Level

The previously mentioned document watermarking methods embed information into the document before it is printed. Several methods exist which embed information into the document at the hardware level of the printer. These techniques exploit the way that the printer puts marks on the paper. This is different than modifying the file to be printed or the printer driver. Embedding at the hardware level allows access to a much larger marking domain and the potential for increased security. Additionally, circumvention of the embedding of such marks is more difficult since the embedding step exists in the hardware of the printing device instead of as a software module or device driver. Changing the document itself will not affect the embedding.

An embedding method developed by printer manufacturers to help trace counterfeit currency uses a pattern of yellow dots overlaid on top of the document content [1]. A rectangular grid of 15 × 8 dots is repeated over the entire page. Since the dots are printed in yellow, they are invisible to the naked eye under normal lighting conditions. Typically about 10 bytes of information is encoded in this grid, including the date and time of printing, serial number of the printer, and parity information for error correction.

This embedding scheme has the advantage that it is independent of image content. However, the method does not protect the content of the document. The dot pattern only identifies the printer that printed the dot pattern and not necessarily the printer that printed the document content. For example, a blank page could be printed on a color printer to lay down the dot pattern, after which the actual page content could be printed by a monochrome printer that does not lay down a dot pattern. If the pattern were somehow content-dependent, then a hash of the document could additionally be embedded to verify the content.

Laser pulse width modulation (PWM) is a technology used in some EP printers to control dot size and placement. This capability can be accessed to embed information into halftone images [65]. Since the electrophotographic process is often unstable for development of isolated single-pixel dots, clustering dots in pairs leads to more stable development. Data is embedded into the halftone by shifting the centroids of dot pairs such that they lie between points on the halftone lattice.

Figure 11 shows the block diagram for extracting the embedded information from a halftone document containing an extrinsic signature based on PWM. The printed document that contains the information is first scanned. The scanned image is preprocessed using morphological filtering to remove noise caused by the printing and scanning processes. Then the preprocessed image is analyzed using centroid, skew-angle, and lattice detection to extract the embedded information.

Another approach [16] is to modulate the laser intensity in a laser printer, which allows per-scan-line changes in dot size. To modulate the laser intensity,

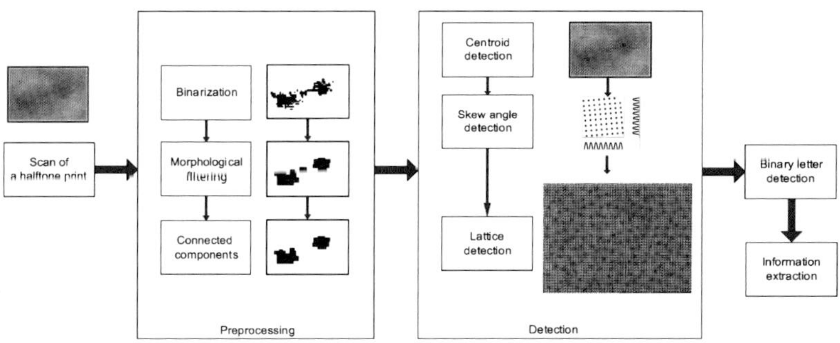

Fig. 11. Process for extracting information from halftone images marked using pulse width modulation (PWM).

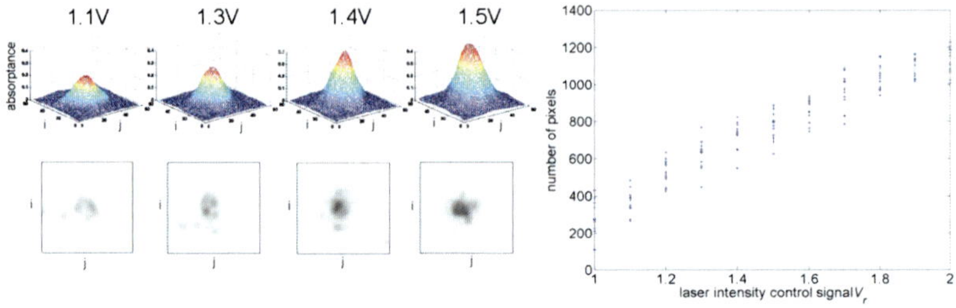

Fig. 12. (a) Average dot profile with input reference voltage 1.1, 1.3, 1.4 and 1.5 volts, respectively. (b) Relationship between the input reference voltages and dot sizes.

the reference voltage input V_{ref} to the laser power control circuit is controlled through the relation

$$V_{ref} = V_0 + V_{mod}, \tag{16}$$

where V_0 is the nominal reference voltage for the laser power and V_{mod} is the modulation signal. An increase in reference voltage V_{ref} causes a proportional increase in laser power amplitude. Modulation of the laser power amplitude, which directly affects the exposure energy and the associated photoconductor contrast voltage, results in different dot sizes. Figure 12(a) shows the average dot profile of 16 dots when the input reference voltage is 1.1, 1.3, 1.4 and 1.5 volt, respectively. Figure 12(b) shows relationship between the input voltages and dot sizes. The dot size is determined by counting the number of pixels with absorptance greater than 0.1 in one dot cell. The dot size increases with increasing input reference voltage as shown in Figure 12(b).

A simple set of signals that can be used for embedding is a set of sinusoids at various frequencies and amplitudes such that they lie below the human visual sensitivity threshold curve. For example, in the bottom two rows of Figure 13, the lines of text and halftone blocks are embedded with two different frequency sinusoids as follows:

$$V_{mod}(y) = \triangle V \cdot sin\left(2\pi f_0 y/R_p\right), \tag{17}$$

where y is the scan-line index in the process direction, f_0 is the spatial banding frequency, R_p is the printer resolution, and $\triangle V$ is the modulation amplitude. The effects of this type of embedding are shown in Figure 13. This figure contains three columns containing text characters, a halftone patch, and an enlarged portion of a vertical edge. No signal is embedded in the first row. The second and third rows have 20 cycles/inch and 40 cycles/inch sinusoids embedded, respectively. Here the embedding power was set very large to emphasize the effect of this type of embedding on the different elements that may appear in a printed document. The embedded signal is most visible in the halftone patch. It can also be seen in vertical edges, which are present in many text characters and in the borders of forms.

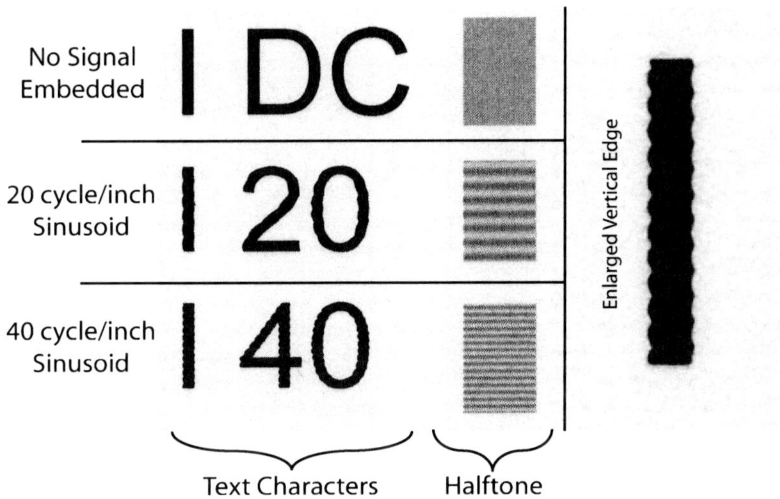

Fig. 13. Effect of exposure modulation. For purposes of illustration, the modulation is much larger here than it would be in practice. The left side of the figure shows several characters and a halftone patch with no modulation and with modulation at 20 and 40 cycles/inch. The right side of the figure shows an enlarged version of the letter 'I' modulated at 40 cycles/inch.

If the embedded document has any large mid-tone gray patches, the signal is easily detectable using Fourier analysis techniques. However, several issues arise when attempting to do this. One is the assumption that every portion of the image consists of a mid-tone gray level. White and black parts of the image—those which contain no toner or are saturated with toner—are not useful for detecting the embedded signal. In addition, synchronization of the signaling periods becomes difficult. The issue of synchronization can be solved by using binary pseudo-random noise (PN) sequences and repeating each symbol multiple times within the image [7]. The Gold sequence [58] is one type of binary PN sequence that can be used [18]. An n-bit Gold sequence has the following properties:

- Period $T = 2^n - 1$
- Each code set contains $2^n + 1$ unique sequences
- Cross-correlation values between any two sequences from the same code set take on one of 3 values $\{-1, t(n) - 2, -t(n)\}$ where $t(n) = 1 + 2\lfloor (n+2)/2 \rfloor$ and $\lfloor x \rfloor$ denotes the integer part of x.

Let $\mathbf{x} = [x_1, \ldots, x_{2^n-1}]$ be the embedded Gold sequence. Each element in the sequence is repeated T_s times to overcome signal loss due to the modulation transfer function (MTF) of the printer [70]. The Gold sequence is essentially a base-band signal. Based on the laser intensity modulation threshold developed in [16], high spatial frequencies provide a large range of intensity modulation while remaining below the human perception threshold. Therefore, the sequence

Table 12. Detection rates for halftone images embedded with one 5-bit Gold sequence, $T_s = 5$, for different modulation amplitudes $\triangle V$ and carrier modulation frequencies F_s.

accurate rate (%)	$F_s = 60$ (cycles/in)	$F_s = 120$ (cycles/in)
$\triangle V = 0.1$ (V)	100%	100%
$\triangle V = 0.05$ (V)	80%	20%

x is modulated to a higher frequency band using a sinusoidal signal. The laser intensity modulation signal, as a function of the **x**, can be represented as

$$V_{mod}(y) = \triangle V \cdot X(y) = \triangle V \cdot \sum_{i=0}^{T-1} x_i \Omega(y - iT_s) \sin\left(2\pi F_s \left(y - iT_s\right)/R_p\right), \quad (18)$$

where F_s is the modulation frequency, R_p is the printer resolution, Ω is a function used to repeat each element of **x**, and T_s is the number of each element repeated.

To detect the embedded signal, a scanned image is first segmented into multiple sub-images. Each sub-image is projected to the cross-scan direction to obtain a 1-D data sequence. After normalization, the correlation between the data sequence and the set of all possible Gold sequences is computed to identify the sequence with the maximum correlation value. A majority vote from all the extracted images of their Gold sequence with maximum correlation value will determine the embedded sequence. Using a 5-bit Gold sequence and a 600 dpi native resolution with printer's default halftone algorithm, approximately 20 bits of data per inch can be embedded in an image [18]. Table 12 shows the code detection accuracy for 50% gray filled images embedded with one 5-bit Gold sequence, and $T_s = 5$, at different modulation amplitudes $\triangle V$ and carrier modulation frequencies F_s. The experimental results indicate that for images embedded with $F_s = 120$ (cycles/in) and $\triangle V = 0.1$ (V), the detection rate can be as high as 100% while preserving printed image quality.

To embed information into a text document using laser modulation, a slightly different approach needs to be taken [52]. Looking again at Figure 13, we notice that the embedded signal affects straight edges. In fact, in the ideal case an estimate of the edge profile will be similar to the embedded signal.

To make use of this observation, each line of text in a document is treated as a signaling period during which one symbol is transmitted. The symbols, each representing n bits of data, can be defined as follows. Let $\mathbf{b}^{(k)} = \{b_0^{(k)}, b_1^{(k)}, ..., b_n^{(k)}\}$ be a sequence of bits to be embedded into the kth line of text. The corresponding symbol, or signal, $B^{(k)}(y)$ can then be defined as

$$B^{(k)}(y) = \sum_{i=0}^{n} b_i A_i sin\left(\frac{2\pi f_i y}{R_p}\right), \quad (19)$$

where

$$\mathbf{f} = \{f_0, f_1, \ldots, f_n\}, \quad (20)$$

is an increasing set of carrier frequencies to be used for each bit in the sequence, and

$$A_i = \frac{n-i}{n} A_{max} + \frac{i}{n} A_{min}. \tag{21}$$

A_{max} is the amplitude to be used for frequency f_0 and A_{min} is the amplitude to be used for frequency f_n. The amplitude varies linearly between these two values for frequencies between f_0 and f_n. Choosing $n = 8$ provides 256 symbols, each corresponding to a different $\mathbf{b}^{(k)}$. Detection of the symbols involves several steps shown in Figure 14. First the document is scanned at a sufficiently high resolution, typically 600 DPI or higher. Next, all characters in each line of text are segmented from the scanned image. Each character is then filtered using a threshold edge detector and morphological operations to find all the left edges of the characters. Only vertical edges that are long enough to contain at least one cycle of the lowest possible embedding frequency are used for decoding. First the edge profile $\hat{B}[y]$ is found for each extracted edge from the line. The power spectral density (PSD) of each profile is obtained using a binned DFT such that the frequencies of interest, f_i, are each contained in 10 cycle/inch wide bins. The frequencies f_i with PSD values greater than a pre-determined threshold determine the embedded symbol $\mathbf{b}^{(k)}$. One such scheme is able to embed 7 bits per line of text or as many as 200-400 bits/page.

Embedding an extrinsic signature into a document containing neither text nor images, such as a form [56], requires a different approach from those presented for text and halftone images. Security documents such as bank notes, statements, and event tickets, are types of forms which typically contain a frame or border around the edge of the document. A similar technique as has been described for text embedding can be used to embed information into the vertical edge of a frame. However, because the frame is typically much longer than a text character and has no breaks, different encoding and detection schemes can be employed.

The length and continuity of borders facilitate the implementation of error-correcting coding techniques that increase the embedding capacity. Two

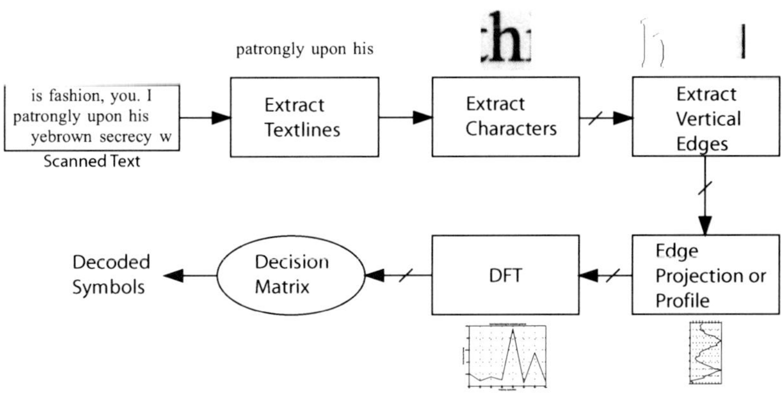

Fig. 14. Process for extracting embedded information from text.

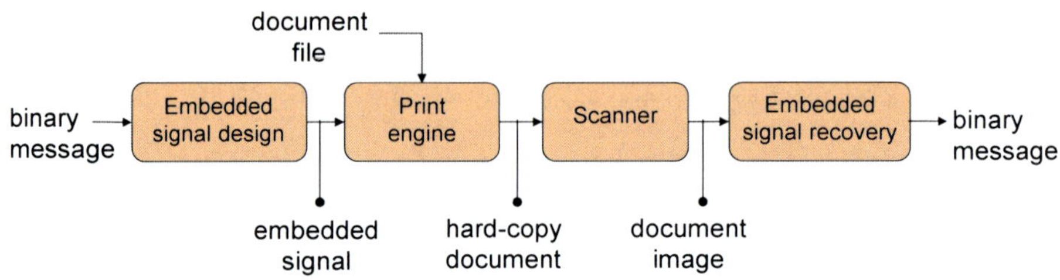

Fig. 15. Block diagram for extrinsic signature embedding in elements of forms, such as borders.

error-correcting codes that exhibit attributes from different coding categories have been chosen: Reed-Solomon (RS) codes and convolutional codes (CC) [66]. The RS code that was implemented is systematic and uses a hard-decision decoder. In contrast, the CC code is non-systematic and employs a soft-decision decoder. A code is said to be systematic if the encoded message contains the information bits unchanged with some parity bits added. Similarly, a non-systematic code is one in which the output does not contain the input bits.

The block diagram of Figure 15 describes the signature embedding and recovery process. First, the message carrier (or embedded signal) is designed according to the encoded binary message. Different signal attributes, including the waveform of the carrier, are determined at this step. Even though different values of amplitude, frequency (in cycles per inch), and code word length (in scan lines) can be considered, the set of options is constrained by the physical limitations of the printer, the instabilities of the EP process, and the perceived raggedness at the edges of the lines in the form. The document is then printed along with the embedded signal. A 600 DPI scanned image of the document is acquired using a flatbed scanner. Finally, the embedded signal is extracted from the edge of the border and decoded to recover the binary message.

Sine, square, and constant amplitude waveforms are the options that have been tested for signature carriers. The code words for a sine or a square wave carrier are defined by a pair of frequencies, where the lower frequency represents the bit '1' and the higher frequency the bit '0'. For the constant amplitude waveform, two opposite amplitudes are used as code words, the negative value for the bit '0' and the positive value for the bit '1'. A summary of the code word alternatives is presented in Figure 16.

The embedded signal $S[y]$ is shown in Equation 22, where T is the length of each code word in scan lines, A is the embedding signal amplitude, and V_o is the bias voltage of the laser beam unit. The signal depends on y, which is the scan line index. The length in bits of the encoded message is L. The encoded bits $c_i = 0$ or 1 determine which code word segment, W_0 or W_1, is to be used in the embedded signal for the ith code word. Once the waveform of the carrier,

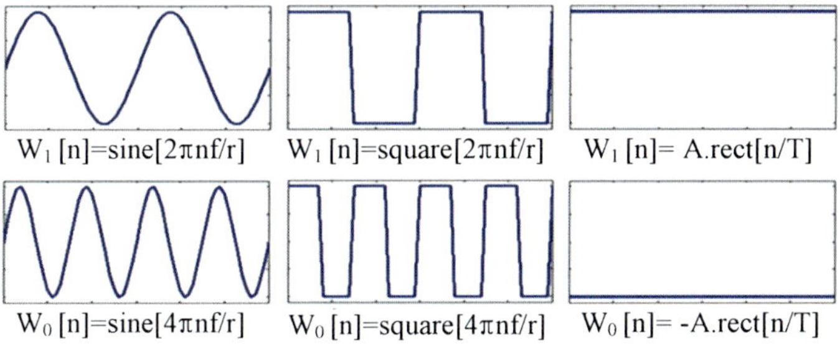

Fig. 16. Code word choices ($n = 0, 1 ... T$ and $r = 600\ DPI$). The first row shows the choices for a '1' bit; and the second one shows the choices for a '0' bit. The abscissa is the scan line index and the ordinate is the code word amplitude. In the plots for sine and square waves, 2 and 4 cycles of the lower and higher frequency signals, respectively, are shown. For the constant amplitude waveform, the number of scan lines per code word determines the support or domain of the function.

the code word length, and the signal amplitude and frequency are chosen, the embedded signal $S[y]$ is completely defined.

$$S[y] = \sum_{i=0}^{L} A \cdot W_{c_i}[y - iT] + V_o. \tag{22}$$

The signal recovery process starts with the segmentation of the border of the form from the scanned image. The recovery process is illustrated in the block diagram of Figure 17. First, the left edge of the border is extracted from the scanned document. Then, the edge profile is rotated using a linear least squares method to correct for misalignment of the document on the scanner bed. Subsequently, the rotation compensated 1-D edge profile is passed through a set of matched-filters. The matched-filtering outputs indicate the positions of the code words through peaks in the correlation between the edge profile and the filters. The correlation peaks are ideally located at multiples of the code word length T. To address sampling synchronization issues due to noise, timing error detection (TED) techniques can be adopted. Finally, the samples are decoded and the binary message is recovered.

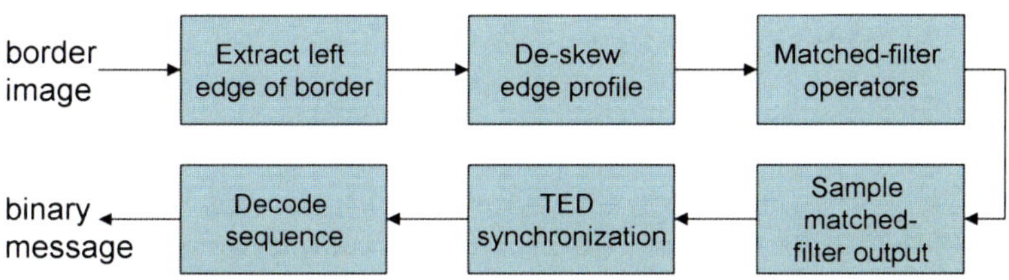

Fig. 17. Block diagram for the embedded signal recovery process.

Table 13. Capacity and error rates for embedded signatures using different carrier waveforms.

Coding method	Carrier (frequency)	Symb. or bit length (scan-lines)	Code word length (scan-lines)	Capacity (bits/inch)	Error rate
-	sine (20,40)	30	30	20	2.3%
-	sine (50,100)	12	36	16	3%
-	square (20,40)	30	30	20	0%
-	square (50,100)	12	36	16	2%
RS	constant	7	28	85	4.8%
RS	constant	8	32	75	0%
CC	constant	7	14	85	3.1%
CC	constant	8	16	75	0%

In Table 13, the results of using different carrier waveforms, code word lengths and coding methods are presented. Notice that when the message is not coded, the embedding capacity is similar to the capacities reported for halftone images and text characters. In contrast, the capacity increases substantially when the proposed coding methods are included.

4 Ink-jet Printers

Ink-jet printers follow the same pipeline illustrated in Figure 5. However, the techniques ink-jet printers use to create a hardcopy document are different from those used by EP printers.

4.1 Ink-jet Printer Architecture

The ink-jet mechanism consists of three principle components – the print-head, the carriage, and the paper advance mechanism. Figure 18 illustrates the operation of a typical ink-jet printer [9]. The paper is picked up and advanced in the process direction under the carriage by the paper advance mechanism. The carriage moves the print-head back and forth across the paper in the scan direction. Drops of ink are fired onto the paper by a print-head consisting of a nozzle plate containing several columns of nozzle openings and reservoirs for one or more different color inks. As illustrated in Fig. 19, the nozzles for each colorant are arranged in one or more staggered columns (two for the case of Fig. 19). By appropriately timing the firing of the nozzles in each column, it is possible to achieve an effective vertical resolution that is equal to the vertical offset between adjacent columns. In this way, the nozzles in each column can be spaced far enough apart and the adjacent columns spaced far enough apart to ease fabrication requirements, provide better structural integrity for the nozzle plate, and provide the necessary room for fluid channels and firing circuitry.

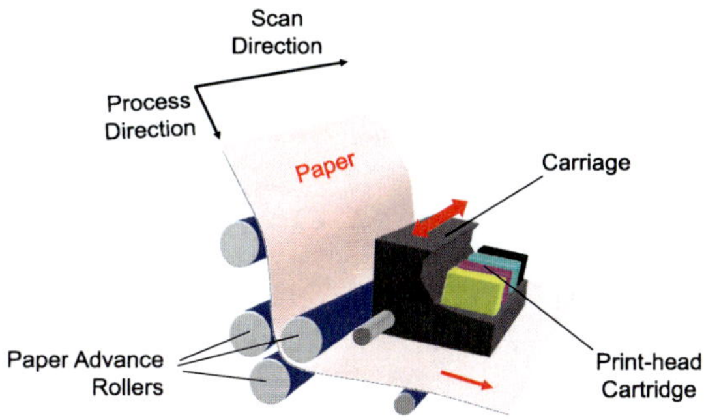

Fig. 18. Architecture for a typical ink-jet printer.

Each column of nozzles allows the simultaneous printing of several rows of pixels during a single pass of the print-head across the paper. Once the print-head has completed an entire pass in the scan direction, the paper is advanced again in the process direction; and a new pass is completed. This process is repeated until the print job is completed, at which time the paper advance mechanism ejects the paper into the output tray.

Several print options exist for ink-jet printers which control print resolution, speed, directionality, and the number of printing passes over each point on the paper. The pixels that are printed in a given pass across the page comprise a subset of the pixels in a horizontal band with height equal to the height of the print-head. This horizontal band of pixels is called a swath. In single-pass print modes, the print-head passes only once over each position on the paper, so the swaths do not overlap. For a multi-pass print mode with N passes, the paper only advances a fraction $1/N$ of the height of the print-head between passes. With the single pass print modes, misalignment between adjoining swaths is more likely to be visible. With multi-pass modes, the misalignment is masked to some extent by the overlapping swaths. In unidirectional print modes, the pen only fires ink while it is traveling in one direction across the page (either while traveling from left to right or from right to left), while in bidirectional print modes, successive swaths are printed in opposite directions.

The process of printing a three-pixel wide vertical line with a two-pass, bidirectional mode is illustrated in Figure 19 for a simplified printer architecture. The two-dimensional image of the line is encoded with a print mask, which consists of a two-dimensional array of 1's and 2's indicating firing the nozzle in the first pass from left to right and second pass from right to left, respectively. With different drop geometries associated with the different printing directions, the edge of the vertical line will exhibit a unique edge profile for different print modes.

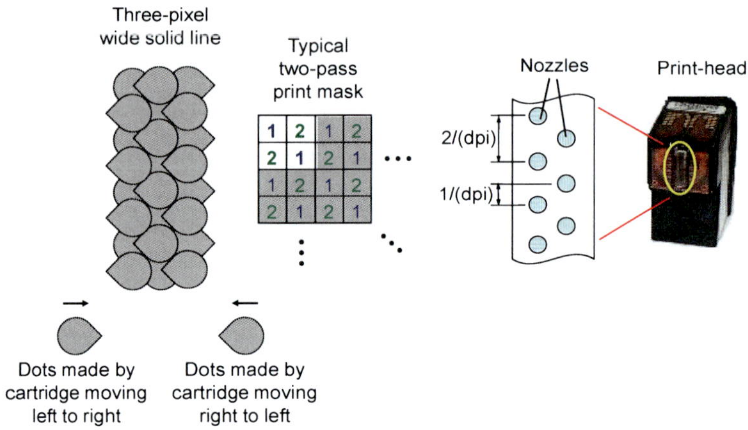

Fig. 19. Illustration of the process of printing a three-pixel wide vertical line in a two-pass, bidirectional print mode. The print mask consists of a two-dimensional array of 1's and 2's. The print mask is tiled over the entire page. Pixels on the page are printed on pass 1 (right-to-left) or pass 2 (left-to-right) according to whether they are labeled with a 1 or 2 from the print mask.

4.2 Ink-jet Printer Signatures

The large number of combinations of ink-jet print options can lead to a very complex intrinsic signature with many different identifiable features. Figure 20 shows the appearance of a typical dot printed with a single pass, 300 DPI resolution print mode with different carriage speeds and printing directions [9]. It illustrates the fact that as print speed increases, the dot shape becomes more asymmetric, and thus more dependent on the printing direction. Other artifacts that are related to print speed are tails and satellites which occur when the drop of ink breaks up as it exits the print nozzle. If the secondary droplet breaks away completely from the main droplet it forms a satellite (see center and right in Figure 20), and if it breaks away only partially, it forms a tail (see left in Figure 20). Tails and satellites usually trail the main dot relative to the direction of travel of the pen.

Another potentially useful ink-jet print artifact is dot placement. Dot placement errors can be caused by paper advance errors, carriage positioning errors, or misaligned nozzles in the print head. Each ink drop travels roughly perpendicular to the surface of the nozzle plate at the nozzle position. Due to structural characteristics of the pen, the nozzle plate may not be flat. This will cause drops from different nozzles to fire in slightly different directions, thereby creating characteristic patterns in the printed content. This is referred to as a toe-in or dimple effect [46]. Other characteristic features of ink-jet printers such as the fluid dynamics of the ink-jet nozzle, ink chemistry, and periodic variation caused by missing jets or paper advance errors are all potential signatures [71].

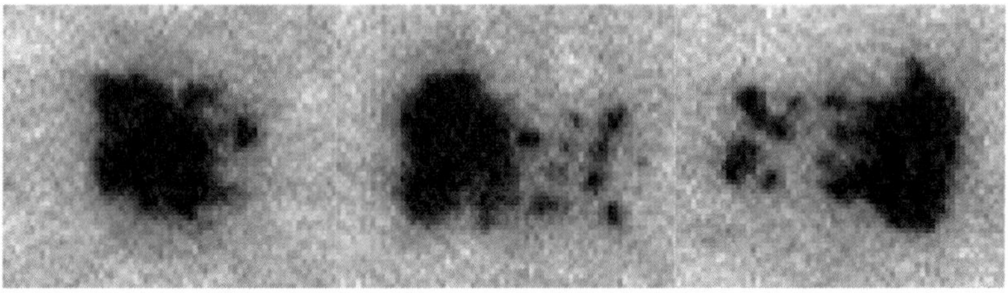

Fig. 20. Typical dots printed by a 300 DPI ink-jet printer: 15 inch/second left-to-right print mode (left), 45 inch/second left-to right print mode (center), and 45 inch/second right-to-left print mode (right). Dots were captured at 7000 DPI using a QEA IAS-1000 imaging system.

To identify the intrinsic signature of an ink-jet printer for security and forensic purposes, image analysis techniques have been used in recent years [6, 21, 55, 67, 71]. The majority of the approaches combine a variety of image analysis techniques with different classification algorithms to either distinguish different printing technologies (ink-jet vs. EP) [55, 63] or distinguish between make and models [6, 28, 55]. It has been shown that a variety of image features and print quality metrics can be used to discriminate among ink-jet printers. For example, quantitative analysis of 14 pt. character "i"s using features such as line raggedness, dot roundness, and background noise have been used to characterize different ink-jet printers [55]. An overview of other characteristics of ink-jet printing and methods of printer identification using these characteristics can be found in [71].

With the different print modes and print masks [72] that ink-jet printers use to improve image quality by hiding intrinsic noise or artifacts generated by the printer, it is difficult to use image analysis alone to deliver the necessary detection rate and payload capacity. The use of GLCM features has been investigated for use with text documents created by a set of six different ink-jet printers [6]. A scatter plot of the data for the six ink-jet printers with respect to the entropy and maximal correlation coefficient features is shown in Figure 21. Using these two features alone separates the six printers into four clusters. The three HP printers fall in the same cluster. This may be due to a similar printer technology being used in these printers.

Another characteristic that can be used to discriminate among ink-jet printers is spur marks. Spur marks are formed on a print by the paper conveyance gears of ink-jet printers, and are clearly distinguished from the background by infrared oblique lighting and gradient image processing [3]. Ink-jet printers can be classified by the type of spurs and their arrangement. This technique allows identification of the model and manufacturer of the printer, and is significant not only for counterfeit detection, but also in the field of forensic document examination.

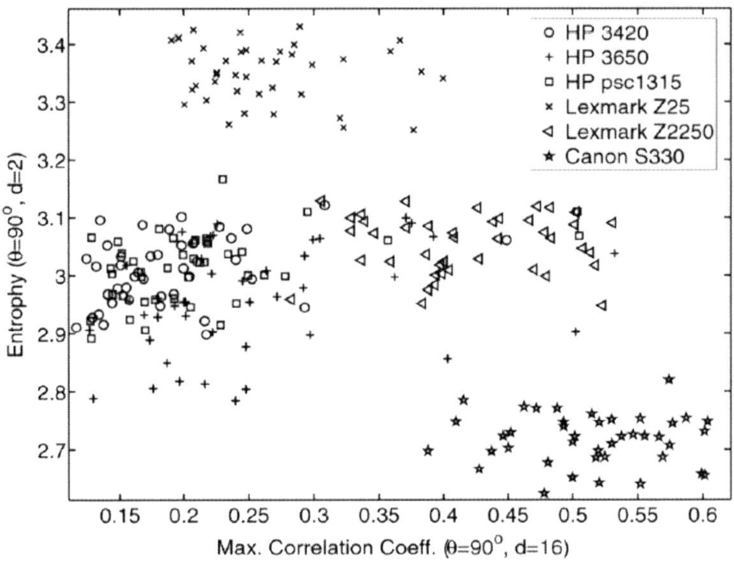

Fig. 21. Scatter plot of the data for the six ink-jet printers with respect to the entropy and maximal correlation coefficient features.

The use of security inks with narrow, reproducible excitation and emission band-gaps has also been investigated [64]. By matching the light source to the emission band-gap of the ink, ink-jet printers with specific inks can be identified. Similarly, by studying the ink solvents, 13 inks from 16 different ink-jet cartridges have been successfully identified [57].

5 Anti-forensics and Spoofing

Every security system has its limitations, and current techniques for securing a printed document or scanned image are no exception. Two types of attacks are possible. In one attack the signature is removed; in the second attack the signature is replaced by another signature (spoofing). Other attacks such as a "scan-print" attack used by counterfeiters are unique to these types of devices. The success of these attacks is dependent upon the document or image content.

In the case of printed text documents, scanning the document using optical character recognition (OCR) software to extract the text, and then re-printing it on another printer would remove any extrinsic or intrinsic signature. However the document may then be labeled as "suspect" if a particular signature that should be in the document is not found in the document. Scan-print attacks could be detected by use of the signatures in the scanner and printer. Similarly, printed halftone images could be scanned, converted to continuous-tone, and then re-printed after using standard watermarking attacks to remove any embedded watermark [68].

Spoofing the original printer or scanner intrinsic signature may be more difficult for signatures that originate at the device mechanism level since attacks would require modification of the device firmware. We have shown that it is possible to "replace" banding patterns in printers used to form intrinsic and extrinsic signatures; but this type of attack would be very difficult for the average user to do. With respect to scanners, some of the attacks reported for digital cameras that exploit the sensor pattern noise are possible [15, 48].

6 Security Protocol and Privacy

There are two major components to any security system: security primitives and protocols. The primitives are tools and techniques, such as the data-hiding and feature extraction methods introduced above. The protocol describes how the security primitives are used to protect content. For example, a watermarking protocol would describe how the message to be embedded is processed, e.g. encrypted, before it is embedded in the document, and where the keys are maintained. Nearly every failure in a security system is due to a protocol failure. For example, a computer system may have a very secure password selection primitive. However, due to the complexity of the password, a user may write it on a slip of paper, and attach it to the computer. In this case, the protocol has failed, rather than the primitive. Printers and sensors are integral components of the digital document world. As described in this paper, many effective security primitives exist for printers and sensors. One of the future challenges is to integrate these primitives into security protocols that are needed for different application scenarios. A successful document security protocol will need to consider the strength and vulnerability of the security primitives, as well as all workflow associated with the generation, distribution, and storage of the document.

The methods presented in this paper have many applications in law enforcement such as tracking, counterfeiting, and child pornography. The downside is that they provide a mechanism for a simple device, a printer or a digital camera, to spy on its user. A typical user cannot turn off these signatures, particularly the intrinsic signature, without very detailed knowledge of how the device operates. This could have dire consequences for many important uses of these devices in our society. For example, a whistleblower who would like to share documents with a regulatory agency could be in danger in that their printer could be identified as the one that produced the documents. A citizen who would like to anonymously report a chemical spill using a digital camera is also threatened with possible discovery, even though these acts are protected by Federal law (31 U.S.C. Sec. 3730(h)) and many state laws. It is also possible to develop signal analysis tools and system protocols that can be used to effectively defeat or turn off a device's intrinsic and extrinsic signatures to protect a users's privacy for legal and ethical uses of the sensor device. It is important to note that the intention is not to create a set of hacking tools that can be used to defeat the ability of the output of the sensor to be tied to a particular device, since it is clear that sensor forensics has many important legitimate uses.

7 Conclusions

The use of printers and scanners in our society as an interface to the digital world will not disappear any time during the foreseeable future. Security mechanisms are required for this unique interface just as security is required for digital media such as digital images, audio, and video. Preventing forgeries and unauthorized duplication, as well as determining the source of illicit content, is just as important in the physical world as it is in the digital one. Printer and scanner forensics is a growing area of research building upon a multitude of disciplines ranging from image processing and communications, to mechatronics and psychophysics.

What is interesting and challenging about printers and scanners it that one can exploit how these devices generate their output to provide security features. But these properties are governed by complex physical processes that are characterized by a great deal of variability that from a signal processing perspective might be viewed as noise. In the realm of extrinsic signature embedding, another challenge is to develop methods that provide the right combination of robustness, fragility, and resistance to attack. From a practical perspective, dealing with the large number of different scanner and printer products in the marketplace and the relatively short time between new product introductions is a challenge for developing algorithms to detect intrinsic signatures and to embed extrinsic signatures. It is important that these algorithms be as broadly and easily generalizable as possible.

Future research topics include the following. First, there is a dearth of research addressing extrinsic signature embedding for inkjet printers. Second, for signature embedding in general, there is a need for end-to-end systems analyses that incorporate channel coding to combat noise, and which offer guaranteed capacity levels. Work has been done in this area; but more remains. The problem is made more challenging by the content-dependence of many of the extrinsic signal embedding methods. As discussed above, embedding in forms offers a way to remove content-dependence; but more work remains here, as well. Finally, as mentioned in the preceding section, there are situations when it is appropriate to disable or defeat intrinsic and extrinsic signatures. The development of methods to achieve these goals is an open issue.

Acknowledgments

This work was supported by the National Science Foundation under Grant No. CCR-0219893 and Grant No. CNS-0524540. Any opinions, findings, and conclusions or recommendations expressed in this material are those of the author(s) and do not necessarily reflect the views of the National Science Foundation.

References

1. Investigating machine identification code technology in color laser printers (July 2005), http://www.eff.org/wp/investigating-machine-identification-code-technology-color-laser-printers
2. Check Clearing for the 21st Century Act (April 24, 2007), http://www.federalreserve.gov/paymentsystems/truncation/default.htm
3. Akao, Y., Kobayashi, K., Sugawara, S., Seki, Y.: Discrimination of inkjet-printed counterfeits by spur marks and feature extraction by spatial frequency analysis. In: van Renesse, R.L. (ed.) Proceedings of the SPIE International Conference on Optical Security and Counterfeit Deterrence Techniques IV, vol. 4677, pp. 129–137 (2002)
4. Ali, G.N., Chiang, P.J., Mikkilineni, A.K., Allebach, J.P., Chiu, G.T.C., Delp, E.J.: Intrinsic and extrinsic signatures for information hiding and secure printing with electrophotographic devices. In: Proceedings of the IS&T's NIP19: International Conference on Digital Printing Technologies, New Orleans, LA, September 2003, vol. 19, pp. 511–515 (2003)
5. Ali, G.N., Chiang, P.J., Mikkilineni, A.K., Chiu, G.T.C., Delp, E.J., Allebach, J.P.: Application of principal components analysis and gaussian mixture models to printer identification. In: Proceedings of the IS&T's NIP20: International Conference on Digital Printing Technologies, Salt Lake City, UT, October/November 2004, vol. 20, pp. 301–305 (2004)
6. Arslan, O., Kumontoy, R.M., Chiang, P.J., Mikkillineni, A.K., Allebach, J.P., Chiu, G.T.C., Delp, E.J.: Identification of inkjet printers for forensic applications. In: Proceedings of the IS&T's NIP21: International Conference on Digital Printing Technologies, Baltimore, MD, October 2005, vol. 21, pp. 235–238 (2005)
7. Barni, M., Bartolini, F.: Watermarking Systems Engineering: Enabling Digital Assets Security and Other Applications. Marcel Dekker, New York (2004)
8. Barni, M., Podilchuk, C.I., Bartolini, F., Delp, E.J.: Watermark embedding: hiding a signal within a cover image. IEEE Communications Magazine 39(8), 102–108 (2001)
9. Bernal, E., Allebach, J.P., Pizlo, Z.: Improved pen alignment for bidirectional printing. The Journal of Imaging Science and Technology 51(1), 1–22 (2007)
10. Brassil, J.T., Low, S., Maxemchuk, N.F.: Copyright protection for the electronic distribution of text documents. Proceedings of the IEEE 87, 1181–1196 (1999)
11. Bulan, O., Monga, V., Sharma, G., Oztan, B.: Data embedding in hardcopy images via halftone-dot orientation modulation. In: Proceedings of the SPIE International Conference on Security, Steganography, and Watermarking of Multimedia Contents X, vol. 6819 (January 2009)
12. Burges, C.J.C.: A tutorial on support vector machines for pattern recognition. Data Mining and Knowledge Discovery 2(2), 121–167 (1998)
13. Chang, C.C., Lin, C.J.: LIBSVM: a library for support vector machines (2001), http://www.csie.ntu.edu.tw/~cjlin/libsvm
14. Chen, C.L.: Modeling, Analysis, and Robust Controls for Electrophotographic Imaging Systems Subject to Spatially Periodic Disturbances and Measurable Nonlinearities. Ph.D. thesis, Purdue University (2003)
15. Chen, M., Fridrich, J., Goljan, M., Lukas, J.: Determining image origin and integrity using sensor noise. IEEE Transactions on Information Forensics and Security 3(1), 74–90 (2008)

16. Chiang, P.J., Ali, G.N., Mikkilineni, A.K., Chiu, G.T.C., Allebach, J.P., Delp, E.J.: Extrinsic signatures embedding using exposure modulation for information hiding and secure printing in electrophotographic devices. In: Proceedings of the IS&T's NIP20: International Conference on Digital Printing Technologies, Salt Lake City, UT, October/November 2004, vol. 20, pp. 295–300 (2004)
17. Chiang, P.J., Khanna, N., Mikkilineni, A.K., Segovia, M.V.O., Suh, S., Allebach, J.P., Chiu, G.T.C., Delp, E.J.: Printer and scanner forensics. IEEE Signal Processing Magazine 26(2), 72–83 (2009)
18. Chiang, P.J., Mikkilineni, A.K., Delp, E.J., Allebach, J.P., Chiu, G.T.C.: Extrinsic signatures embedding and detection in electrophotographic halftone images through laser intensity modulation. In: Proceedings of the IS&T's NIP22: International Conference on Digital Printing Technologies, Denver, CO, September 2006, pp. 432–435 (2006)
19. Conners, R.W., Trivedi, M.M., Harlow, C.A.: Segmentation of a high-resolution urban scene using texture operators. Computer Vision, Graphics, and Image Processing 25, 273–310 (1984)
20. Cristianini, N., Taylor, J.S.: An Introduction to Support Vector Machines and Other Kernel-Based Learning Methods. Cambridge University Press, Cambridge (2000)
21. Doherty, P.: Classification of ink jet printers and inks. J. Am. Soc. of Quest. Doc. Exam. 1(1), 88–106 (1998)
22. Duda, R.O., Hart, P.E., Stork, D.G.: Pattern Classification, 2nd edn. Wiley-Interscience, Hoboken (2000)
23. Foi, A., Katkovnik, V., Egiazarian, K., Astola, J.: A novel local polynomial estimator based on directional multiscale optimizations. In: Proceedings of the 6th IMA Int. Conf. Math. in Signal Processing, vol. 5685, pp. 79–82 (2004)
24. Fu, M.S., Au, O.: Data hiding in halftone images by stochastic error diffusion. In: Proceedings of IEEE International Conference on Acoustics, Speech, and Signal Processing, ICASSP 2001, vol. 3, pp. 1965–1968 (2001)
25. Fukunaga, K.: Introduction to Statistical Pattern Recognition. Academic Press, San Diego (1990)
26. Gloe, T., Franz, E., Winkler, A.: Forensics for flatbed scanners. In: Delp III, E.J., Wong, P.W. (eds.) Proceedings of the SPIE International Conference on Security, Steganography, and Watermarking of Multimedia Contents IX, vol. 6505, 65051I. SPIE, San Jose (2007)
27. Gou, H., Swaminathan, A., Wu, M.: Robust scanner identification based on noise features. In: Delp III, E.J., Wong, P.W. (eds.) Proceedings of the SPIE International Conference on Security, Steganography, and Watermarking of Multimedia Contents IX, vol. 6505, 65050S. SPIE, San Jose (2007)
28. Gupta, G., Saha, S.K., Chakraborty, S., Mazumdar, C.: Document frauds: Identification and linking fake document to scanners and printers. In: International Conference on Computing: Theory and Applications (ICCTA 2007), March 2007, pp. 497–501 (2007)
29. Haralick, R.M., Shanmugam, K., Dinstein, I.: Textural features for image classification. IEEE Transactions on Systems, Man, and Cybernetics SMC-3(6), 610–621 (1973)
30. Holst, G.C.: CCD Arrays, Cameras, and Displays, 2nd edn. JCD Publishing & SPIE Press, USA (1998)

31. Hsu, C.W., Chang, C.C., Lin, C.J.: A practical guide to support vector classification. Department of Computer Science and Information Engineering, National Taiwan University (July 2003), `http://www.csie.ntu.edu.tw/~cjlin/papers/guide/guide.pdf`
32. Janesick, J.R.: Scientific Charge-Coupled Devices. SPIE, San Jose (January 2001)
33. Joachims, T.: Making large-scale support vector machine learning practical. In: Schölkopf, B., Burges, C., Smola, A. (eds.) Advances in Kernel Methods: Support Vector Machines. MIT Press, Cambridge (1998)
34. Kacker, D., Allebach, J.P.: Joint halftoning and watermarking. IEEE Transactions on Signal Processing 51(4), 1054–1068 (2003)
35. Kacker, D., Camis, T., Allebach, J.P.: Electrophotographic process embedded in direct binary search. IEEE Transactions on Image Processing 11(3), 234–257 (2002)
36. Khanna, N., Chiu, G.T.C., Allebach, J.P., Delp, E.J.: Forensic techniques for classifying scanner, computer generated and digital camera images. In: Proceedings of the 2008 IEEE International Conference on Acoustics, Speech, and Signal Processing, Las Vegas, NV (March 2008)
37. Khanna, N., Chiu, G.T.C., Allebach, J.P., Delp, E.J.: Scanner identification with extension to forgery detection. In: Proceedings of the SPIE International Conference on Security, Steganography, and Watermarking of Multimedia Contents X, San Jose, CA, January 2008, vol. 6819, 68190G (2008)
38. Khanna, N., Mikkilineni, A.K., Chiang, P.J., Ortiz, M.V., Shah, V., Suh, S., Chiu, G.T.C., Allebach, J.P., Delp, E.J.: Printer and sensor forensics. In: IEEE Workshop on Signal Processing Applications for Public Security and Forensics (SAFE 2007), April 2007, pp. 1–8 (2007)
39. Khanna, N., Mikkilineni, A.K., Chiang, P.J., Ortiz, M.V., Suh, S., Chiu, G.T.C., Allebach, J.P., Delp, E.J.: Sensor forensics: Printers, cameras and scanners, they never lie. In: 2007 IEEE International Conference on Multimedia and Expo., July 2007, pp. 20–23 (2007)
40. Khanna, N., Mikkilineni, A.K., Chiu, G.T.C., Allebach, J.P., Delp, E.J.: Forensic classification of imaging sensor types. In: Proceedings of the SPIE International Conference on Security, Steganography, and Watermarking of Multimedia Contents IX, January 2007, vol. 6505, 65050U. SPIE, San Jose (2007)
41. Khanna, N., Mikkilineni, A.K., Chiu, G.T.C., Allebach, J.P., Delp, E.J.: Scanner identification using sensor pattern noise. In: Proceedings of the SPIE International Conference on Security, Steganography, and Watermarking of Multimedia Contents IX, January 2007, vol. 6505, 65051K. SPIE, San Jose (2007)
42. Khanna, N., Mikkilineni, A.K., Chiu, G.T.C., Allebach, J.P., Delp, E.J.: Survey of scanner and printer forensics at purdue university. In: Srihari, S.N., Franke, K. (eds.) IWCF 2008. LNCS, vol. 5158, pp. 22–34. Springer, Heidelberg (2008)
43. Khanna, N., Mikkilineni, A.K., Delp, E.J.: Forensic camera classification: Verification of sensor pattern noise approach. Forensic Science Communications (FSC) 11(1) (January 2009)
44. Khanna, N., Mikkilineni, A.K., Delp, E.J.: Scanner identification using feature-based processing and analysis. IEEE Transactions on Information Forensics and Security 4(1), 123–139 (2009)
45. Khanna, N., Mikkilineni, A.K., Martone, A.F., Ali, G.N., Chiu, G.T.C., Allebach, J.P., Delp, E.J.: A survey of forensic characterization methods for physical devices. Digital Investigation 3, 17–28 (2006)
46. Lee, J.H., Allebach, J.P.: Inkjet printer model-based halftoning. IEEE Transactions on Image Processing 14(5), 674–689 (2005)

47. Lukas, J., Fridrich, J., Goljan, M.: Determining digital image origin using sensor imperfections. In: Said, A., Apostolopoulos, J.G. (eds.) Proceedings of the SPIE International Conference on Image and Video Communications and Processing, vol. 5685, pp. 249–260. SPIE, San Jose (2005)
48. Lukas, J., Fridrich, J.J., Goljan, M.: Digital camera identification from sensor pattern noise. IEEE Transactions on Information Forensics and Security 1(2), 205–214 (2006)
49. Mikkilineni, A.K., Arslan, O., Chiang, P.J., Kumontoy, R.M., Allebach, J.P., Chiu, G.T.C., Delp, E.J.: Printer forensics using svm techniques. In: Proceedings of the IS&T's NIP21: International Conference on Digital Printing Technologies, Baltimore, MD, October 2005, vol. 21, pp. 223–226 (2005)
50. Mikkilineni, A.K., Chiang, P.J., Ali, G.N., Chiu, G.T.C., Allebach, J.P., Delp, E.J.: Printer identification based on textural features. In: Proceedings of the IS&T's NIP20: International Conference on Digital Printing Technologies, Salt Lake City, UT, October/November 2004, vol. 20, pp. 306–311 (2004)
51. Mikkilineni, A.K., Chiang, P.J., Ali, G.N., Chiu, G.T.C., Allebach, J.P., Delp, E.J.: Printer identification based on graylevel co-occurrence features for security and forensic applications. In: Proceedings of the SPIE International Conference on Security, Steganography, and Watermarking of Multimedia Contents VII, San Jose, CA, March 2005, vol. 5681, pp. 430–440 (2005)
52. Mikkilineni, A.K., Chiang, P.J., Chiu, G.T.C., Allebach, J.P., Delp, E.J.: Channel model and operational capacity analysis of printed text documents. In: Proceedings of the SPIE International Conference on Security, Steganography, and Watermarking of Multimedia Contents IX, 65051U, San Jose, CA (January 2007)
53. Mukhtar, M., Meckl, P., Chiu, G.T.C.: Color registration error reduction in document scanner using iterative velocity command synthesis. In: Proceedings of the 2008 ASME Dynamic Systems and Control Conference, Ann Arbor, Michigan (October 2008)
54. Norris, M., Barney, E.: Printer modeling for document imaging. In: Proceedings of CISST: International Conference on Imaging Science, Systems and Technology, Las Vegas, NV, June 2004, pp. 14–20 (2004)
55. Oliver, J.F., Chen, J.X.: Use of signature analysis to discriminate digital printing technologies. In: Proceedings of the IS&T's NIP18: International Conference on Digital Printing Technologies, San Diego, California, September 2002, pp. 218–222 (2002)
56. Ortiz Segovia, M.V., Chiu, G.T.C., Delp, E.J., Allebach, J.P.: Information embedding in electrophotographic document forms through laser intensity modulation a communications systems perspective. In: Proceedings of the IS&T's NIP25: International Conference on Digital Printing Technologies, Louisville, KY (September 2009)
57. Partouche, F., Espanet, B., Villena, C., Murie, C.: Forensic analysis of inkjet printings by pyrolysis gc/ms. In: Proceedings of NIP21: The 21st International Conference on Digital Printing Technologies, Baltimore, MD, pp. 216–219 (2005)
58. Pursley, M.B.: Introduction to Digital Communications. Pearson Education, Inc., London (2005)
59. Sencar, H.T., Memon, N.: Overview of state-of-the-art in digital image forensics. In: Part of Indian Statistical Institute Platinum Jubilee Monograph series titled 'Statistical Science and Interdisciplinary Research' (2008) (to appear)

60. Sharma, G., Wang, S.: Show-through watermarking of duplex printed documents. In: Proceedings of the SPIE International Conference on Security, Steganography, and Watermarking of Multimedia Contents VI, January 2004, vol. 5306 (2004)
61. Shein, L.B.: Electrophotography and Development Physics, 2nd edn. Springer, New York (1992)
62. Shein, L.B.: Handbook of Electrostatic Processes, pp. 321–348. Marcel Dekker, New York (1995)
63. Simske, S.J., Aronoff, J.S., Sturgill, M., Collins, F., Golodetz, G., Israel, R.: Security printing deterrents: A comparison of TIJ, DEP and LEP printing. In: Proceedings of the IS&T's NIP23: International Conference on Digital Printing Technologies, Anchorage, AK, September 2007, pp. 543–548 (2007)
64. Simske, S.J., Hattersley, J.R., Golodetz, G., Stssiak, J., Aronoff, J.S.: Ink-specific handheld readers and security variable data printing (svdp). In: Proceedings of NIP22: The 22nd International Conference on Digital Printing Technologies, Denver, CO, vol. 2006, pp. 441–443 (2006)
65. Suh, S., Allebach, J.P., Chiu, G.T.C., Delp, E.J.: Printer mechanism-level information embedding and extraction for halftone documents: New results. In: Proceedings of the IS&T's NIP23: International Conference on Digital Printing Technologies, Anchorage, Alaska, September 2007, pp. 549–553 (2007)
66. Sweeney, P.: Error Control Coding: From Theory to Practice. Wiley, New York (2002)
67. Talbot, V., Perrot, P., Murie, C.: Ink jet printing discrimination based on invariant moment. In: Proceedings of the IS&T's NIP22: International Conference on Digital Printing Technologies, Denver, CO, September 2006, pp. 427–431 (2006)
68. Voloshynovskiy, S., Pereira, S., Pun, T., Eggers, J.J., Su, J.K.: Attacks on digital watermarks: classification, estimation based attacks, and benchmarks. IEEE Communications Magazine 39(8), 118–126 (2001)
69. Williams, E.M.: The Physics and Technology of Xerographic Processes. Wiley, New York (1984)
70. Williams, E.M.: The Physics and Technology of Xerographic Processes. Wiley, New York (1984)
71. Wolin, D.: Document verification and traceability through image quality analysis. In: Proceedings of the IS&T's NIP18: International Conference on Digital Printing Technologies, San Diego, California, September 2002, pp. 214–217 (2002)
72. Yen, J., Carlsson, M., Chang, M., Garcia, J.M., Nguyen, H.: Constraint solving for inkjet print mask design. The Journal of Imaging Science and Technology 44(5), 391–397 (2000)
73. Yi, J.: A Xerographic simulation model. Master's thesis, University of Idaho, Boise, ID (1999)

Privacy Enhancing Solutions for Personal Information Based Multimedia Content Sharing

Hosik Sohn[1], Yong Man Ro[1], and Konstantinos N. Plataniotis[2]

[1] Department of Electrical Engineering, Korea Advanced Institute of Science and Technology, South Korea
sohnhosik@kaist.ac.kr, ymro@ee.kaist.ac.kr
[2] Department of Electrical and Computer Engineering, University of Toronto, Canada
Kostas@comm.utoronto.ca

Abstract. User generated content (UGC) is shared in an open space like social media where users can upload and consume contents freely. Since the access to the contents is not restricted, the contents could be delivered to unwanted users or misused sometimes. Although it is contrary to the philosophy of UGC, the contents can be encrypted by a key and the key can be shared among users to protect the contents of a UGC creator. However, it is quite challenging to distribute the key to anonymous users since there is no relationship or acquaintanceship between the content consumers and creators. To solve this problem, user profile or content preference can be utilized. By allowing accessibility to only users who have similar personal information, the contents creator or distributor can select the target group of consumers. In this chapter, we present a method for sharing contents securely based on the personal information of users that includes user profiles and contents preference. With the proposed method, virtual secure space is created for contents delivery. The virtual secure space allows the UGC creator to deliver contents to users who have similar personal information and they can consume the contents without any leakage of personal information during similarity measurement. To realize the virtual secure space, the personal information is applied to fuzzy vault scheme, which enables the generation of secure biometric template.

1 Introduction

The number of user generated content on the Web has been rapidly increasing during the last several years [1, 16, 22]. Ordinary people have become producers of content as well as consumers so called 'prosumers', being capable of publishing their own contents on the Web such as Flickr, MySpace, FaceBook and Youtube [5, 6, 19, 33]. These social networking sites provide an open space where every user is allowed to upload and consume contents freely. Since this open space offers better accessibility and unlimited sharing of contents among users, uploaded content can be easily reviewed by anyone.

As the UGC is becoming increasingly popular, the need to protect these personal contents has also arisen. One of the examples of limited restriction of the access to personal contents is to utilize the list of acquaintances to be trusted for sharing contents. More specifically, personal contents can be protected by restricting the access to a selected group of users by creating the 'circle of trust' [18], such as friends list. Within the circle of trust, users can invite friends and acquaintances to be added to

H.T. Sencar et al. (Eds.): Intel. Multimedia Analysis for Security Appli., SCI 282, pp. 189–208.
springerlink.com

their list and share personal contents among them. Most existing solutions are based on authentication mechanism or encryption to protect users' information from being accessed by unexpected users [2, 4, 7, 18]. In [18], Instant Messaging (IM) based content sharing was proposed in the IM network environment. The authenticated encryption scheme was also proposed in [2] to guarantee secure data sharing.

Anonymous users, on the other hand, distribute contents without any relationship or acquaintanceship between the content consumers and content creators. So, content sharing among anonymous users contain a potential risk that the contents can be delivered to the unwanted group of people and misused if there is no restriction of the access to the contents. Therefore, the content creator would like to restrict access to content for anonymous users and partially share his/her contents with only the group of people who have similar personal preferences such as user profile and content preference e.g., people who have same occupation, hobby, or favorite movie.

Although it is contrary to the philosophy of UGC, if the contents are encrypted by a key and the key is distributed to the preferred group of people, content creators would not be concerned about the misusage of their contents. However, it is quite challenging to distribute the key to anonymous users who have the similarity in terms of user profile and content preference. Concerns of privacy might come up: During the process of measuring the similarity of personal information, user's personal information could be revealed to another user. There should not be any revelation of personal information during the comparison of personal information between anonymous users and content creators. The personal information of users such as name, gender, date of birth, religion, occupation and hobby should be kept private. Consequently, users would like to hide their personal information in public. In fact, revealing personal information on the Web could cause various security related attacks [7].

In this chapter, we present a virtual secure space that affords users the opportunity to generate and consume contents securely without revealing their personal information. By the proposed method, virtual secure space, which is generated by the creator's personal information, is available to the only user whose personal information is well matched. In order to create the virtual secure space, we utilized the vault set that is an encoding result of fuzzy vault scheme [12]. By applying the personal information to fuzzy vault scheme, the personal information is secured when measuring the similarity between users.

The advantage of such method is that content creators would not be concerned about the misusage of their contents by being assured that their contents are delivered to users with similar preference. Note that the proposed method is not a fundamental solution that guarantees the security of content or deters its misuse. However, by restricting the group of content consumers based on the preference, we can mitigate the potential risk that the content is being misused. Another advantage is that if uploading and consumption of contents are done based on the preferences, the contents can be automatically classified according to the user profile and contents preference, thereby, assisting data and server management. Moreover, by making use of the proposed scheme, it is possible to create the 'circle of trust' with even anonymous users.

Fuzzy vault scheme had been introduced to provide security for biometric data, such as fingerprint, face, and iris features, represented by discrete features [14, 20, 32]. The basic concept of fuzzy vault is that, the vault set is generated through the binding of both biometric data and a security key, and this generated vault set provides the secret key only if it is decoded with the biometric data of the same identity. Additionally, several efforts have been dedicated to generate revocable biometrics templates based on non-invertible transforms; If the template is compromised, the transform function is changed and the same biometrics are mapped to a new template. In [27], pseudo-random matrix was utilized to project face feature vector onto random subspace. Since the transform is non-invertible, the privacy of the biometric data can be preserved even when a secret key is compromised. Also, in [11], the generation of binarized biometric templates, so-called bioHashing, based on the random projection has been reported. This scheme utilizes iterated inner product between tokenized pseudo-random number and biometric features. Although the non-invertible transform based schemes provide better privacy protection than fuzzy vault scheme since the original data cannot be recovered even in the stolen-token scenario, they are two-factor authenticator schemes that require a secret key to measure the similarity between genuine and query biometric feature. Therefore, they are not appropriate for our concept of a secret key sharing method for contents distribution.

Biometric traits for human recognition largely consist of primary and soft biometric traits [10]. The primary biometric data represents unique features that distinguish an individual, while the soft biometric data represents relaxed characteristics of a human such as height, gender, and ethnicity etc. Although the characteristics of the soft biometric data lack the uniqueness and permanence to identify an individual reliably, they offer additional information about the user identity. Therefore, in the perspective of biometrics, the personal information, which contains user profile and content preference, can be regarded as soft biometric traits of a human.

Biometric data and personal information have two common characteristics: uniqueness and fuzzy match. Both data have very unique characteristics that represent an individual's identity, such that they should not be revealed to other people. Moreover, biometric authentication is based on fuzzy match due to the noisy nature of biometric data. Likewise, the personal information of users may be similar to each other, but the probability of exact match will be low. Ultimately, our content sharing scheme mechanism can be seen as a sole soft biometric authenticator that requires relaxed matching among users.

In this chapter, we present the concept of personal information based content sharing among anonymous users and its application scenario in the social network as well. The contents of this chapter are organized as follows: Section 2 introduces content sharing using personal information and the philosophy of the virtual secure space as well as detailed explanation of procedure to generate the virtual secure space with its conceptual scenario. We extend the concept of virtual secure space to the home network scenario in Section 3. Also in Section 3, simulation results and security issue are presented. Finally, Section 4 provides some concluding remarks.

2 Content Sharing Using Personal Information

2.1 User Scenario for Virtual Secure Content Sharing System

In this Section, the concept of a virtual secure content sharing system is presented in the user scenario. For contents sharing, there are two choices a user can make. One is to upload contents without encryption for unlimited consumption to any users, while the other is our proposed method which encrypts contents with a secret key. In this case, the content creators allow users to decrypt and consume the contents only if they have similar personal information including user profile and contents preference such as gender, hobby, and favorite movie etc. Note that restriction to content consumption is not achieved only through encryption. For instance, it can be done by only providing preview.

In Fig. 2.1, the open space refers to the web space such as Flickr, MySpace, FaceBook and Youtube. Since access to the contents in the open space is not restricted, users can consume contents anytime anywhere. In this space, both encrypted contents (C^e_{Alice} and C^e_{Bob}) and unencrypted contents ($C_{Charlie}$ and C_{Dave}) can be uploaded according to the choices that content creators made. For the encrypted content, corresponding vault sets (VS_{A_P} and VS_{B_P}), which are encoding results from the fuzzy vault encoder, are uploaded together. The access to the encrypted contents is given to users according to their personal information, and users with similar personal information can be defined as a virtual group. This space, generated by the vault set, is opened to the only users with similar personal information. Thus, the virtual space can be secured by grouping the contents according to personal information. In the view of content categorization, each virtual secure space is a category generated by the combination of personal information. By making use of the proposed content sharing scheme, we can generate more varying and flexible categories in comparison to the current categorization of contents; Each user can have his own personalized group to share contents with other users. On the other hand, current social networking sites only allow a limited number of categories which are defined by a central authority. The following show possible scenarios that can occur.

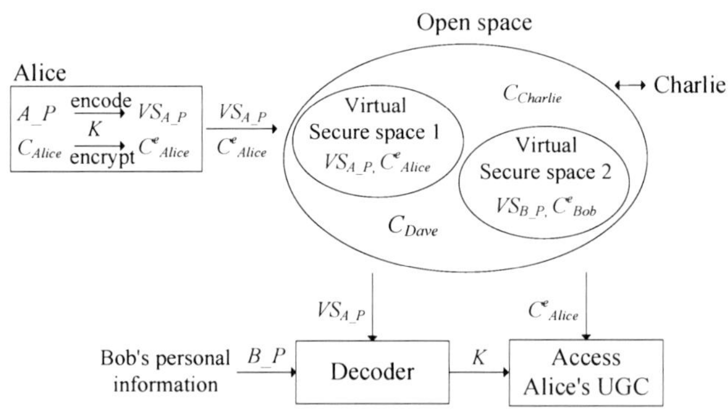

notation

K	secret key
L_P	user's personal information (L = initial letter of user name)
VS_{L_P}	vault set generated by L_P
$C_{user\ name}$	user's content
$C^e_{user\ name}$	user's encrypted content

Fig. 2.1. The consumption process of UGC in the open space [25]

Alice is a user (or UGC creator) who wants to provide her content to a group of people who have similar personal information. To do so, she uploads UGC (C^e_{Alice}) that is encrypted by her secret key (K) and the vault set (VS_{A_P}) generated by her personal information. A possible example of her personal information could be that, she is in 'twenties', 'not married' and her favorite genre of movie is documentary movie. Her profile and content preference are the features of her personal information. Uploaded Alice's vault set creates a virtual secure space and allows accessibility to only users who have similar personal information.

Bob is another user who wants to consume Alice's content. In order to consume the content, Bob downloads Alice's vault set (VS_{A_P}) and decrypts it with his personal information. For example, assume that Bob's personal information is, 'twenties', 'not married', favorite content is documentary movie and comedy show. Since his personal information has significant overlapping with Alice's, he can decode the vault set and acquire the secret key which allows the decryption of the content. Hence, Bob can freely consume the contents inside the virtual secure space generated by Alice where this space is opened to only users who have similar personal information with Alice, like Bob. From Alice's perspective, this can be seen as a method for limiting the accessibility of the contents to the consumers by creating the virtual secure space.

Lastly, Charlie and Dave are consumers who want to provide their contents to the public. The contents of Charlie and Dave ($C_{Charlie}$ and C_{Dave}) are uploaded without any encryption. However they cannot access the contents in the virtual space generated by Alice and Bob. This method is currently well used in most of UGC sites, where consumption of the contents is unlimited but cannot protect the contents being delivered to unwanted users.

For the scenario mentioned above, the benefit of applying the fuzzy vault scheme is that, content sharing is possible even though personal information between users is not perfectly matched. And, unlike other general web sites, the proposed system does not require a centralized authority. The only function of authority in open space is to distribute the personal information table (more on this will be mentioned in the next Section) and store contents along with vault set. If the personal information is encrypted without utilizing the fuzzy vault scheme, the centralized authority is required to handle key management and distribution. Also, security of centralized authority should be guaranteed. In practice, key management is one of the most difficult aspects of cryptography.

Moreover, in order to compare user information among other users, encrypted personal information should be transformed to the same domain or decrypted to original information. For the case of transforming to the same domain, an authority that has the knowledge of the rules used in transforming the encrypted personal information into the identical domain for comparison is required. If the original information is decrypted from the encrypted personal information, the personal information is revealed at the other user's side or in the open space. If the centralized authority has perfect security, the utilization of the fuzzy vault scheme is not necessary. However, from the practical standpoint, perfect security cannot be guaranteed. Vulnerable authority also has a risk of leaking all personal information at once. In the next Section, the method for generating the virtual secure space is presented, where secure delivery and consumption of contents are guaranteed.

2.2 Personal Information Model

As explained in the previous Section, vault set is generated by binding the personal information of the content creator and a secret key which is used for protecting the content. In order to consume the contents, a consumer should decode the vault set by his/her personal information. If a certain number of the personal information matches personal information of consumers, then the proper secret key is acquired, and finally consumers can decrypt or access the content. Before explaining the procedure for encoding and decoding, we will first define the personal information item of a user, *i.e.*, binarized personal information, and describe how to generate it.

In order to use personal information as an input of fuzzy vault scheme, we defined personal information items such as age, gender, and favorite contents. Table 2.1 presents the defined personal information items. The personal information consists of user profile and content preference. For the construction of the personal information, the User Preference Description Scheme (DS) in MPEG-7 Multimedia Description Schemes (MDS) [9] was referred. MPEG-7, which is formally named "Multimedia Content Description Interface", is standardized for describing the content of multimedia data. MPEG-7 description tools allow for generating descriptions of specific applications [17]. Descriptions include information of creation, production process, usage of content, storage features, and interaction of the user with the content etc. In MPEG-7 MDS, two tools are defined for describing user interaction: User Preferences DS and the User History DS.

The User Preferences DS contains all the tools for describing the user in terms of the preferences for information contained in other tools within MPEG-7 as well as the usage environment of the user. The user model mainly consists of three components; preferences for content (e.g. music, movies, and books, etc), demographic information (e.g. language, gender, etc), and usage environment (*i.e.*, place, time, and recurrence etc). On the other hand, the User History DS is designed as a tool for logging user interactions; it logs the interaction with the user in terms of preferred content and usage information (e.g. where and when interactions of the content with user took place). In our research, we utilized the Genre CS for sub-categories of contents preference and the ParentalRating CS for age in user profile in order to construct personal information. Each CS contains set of several terms to define concepts in the domain covered by the CS. In addition, hobby [31] and occupation [28] sections were appended to define the personal information.

As seen in Table 2.1, user profile and content preference have sub-items. The personal information consists of sub-categories (S_i, i=1~12). And each sub-category consist of personal information items (P_i, i=1~288). User profile consists of items related to personal information, such as age, gender, occupation and hobby. Content preference contains items that can represent the user's favorite contents, such as drama, music and movies. In order to construct the personal information of each user, one personal information item is selected from each sub-category.

Table 2.2 shows an example of personal information items of two sub-categories of user profile (hobby) and contents preference (movie). Each sub-category contains several personal information items that describe the concept of each sub-category. To use the personal information items as an input of the fuzzy vault scheme, each

Table 2.1. User profile/Contents preference table

User information	Sub-category	Personal information items
User profile	Age (S_1)	$P_1 \sim P_6$
	Gender (S_2)	$P_7 \sim P_8$
	Marriage status (S_3)	$P_9 \sim P_{12}$
	Hobby (S_4)	$P_{13} \sim P_{35}$
	Occupation (S_5)	$P_{36} \sim P_{59}$
	Language (S_6)	$P_{60} \sim P_{198}$
Contents preference	Information (S_7)	$P_{199} \sim P_{211}$
	Drama (S_8)	$P_{212} \sim P_{225}$
	Entertainment (S_9)	$P_{226} \sim P_{250}$
	Music (S_{10})	$P_{251} \sim P_{260}$
	Enrichment (S_{11})	$P_{261} \sim P_{271}$
	Movie (S_{12})	$P_{272} \sim P_{288}$

Table 2.2. Example of personal information items for hobby (S_4) and movie (S_{12}).

Sub-category	P_i	Personal information item	Sub-category	P_i	Personal information item
Hobby (S_4)	P_{13}	Amateur science related	Movie (S_{12})	P_{272}	Eastern
	P_{14}	Animal-related		P_{273}	Effect movies
	P_{15}	Arts and crafts		P_{274}	Fiction
	P_{16}	Collecting		P_{275}	Science fiction
	P_{17}	Computer-related		P_{276}	Horror
	P_{18}	Cooking		P_{277}	Romance
	P_{19}	DIY (Do It Yourself)		P_{278}	Romance comedy
	P_{20}	Electronics		P_{279}	Thriller
	P_{21}	Film-making		P_{280}	War
	P_{22}	Games		P_{281}	Western
	P_{23}	Historical reenactment		P_{282}	Western comedy
	P_{24}	Interactive fiction		P_{283}	Detective
	P_{25}	Internet-based hobbies		P_{284}	Adventure
	P_{26}	Literature		P_{285}	Fantasy
	P_{27}	Model building		P_{286}	Comedy
	P_{28}	Music		P_{287}	Historical
	P_{29}	Observation		P_{288}	Adult

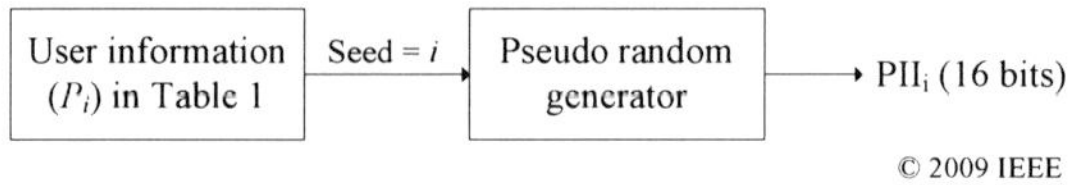

Fig. 2.2. Binarization of user information [26].

personal information item should be binarized. More specifically, in order for sub-items in Table 2.1 to be used as an input of the fuzzy vault scheme, we assigned a 16 bit-pseudo-random number to each personal information item as presented in Fig. 2.2.

We defined this value as personal information item (PII) value. N number of PII values corresponding to his profile and content preference at the encoder and decoder side is used as an input. Let the set of PII values at the encoder side be the $T=\{t_1, t_2, ..., t_N\}$, while at the decoder side be the $Q=\{q_1, q_2,..., q_N\}$. Note that the seed value used for the generation of pseudo-random numbers is determined by central authority. In the proposed content sharing scheme, the central authority only generates and distribute the personal information table. The central authority does not involve in any management regarding content sharing as well as secret key distribution which allows access to the contents.

2.3 Fuzzy Vault Scheme Using Personal Information

Fuzzy vault scheme is a polynomial reconstruction bases scheme [12]. It has been introduced to provide security using biometric data such as fingerprint and face features [14, 20, 32]. The basic concept of fuzzy vault is to bind both biometric data and secret key so that the vault set is generated at the side of encoder, and secret key can be reconstructed only if the generated vault set is decoded by the biometric data originated from the same identity. In our research, by using the personal information as a biometric feature, we utilized the fuzzy vault scheme to provide secure match of personal information among anonymous users for the purpose of contents sharing. Since the access to the contents is restricted by a secret key and given to users according to their personal information, users with similar personal information can be defined as a virtual group. This space is opened to the only users with similar personal information. Thus, the virtual space can be secured by grouping the contents according to personal information. In this Section, detailed procedures for generating the vault set at the side of encoder and reconstructing the secret key to access contents at the side of decoder are presented.

2.3.1 Binding of Personal Information and Secret Key

Fig. 2.3 represents the fuzzy vault encoder using personal information item, PII. As seen in the figure, the fuzzy vault encoder generates a vault set with the list of PII and secret key as an input. User's PIIs are generated through the procedure of Fig. 2.2 with N number of items P_i in Table 2.1. The set of user's PII is $T=\{t_1, t_2, ..., t_N\}$ which is identical to template personal information item list, T in Fig. 2.3. Note that 128-bit Advanced Encryption Standard (AES) [21] key is used as a secret key to protect the contents. At first, 16-bit redundancy is concatenated to 128-bit AES key so that total of 144-bit data SC is constructed through Cyclic Redundancy Check (CRC) [23]. For the generation of cyclic redundancy, CRC-16 scheme which has 16-bit primitive polynomial is utilized. The polynomial used is as follow:

$$\mathrm{P}_{\mathrm{CRC}}(x) = x^{16} + x^{15} + x^{2} + 1. \tag{2.1}$$

In order to construct polynomials of (2.2), *SC* is divided into non-overlapping 16-bit unit in order to be used for generating coefficients ($C_8 \sim C_0$) of polynomial. Note that every operation after the construction of polynomial is under the Galois field ($GF(2^{16})$).

$$F(u) = c_8 u^8 + c_7 u^7 + \cdots + c_1 u + c_0. \tag{2.2}$$

A vault set, which is the encoding result, consists of genuine set *G* and chaff set *C*. The elements of genuine set is a pair of values, which is a template personal information item list $T=\{t_1, t_2, \ldots, t_N\}$ and its projected value to polynomial $F(u)$ of equation (2.2). Genuine set *G* is expressed as equation (2.3).

$$G = \{(t_1, F(t_1)), (t_2, F(t_2)), \ldots, (t_N, F(t_N))\}. \tag{2.3}$$

And chaff set *C* is defined as follows:

$$C = \{(u'_1, r_1), (u'_2, r_2), \ldots, (u'_M, r_M)\}. \tag{2.4}$$

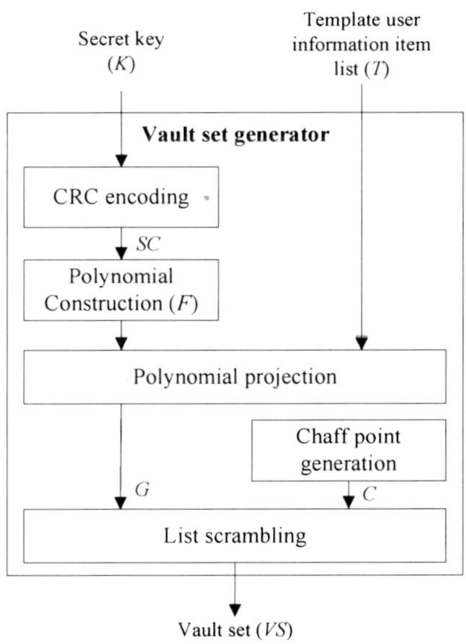

Fig. 2.3. Vault set generation using personal information [25].

The purpose of utilizing the chaff set C is to securely protect the genuine set by maxing with genuine set G. Note that random points are used as chaff points in original scheme. However, in our case, if we use the same method, an adversary can easily separate chaff points and genuine points using personal information table, as in Table 2.1. Therefore, we defined chaff set C as a set that is composed of M number of PII value ($u'_1, u'_2,\ldots,u'_M$) from PII values in the personal information table ($u'_i \neq t_j$, for $1 \leq i \leq M,\ 1 \leq j \leq N$), which are not used as personal information. For r_1~r_M, the values satisfying equation (2.5) are chosen. Then, even if the adversary (malicious attacker) knows the personal information table, he/she cannot distinguish between genuine set and chaff set.

$$F(u_i^{'}) \neq r_i \quad 1 \leq i \leq M. \tag{2.5}$$

Finally, the vault set VS is generated by scrambling the genuine set G and chaff set C.

$$VS = \{(x_1, y_1), (x_2, y_2), \ldots, (x_{N+M}, y_{N+M})\}. \tag{2.6}$$

2.3.2 Secret Key Reconstruction Based on Personal Information

The fuzzy vault decoder uses query PIIs as an input to decode the vault set. If a query user wants to successfully reconstruct the secret key and access the desired contents to consume, more than D+1 number of PIIs should be the same as that of encoding side. By decoding the vault set, the original secret key can be acquired which guarantees the successful decryption of the content. Here, D denotes the degree of polynomial in (2.2). Query PIIs are generated through the procedure of Fig. 2.2 with N number of PII from items (P_i). The set of user's PII is Q={$q_1, q_2,\ldots,q_N$}, which is identical to the query personal information item list Q in Fig. 2.4.

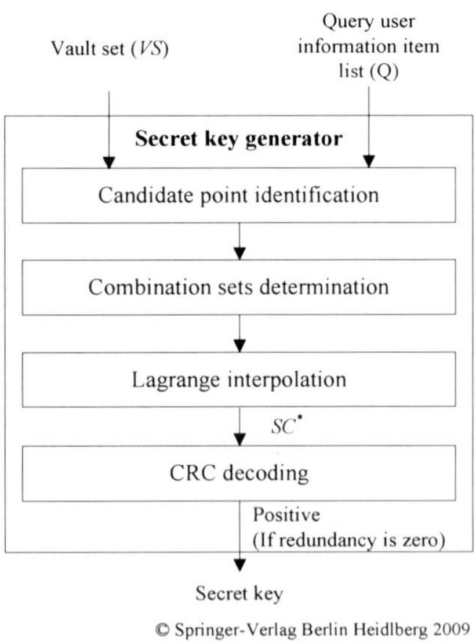

Fig. 2.4. Secret key reconstruction from the vault set and personal information of user [25].

In 'Candidate point identification' module, the element of *VS*, (x_j, y_j) which satisfies the equation (2.7), is selected as a candidate point after comparing query PII values and elements of the vault set.

$$q_i = x_j,\ 1 \le i \le N,\ 1 \le j \le N + M. \tag{2.7}$$

Then, if the number of candidate points is *k*, the next module 'Combination sets determination' generates every possible set that can select *D*+1 number of points from *k* number of candidate points, *i.e.*, *C*(*k*, *D*+1). Let each candidate set be the $L=\{(a_0, b_0), (a_1, b_1),\ldots,(a_D, b_D)\}$, then the polynomial is reconstructed using equation (8) for *C*(*k*, *D*+1) cases in 'Lagrange interpolation' [8] module.

$$F^*(u) = \sum_{j=0}^{D} b_j f_j(u) = c_8 u^8 + c_7 u^7 + \cdots + c_1 u + c_0,$$
$$\text{where } f_j(u) = \prod_{i=0, i\ne j}^{D} \frac{u - a_i}{a_j - a_i}. \tag{2.8}$$

In order to construct *SC** of 144 bits data, the coefficients $C_8 \sim C_0$ obtained by equation (2.8) are concatenated. Then, each *SC** is checked through CRC decoder. And if the redundancy is not zero, the result of CRC decoding is negative. Thus, it is impossible to acquire the proper secret key. On the other hand, if the redundancy is zero, the result of CRC decoder is positive and the proper secret key can be obtained. Consequently, in the case that the result is positive, only 128 bits of the *SC**, excluding LSB 16 bits, are used for decrypting the content.

3 Application of Virtual Secure Space in Home Network Environment

In this Section, we extend the concept of the personal information based content sharing scheme to home network system and present a detailed explanation as well as several simulation results. A home network is a residential local area network that is used to connect several devices inside or even outside the home. Within a home, multiple devices are connected to a local gate way known as the home server [15].

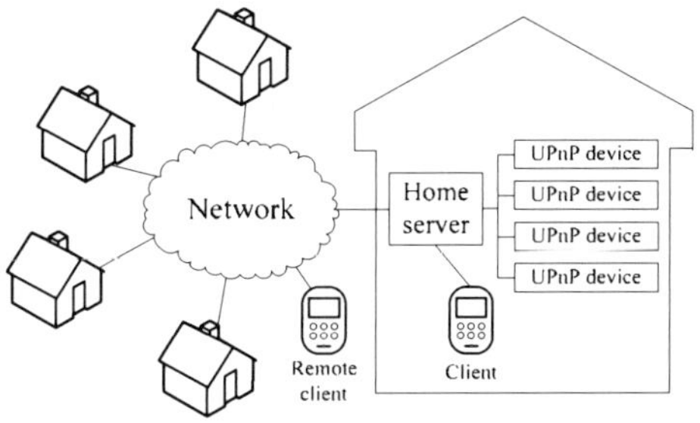

Fig. 3.1. Simplified illustration of a home network system.

Fig. 3.1 shows a simplified illustration of a home network system. In the home network system, the home server functions as a communication gateway between the inside and the outside of a home. The types of connections to the home server are classified into local and remote connections based on the connection spot. The local connection to home is established using UPnP [29] audio/video (AV) control point (CP). Control Point and UPnP devices are compared to the client and server. The CP controls the UPnP devices to transmit contents and find UPnP devices. In other words, the CP refers devices or applications that are used by particular client user. The remote connection to the home can be feasible using virtual private network (VPN) [3]. VPN guarantees the secure tunnel between a visitor and home server. In the context of our work, the home server acts as an authentication server, while content consumer's device acts as a client.

The personal information based content sharing scheme can be utilized as a method for sharing contents among anonymous home networks [26]. The increasing availability of high speed network access and the tendency of decreasing cost of personal storage triggers the need for content sharing among different home networks [24] as well as within a home [13, 30]. For the content sharing between anonymous users in different home networks, there is no relationship or acquaintanceship between the content consumers and content creators. Therefore, the content creators would like to give a restriction to the access of contents in their home network and share his/her contents with only the group of people who have similar personal preferences about the contents. In such an environment, the proposed personal information based content sharing system can be utilized as a solution for guaranteeing the security of both content creators' contents and their personal information.

3.1 Personal Information Based Content Sharing among Anonymous Home Network

Content sharing is getting popular in home network as well as in social media. The increasing availability of high speed network access and the tendency of decreasing cost of personal storage, triggers the need for content sharing among different home networks as well as within a home [13, 30]. Content sharing within these homes is usually done among family members, while that between different home networks is done by friends and even anonymous users. Especially, anonymous users could access contents in home networks which could act like home portals. For the acquaintances, the full right to access contents could be granted so that they can freely consume the contents inside or outside the home. However, for the case of anonymous users, since there is no relationship or acquaintanceship between the users, i.e., content consumers, and content creators, the content creators will be concerned about delivery to unwanted groups of people and the misusage of contents. Therefore, the content creator would like to share his/her contents with only group of people who have a similarity such as user profile or content preference. However, it is not easy to securely distribute the key to preferred group of anonymous users.

In the proposed system, contents are encrypted with a secret key in order to restrict the access of unpreferred group of people. The secret key is granted to trusted group of people, such as family and friends, for free contents consumption while the visitors can be granted the secret key based on comparison of personal information. Note that the visitors represent the anonymous users who want to access and consume the content. Fig. 3.2 represents visitor's remote connection to home server. This figure shows how the visitor accesses from outside the home. Content creator (home server side) is a user who wants to open the content to the group of people who have similar personal information. The creator encodes a vault set with personal information of preferred visitor (PI_p) and secret key (K). The vault set (VS) is generated by binding the personal information and secret key. The contents are encrypted by the secret key and opened to the public along with the vault set. C^e in the figure represents encrypted content.

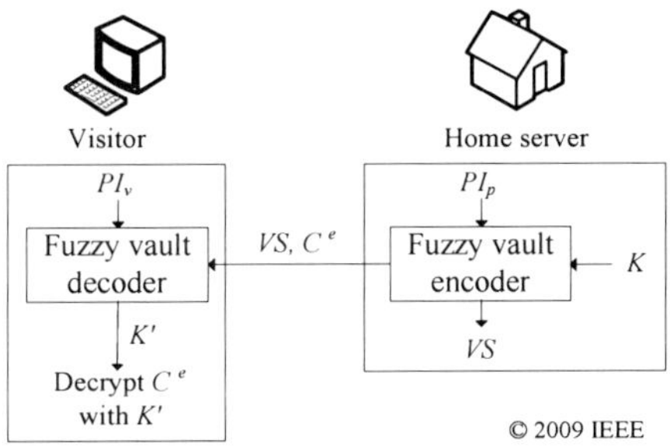

Fig. 3.2. Content consumption process with vault set in home network [26].

A visitor is a user who wants to consume the content of home server. In order to consume the content, the visitor downloads vault set (VS) and decodes it with his/her personal information (PI_v). The decoding result of vault set is secret key (K'). Therefore, if the visitor's personal information is overlapped as much as the content creator wants, he/she can acquire identical secret key (K') with original secret key (K). Then, visitor can decrypt the encrypted content and consume it.

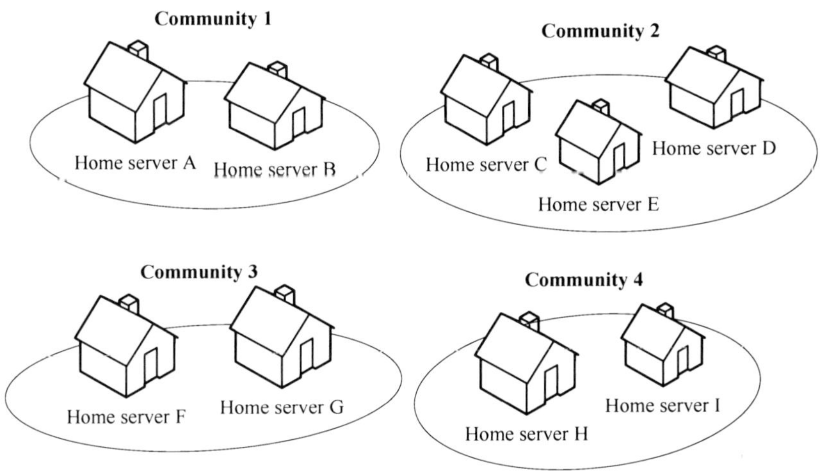

Fig. 3.3. Communities generated by personal information [26].

Since the access to the encrypted contents are given to visitors according to their personal information (user profile and content preference), home servers with similar personal information can be defined as a group, i.e. community. This space, generated by the vault set, is only opened to visitors with similar personal information. Thus, the community can be secured by grouping the contents according to personal information. Grouped communities according to similar preferred personal information is depicted in Fig. 3.3. From the perspective of home server, this could provide limited accessibility to the unpreferred visitors.

3.2 Simulation Results

Simulations were performed to verify the proposed content sharing between home networks by using personal information. In the simulation, the consumption process for protected contents with vault set developed by personal information was evaluated. Through content sharing using personal information, we demonstrated a content sharing community among home servers in which protected contents could be shared. Fig. 3.4 shows the detailed scenario of the simulation.

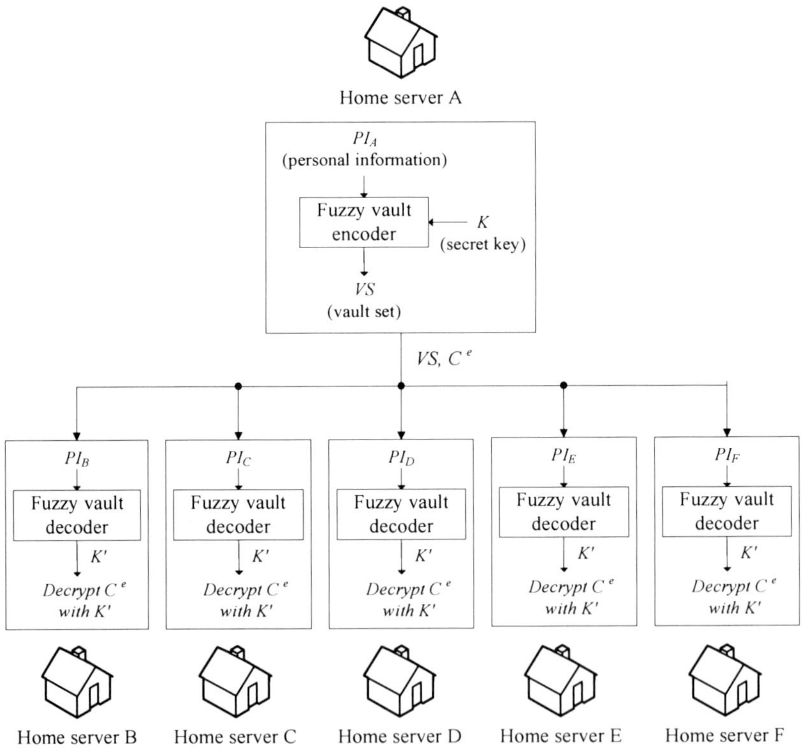

Fig. 3.4. Simulation scenario. Notes Home server A produces encrypted content C^e and associated vault set by using personal information of home server A user [26].

We established six home servers. Six home server users were named as A, B, C, D, E, and F, respectively. Users of home server B, C, D, E, and F accessed the encrypted content of home server A. The home server user A generated encrypted contents with the vault set and opened them to the public so that the rest of home server users could consume them. To simplify the simulation, we assume that each home server user has 12 PII values. If nine PII values are identical with each other, i.e. PII values used to encode a vault set and query user's PII values are identical, the encrypted content can be consumed. To realize this, the number of PII value is 12 in the fuzzy vault encoder, the degree of encoder polynomial is eight and the number of chaff point is 276. Since the degree of the polynomial is eight, 9 out of 12 of user's PII values should be identical to that of home server A in order to decode the vault set.

Detailed personal information of six home server users used in the simulation is presented in Table 3.1. As seen in Table 3.1, user B has 7, user C has 8, user D has 9, user E has 10, and user F has 11 identical personal information with home server user A. Note that home server A has encrypted contents and vault set generated by the

Table 3.1. Personal information of user A, B, C, D, E, and F.

User information	Sub-category	Home server A	Home server B	Home server C	Home server D	Home server E	Home server F
User profile	Age	Young Adult	Teenager	Senior citizen	Young adult	Senior citizen	Young adult
	Gender	Male	Female	Male	Female	Female	Female
	Marriage status	Single	Single	Single	Single	Single	Single
	Hobbies	Photography	Film-making	Collecting	Photography	Photography	Photography
	Occupation	Computer and Mathematical	Computer and Mathematical	Computer and Mathematical	Legal	Computer and Mathematical	Computer and Mathematical
	Language	English	English	English	English	English	English
Contents preference	Information	Social/Political	Leisure	Sport events	Social/Political	Social/Political	Social/Political
	Drama	Docudrama	Docudrama	Docudrama	Docudrama	Docudrama	Docudrama
	Entertain-ment	Quiz/Contest	Panel-show	Travel variety	Quiz/Contest	Quiz/Contest	Quiz/Contest
	Music	Jazz	Jazz	Jazz	Jazz	Jazz	Jazz
	Enrichment	Language studies	Language studies	Language studies	Language studies	Language studies	Language studies
	Movies	Effect movies	Effect movies	Effect movies	Horror	Effect movies	Effect movies

Table 3.2. Decoding results [25].

D	Consumer	Identical PII to user A	Candidate set	Positive set	Negative set	Authentication
	user B	7	792	1	791	Yes
	user C	8	792	8	784	Yes
6	user D	9	792	36	756	Yes
	user E	10	792	120	672	Yes
	user F	11	792	330	462	Yes
	user B	7	495	0	495	No
	user C	8	495	1	494	Yes
7	user D	9	495	9	486	Yes
	user E	10	495	45	450	Yes
	user F	11	495	165	330	Yes
	user B	7	220	0	220	No
	user C	8	220	0	220	No
8	user D	9	220	1	219	Yes
	user E	10	220	10	210	Yes
	user F	11	220	55	165	Yes
	user B	7	66	0	66	No
	user C	8	66	0	66	No
9	user D	9	66	0	66	No
	user E	10	66	1	65	Yes
	user F	11	66	11	55	Yes
	user B	7	12	0	12	No
	user C	8	12	0	12	No
10	user D	9	12	0	12	No
	user E	10	12	0	12	No
	user F	11	12	1	11	Yes

personal information of home server A user. We want to observe whether home server users B, C, D, E, and F who have different personal information, could decrypt and consume the encrypted content of home server A.

Table 3.2 shows the decoding result of users with a varying degree of the polynomial (D). In the case when D is eight, since the number of elements in the vault set is the same as the number of sub-items (P_i, $1 \leq i \leq 288$) in Table 2.1, PII values which are different from user A's are always matched to the chaff points. And since it was assumed that the number of all user's PII used to generate a vault set is 12, the number of candidate set generated at the decoder size is always 220 ($C(12, 9)$) which is a combination of selecting 9 from 12. For user B and C, since the number of PII values identical to user A is less than 9, it is impossible to reconstruct a polynomial. Thus, a proper secret key cannot be acquired. For user D, 9 PII values are identical to user A. Thus, only one set ($C(9, 9)$) out of 220 is positive. And for user E, since ten PII values are identical to user A, ten sets ($C(10, 9)$) are positive. Finally, for user F, since eleven PII values are identical to user A, fifty-five sets ($C(11, 9)$) are positive. Therefore, user D, E, and F decrypted content of user A with secret key which is a 128-bit from MSB of 144-bit SC^* in Fig 2.4.

Note that the positive represents the number of sets when the redundancy is zero after CRC decoding at the decoder side in Fig. 2.4. Likewise, the negative represents the number of sets when the redundancy is not zero. If the CRC decoding result is positive, the probability of a set containing the proper secret key is very high. In the case of the result being negative, the set has no secret key. The decoding process is completed as soon as the secret key is found, and encrypted content is decrypted by the key.

This proves and validates that the virtual secure space generated by personal information of user A is opened to user D, E and F, but not to user B and C. Since user A, D, E and F have more than eight identical personal information items, the encrypted content and the vault set generated by their own personal information create a secure space where user B and C have no accessibility as shown in Fig. 3.5. In this virtual space, only user A, D, E and F can upload their contents freely and share them securely.

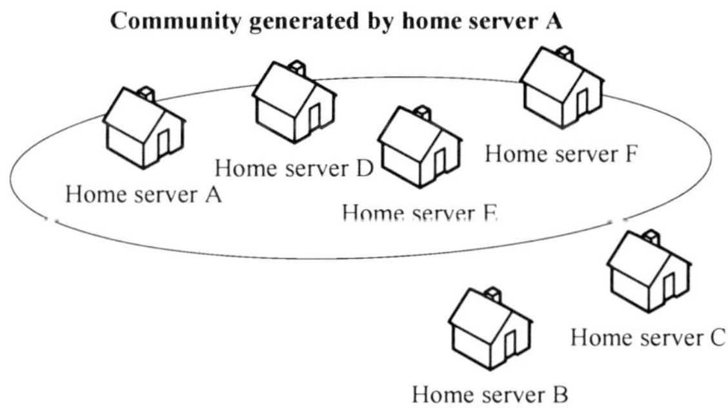

Fig. 3.5. A content sharing community generated by personal information of home server A [26].

3.3 Security Issue

In this Section, we consider the possibility of personal information leakage. For security considerations, let us assume that the adversary wants to break the virtual secure space using brute force attack simulated by our simulation environment. That is to iterate over all combinations of attributes and try to access the virtual space by randomly selecting 9 PII values from 12 sub-categories where only one PII value is selected for each sub-category.

For a given personal information table, the adversary can break the virtual secure space by evaluating all combination sets ($8.80\text{x}10^{12}$ sets) at maximum which takes about $7.25\text{x}10^4$ years when the degree of polynomial is 8 (Note that the evaluation time for one set is about 0.26 sec at 3.4GHz CPU). The required number of evaluations to break the virtual secure space according to degree of the polynomial is

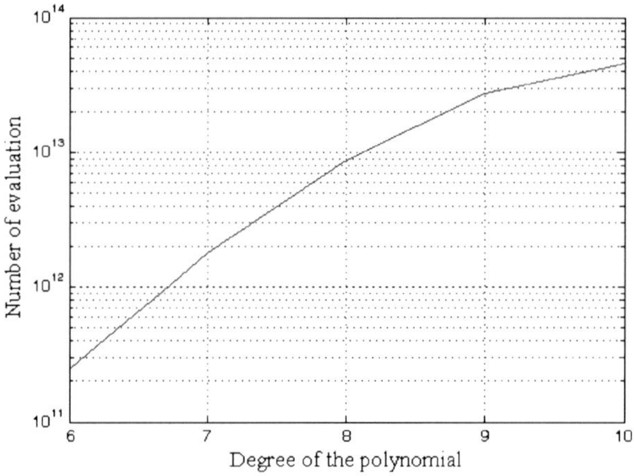

Fig. 3.6. Evaluations required for an adversary to brute force attack.

presented in Fig. 3.6. In this figure, the horizontal axis is the degree of the polynomial and the vertical axis indicates the number of evaluation to break virtual secure space. The graph shows that as the degree increases, the security against brute force attack also increases. Moreover, since the security level is dependent on the size of the personal information table, the security can be enhanced by increasing the number of sub-category or corresponding items.

4 Conclusions

Most of the well-known open spaces, such as social network sites, allow for unlimited sharing of contents among all users. In such environments, content creators do not have any options to restrict access to consumers. Therefore, content sharing among anonymous users in the open space has potential risk since contents could be delivered to unwanted users or misused sometimes. To solve this problem, we presented the content sharing method based on the personal information of users. The personal information based content sharing method is to restrict the access to content by using a secret key and uploading it to open space with a vault set that is generated by personal information of users. Since only consumers with personal information similar to that of the content creator can acquire the proper secret key, content creator can limit the consumption of contents over groups of consumers by his/her intention. The first priority to take into consideration in personal information based content sharing is the risk of personal information leakage. During the process of comparing personal information among anonymous users, no information is revealed to the public by making use of the fuzzy vault scheme. With the proposed method, content creators can not only protect their content, but build more trust in the consumers of the content.

References

[1] Ames, M., Naaman, M.: Why we tag: motivations for annotation in mobile and online media. In: The Proceedings of the SIGCHI conference on Human factors in computing systems, pp. 971–980 (2007)

[2] Bellare, M., Namprempre, C.: Authenticated encryption: Relations among notions and analysis of the generic composition paradigm. In: Okamoto, T. (ed.) ASIACRYPT 2000. LNCS, vol. 1976, p. 531. Springer, Heidelberg (2000)

[3] Cisco, How virtual private networks work, `http://www.cisco.com/en/US/tech/tk583/tk372/technologies_tech_note09186a0080094865.shtml`

[4] Dwyer, C., Hiltz, S.R.: Trust and privacy concern within social networking sites: A comparison of facebook and myspace. In: Proceedings of the Thirteenth Americas Conference on Information Systems, Keystone, Colorado (2007)

[5] Facebook, `http://www.facebook.com/`

[6] Flickr, `http://www.flickr.com/`

[7] Gross, R., Acquisti, A.: Information revelation and privacy in online social networks. In: Proceeding of ACM Workshop on Privacy in the Electronic Society (WPES), pp. 71–80 (2005)

[8] Hildebrand, F.B.: Introduction to numerical analysis, New York (1974)

[9] ISO/IEC JTC1/SC29/WG11, Information technology - multimedia content description interface - Part 5: Multimedia description schemes. FDIS (N4242) (2001)

[10] Jain, A.K., Dass, S.C., Nandakumar, K.: Soft biometric traits for personal recognition systems. In: Zhang, D., Jain, A.K. (eds.) ICBA 2004. LNCS, vol. 3072, pp. 731–738. Springer, Heidelberg (2004)

[11] Jin, A.T.B., Ling, D.N.C., Goh, A.: BioHashing: two factor authentication featuring fingerprint data and tokenised random number. Pattern Recognition 37, 2245–2255 (2004)

[12] Juels, A., Sudan, M.: A fuzzy vault scheme. In: IEEE International Symposium on Information Theory, p. 408 (2002)

[13] Lee, H.Y., Kim, J.W.: An approach for content sharing among UPnP devices in different home networks. IEEE Transactions on Consumer and Electronics 53(4), 1419–1426 (2007)

[14] Lee, Y.J., Bae, K., Lee, S.J., Park, K.R., Kim, J.: Biometric key binding: fuzzy vault based on iris images. In: Lee, S.-W., Li, S.Z. (eds.) ICB 2007. LNCS, vol. 4642, pp. 800–808. Springer, Heidelberg (2007)

[15] Lee, Y., Ju, H., Park, J., Han, J.: User authentication mechanism using authentication server in home network. In: International Conference of Advanced Communication Technology (ICACT), vol. 1, pp. 503–506 (2006)

[16] Loia, V., Pedrycz, W., Senatore, S.: Semantic web content analysis: a study in proximity-based collaborative clustering. IEEE Transactions on Fuzzy Systems 15(6), 1294–1312 (2007)

[17] Manjunath, B.S., Salembier, P., Sikora, T.: Introduction to MPEG-7: multimedia content description interface. John Wiley & Sons, Chichester (2002)

[18] Mannan, M., Oorschot, P.C.V.: Privacy-enhanced sharing of personal content on the web. In: Proceeding of the 17th international conference on World Wide Web (WWW), pp. 487–496 (2008)

[19] Myspace, `http://www.myspace.com/`

[20] Nandakumar, K., Jain, A.K., Pankanti, S.: Fingerprint-based fuzzy vault: implementation and performance. IEEE Transactions on Information Forensics and Security 2(4), 744–757 (2007)
[21] NIST. Advanced Encryption Standard (AES), http://csrc.nist.gov/publications/fips/fips197/fips-197.pdf
[22] OECD study on the participative web: user generated content, http://www.oecd.org
[23] Press, W.H., Teukolsky, S.A., Vetterling, W.T., Flannery, B.P.: Numerical recipes in C. Cambridge University Press, Cambridge (1992)
[24] Rose, B.: Home networks: a standards perspective. IEEE Communications Magazine 39(12), 78–85 (2001)
[25] Sohn, H., Ro, Y.M., Plataniotis, K.N.: Content sharing based on personal information in virtually secured space. In: International Workshop on Digital Watermarking (IWDW), pp. 401–413 (2008)
[26] Sohn, H., Ro, Y.M., Plataniotis, K.N.: Content sharing between home networks by using personal information and associated fuzzy vault scheme. IEEE Transactions on Consumer and Electronics 55(2), 431–437 (2009)
[27] Teoh, A.B.J., Yuang, C.T.: Cancelable biometrics realization with multispace random projections. IEEE Transactions on Systems, Man, and Cybernetics-part B: Cybernetics 37(5), 1096–1106 (2007)
[28] U.S. Department of Labor, Bureau of Labor Statistics. Standard Occupational Classification (SOC) Major Groups, http://www.bls.gov/soc/soc_majo.htm
[29] Universal Plug and Play, http://www.upnp.org/
[30] Walker, J., Morris, O.J., Marusic, B.: Share It! - the architecture of a rights-manages network of peer-to-peer set-top-boxes. In: The Proceedings of EUROCON (2003)
[31] Wikipedia, List of hobbies, http://en.wikipedia.org/wiki/List_of_hobbies
[32] Yong, W., Plataniotis, K.N.: Fuzzy vault for face based cryptographic key generation biometrics symposium, pp. 1–6 (2007)
[33] Youtube, http://www.youtube.com

Image-Based Sensitive Information Filtering for Teleradiology

Ahmet Ekin

Philips Research Europe, Video Processing and Analysis Group,
Eindhoven, The Netherlands
ahmet.ekin@philips.com

Teleradiology is the transmission of radiological information from one place to another for interpretation and/or consultation. The content of the transmitted data, the medium of transmission, and the manner of interpretation significantly affect the application and call for efficient automated solutions. Before transmission, the medical data must be processed to remove the content features that may lead to the identification of the patient. Depending on the teleradiology application and the suspected pathology, instead of the raw data, only computer-extracted features or only some critical part of the data may be shared; hence, some automated processing may be needed at the transmitting medical site. The transmission method has to ensure the security of the data; but, that is not sufficient and it should be complemented with efficient coding for fast transmission of large amounts of data because more experts are beginning to access the data with handheld devices over lower-bandwidth channels. At the receiver site, the radiologists may have different types of access to the data depending on the predetermined policies between the clinical sites. For example, only some computer-extracted features can be shown to a resident whereas only the expert radiologist may be able to read the raw data. To deal with these and other issues in teleradiology, this chapter will start with a brief overview of the literature and then will introduce fully automatic content analysis methods. Specifically, we will provide solutions in the field of neuroradiology by using MR (magnetic resonance) brain images.

1 Introduction

Teleradiology is the electrical transmission of radiological information, such as X-rays, MR, and computed tomography, from one place to another for interpretation and/or consultation [1]. It has first gained increased usage in the hospitals because of the need to interpret emergency radiological images at nights when there are insufficient number of on-site radiologists [2, 3]. To this end, the medical data have been transferred to a remote site in another time zone where it is daytime and the radiologists there have provided their critical interpretation. The favorable cost of such services have offered additional motives for their use and, as a result, more than 300 hospitals in the US are currently using some

H.T. Sencar et al. (Eds.): Intel. Multimedia Analysis for Security Appli., SCI 282, pp. 209–224.
springerlink.com

form of teleradiology [4]. The success of night-time application has initiated discussions on the daytime and non-emergency use of such services. This extension has a potential to modify the traditional radiology workflow to the degree of, some argue, commoditizing the radiology services [5, 6]; but, on the positive side, this may help shifting the decision power to the patient from the medical specialist [2].

The digitization of the medical data, increased usage of PACS (Picture Archiving and Communications Systems), reduced cost of bandwidth, data storage and transfer, larger volumes of data awaiting radiologists' reading, and the need to have timely and cost-efficient expert opinion even in the remote areas of the world have all contributed to the above-outlined growing interest in teleradiology. Such demand has also intensified both legal and algorithmic discussions on the safety, privacy, and the quality of the transmitted medical data. In addition, different flavors of teleradiology applications, from accessing the data with handheld devices to transferring it to a far-away country, have required efficient data description and coding not only for fast transmission but also for better diagnosis. In this chapter, we will first provide an overview of the field of teleradiology, and then propose algorithms in the domain of brain MR images to alleviate some of these concerns. Specifically, we first discuss automated algorithms for their use in preserving privacy and limiting access to the medical data. Then, we will propose a reference(template)-based description and coding of the data for efficient and secure transmission to the receiver side. Finally, we will propose a data organization framework to set access conditions to the medical data.

2 Teleradiology

Teleradiology is one of the most advanced and widely reported application areas of telemedicine [7]. In 2003, 67% of all radiology practices in the US having 78% of all radiologists reported using teleradiology for (at least one of) the purposes of sending data to the radiologists at home, to a different facility, to a different location than the reading room in the same facility, and to a radiologist in another state or country [3].

Uses, Benefits and Drawbacks, and Impact

Teleradiology brings about beneficial and sometimes controversial changes to how the medical practice is conducted [6, 8, 9]. Foremost, it makes the expert opinion readily available even in remote areas of the world. This ability to access the radiological data immediately from a distance improves the diagnosis and may even be life-critical, such as image interpretation for stroke [10]. This can also be in the form of supporting students completing their studies on remote sites [11] or making it possible for the less experienced specialists to consult with an expert on the diagnosis of difficult cases [12]. The possibility of having the expert opinion from the local site is also immediately noticeable from the patients' side as their travel times are diminished or completely eliminated. It is reported that such services can even improve the patient-physician relationship at the local site [13]. Furthermore, the use of teleradiology has been leading to a more affordable radiological interpretation; thereby, reducing the cost of clinical care imposed on insurance companies and governments.

More often than any other use, teleradiology has been utilized in the outsourcing of radiologists' out-of-hours emergency imaging to off-site teleradiology companies [5]. Various factors, such as increasing need for emergency imaging, the paucity of the radiologists in the facilities at night, and the decrease in the cost of bandwidth for digital transmission, have contributed to this type of teleradiology application. Additional benefits include fast and easy transfer of data between centers because of an established teleradiology network. This way, the radiologist may have access to the past radiological images of the patient and can make a diagnosis with the aid of prior scans [8]. Another advantage of teleradiology is that it relieves the burden on the radiology departments by reducing the waiting times for the imaging procedures because imaging does not have to be done where the expert radiologist is located.

Despite the above-outlined benefits, teleradiology also draws some criticism. It is argued that teleradiology may result in distancing the clinical radiologist from the patient impacting upon the direct clinical care [14]. The effect of this is difficult to quantify. Another aspect to the teleradiology is that whether or how often a radiologist who is making a diagnosis from a distance will miss cases compared to the local environment [8]. This possible decision quality discrepancy may stem from the differences in the quality of the original and transmitted data [8], the distance of the radiologist from the patient and the other metadata information, and the differences in the training and the expertise of the off-site radiologist from those on-site [5]. There are solutions to some of these concerns. The compression of the data for transmission may be needed for some cases, such as accessing the data over lower bandwidth channels; but, the associated consequences must be considered thoroughly. For some clinical cases, accessing to metadata may be needed and, when it is not supported, the teleradiology result should not be given the highest weight. The training of the teleradiologist may be inferior to the one on-site that can result in worsening of the diagnostic decisions [5]; however, with certain conditions on credentials and licenses, some of these concerns can be alleviated.

Teleradiology results in a power shift from the medical specialist to the patient. Microsoft's HealthVault [15] and Google Health [16] are personally controlled products of health records [17]. Although the use of teleradiology may have some risks, such as privacy and security, the non-use of it may prevent the patient from receiving the beneficial expert opinion. In such cases, the courts may consider the failure to use such services as evidence of negligence on the non-referring physician's part [8]. Teleradiology may also change the way radiologists work, and this change may not always be deemed favorable [18]. Distance consultation by teleradiology will allow referring doctors to demand that a specialist radiologist should report on, for example, brain MRI, and will force radiologists to move more rapidly into providing subspecialist radiological opinions for all clinical specialists.

Examples and Further User Needs

Teleradiology finds uses in the management of several diseases in various locations in the world. In Ireland, which is characterized by high rural population, a teleradiology service between rural town community hospital X-ray centers and radiologists in regional hospitals has been in use since 2001 [13]. 1500-1800 X-ray images are transferred annually within this system. A similar example has been reported for the US [19]. As a specialty in teleradiology, telestroke within the overall acute stroke management can provide major improvements in the on-site stroke therapy [10]. One such telestroke system, REACH (remote evaluation of acute ischemic stroke), is being use to guide stroke care in eight rural community hospitals in the state of Georgia in the US [20].

In a study, Kern [21] investigated the criteria used for evaluation of the teleconsultation systems. He identified hardware and technical aspects as the leading criteria whereas satisfaction of both healthcare provider and patient have shown the higgest increase in the studied time intervals. Being aware of the differentiating feature of technical aspects, hardware and software companies are building teleradiology tools. For example, the distant data viewing systems of several companies, such as Philips, GE, Siemens, Agfa, and Kodak, are being used in the domain of teleradiology [22]. The field boasts a number of key commercial players including Agfa-Gevaert Group, BRIT Systems, Canon Medical Systems, DR Systems, Carestream Health Inc., FUJIFILM Medical Systems, General Electric Medical Systems, Hitachi Medical Systems America, IBM, IMCO Technologies, Intelerad Medical Systems Inc, McKesson's Medical Imaging Group, Philips Healthcare, and Siemens Medical Solutions [7].

Teleradiology can be improved by augmenting it with different flavors. In [23], a web-based 3D teleradiology has been made possible by web-based medical data visualization tools. An Internet-based system is presented in [24]. Poon et al. [25] have found out that teleradiology with video consultation where neurological examination of the patient can also be carried out simultaneously brought superior performance than regular teleradiology in the referral of neurosurgical patients from a large district general hospital. Muench et al. [26] have proposed to integrate the medical images with the electronic patient records (EPRs) for an improved web-based teleradiology application. In [27], multimedia messaging service is used to get opinions from senior neurosurgical staff in the handling of neurosurgical emergencies. For wireless teleradiology, a user study among the Spanish and German physicians found out that neurosurgical emergency consultation is considered as the most needed teleradiology application [28]. A mobile teleradiology application for evaluation of neurosurgical CT images has been used in [29]. Stahl et al. [30] provide a customized solution for visualization of dynamic images of cardiac angiography and intravascular ultrasound for clinical conferencing.

Open Issues

In teleradiology applications, regulation has been an issue. Surrogate interpretation of teleradiology images is possible and refers to the interpretation by unlicensed individuals under the supervision of someone with a license. Then, a licensed radiologist could theoretically sign off those reports seen by others [17]. The regulations usually differ from one region to another. For example, 19 of the US states require doctors to have special license to practise telemedicine whereas the EU focuses on data protection, patients' confidentiality, and consent [17]. In the US, the American College of Radiologist (ACR) has been issuing guidance on the use and practice of teleradiology since 1994 [14]. In addition to the licensing-related regulations, the ACR also recommends that the interpreting radiologists should be responsible for the quality of images being interpreted.

Tampering with a patient's image and the medical record may be a concern. In film radiography, a patient's image and information is directly overlaid on the image exposing the identity of the person. In teleradiology, the reporting radiologist may be located in a different country. Despite this, the regulations in the transferring region applies and the reporting site should be able to conform to different regulations.

3 Content Processing of Brain MR to Limit Shared Information

In this section, we explain a number of methods that can be applied to process the brain MR images before sharing the data for teleradiology application. These methods can operate in two ways: 1) They directly eliminate the sensitive information from the raw data and generate filtered versions of it. 2) They can extract features from the raw medical data, such as activation maps in fMRI (functional MRI), and then, only those features can be shared with the receiving side. This type of application makes use of standardized brain templates, explained in Section 4, to visualize the features on the receiver side.

Filtering the Raw Data

In this section, we consider processing of medical data to filter some part of the intensity information. In the end, the shared information with the receiver side will be intensity-based; however, some privacy-related or clinically-irrelevant information will be removed.

Privacy-related information may be in the form of overlaid text on the content or a part of the natural content. For example, in some X-ray images, the name of the patient and the scan date may be available as embedded on the image. Before using this data for the teleradiology application, if desired, such data may be removed by running a text detection algorithm and removing text characters [31]. In the brain MR images, the image content contains information about brain tissue and the skull. In many cases, brain tissue characteristics are essential for

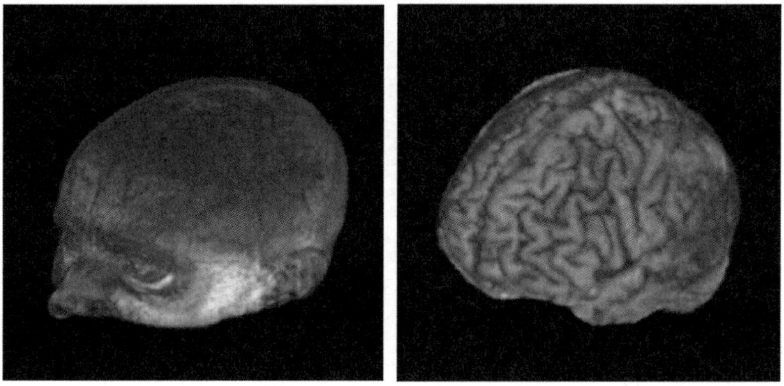

Fig. 1. A brain MR volume with skull (left) shows facial features whereas skull-stripped volume (right) only shows the brain tissue.

a diagnosis, whereas skull information, referring to muscles, head, nose, eyes etc., may not have much diagnostic value. Although it may seem harmless to have this information embedded in the data, the content shown in 3D in Figure 1 on the left makes some facial features visible that may cause some disturbance. Removing the skull information retains only the brain tissue information that has no identifying facial features.

There are several skull-stripping algorithms in the literature, such as publicly available Brain Extraction Tool [32]. Most of these algorithms show the best performance on the T1 contrast of MR. Even on such high quality data, there may be holes in the determined mask and the accuracy on the borders may not be satisfactory. The lower tissue class intensity differences in other MR contrasts, such as T2 and PD, are often problematic resulting in unacceptable skull maps. In contrast to intensity-based approaches, our approach tries to solve the problem by using more-robust registration methods. In this approach, a reference brain volume and its skull mask are stored in a database. To detect skull mask of the input, this reference atlas is registered rigidly onto each new MR volume (or vice versa). The obtained transformation map is applied on the skull mask of the reference volume to find the skull mask of the new volume. Figure 2 shows the flowchart of the proposed method where the input is registered onto the reference. We have tested this method on more than thousand volumes achieving a very high (greater than 95%) accuracy rate. In Figure 3, skull masks obtained by the Brain Extraction Tool and the proposed method are shown for T1 contrast. Not only does the proposed method extract the skull more accurately but also it is applicable to different MR contrasts with similar performance to that obtained on the T1 contrast.

The raw data may also be filtered further by detecting clinically-relevant parts of the data and transmitting only that limited part to the external site. This is possible because for some diseases, such as differential diagnosis of dementia,

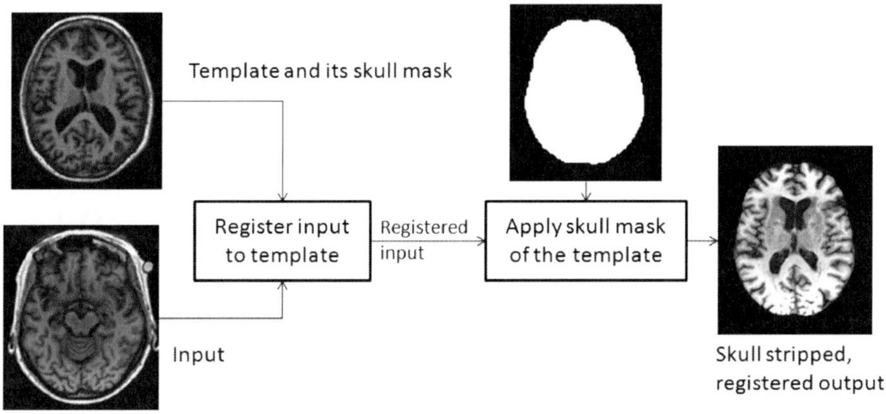

Fig. 2. The flowchart of the proposed skull stripping scheme where input is registered onto the template. The opposite case is also possible.

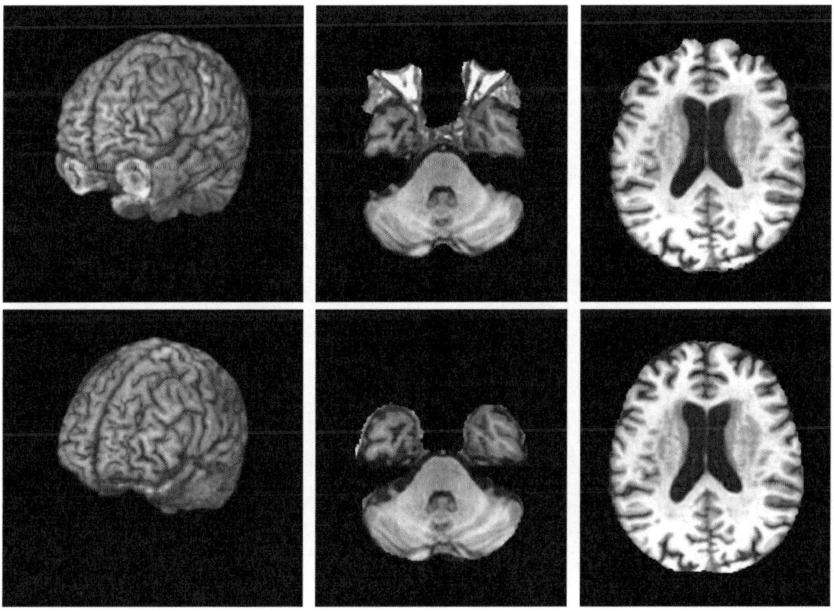

Fig. 3. The proposed method, whose results are shown in the second row, provides more accurate results than Brain Extraction Tool that keeps some parts of eyes and returns jaggy borders (first row).

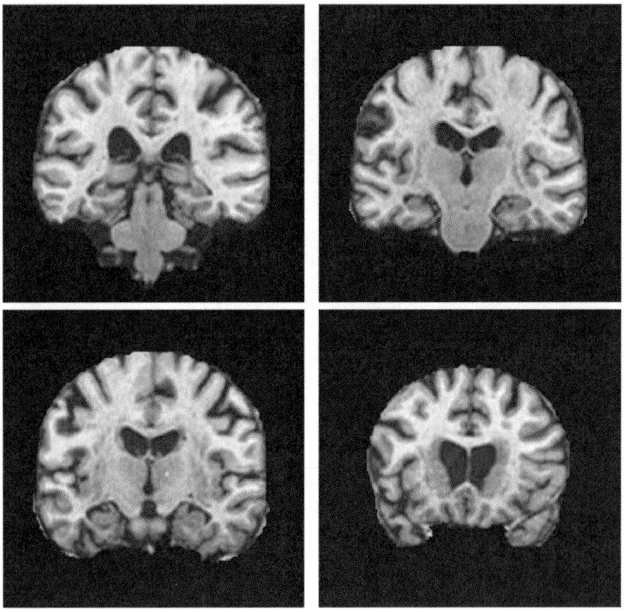

Fig. 4. The key slices proposed in [33] for differential diagnosis of dementia

visual atrophy rating of several slices can provide statistically significant differences between the compared classes. Davies et al. [33] have chosen four coronal slices defined by the visibility of some anatomical parts. It is possible to define such slices on a template volume so that after registering the input volume to the template, those slices are easily selected and sent to the external medical site. The slices are shown in Figure 4.

Extracting Clinical Features

In this section, we explain the use of automated processing of brain MR images to extract clinical features. Using only the extracted features for the teleradiology application may prevent any potential privacy concerns. A standardized coordinate system helps efficiently describe the detected features for transmission. It also provides information on the visualization of them on the receiver side. The features can be extracted for structural MR data as well as for functional MR images; but, it is also possible to add examples from image modalities that measure metabolism, such as PET (positron emission tomography) and SPECT (single photon emission computed tomography).

Structural features that can be extracted from brain MR data may be related to specific brain tissue, brain structure, lobe, or the whole volume. The size and shape features of a brain structure, a brain tissue map, atrophy-indicating features, and abnormality, such as a tumor and microbleeds, and hyper-, or hypo-intense regions than normal can be candidates to such clinically-relevant structural features.

Functional features can be extracted from functional MR data. The use of fMRI is becoming more prevalent and may be candidate for the use in teleradiology in the future. In the literature, the fMRI data are already being reported on standardized coordinates to be able to compare different studies. The raw data is neither important nor sufficient for interpretation. The activation maps obtained from task-induced fMRI data can also be sent to other sites for interpretation.

4 Reference-Based Content Description for Secure and Fast Transmission

In the previous section, we have discussed the content analysis operations to remove sensitive content and to extract quantitative features from the data. The intensity data of the whole or part of the volumetric data and/or extracted features can be desired at the receiver side. In this section, we will introduce the use of standardized templates, or reference data, to provide robust, secure, and efficient transmission of the content.

A reference volume is a brain volume that determines the coordinates other brain volumes are spatially registered. It can be obtained by averaging the intensity of the brain data of a number of individuals at each spatial location. This provides information on general appearance of brain in the selected age group and washes away individual brain differences. It is also common that for different pathologies, reference volumes are constructed and used to detect the modelled pathology. In Figure 5, an input volume, a reference volume, and the registered volume obtained by rigidly registering the input to the reference are shown. The main benefit of using reference volumes is the standardization of spatial locations that allows for easy comparison of different volumes. In Figure 6, same slices of the two brain volumes are shown before and after registration to a reference volume. It is clearly noticeable that such transformation provides visualization possibilities that can enhance the clinical interpretation.

Another benefit of using reference is related to the description of the data. When a reference volume is determined, the data can be more easily exchanged between centers. For example, as explained in the earlier section, a set of slices can be defined before the teleradiology application. The patient data can be registered to the reference and only the requested slices can be sent. This type of region-of-interest detection and transmission is very useful for accessing the data from lower-bandwidth channels, e.g., by using handheld devices. If a reference volume has 100 to 150 axial slices 1mm slice thickness, then using 5 regions-of-interest instead of the whole data can reduce the data transmission load 20 to 30 times.

A reference volume can also be personalized by defining unique reference volumes for distinct patients. This is practical especially when a patient has medical data corresponding to multiple time points. The first data of the patient in time

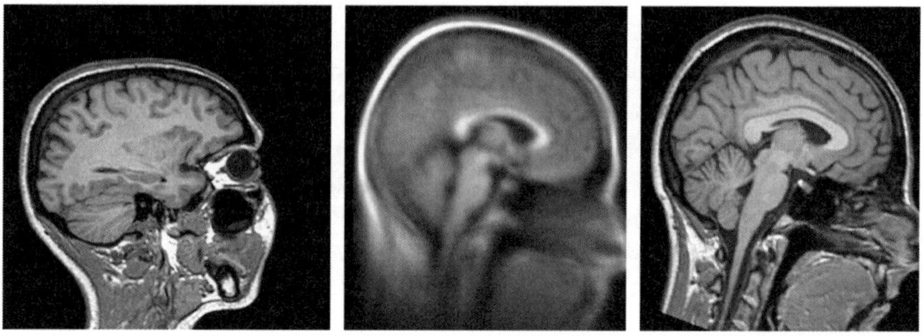

Fig. 5. The process of registration: a slice of input volume, a slice of the template, and a slice of the registered volume.

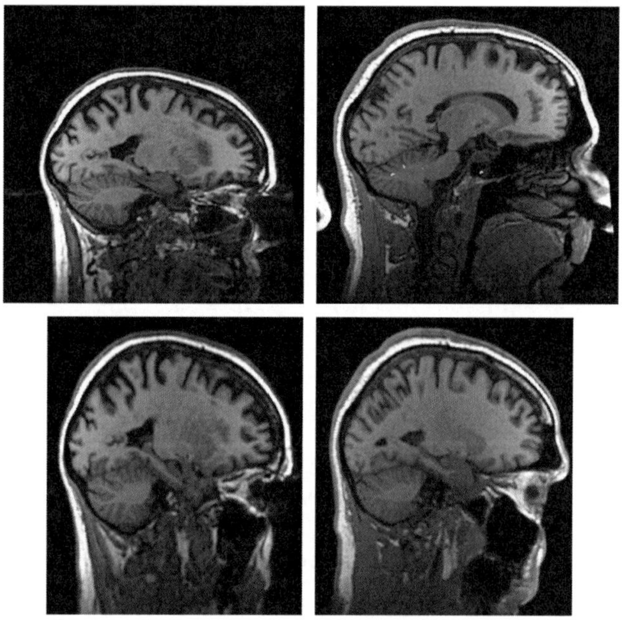

Fig. 6. The effect of registration: Two equally-indexed slices of two volumes before registration (top row) and after registration (bottom row)

is called baseline data and it can be defined as reference for that patient after being registered to a pre-defined coordinate locations. Then, the data corresponding to the follow-up visits can be registered onto the baseline data. By doing that, the changed structures between different visits can be more accurately captured. The use of baseline data of the same patient as reference can also improve the compression efficiency because of more similarity between the

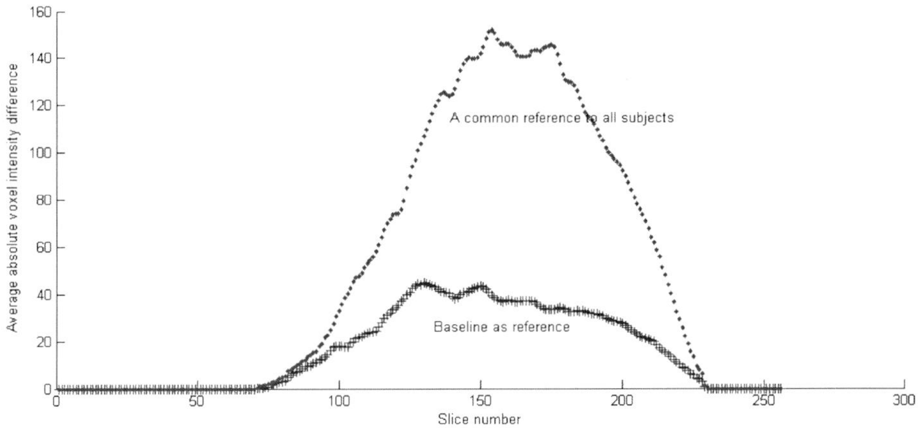

Fig. 7. The average voxel intensity difference between reference and MR volume decreases when the reference is baseline volume (shown with '+') compared with when it is a generic volume.

reference and the data to be compressed. In Figure 7, the graph of the average absolute voxel intensity difference between registered volumes is plotted for cases when the reference is baseline, resulting in smaller difference, and when the reference is a generic volume. Transmitting only the changed areas between the current time point and the reference also limits the amount of shared data between the sites. This also provides an alternative representation of the data instead of raw intensity information; hence, it also helps reduce the risks related to data security.

In MR, by changing the device parameters, different contrasts, such as T1, T2, and FLAIR, can be obtained. Each contrast provides additional information that can improve the diagnostic decision. Therefore, in almost all MR sessions, multiple contrasts are obtained. Because, in most cases, a diagnostic information is sought from the teleradiology application, it is desirable to transfer the whole data rather than just one contrast. Similar to the time series case, a reference data for the subject can be chosen from his own MR data. In this case, one contrast can be assigned as reference and each of the other contrasts can be registered to that reference contrast. The use of one contrast as reference can also eliminate transfer of redundant information by enabling a multi-step, expert-controlled data transfer session. In this case, the expert first gets the reference contrast, which was chosen to be the most clinically relevant one, and then, determines the suspicious, abnormally-appearing brain regions from the reference contrast. To further enhance the diagnostic confidence, he requests the other contrasts only for those spatial locations.

In summary, reference data can apply to a whole set of subject population as well as differently to individual subjects. In any of these cases, the use of such reference MR data can improve compression efficiency, enable interactive teleradiology application, and reduce the amount of sensitive information shared between the sites.

5 Organizing Content for Controlling Access in Teleradiology

In teleradiology, it is important to be able to limit the access to the data. This is necessary to protect the privacy of the data by allowing access to only those with necessary credentials and to assure certain quality level in the interpretation. For example, one of the most criticized aspects of teleradiology is the possible reduction in the quality of radiological interpretation of the data at the distant center. This type of complaint is partly resulting from the doubts on the interpretation of the data by non-qualified radiologists who may have replaced a qualified personnel temporarily (surrogate interpretation). We believe that organization of the data is essential to set policies for access. In this section, we will propose a hierarchical description of the data, whereby access to the content can be constrained.

Medical data is composed of medical image information, such as MR images, medical non-image information, such as lab tests, and metadata information, such as age, gender, and education. The imaging data is stored in PACS, while the other information including the reports are usually stored in EPRs (electronic patient records) [34]. In addition to the PACS and EPRs, an hospital information system (HIS) also stores the non-clinical administrative information, such as financial data, visit dates, insurance and billing. In many cases, PACS and EPRs are separate, and the EPR information may not be easily accessible to the radiologists. Furthermore, in teleradiology, only imaging information may be considered useful to transmit to the distant site. Because linking these two records may improve clinical decision, some studies have considered seamless integration of such information over the web [34] or locally for the teleradiology application. In this section, we assume that both types of information are available, and we propose a data organization and presentation scheme on top of these systems.

We will explain the proposed scheme by using MR data as image information; but, other imaging modalities are equally applicable. Figure 8 shows that the visit of the patient is an integral part of the proposed scheme. Multiple visits can be added very easily to the system and the imaging information from different visits can be compared. As explained in the previous section, the first visit is usually called the baseline and the rest are called the follow-ups. The access policy for teleradiologist can be dependent on the visit. For example, only some radiologists can be given permission to see all visits whereas even in a single visit, some data may not be available to all. The policy for accessing multiple visits is also dependent on the requested consultancy service. As shown in Figure 8, the setting of access conditions to the data and granting access rights to the radiologists are handled by the teleradiology application in the proposed scheme. A clinical opinion may

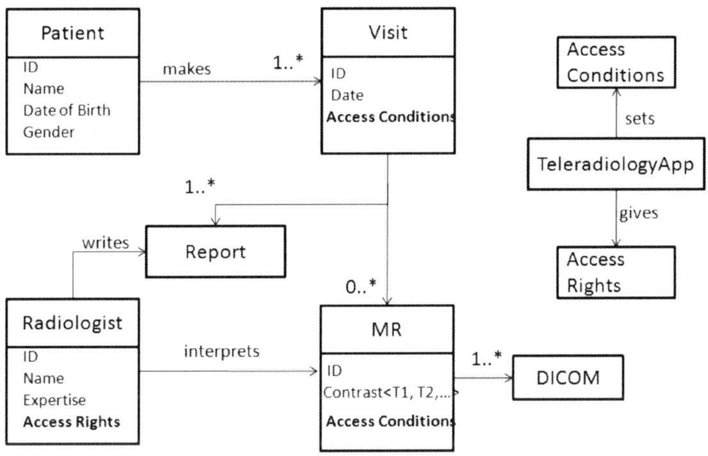

Fig. 8. The UML (Unified Modeling Language) diagram of the proposed data description model.

require comparing data from multiple visits whereas another may just need one contrast. Furthermore, because some metadata information, such as age and gender, affects the clinical interpretation, this type of information should be visible to the radiologist. Depending on the clinical question, family history and genetic information can also be added to the radiologists' viewing form.

A visit will generally include an MR session, where multiple contrasts for structural and functional information are obtained (MR entity in the UML diagram). Conventional MR contrasts, such as T1 and T2, provide structural information while functional information can be obtained from functional MR data. Therefore, each scan in the visit can have such identifying features as structural or functional image. Furthermore, the explanation on how these images are obtained should be available for a correct interpretation. For example, sometimes contrast agents are injected to image brain tumors; the information on the amount of injection and the timing of the imaging can be crucial. In stroke, the information on the time passed after the event is crucial for the interpretation. In fMRI, a crucial side information is the cognitive tasks that have been used as stimuli during the functional study. The access to the data can be determined by setting which contrasts are visible to the qualified radiologist.

In Section 3, we have explained the processing of MR images. After some of these processing steps, clinically-relevant features are extracted from the data. These can be segmentation maps, abnormal intensity maps, detected brain tumors from computer-aided detection systems, activation maps of functional image data, and other related features. In teleradiology, in addition to the image information, data on automatically extracted features can be available. For functional images, this is required because the image by itself is not suitable for interpretation. Having both raw data and quantified data allows for setting further access controls. For example, only some radiologists can see the raw data. For some teleradiology applications, extracted features can be checked by a radiologist and only those very suspicious cases can be reported for the second reading. The quality of the extracted features and the performance of the automated tool can also be checked by a radiologist at the distant site. Because of the verification aspect of this task, the access conditions may not be as strong as interpreting from the original data.

Each volumetric image is composed of 2D slices. As explained before, only certain slices can be of interest to the clinical question. The data may be described in a reference scheme and slices can be made available depending on the suspected pathology. An access protocol based on slices can be quite limiting compared to the other access conditions; however, in some cases, the application may just require evaluation of atrophy in established key slices.

To keep a certain quality level on the interpretation, a diagnostic quality control may be inserted before giving an access permission to the data. In this quality control scheme, a distant radiologist can be presented a simulated or real test images for interpretation, and is assigned a performance value dependent on the success in that test set. This can also prevent surrogate interpretation, where licensed radiologist signs off the report of unlicensed one. Furthermore, it provides a figure for the quality of the service for the main site.

All in all, the proposed data organization scheme offers an intuitive way to set the access policies in various ways depending on the clinical application utilized in the teleradiology.

6 Summary

In this chapter, we provided an overview of teleradiology, including the benefits, drawbacks, needs, and open issues, and offered image analysis algorithms and data description model to remove the sensitive content and to control access to the sensitive data. Although we considered the brain MR images as the application domain, all of the proposed methods, except for the skull stripping, are generalizable to other organs and modalities. As a novel approach to the literature, we also considered the use of automatically-extracted features as part of teleradiology.

In the future, as the data size and patient needs continue to rise, the expert opinion will be sought at any time and from all around the world to make the teleradiology a more common application. The main issues for the future research in teleradiology remain the measurement and assurance of diagnostic quality, the integration of PACS, radiology information systems, and hospital information systems to be able to support more involved diagnostic questions, e.g., "how much has the tumor grown from the last time," than binary questions, such as "is there a tumor," the visualization of the data in the best possible way to enhance the diagnostic decision, and the utilization of the links to personally controlled health records that will be more commonly used in the future. In all of these, solutions that preserve the security of the data and the privacy of the personal information are needed.

References

1. Engelmann, U., Schroeter, A., Werner, O., Schwab, M., Mueller, H., Meinzer, H.-P.: A three-generation model for teleradiology. IEEE Trans. Info. Tech. in Biomedicine 2(1), 20–25 (1998)
2. Boland, G.W.L.: Teleradiology Coming of Age: Winners and Losers. Am. J. Roentgenol. 190, 1161–1162 (2008)
3. Steinbrook, R.: The Age of Teleradiology. New England Journal of Medicine 357, 5–7 (2007)
4. Singh, S.N., Wachter, R.M.: Perspectives on Medical Outsourcing and Telemedicine – Rough Edges in a Flat World? New England Journal of Medicine 358, 1622–1627 (2008)
5. Boland, G.W.L.: Visibility of Radiologists: Helping to Secure Your Future. Am. J. Roentgenol. 192, 1373–1374 (2009)
6. Boland, G.W.L.: The impact of teleradiology in the United States over the last decade: driving consolidation and commoditization of radiologists and radiology services. Clinical Radiology 64, 457–460 (2009)
7. PACS and Teleradiology Systems: A Global Strategic Business Report, Global Industry Analysts, Inc., `http://www.strategyr.com/PACS_and_Teleradiology_Systems_Market_Report.asp`
8. Berger, S.B., Cepelewicz, B.B.: Medical-legal issues in teleradiology. American Journal of Radiology 166, 505–510 (1996)
9. Nicolini, D.: The work to make telemedicine work: a social and articulative view. Social Science and Medicine 62, 2754–2767 (2006)
10. Audebert, H.: Telestroke: effective networking. Lancet Neurology 5, 279–282 (2006)
11. Mukundan, S., Vydareny, K., Vassallo, D.J., Irving, S., Ogaoga, D.: The telemedicine system for supporting medical students on elective in the developing world. Academic Radiology 10(7), 794–797 (2003)
12. Balasingham, H., Ihlen, H., Leister, W., Samset, E.: Communication of medical images, text, and messages in inter-enterprise systems: a case study in Norway. IEEE Trans. Info. Tech. in Biomedicine 11(1), 7–13 (2007)
13. MacFarlane, A., Murphy, A.W., Clerkin, P.: Telemedicine services in the Republic of Ireland: an evolving policy context. Health Policy 76, 245–258 (2006)
14. Jarvis, L., Stanberry, B.: Teleradiology: threat or opportunity? Clinical Radiology 60, 840–845 (2005)
15. Microsoft HealthVault, `http://www.healthvault.com`
16. Google Health, `http://www.google.com/health`
17. Fitzgerald, R.: Medical regulation in the telemedicine era. The Lancet 372, 1795–1796 (2008)
18. Hynes, D.M., Stevenson, G., Nahmias, C.: Towards filmless and distance radiology. The Lancet 350, 657–660 (1997)
19. Stamford, P., Bickford, T., Hsiao, H., Mattern, W.: The significance of telemedicine in a rural emergency department. IEEE Engineering in Medicine and Biology, 45–52 (1999)

20. Hess, D.C., Wang, S., Gross, H., Nichols, F.T., Hall, C.E., Adams, R.J.: Telestroke: extending stroke expertise into underserved areas. Lancet Neurology 5, 275–278 (2006)
21. Kern, J.: Evaluation of teleconsultation systems. Medical Informatics 75, 330–334 (2006)
22. Mathiesen, F.K.: Web technology - the future of teleradiology. Computer Methods and Programs in Biomedicine 66, 87–90 (2001)
23. John, N.W., Riding, M., Sadarjoen, A., Blumrozen, L.: Bringing 3D to teleradiology. In: Proc. of the Int. Conf. on Information Visualisation, vol. 4, pp. 4–9 (2000)
24. Laird, S.P., Wong, J.S.K., Schaller, W.J., Erickson, B.J., de Groen, P.C.: Design and implementation of an Internet-based medical image viewing system. The Journal of Systems and Software 66, 167–181 (2003)
25. Poon, W.S., Leung, C.H.S., Lam, M.K., Wai, S., Ng, C.P., Kwok, S.: The comparative impact of video consultation on neurosurgical health services. Medical Informatics 62, 175–180 (2001)
26. Muench, H., Engelmann, U., Schroeter, A., Meinzer, H.P.: The integration of medical images with the electronic patient record and their web-based distribution. Academic Radiology 11(6), 661–668 (2004)
27. Ng, W.H., Wang, E., Ng, I.: Multimedia messaging service teleradiology in the provision of emergency neurosurgery services. Surgical neurology 67, 338–341 (2007)
28. Engelmann, U., Schroeter, A., Boralv, E., Schweitzer, T., Meinzer, H.-P.: Mobile teleradiology: all images everywhere. International congress series 1230, 844–850 (2001)
29. Reponen, J., Niinimaki, J., Kumpulainen, T., Ilkko, E., Karttunen, A., Jartti, P.: Mobile teleradiology with smartphone terminals as part of a multimedia electronic patient record. International Congress Series 1281, 916–921 (2005)
30. Stahl, J.N., Zhang, J., Zellner, C., Pomerantsev, E.V., Chou, T.M., Huang, H.K.: Teleconferencing with dynamic medical images. IEEE Trans. Info. Tech. in Biomedicine 4(2), 88–96 (2000)
31. Ekin, A.: Local information based overlaid text detection by classifier fusion. In: Proc. IEEE ICASSP, Toulouse, France (2006)
32. Smith, S.M.: Brain Extraction Tool, FMRIB technical Report TR00SMS2b (2005)
33. Davies, R.R., Scahill, V.L., Graham, A., Williams, G.B., Graham, K.S., Hodges, J.R.: Development of an MRI rating scale for multiple regions: comparison with volumetrics and with voxel-based morphometry. Neuroradiology 51(8), 491–503 (2009)
34. Muench, H., Engelmann, U., Schroeter, A., Meinzer, H.P.: Web-based distribution of radiological images from PACS to EPR. International Congress Series 1256, 873–879 (2003)

Detecting Nakedness in Color Images

Jiann-Shu Lee[1], Yung-Ming Kuo[2], and Pau-Choo Chung[2]

[1] Department of Computer Science and Information Engineering, National University of Tainan, 33, Sec. 2, Shu-Lin St. Tainan, 700, Taiwan
Tel.: 886-6-2606123 ext:7717
Fax: 886-6-2144409
cslee@mail.nutn.edu.tw

[2] Department of Electrical Engineering, National Cheng Kung University

Abstract. This chapter presents a new naked image detection algorithm. A learning based chromatic distribution-matching scheme is proposed to determine the image's skin chroma distribution online such that it can tolerate the chromatic deviation coming from special lighting without increasing false alarm.

Keywords: Skin detection, naked image detection, chromatic distribution-matching scheme.

1 Introduction

In a relatively short period of time, the Internet has become readily accessible in most organizations, schools and homes. Meanwhile, however, the problem of pornography through the Internet access in the workplace, at home and in education has considerably escalated. Therefore, how to effectively block or filter out pornography has been arousing a serious concern in related research areas.

The mostly used approach to blocking smut from the Internet is based on contextual keyword pattern matching technology that categorizes URLs by means of checking contexts of web pages and then traps the websites assorted as the obscene. Although this method can successfully filter out a mass of obscene websites, it is unable to deal with images, leading to its failure to detect those obscene web sites containing naked images instead of smut texts. Hence, the development of naked image detection tech nology is urgently desired to prevent juveniles from getting access to pornographic contents from the Internet more thoroughly.

Naked image detection using semantic matches is a hard task because it has to deal with jointed objects of highly variable shapes, in a diverse range of poses, seen from many different views. Furthermore, lighting and background are uncontrolled, making segmentation very difficult. In [1-2] Forsyth and Fleck proposed an automatic system that marks skin-like pixels using color and texture properties to tell whether there are naked humans present in an image. These skin regions are then fed to a specialized grouper, which attempts to group a human figure using geometric constraints on human structure. If the grouper finds a predefined structure, the system decides a naked image is present. Jiao et al. [3] presented an adult image detection method. They first use the

H.T. Sencar et al. (Eds.): Intel. Multimedia Analysis for Security Appli., SCI 282, pp. 225–236.
springerlink.com

skin color model to detect naked skin areas roughly. Then the Sobel operator and the Gabor filter are applied to remove those non-skin pixels. Finally, the color coherence vector and the color histogram are employed to determine which image contains naked people. J. Z. Wang et al. [4] combined an icon filter, a graph-photo detector, a color histogram filter, a texture filter and a wavelet based shape-matching algorithm to detect the objectionable images. For the color histogram filter, they use a total of 512 bins to compute the histogram. Jones and Rehg [5] proposed a skin color detection technique by estimating the distribution of skin and non-skin color in the color space using labeled training data. Some simple features are extracted to detect adult images. A. Bosson et al. [6] presented a method to block the pornographic images. The authors compute the likelihood ratio for a quantized color space. Then, the blobs' features of the image, detected by likelihood histogram in RGB space, are computed and presented as a vector. Finally, the artificial neural network is utilized to classify the image is pornographic or not. B. Jedynak et al. [7] proposed a statistical model for skin detection. The Maximum Entropy Model is used to infer the skin models from the data set. Then, the Bethe Tree approximation and Belief Propagation algorithm are utilized to approximate the probability for skin at pixel locations. Cao et al. [8] proposed a learning scheme for judging if there are any naked people in an image. The Learning Vector Quantization is first utilized to build several classifiers based on low-level features, such as color histogram, texture etc. Subsequently, these classifiers are combined to detect naked images. H. Zheng et al. [9] proposed a method to detect the adult images. The architecture is divided into two parts. The skin detecting model, similar to [7], is applied to detect the skin blocks in the image first. Next, the features of skin blocks in the testing image are fed into the Multi-Layer Perceptron Classifier to identify that is an adult image or not.

As can be seen in these methods, none of them consider the inference coming from special lighting and color altering. There exist a large number of naked pictures taken under special lighting. Usually, warm lighting is applied to make skin tone look more attractive, while human skin color deviates from the normal case at the same time. If the skin color model cannot tolerate the deviation, it will tend to miss a lot of naked pictures. On the contrary, if the skin color model accommodates the deviation, an abundance of non-skin objects like wood, desert sand, rock, foods, and the skin or fur of animals would be detected in the skin detection phase and deteriorates the system performance. Accordingly, the above mentioned approaches suffer from the skin color deviation resulting from special lighting, which is often seen in the naked images. Dealing with the special lighting effect in the naked images is a difficult task. If the skin color model tolerates the deviation, lots of non-skin objects would be detected simultaneously. A feasible solution for the problem is to adapt the adopted skin chroma distribution to the lighting of the input image. Based on this concept, a new naked image detection system is proposed. We develop a learning based chromatic distribution-matching scheme that consists of the online sampling mechanism and the one-class-one-net neural network. Based on this approach, the object's chroma distribution can be online determined so that the skin color deviation coming from lighting can be accommodated without sacrificing the accuracy. The roughness feature is further applied to reject confusion coming from non-skin objects, so the skin area can be more effectively detected. Several representative features induced from the naked images are used to verify these skin areas. Subsequently, the face detection process is employed to filter out those false candidates coming from mug shots.

The chapter is structured as follows. In section two, the adaptive skin segmentation procedure is introduced. The post-processing employed to enhance the system accuracy is presented in section three. Section four demonstrates the experimental results. Finally, we make conclusions in section five.

2 Skin Segmentation

Adapting the skin chroma distribution to lighting is difficult if without any priori information. Here, we propose an adaptive scheme, which can adapt the skin color model to the corresponding lighting condition of the input naked image. From the viewpoint of photography, the shot subject should be positioned around the center of an image. Based on this property, we propose an on-line sampling strategy, i.e. the skin chroma distribution of the input image is decided by directly sampling around the image center for skin areas. Subsequently, a neural network is used to determine the corresponding chroma distribution that is then applied to detect the skin area.

The removal of luminance component helps constructing skin classifier that will work well for images with different skin brightness. Also, the reduction of space dimensionality profits decreasing the computational complexity. Because it provides an effective separation into luminance and chrominance channel and generates a compact skin chroma distribution, we adopt the C_b-C_r space as the skin color space. We classify the popular skin colors of naked images, including various lighting conditions, into several categories. The corresponding compactly chroma histogram of each category is compiled in advance. The existing skin detection methods usually generate the corresponding chroma histogram by straightforward compiling the whole skin samples. If we follow that approach, we need to manually categorize the skin samples in advance. It is difficult to do accurate categorization because of limited chromatic sensitivity and subjective bias. To avoid this drawback, we design a systematic approach to the automatic and objective classification of the whole skin samples into several categories. This approach consists of two major steps: **(a)** merging and **(b)** grouping.

(a). At the merging step, every sample skin image is first transformed from the RGB color space to the YC_bC_r space by using the following formula:

$$\begin{bmatrix} Y \\ C_b \\ C_r \end{bmatrix} = \begin{bmatrix} 16 \\ 128 \\ 128 \end{bmatrix} + \begin{bmatrix} 65.481 & 128.533 & 24.966 \\ -37.797 & -74.203 & 112 \\ 112 & -93.786 & -84.214 \end{bmatrix} \begin{bmatrix} R \\ G \\ B \end{bmatrix}. \tag{1}$$

Each transformed result forms a chroma histogram denoted as CH. Then, dichotomizing the CH followed by the closing and the filling operations to get the solid skin chroma region, denoted as CR. For two CHs, if the overlapping part exceeds eighty percent of the individual CH, they are merged into a larger CH. For more explicit expression, let CH_1 and CH_2 represent the two CHs, CR_1 and CR_2 denote the corresponding CRs, and Z represent the overlapping area, i.e. $CR_1 \cap CR_2$. Once the two CHs satisfy both the equations (2) and (3), they are merged into a single CH.

$$\sum_{C_b \in Z} \sum_{C_r \in Z} CH_1(C_b, C_r) / \sum_{C_b \in CR_1} \sum_{C_r \in CR_1} CH_1(C_b, C_r) \geq 0.8 \tag{2}$$

$$\sum_{C_b \in Z} \sum_{C_r \in Z} CH_2(C_b, C_r) / \sum_{C_b \in CR_2} \sum_{C_r \in CR_2} CH_2(C_b, C_r) \geq 0.8 \tag{3}$$

(b). After the merging step, the merged CHs can be further merged again. For example, if CH_1 and CH_2 are merged into $CH_{1,2}$ and CH2 and CH3 are merged into $CH_{2,3}$, then $CH_{1,2}$ and $CH_{2,3}$ can be further merged into $CH_{1,2,3}$. Repeat this process until no more mergence occurs. Finally, we acquire five major skin chroma clusters.

The chroma histogram is denoted as H_x, where the subscript x means the x-th skin chroma cluster. The corresponding element at position (i, j) is represented as $H_x^{(i,j)}$. The learning-based chroma distribution matching mechanism comprises a multilayer feed-forward neural network used to learn and judge whether the input image contains skin, as shown in Fig. 1. Each skin module consists of a three-layer feed-forward neural network and is responsible for learning and classifying the corresponding skin chroma. The advantage of this architecture is that the system can adjust the number of skin modules to meet the practical requirement. For example, once a new skin chroma, not included in the existing skin modules, frequently appears in the naked images we can establish the corresponding new skin module easily.

The macro block is the basic unit for matching mechanism. The macro block size depends on the size of the input image. For the real naked images, the sizes are not uniform and usually larger than 256 by 256. To save computational time, the input

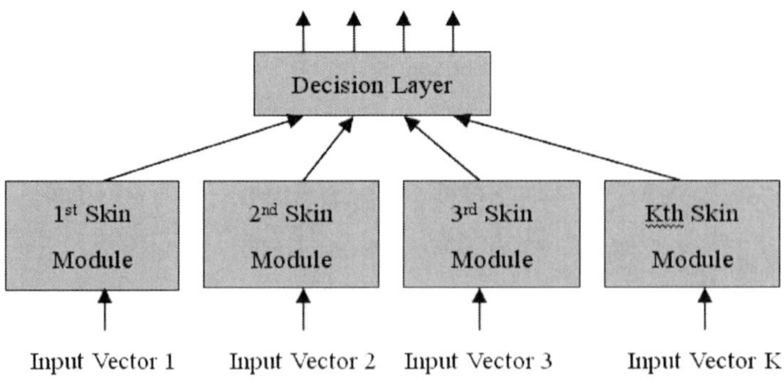

Fig. 1. The learning-based chroma distribution matching mechanism comprises K multilayer feed-forward neural networks.

image is first normalized before further processing. Let the width and the height of the input image is *W* and *H*, respectively. The bigger one is scaled to 256 and the other dimension is scaled proportionably to retain the aspect ratio. The normalized image is divided into 16 by 16 blocks. Each block is called a macro block. The matching mechanism can be divided into the training phase and the testing phase. In the training phase, the *K* skin modules are trained, respectively. For each skin module, both the positive and the negative training samples are interlaced fed into the network until convergence. The input vector is composed of the foreground feature vector derived from the foreground blocks (*FB*). And, four macro blocks construct these foreground blocks. Each macro block comprises *M* by *N* pixels. To ensure the training samples can be captured from the right places, the configuration of the macro blocks for the positive training sample is distinct from that for the negative training sample. In our investigation, the skin region in a naked image is usually so close to the image center that it can easily catch the viewer's eye. Hence, for the case of positive training, a central area (*CA*) with size 4*M* by 4*N* pixels is first selected from the training image. In other words, the *CA* comprises 16 macro blocks arranged to a macro block array with four rows by four columns. The larger number of macro block is used to ensure the coverage of the skin region. Four most representative macro blocks are chosen from the 16 blocks. For example, we want to train the *x*-th skin module. A positive sample image *P* belonging to the *x*-th skin chroma cluster is fed to train this module. First, the *CA* is picked out from *P*. Let $MB^{(i,j)}$ denote the macro block at the intersection of the *i*-th raw and the *j*-th column macro block in *CA*. Let $MB^{(i,j)}_{(u,v)}$ represent the position (*u*, *v*) of $MB^{(i,j)}$. The (*Cb*, *Cr*) components at $MB^{(i,j)}_{(u,v)}$ are denoted as $Cb_{MB^{(i,j)}_{(u,v)}}$ and $Cr_{MB^{(i,j)}_{(u,v)}}$, respectively. Let $MS^{(i,j)}_x$ mean the membership strength of $MB^{(i,j)}$ to the chroma histogram H_x and is defined as:

$$MS^{(i,j)}_x = \sum_{u=1}^{M}\sum_{v=1}^{N} H_x^{(Cb_{MB^{(i,j)}_{(u,v)}},Cr_{MB^{(i,j)}_{(u,v)}})} \tag{4}$$

The four macro-blocks with the top four biggest values of $\{MS^{(i,j)}_x \mid 1 \le i, j \le 16, i, j \in N\}$ are chosen as the *FB*. As to the case of negative training, a central area *CA* with size 2*M* by 2*N* pixels, i.e. four macro blocks, is selected as the *FB*. Fig. 2 is illustrated to explain how to get the *K* input vectors from the *FB*. Let F_i, where $i = 1, 2, 3, 4$, denote the chroma distribution of each macro block of *FB*. By accumulating those F_i, we can get the chroma distribution of the *FB*, that is, *F*.

$$F = \sum_{i=1}^{4} F_i \tag{5}$$

The input feature is defined as the product between the pre-categorized skin chroma distribution H_x and *F*. Assume P_x represents the product between *F* and H_x. Then, we can get

$$P_x = F \times H_x \tag{6}$$

where '×' means the element-to-element product. Because the size of P_x is 256x256, it is not practical for neural network training. We reduce P_x to $\tilde{P}_x$ by down sampling, i.e. the original 32 by 32 elements are summed to generate a new element. Then, $\tilde{P}_x$ is treated as the input vector for the *x*-th skin module.

In the testing phase, the uncertainty of the skin region location is higher than that in the training phase. To increase the hit rate of capturing the skin region, the online sampling strategy is proposed. We utilize five *FB*s arranged as Fig. 3 to sample the skin region. Every *FB* generates *K* input vectors. These vectors are fed into the corresponding skin modules. Subsequently, the outputs of the *K* skin modules are fed into the decision layer. The decision layer is a competitive network, with function

$$Y_k = \begin{cases} f_k, if \ \ f_k \geq f_j \ \ and \ f_k > 0.5, \forall j \neq k \\ 0, \qquad\qquad\qquad\qquad otherwise \end{cases} \tag{7}$$

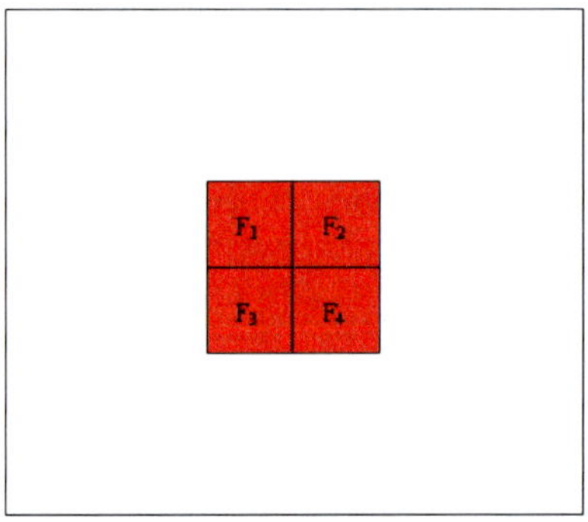

Fig. 2. The foreground blocks chosen for the learning phase.

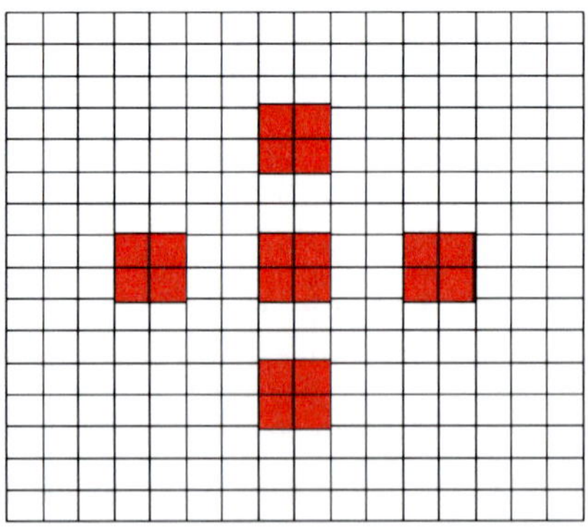

Fig. 3. The foreground blocks chosen for the testing phase.

If $Y_k, k \in \{1,2,3,..., K\}$, is the winner, it means that the skin chroma distribution of the test image may belong to the *k*th skin category. Because there are five *FB*s, the above procedures are executed five times. If there is no winner, it implies that the input image does not contain naked people. If the winner is not unique, further competition is needed. Assume Y_a and Y_b are the winners generated by the *m* and *n FB*, respectively. Then, the bigger one is viewed as the real winner. Assuming Y_a is the winner, the test image is then segmented by setting the pixels with chroma belonging to the *a*-th skin chroma cluster to one and others to zero. And, the acquired binary image, i.e. skin image, is denoted as *S*.

3 The Post-processing

The performance of the naked image detection is extremely dependent on the accurate skin segmentation. There exist a lot of objects, possessing skin like chroma in an image, e.g. wood, foods, rock, desert sand, and animal fur. It is almost impossible to distinguish them from the human skin solely by using the chroma property. According to our observation, smoothness is a very important feature for skin. Therefore, we utilize the roughness feature to further reject the confusion coming from the non-skin objects with skin like chroma. There exist many approaches to measuring the texture roughness in image processing, e.g. Fourier power spectrum [10], fractal dimension [11], co-occurrence grey-level matrix [12], neighborhood grey-tone difference matrix [13], auto-covariance function and edge density [14] etc. Intuitively, the rougher the object's surface the more its extreme number in image intensity. Based on the concern about computational complexity, the window based extreme density approach is proposed to quantify the roughness. First, the input image is transformed to an intensity image I. And, we create an extreme image X with the same size of I and each element is set to zero. A pixel I(x,y) of I at position (x,y), if it is an extreme, i.e. $I(x, y) > I(x + \Delta x, y + \Delta y)$ where both Δx and $\Delta y \in \{-1,0,1\}$ and $|\Delta x| + |\Delta y| \neq 0$, X(x,y) is set to one. Subsequently, the X is divided into 8x8 blocks. If the extreme number of a block is less than a threshold value te, that block is set to one. Otherwise, that block is reset to zero. Therefore, the zero-block in X represents the rough region, whereas the one-block in X corresponds to the smooth region. The smooth skin region SS can be obtained by multiplying S and X.

4 Naked Image Detection

We induce three common properties for naked images through investigating a mass of naked images. First, for delighting viewers, the naked body usually occupies a significant portion of an image. Second, the aspect ratio of the naked body is usually in a reasonable range. Third, the position of the naked body is close to the image center to harmonize with the frame. The features based on these properties are extracted to judge if the smooth skin region *SS* contains naked bodies.

4.1 Feature 1: Area

The area feature is used to quantify the size of the examined area. To further separate the body from the skin like background in the *SS* we apply the opening operation to the *SS* and get the outcome *OSS*. We locate the body area by picking out the maximum object, denoted as *MO*, from the *OSS*. The area of *MO* is denoted as *AMO*. The occupation ratio of *MO*, i.e. $\frac{AMO}{Area(I)}$, represented as *OR* is viewed as the first feature. The maximum object area *MO* corresponding to Fig. 10(a) is shown in Fig. 11(b).

4.2 Feature 2: Shape

The shape feature is used to quantify how the profile of the examined area is like the naked body. For the convenience of explanation we use two figures as examples. Fig. 4(a) shows an *MO*. By applying the principal component analysis (PCA) to *MO*, we can obtain two orthogonal eigenvectors. By projecting *MO* to the two eigenvectors, respectively, we can obtain a minimum rectangle *R* containing *MO*. Assume the lengths of the two sides of the rectangle are *L* and *W* (and $L \geqq W$). If we directly adopt the ratio of *L* and *W* as the aspect ratio, an elongate and bended area may correspond to a significant aspect ratio like the case in Fig. 4(b). To avoid this problem, we take the equivalent aspect ratio as the shape feature. The equivalent aspect ratio *Q* is defined as the ratio of *L* to the equivalent width *W_eq*.

$$Q = L/W\_eq \tag{8}$$

$$W\_eq = AMO/L \tag{9}$$

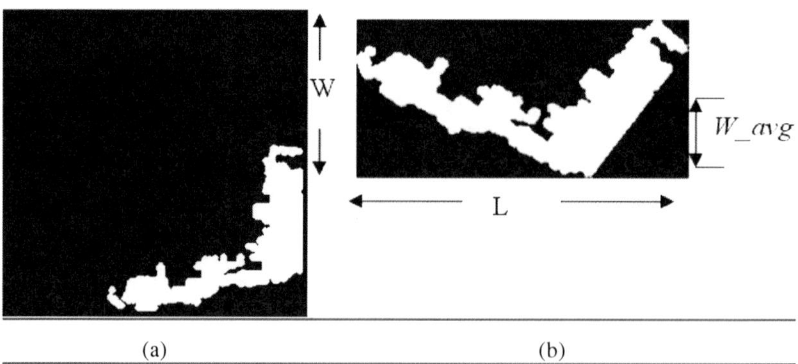

Fig. 4. (a) An MO used to explain how to quantify the shape feature. **(b)**The minimum rectangle containing the maximum object in **(a)**.

4.3 Feature 3: Location

The location feature is used to quantify if the examined area is close to the image center. The normalized horizontal and vertical distances from the gravity center (*GC*) of *MO* to the image center (*IC*) are adopted as the location features *NHD* and *NVD*, respectively. The normalization is used because the naked image's size is not always square. The location features are defined as the following equations.

$$NHD = \left|IC_x - GC_x\right| / IW \tag{10}$$

$$NVD = \left|IC_y - GC_y\right| / IL \tag{11}$$

$$GC_x = \sum_{(x_i, y_i) \in MO} x_i \Big/ AMO \tag{12}$$

$$GC_y = \sum_{(x_i, y_i) \in MO} y_i \Big/ AMO \tag{13}$$

$$(IC_x, IC_y) = \left(\frac{IW}{2}, \frac{IH}{2}\right) \tag{14}$$

where *IW* and *IH* represent the width and the height of the input image, respectively. Subsequently, these features are applied to judge if the *MO* corresponds to the naked body. However, the effective combination of these features into one classifier must be taken into serious account. Boosting is a powerful method to combine a collection of simple classification functions to form a stronger one. The smart combination often outperforms most monolithic strong classifiers such as SVMs and Neural Networks. Adaboost, making the simple classification functions work, is a kind of large margin classifiers [15-16]. It is utilized to serve as the feature classifier in our work. If the *MO* is classified as no naked bodies, our system will conclude that the examined image is not a naked image. On the contrary, if the *MO* is classified as a naked body, further examination is needed because most mug shots also meet the above properties of a naked body. The most important feature for a mug shot is the face area occupies a significant portion of the skin area. So we utilize the face-to-skin ratio (*FSR*) to exclude the false positives coming from mug shots. There exist a lot of face detection algorithms in the literature. For practical applications, we need a face detection algorithm in which faces can be detected effectively as well as efficiently. Recently, Lienhart et al. [17] proposed a new face detection algorithm that minimizes computation time while achieving high detection accuracy. We apply this approach to detecting faces in the so-called *ROI*, obtained by referring to the minimal rectangle containing the *MO*. The *FSR* is derived from calculating the ratio of the predicted face area to the *AMO*. If the *FSR* is larger than a threshold value, the test image is viewed as a mug shot. Otherwise, the test image is considered a naked image. In terms of computational complexity, the face detection stage indeed increases the system's computation load. However, it can effectively filter out those false positives resulting from mug shots in the classification stage. In addition, the face detection routine is invoked only when the *MO* is classified as a naked body that only occupies a little portion in real applications. Therefore, the face detection stage is necessary and worthwhile when the system robustness and complexity are both taken into consideration.

5 Experimental Results

The training images and the testing ones are collected from the Internet and the album. There are 508 naked images, including 312 fully naked and upper-body naked images and 196 intercourse images. The former are categorized as naked-1 images while the latter are categorized as naked-2 images. There are 482 non-naked images, including 127 clothed people and people wearing swimming suits, 41 mug shots and 314 other miscellaneous images (nature scenes, buildings, wood, foods, rock, desert sand, and animals). The naked roles comprise Caucasians, Blacks and Asians. The training set contains 50 naked images and 40 non-naked images.

The confusion matrix of the experimental results is shown in Table 1. The detection rates of the naked images and the non-naked images are 86.4% and 94.8%, respectively. The ROC (receiver operating characteristics) curve of the proposed system is shown in Fig. 5. It is noted that the results meet the requirements for the practical application of a naked image detection system, i.e. the moderate detection rate and low false alarm. From these experiments, our method proves to be able to correctly online determine the skin chroma and effectively distinguish naked images from non-naked images by integrating texture, geometric features and face detection.

To test the superior performance of our method, an existing method proposed by Cao et al. [8] was implemented to compare with ours. This choice was made on the basis of fair and objective considerations: it also adopts YCbCr color space and integrates features by utilizing the Adaboost learning algorithm. The same test images are

Table 1. The confusion matrix of the proposed method.

	Test images	
Classification results	Naked image	Non-naked image
Naked image	439(86.4%)	25(5.2%)
Non-naked image	69(13.6%)	457(94.8%)

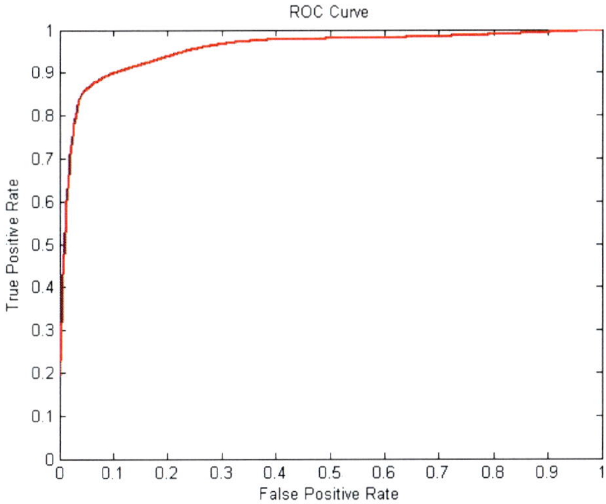

Fig. 5. The ROC curve of the proposed system.

Table 2. The confusion matrix of the compared method.

	Test images	
Classification results	Naked image	Non-naked image
Naked image	364 (71.7%)	76 (15.8%)
Non-naked image	144 (28.3%)	406 (84.2%)

employed to perform comparison. The confusion matrix of the compared method is shown in Table 2. The detection rates of the naked images and the non-naked images are 71.7% and 84.2%, respectively. These results show our method can achieve superior performance for naked images detection.

6 Conclusions

Due to the popularization of the Internet access, the problem of pornography inundation has become considerably serious. Hence, how to block or filter out pornography has come into notice in related research areas. Most of the existing pornography detection methods utilize skin information as the premier clue for detecting naked bodies. Usually, the naked bodies are shot in changeful environments. Therefore, how to accurately detect the skin regions in an image in spite of the interference coming from various lighting conditions is still an open research issue.

In this chapter, a new naked image detection algorithm is presented. We develop a learning based chromatic distribution-matching scheme that consists of the online sampling mechanism and the one-class-one-net neural network. Based on this scheme, we can deal with the chromatic deviation coming from special lighting without increasing false alarm. The utilization of the roughness feature further rejects confusion coming from flesh tone objects such that our skin detection method can get not only high detection rate but also high detection precision. Low-level while reliable geometric constraints of naked bodies are effectively combined by utilizing the Adaboost algorithm. Finally, the mug shot exclusion process makes the overall system achieve satisfactory detection performance. Experiments show our method can correctly online determine the skin chroma and effectively distinguish naked images from non-naked images by integrating texture, geometric features and face detection.

References

[1] Forsyth, D.A., Fleck, M.M.: Body plans. In: Conf. Computer Vision and Pattern Recognition, pp. 678–683 (1997)

[2] Forsyth, D., Fleck, M.: Automatic detection of human nudes. International Journal of Computer Vision 32(1), 63–77 (1999)

[3] Jiao, F., Gao, W., Duan, L., Cui, G.: Detecting adult image using multiple features. In: Proc. of IEEE Int. Conference on Info-tech and Info-net, vol. 3, pp. 378–383 (2001)

[4] Wang, J.Z., Li, J., Wiederhold, G., Firschein, O.: System for Screening Objectionable Images. Computer Communications 15(21), 1355–1360 (1998)

[5] Jones, M.J., Rehg, J.M.: Statistical color models with application to skin detection. Computer Vision and Patter Recognition (1999)

[6] Bosson, A., Cawley, G.C., Chian, Y., Harvey, R.: Non-retrieval: blocking pornographic images. In: Lew, M., Sebe, N., Eakins, J.P. (eds.) CIVR 2002. LNCS, vol. 2383, pp. 50–60. Springer, Heidelberg (2002)

[7] Jedynak, B., Zheng, H., Daoudi, M.: Statistical Models for Skin Detection. In: IEEE Workshop on Statistical Analysis in Computer Vision, in conjunction with CVPR 2003, Madison, Wisconsin, June 16-22 (2003)

[8] Cao, L.L., Li, X.L., Yu, N.H., Liu, Z.K.: Naked people retrieval based on adaboost learning. In: Proc. of the 1st Int. Conference on Machine Learning and Cybernetics, pp. 1133–1138 (2002)

[9] Zheng, H., Daoudi, M., Jedynak, B.: Blocking Adult Images Based on Statistical Skin. Electronic Letters on Computer Vision and Image Analysis 4(2), 1–14 (1996)

[10] Castleman, K.R.: Digital Image Processing. Prentice-Hall, Englewood Cliffs (1996)

[11] Chaudhuri, B., Sarkar, N.: Texture segmentation using fractal dimension. IEEE Trans. Pattern Analysis and Machine Intelligence 17(1), 72–77 (1995)

[12] Haralick, et al.: Texture features for image classification. IEEE Trans. SMC (3), 610–621 (1973)

[13] Amadasun, M., King, R.: Textural features corresponding to textural properties. IEEE Trans. SMC 19(5), 1264–1274 (1989)

[14] Tomita, F., Tsuji, S.: Computer Analysis of Visual Textures. Kluwer, Norwell (1990)

[15] Tieu, K., Viola, P.: Boosting image retrieval. In: IEEE Conf. on ICCV (2000)

[16] Guo, G.D., Zhang, H.J., Li, S.Z.: Pairwise face recognition. In: IEEE Conf. on ICCV, Vancouver, Canada (2001)

[17] Lienhart, R., Kuranov, A., Pisarevsky, V.: Empirical analysis of detection cascades of boosted classifiers for rapid object detection. Technical report, MRL, Intel Labs. (2002)

An Introduction to the Principles and Requirements of Robust Hashing

Félix Balado

School of Computer Science and Informatics
University College Dublin
felix@ucd.ie

1 Introduction

A robust hash is essentially a low-dimensional representation of a multimedia signal which is linked to its perceptual content. This representation acts as a descriptor which allows for flexible identification of this type of signal. Flexibility is needed because any multimedia signal —that is, image, audio, or video— may appear in slightly different but perceptually equivalent forms owing to lossy compression, transcoding, clipping/cropping, or a number of other reasons. Therefore, identifiers strictly associated to a particular bit representation of a multimedia signal —such as a cryptographic hash [1]— are far too strict. If robust hashing is to become something like "cryptographic hashing for multimedia" then it is reasonable to expect from it key properties of cryptographic hashing while simultaneously trying to overcome its intrinsic limitation with respect to multimedia signals. Exploiting source coding with distortion constraints is the natural way to achieve this end. The connection between these two areas is explored further on in this chapter in an attempt to produce an accurate definition of robust hashing and its requirements conforming to the point of view expressed above. This point of view is perhaps not the most common in the literature of the subject, which has mainly dealt with the production of robust low-dimensional descriptors but not so much with their connection with cryptographic hashing.

Any novel discipline is bedevilled by hesitant nomenclature, and, as such, robust hashing is no exception. Some remarks are necessary firstly on the alternative names for robust hashing sometimes found in the literature: *perceptual hashing*, *visual hashing*, *soft hashing*, *perceptual fingerprinting*, *fingerprinting*, and others. A brief query in some current databases will reveal that the terms including "hashing" seem to be gaining higher acceptance. Moreover, it is perhaps best to avoid the use of "fingerprinting" in the robust hashing sense. This is because of its widespread use with a rather different meaning in data hiding applications (see for instance [2] for one of the earliest works in this area). To be more precise, our criterion in making the aforementioned naming choice is that whereas a robust hash associated to a given signal has to be obtained from that signal, a fingerprint associated to it is usually independently produced.

H.T. Sencar et al. (Eds.): Intel. Multimedia Analysis for Security Appli., SCI 282, pp. 237–252.
springerlink.com

Consequently, a robust hash is unique for a group of perceptually equivalent signals (for a fixed algorithm and parameters), whereas a fingerprint is an arbitrary code not intrinsically linked to a particular signal. Also most robust hashing applications do not involve modifying the signal associated with the hash —as we will discuss in the next section, only authentication applications require this—, while all fingerprinting applications do since they are a subset of data hiding. Finally, although robust hashing applications could in principle be tackled solely by means of watermarking/fingerprinting procedures, the main practical advantage of robust hashing lies in its ability to deal with legacy environments or with situations where it is not practical to modify all the instances of the signals to be identified.

1.1 Some Background to Robust Hashing

Arguably [3] or [4] are among the first papers on robust hashing topics, but it is fair to say that it was the data hiding community who gave a boost to the subject starting from 1999 [5, 6, 7, 8]. Ever since those early days, the number of publications in the field has steadily grown until attaining more than 60 papers per year a decade later, without signs of abatement. This level of activity can be explained by the fact that robust hashing is mainly driven by two broad application areas that have wide implications:

- Multimedia identification, where robust hashing is exploited to reveal the identity of multimedia files which may have been processed without alteration of their perceptual features, or whose metadata (for example, their file names) may have been modified. Sometimes this is also called copy detection or replica detection. The main advantage of using robust hashing in this context with respect to alternative procedures is that robust hashes are typically smaller than their corresponding signals. This allows for small databases with respect to the size of the data to be hashed, and also for fast query times. There are several scenarios where multimedia content identification crops up, which usually correspond either to broadcast monitoring and content filtering in multimedia networks, or to indexing in multimedia databases. Robust hashing overlaps in this last context with techniques such as context-based indexing and retrieval (CBIR), although it is necessary to stress that CBIR is not concerned with cryptography-related security features.
- Multimedia authentication (or multimedia self-identification), where robust hashing plays a role as one of the two stages in the so-called quantize-and-embed schemes [9]. These practical methods, which are suboptimal, entail embedding a distortion-resistant low-dimensional descriptor of a signal —that is, essentially a robust hash— within that very same signal by means of data hiding. Authentication hinges on the fact that a host should remain perceptually untouched by the application of a digital watermark. Owing to this fact, one can retrieve the robust hash in two independent ways, which should both give the same answer if the signal has not been tampered with: 1) by hashing the watermarked signal, without recourse to the watermark;

and 2) by decoding the watermark. The size of the hash value should again be as small as possible in this application.

In recent years there has been plenty of activity in the proposal of practical robust hashing methods (see for instance [10] for a review of audio hashing algorithms), some of them with remarkable commercial impact [8, 11]. Basic groundwork dealing with the general formulation of robust hashing has been more scarce to date. The first paper discussing general aspects of the problem is probably [7]. Other such attempts were given in [12] and [13]. These attempts have combined in different ways techniques such as feature extraction, quantization and error correction codes, in a bid to get some insight in the core elements of robust hashing. The view of robust hashing as a mishmash of these techniques is perhaps behind the difficulty that the discipline has had to carve out an identity of its own. Indeed, it is not uncommon to hear the opinion that robust hashing is "nothing new under the sun" with respect to existing techniques devoted to classification, feature extraction, or quantization. These opinions are doubtless also partially due to the difficulties found in unequivocally defining a completely satisfactory robust hashing framework: what are the central elements of the problem?

In this chapter I will address this question, focusing on the assumption that robust hashing should inherit key security properties which are a staple of cryptographic hashing.

Notation

Uppercase (X, $\mathbf{X}$) and lowercase (x, $\mathbf{x}$) letters denote random and deterministic variables, respectively, where boldface letters denote row vectors. We denote by $X \sim p_X$ that $p_X(x)$ is the probability mass function (pmf) or probability density function (pdf) of X, depending on whether the variable is discrete or continuous. $E[X]$, $\mathrm{Var}[X]$, and $H(X)$ are the expectation, variance, and entropy of X, respectively, and $I(X;Y)$ is the mutual information between X and Y. Calligraphic letters ($\mathcal{A}$) denote discrete sets, and the elements of a discrete set are given enclosed within brackets, that is, $\mathcal{A} = \{a_1, \cdots, a_N\}$, where N is the cardinality of $\mathcal{A}$, denoted by $|\mathcal{A}|$. $\mathcal{A}^n = \mathcal{A}\times\cdots\times\mathcal{A}$ is the n-dimensional Cartesian product of $\mathcal{A}$.

2 The Sources of Robust Hashing

In order to understand what should be required from robust hashing techniques it will be necessary to reflect, even if in a cursory manner, on their main influences. This is something that is sometimes glossed over in the robust hashing literature, and it will hopefully shed light on determining what prior research is applicable to the problem at hand, and what further work is required.

2.1 Cryptographic Hashing

Simplifying matters a lot, a cryptographic hash, such as those produced by the popular MD5 [14] (now obsolete) and SHA [15] algorithms, is basically a low-dimensional representation of a binary signal. Further features —that we will discuss later— are also key to understanding cryptographic hashing techniques. The parallelism of this informal definition of cryptographic hashing with the also informal one of robust hashing given at the start of this chapter is not by chance, since, as we have discussed, the origin of robust hashing is found in the attempt to overcome the limitations of cryptographic hashing methods when dealing with multimedia inputs. One could say that robust hashing unabashedly aims at being "cryptographic hashing for multimedia".

In order to study the relationship between cryptographic and robust hashing, let us assume a generic public cryptographic hashing function of the form

$$\underline{h}(\cdot) : \{0,1\}^n \rightarrow \{0,1\}^m. \tag{1}$$

This is just a function with an n-dimensional binary input and an m-dimensional binary output. Unfortunately, there is no full consensus about the general requirements for this function [16]. We will be assuming here that m is fixed and "small" so that we have that $m \ll n$ for "large" n, but the reader must be aware that some authors consider cryptographic hashing methods with $m = n$. In the requirements definition it will be convenient to use the Hamming distance function, which is of the form

$$d_H(\cdot,\cdot) : \{0,1\}^n \times \{0,1\}^n \rightarrow \{0,1,2\ldots,n\}, \tag{2}$$

and which gives the number of positions in which two n-length (binary) vectors differ. As a minimum common denominator we typically find the following properties in most cryptographic hash functions [1]:

1. Preimage resistance (one-way function): given $\mathbf{h} \in \{0,1\}^m$, it must be computationally infeasible to find $\mathbf{x} \in \{0,1\}^n$ such that $\underline{h}(\mathbf{x}) = \mathbf{h}$.
2. Collision resistance: it must be computationally infeasible to find any $\mathbf{x}$ and $\mathbf{x}'$ such that $d_H(\mathbf{x},\mathbf{x}') > 0$ and $\underline{h}(\mathbf{x}) = \underline{h}(\mathbf{x}')$.
3. Second preimage resistance: given $\mathbf{x}$ and $\mathbf{h}$ such that $\underline{h}(\mathbf{x}) = \mathbf{h}$, it must be computationally infeasible to find $\mathbf{x}'$ such that $d_H(\mathbf{x},\mathbf{x}') > 0$ and $\underline{h}(\mathbf{x}') = \mathbf{h}$. Sometimes, but not always, this is assumed to be implied by collision resistance. For simplicity we will assume this implication to be true in the following.

The definitions of the three properties above are imprecise on the same single point: how does one measure computational infeasibility? This is the reason why "cryptographic hashing" bears the "cryptographic" label: key-dependent functions which are empirically known to be hard to invert without knowing the secret —such as the ones used in cryptography— lie at the heart of cryptographic hashing techniques. Dimensionality reduction helps to make invertibility even harder since in that case $\underline{h}(\cdot)$ cannot be a bijection, but it will inevitably decrease preimage and collision resistance.

The probabilistic view of cryptographic hashing

There is another insightful view of cryptographic hashing, which will prove more useful to define the properties that we want to achieve with robust hashing. Mıhçak and Venkatesan [7] were probably the first to use this view in order to translate cryptographic properties into the robust hashing problem. The probabilistic perspective is afforded by observing a hashing function as a random oracle [1], in which the output of the hashing function is a random variable $\mathbf{H}$ produced by feeding a random input $\mathbf{X}$ with pmf $p_{\mathbf{X}}$ into a deterministic hashing function $\underline{h}(\cdot)$. Using this probabilistic approach we may rewrite the first property above as follows:

1. Preimage resistance: $\underline{h}(\mathbf{X})$ is a uniform random variable not correlated with $\mathbf{X}$; furthermore, for any $\mathbf{X}' \sim p_{\mathbf{X}}$, $\underline{h}(\mathbf{X})$ is independent of $\underline{h}(\mathbf{X}')$ (pairwise independence).

A uniform distribution maximizes the entropy of a discrete random variable, therefore maximising uncertainty for an observer of $\mathbf{H} = \underline{h}(\mathbf{X})$. The lack of correlation between input and output makes it difficult to guess the function by observing several outcomes of the oracle, which makes it difficult to invert. Note that in practice this requirement provides pairwise independence already, and that it is stronger than just requiring $\underline{h}(\mathbf{X})$ to be uniform.

We may now see properties 2 and 3 as consequences of property 1:

2. Collision resistance:

$$P_c \triangleq \Pr[\underline{h}(\mathbf{X}) = \underline{h}(\mathbf{X}')] = 2^{-m}, \quad \text{for } \mathbf{X}, \mathbf{X}' \sim p_{\mathbf{X}}. \tag{3}$$

   This is equivalent to minimizing the probability of collision P_c between two random input signals with the same distribution.
3. Second preimage resistance: for any given $\mathbf{h}$ it should hold that

$$\Pr[\underline{h}(\mathbf{X}) = \mathbf{h}] = 2^{-m}. \tag{4}$$

Discussion

One would like, of course, to translate these properties into the domain of robust hashing. The most serious difficulty is found with properties 2 and 3, since they imply that two input signals differing in one single bit must lead to radically different cryptographic hashes. In fact, when defining these properties we have emphasized the fact that the null Hamming distance is implicit when enforcing equality in cryptographic hashing. However, as we have discussed in the introduction, this behaviour will be utterly misleading for hashing of multimedia signals, for which the difference of one single bit will typically be completely irrelevant from a perceptual viewpoint.

Clearly, the way distances are measured in cryptographic hashing has to be modified in order to achieve our goal. We might think that a nonzero threshold on the Hamming distance might solve our problems, but unfortunately it is difficult to capture perceptual features just through this measurement.

2.2 Source Coding with Distortion Constraints

In order to explore the metamorphosis of cryptographic hashing into something more akin to what robust hashing should be, we need to rely on procedures that allow for signal representation in a less strict way. This immediately leads to a number of candidate techniques, which can all be traced back to some articulation of concepts belonging to source coding with distortion constraints. Therefore we will call them source coding methods in a somewhat loose manner. Since the theory that underpins source coding is expressed probabilistically, it will become clear that the probabilistic rephrasing of the properties of cryptographic hashing given in the previous section will make it easier its integration with source coding.

Some examples of techniques which ultimately implement concepts of source coding with distortion constraints are:

- quantization
- lossy compression
- clustering and classification

All of them provide straightforward ways to derive a low-dimensional descriptor from an input signal, in such a way that signals which are equivalent according to some fidelity criterion will lead to equal or perhaps similar descriptors. Although the actual implementations will differ, in all these cases a function resembling (1) may be established, having the form

$$h(\cdot) : \mathbb{R}^k \to \{0,1\}^m. \tag{5}$$

This function divides the input space into 2^m disjoint regions $\mathcal{R}_i$, each of which maps into the same value (quantization regions, clustering regions, etc). For convenience, and unlike in (1), we have now required the input signal to have support in $\mathbb{R}^k$, since in order to measure the variability inherent to multimedia signals it makes good sense to approximate them as real-valued. Notice nonetheless that the dimensionality of the input space of (5) is actually the same as that of (1) if b bits per sample are used to represent the input, which implies that $n = b\,k$ for the corresponding n-length binary vector. Unlike in standard cryptographic hashing, the output range in (5) is typically not fixed. Rather the size of the output may depend on k through some rate $R < 1$, that is, $m = R\,k$.

The fidelity criterion that we previously mentioned is established by means of a distortion function of the form

$$d(\cdot,\cdot) : \mathbb{R}^k \times \mathbb{R}^k \to \mathbb{R}^+, \tag{6}$$

which establishes how well we can reconstruct a signal $\mathbf{x}$ from its low-dimensional descriptor $\mathbf{h} = h(\mathbf{x})$ using a certain reconstruction function of the form

$$r(\cdot) : \{0,1\}^m \to \mathbb{R}^k, \tag{7}$$

such that the reconstruction is obtained as $\widehat{\mathbf{x}} = r(\mathbf{h})$. The distortion function (6) may be stated explicitly, such as in quantization or lossy compression, or implicitly, as it happens in classification/clustering problems.

A threshold D on the distortion between the reconstruction and the original allows to capture the perceptual "granularity" of the lossy representation of a signal. Optimality in source coding scenarios is achieved when the descriptor requires as few bits as possible for achieving a distortion constraint. This constraint is typically expressed as

$$E[d(\mathbf{X}, \widehat{\mathbf{X}})] \leq D. \tag{8}$$

The absolute minimum source coding rate for a practical method is given by the rate-distortion function $R^*(D) \triangleq \min I(X; \widehat{X})$ bits/signal sample [17], that is, the minimum mutual information between the original signal and its reconstruction.

It is also worth recalling that the reconstruction function can be aided by side information at the decoder [18], which can be exploited to further decrease the rate R and then the size of the descriptor $\mathbf{h}$. Intuitively, if the decoder can access a signal $\widetilde{\mathbf{x}}$ related to the original $\mathbf{x}$ (for instance, a noisy version of $\mathbf{x}$), then the encoder can afford to use less information to describe it since $\widetilde{\mathbf{x}}$ contains information about $\mathbf{x}$ already. In this case the reconstruction function is of the form

$$r(\cdot, \cdot) : \{0, 1\}^m \times \mathbb{R}^k \rightarrow \mathbb{R}^k. \tag{9}$$

We can then obtain the reconstructed signal as $\widehat{\mathbf{x}} = r(\mathbf{h}, \widetilde{\mathbf{x}})$, with the same distortion constraint as before. Optimality in the side-informed setting is achieved when the rate of a method achieves $R^*_{\text{si}}(D) \triangleq \min I(U; \widehat{X}) - I(X; U)$ under (8), with U an auxiliary random variable that acts as a proxy for X.

Discussion

So far so good with respect to the "goodness of fit" of source coding with distortion constraints for robust hashing. The low-dimensional descriptor $\mathbf{h}$ obtained from a given signal $\mathbf{x}$ allows for a degree of tolerance in variations of $\mathbf{x}$ which is absent in cryptographic hashing, and therefore plays a similar role to that of a robust hash corresponding to $\mathbf{x}$. A conspicuous difference is that, unlike in cryptographic hashing, it is the actual reconstruction representative $\widehat{\mathbf{x}}$ that matters in source coding, and not the actual $\mathbf{h}$ value attached to it. This value simply acts as a *label* for a quantization or clustering region. More importantly, not only all source coding approaches are oblivious to cryptographic hashing concepts such as preimage and collision resistance, but they are in direct opposition to them. For instance, a too lax approach to the definition of similarity will make collisions highly likely and also turn preimage resistance into wishful thinking. Also adjacent classes, clusters, or quantization regions (that is, regions which map into unique descriptors) may or may not be suitable for collision resistance depending on the labels attached to these regions. That is to say, the robust hashing scenario requires *semantics* which source coding lacks. Finally, the uniformity of the descriptors produced is not guaranteed *a priori* when regions are designed just to meet a distortion constraint.

Although flexibility in representing similar signals is a very appealing property for robust hashing, source coding with distortion constraints still has to be reconciled somehow with the set of properties discussed in Section 2.1. These properties are also key to obtaining robust hashing methods truly resembling cryptographic hashing.

Furthermore, even if we would conceivably drop properties typical of cryptographic hashing such as preimage and collision resistance, standard source coding techniques would still not be suitable for two relevant multimedia identification scenarios: 1) desynchronized signals, that is, differently aligned instances of the same input signal, are an important concern for robust hashing; and 2) matching of a signal against a small portion of it —which we may term clipped matching— is an important challenge in some applications. Notice that both issues are related, since clipped matching implies misalignment and therefore lack of synchronization. Therefore they both run into the same problem: very different descriptors for signals with many similarities. We will defer the discussion of synchronization issues until Section 4.

3 Fundamentals of Robust Hashing

The very fact that robust hashing —with the properties previously discussed— exists at the sometimes contradictory intersection between the two fields in Sections 2.1 and 2.2 warrants its study as an area of its own right. In any case, the previous discussion allow us, at the very least, to see that none of these fields suffices *on their own* to tackle all robust hashing goals. Whether the time has come to rebrand robust hashing as something akin to *source-encoded cryptographic hashing* is something yet to be decided.

The issues faced by robust hashing of real multimedia signals are complex. It is unlikely that these issues can be disentangled if we are not able to clearly define first what robust hashing is, and to understand simple cases adhering to well-established design principles. Consequently, the main purpose of this section is to establish a framework for the robust hashing problem that is able to conjugate the properties of the cryptographic hashing and source coding with distortion constraints scenarios, and to try to find its implications. The ultimate goal is to unambiguously establish the core requirements of robust hashing, so as to shed light on its relationships with cognate techniques.

Before proceeding, it is important to remark that all robust hashing systems operate according to two basic stages:

1. Enrollment (acquisition) stage, in which a set of signals is hashed and the robust hashes are stored for future reference.
2. Matching (comparison, identification) stage, in which arbitrary signals are presented to the system, and a match of their hash against the hash values stored in the enrollment stage is attempted.
   We will only deal here with exact matches. It is also possible to consider matching based on multiple hypotheses, with associated reliabilities or scores, such as in CBIR scenarios. Nevertheless, this type of graceful matching degradation (as two signals become less and less similar) is clearly

at loggerheads with a system possessing preimage resistance, which is something that we absolutely need in order to echo cryptographic hashing in robust hashing. As we will see next, in order to achieve this goal we will assume that similarity is solely encapsulated by source coding with distortion constraints, and not by the hash values themselves.

3.1 A General Basic Setting

We will discuss in this section the simplest possible robust hashing setting, in which we assume: 1) entire signals are presented at the input of the hashing system; 2) a pdf $p_{\mathbf{X}}(\mathbf{x})$ of the input signals is available, and the elements X of $\mathbf{X}$ are independent and identically distributed (i.i.d.); and 3) there are no synchronization issues, meaning that the geometrical integrity and scale of the signals is preserved between the enrollment and matching stages. For instance, if we enroll a signal represented by $\mathbf{x}$, any processing carried out on $\mathbf{x}$ is assumed not to scale (geometrically or in amplitude), shift, or clip the signal, and therefore versions of $\mathbf{x}$ subsequently presented to the system are essentially of the form $\mathbf{y} = \mathbf{x} + \mathbf{n}$.

It is worthwhile remarking that a transform and/or feature extraction stage (see Section 4) will be typically required in order to approximate the strict requirements on the distribution of $\mathbf{X}$ in practical robust hashing systems. Despite their ubiquity in practice, we do not consider feature extraction to be a truly core issue in establishing a robust hashing framework informed by the guidelines in Section 2. Notice the parallel with data hiding theory, whose core principles are also independent of preprocessing stages such as the ones mentioned, although they also usually benefit in practical settings of transforms or projections.

The parameter space of the robust hashing function is the same as in function (5) for source coding techniques, which in turn is similar to (1). In order to achieve our goal of conjugating cryptographic hashing and source coding, it is convenient to consider robust hashing functions as the composition $h = f \circ g$ of two distinct functions, that is,

$$\mathbf{h} = h(\mathbf{x}) = g(f(\mathbf{x})), \tag{10}$$

where

- $f(\cdot)$ is a *mapping* function, which embodies the source coding with distortion constraints aspect underlying robust hashing, and which is of the form

$$f(\cdot) : \mathbb{R}^k \rightarrow \mathcal{H}, \tag{11}$$

where $\mathcal{H}$ is a set of integers representing $|\mathcal{H}|$ regions $\mathcal{R}_1, \cdots, \mathcal{R}_{|\mathcal{H}|}$ into which $\mathbb{R}^k$ is divided. For convenience we will choose $|\mathcal{H}| = 2^t$. These *hashing regions* are disjoint and tile $\mathbb{R}^k$, that is, $\cap_{i=1}^{|\mathcal{H}|}\mathcal{R}_i = \emptyset$ and $\cup_{i=1}^{|\mathcal{H}|}\mathcal{R}_i = \mathbb{R}^k$. The function (11) is defined as

$$f(\mathbf{x}) = i, \quad \text{if } \mathbf{x} \in \mathcal{R}_i. \tag{12}$$

The regions $\mathcal{R}_i$ play a similar role as those in a source coding context, and thus they are subject to certain constrains that we will discuss later.

- $g(\cdot)$ is a *labelling* function, which embodies the cryptographic hashing part of the robust hashing problem, and whose pattern is

$$g(\cdot) : \mathcal{H} \rightarrow \{0, 1\}^m, \tag{13}$$

with $m \leq t$. Therefore (13) labels each hashing region with a binary symbol, which constitutes the final robust hash value. We will assume $g(\cdot)$ to be evenly spread, in the sense that each output maps back to exactly 2^{m-t} input values.

It must be noted from the two functions above that, according to this framework, dimensionality reduction in robust hashing is achieved in a two-step process (unless $t = m$). Also the labelling function effectively blends together hashing regions (for $t > m$).

Side information

Before examining the problem requirements under the "division of labour" proposed above, a remark is due on the use of side information in the source coding part of the robust hashing problem. Some authors have posited side-informed robust hashing along the lines of side-informed source coding with distortion constraints (see for instance [19, 13]). Side information at the decoder is provided by the distorted version of $\mathbf{x}$ that the decoder is trying to match, that is, $\widetilde{\mathbf{x}} = \mathbf{x} + \mathbf{n}$. Exploiting side information does allow to decrease hash size (an important concern, as we discussed in Section 1.1), for a given perceptual distortion constraint implied by the hashing regions. Nonetheless reliance on side information may not be desirable in all applications of robust hashing. This is because of the fact that the use of side information will create unconnected hashing regions through the use of binning schemes [18]. The implications of this are as follows:

- In authentication applications, side information is indeed useful because the hash value needs to be as small as possible (since it needs to be embedded in the hashed signal) and only a binary authentication decision has to be made. The decoder verifies whether the hash value matches the one embedded, and, if the system is well designed, it is unlikely that the signal can be tampered with in such a way that the robust hash is preserved.
- In content identification applications, matching amounts to choosing a hash value among all the enrolled ones. As with side information the same index will be assigned to hashing regions that are perceptually different, this will unavoidably increase the probability of collision with respect to a equivalent non-side-informed approach. Bypassing this shortcoming requires that comparisons are made not through robust hashes, but through reconstructions based on robust hashes —recall our discussion on side-informed source coding. This arguably defeats one of the main purposes of robust hashing: to achieve identification *solely* through low-dimensional descriptors.

3.2 Requirements

Let us see next how to conjugate the requirements of cryptographic hashing with the properties of source coding.

- Preimage resistance. If $\mathbf{X}$ is i.i.d., both uniformity of the hash values distribution and independence between hash values are guaranteed by the condition

$$\Pr[f(\mathbf{X}) = i] = 2^{-t}, \quad \text{for all } i \in \mathcal{H}. \tag{14}$$

This can be straightforwardly achieved by designing $f(\cdot)$ such that the regions $\mathcal{R}_i$ fulfil the following equation:

$$\int_{\mathcal{R}_i} p_{\mathbf{X}}(\mathbf{x})\, d\mathbf{x} = 2^{-t}, \quad \text{for all } i \in \mathcal{H}. \tag{15}$$

Note that the choice of these 2^t regions is not unique; in fact, (15) does not even imply that they have to be connected. Nonetheless an additional constraint has to be put in place because, as in source coding with distortion constraints, we know that the hashing regions must also conform to a perceptual constraint such as (8). In this way signals belonging to the same region, which get mapped onto the same hash value, are also guaranteed to be perceptually similar. The fact that we have to impose an additional constraint means that, the optimum rate-distortion function for describing the hashing regions will be lower bounded by $R^*(D)$ (see Section 2.2), all other conditions being the same.

One possible region design approach consists of choosing connected regions compliant with (15) —which just requires inverting the cumulative distribution of $\mathbf{X}$— and verifying afterwards that the distortion associated to assigning one reconstruction representative to each of these regions complies with (8). If not, the rate has to be increased, that is, we need more hashing regions, and hence bigger hash values. For symmetric $p_{\mathbf{X}}$ and $R(D) = 1$ bit/signal sample, the regions for source coding and robust hashing coincide, using the aforementioned design approach. This simple arrangement is used in a number of methods (see for instance [8]) which obtain robust hashes as thresholded scalar values obtained using transforms of the signal to be hashed.

Unlike in source coding with distortion constratints, assigning reconstruction representatives to the hashing regions is not strictly required. Robust hashing is not concerned problem with rebuilding a signal from a descriptor according to a distortion criterion (unless side information is exploited, as we have discussed), but with endowing that descriptor with an altogether different set of properties (see Section 2.1). Therefore we may establish the following set of similarity constraints:

$$E[d(\mathbf{X}, \mathbf{X}') \,|\, \mathbf{X}, \mathbf{X}' \in \mathcal{R}_i] \leq D, \qquad \text{for all } i \in \mathcal{H}. \tag{16}$$

Measuring similarity with a measure like (8) may be more amenable to analysis, glossing over the fact that no reconstructions are needed. In any case, this discussion on the achievement of preimage resistance in robust hashing is illustrative of the fact that region design in this problem constitutes a major departure with respect to the way region design is tackled in source coding settings.

Finally, in order to guarantee that the output of $h(\cdot)$ is not correlated with its input we also need a labelling function $g(\cdot)$ that not only is uniform, as we have stated already, but also able to randomize the mapping between the 2^h regions and 2^m labels. This requires that the function (13) is also unknown by third parties, that is, key dependent. We will return to the issue of the key in Section 3.3.

- Collision resistance. The cryptographic hashing requirement (3) is similarly expressed in this context as the minimization of

$$P_c = \Pr[h(\mathbf{X}) = h(\mathbf{X}')], \tag{17}$$

which amounts to achieving $P_c = 2^{-m}$. For i.i.d. $\mathbf{X}$, preimage resistance implies collision resistance, in the same way as in the cryptographic hashing case. If P_c is minimized, the Hamming distance between the hash values $d_H(h(\mathbf{x}), h(\mathbf{x}'))$ of two arbitrary signals $\mathbf{x}, \mathbf{x}'$ will not follow any obvious pattern, that is, it will not correlate on average neither with the distance $d(\mathbf{x}, \mathbf{x}')$ nor with the direction $\mathbf{x} - \mathbf{x}'$, unless the distance between them is below D on average. In this way members of a region map into the same robust hash value, which in turn will be uncorrelated with hash values corresponding to other regions.

3.3 Additional Performance Measures

Preimage and collision resistance are the core requirements for a robust hashing method that aims at emulating cryptographic hashing. As such, measuring how well these requirements are implemented by a practical robust hashing method should be the main way to gauge its suitability for that task. Nevertheless there are additional performance measures for analyzing and comparing robust hashing systems that are also worth discussing.

- Robustness. Since robust hashing should withstand small distortions that do not radically alter the perceptual content of a signal, it makes sense to benchmark the probability of error when subjecting signals to be hashed to distortions. The parameter to be measured in this situation is

$$P_e \triangleq \Pr[h(\mathbf{X}) \neq h(\mathbf{X} + \mathbf{N})]. \tag{18}$$

The fact that the system is designed for $p_{\mathbf{X}}$ and not for $p_{\mathbf{X}+\mathbf{N}}$ —which is unknown to the designer— is the source of errors in this scenario. As similarity depends on a given D, we would like to obtain a low P_e for a variance of $\mathbf{N}$ comparable to this threshold.

Notice that (18) is independent of the labelling applied. An alternative measure of robustness which does depend on $g(\cdot)$ is the probability of bit error, given by

$$P_b \triangleq \frac{1}{m} \sum_i \Pr[h_i(\mathbf{X}) \neq h_i(\mathbf{X} + \mathbf{N})], \tag{19}$$

were $h_i(\cdot)$ refers to the i-th bit of the hash value. It is possible to sacrifice some basic cryptographic properties of robust hashing in order to improve robustness as measured by P_b. This can be done by using a labelling function that, instead of trying to decorrelate the physical proximity of the hashing regions in order to achieve preimage resistance, exploits it instead in order to obtain a graceful degradation the hash under distortion (such as in CBIR contexts). One example of this procedure for scalar X and $t = m$ consists of using a Gray code as the labelling function, in such a way that robust hashes corresponding to neighbouring regions differ only in one bit. In this case the bit error rate under distortion will decrease gracefully with the variance of the distortion beyond the similarity target established by D.

- Security of keyed robust hashing. In applications such as indexing it is acceptable to publicly disclose the hashing function $h(\cdot)$. This is not wise for authentication-related scenarios, but keeping the robust hashing function secret amounts to violating the Kerckhoffs principle —all the details of the system must be disclosed. Therefore we will have to assume that security solely relies on a secret key K parametrizing the system, that is, $h_K(\cdot)$. Although the hashing regions defined by the mapping function $f(\cdot)$ might be made key-dependent, this would only make their design even harder, since as we saw in Section 3.1 they involve two fundamental constraints already. For this reason, the simplest way to introduce a secret key in the robust hashing process is through a keyed $g_K(\cdot)$ function. For $t = m$ this could be simply achieved by choosing one among the $2^m!$ permutations possible. For more typical cases in which $t > m$ the best approach would be to rely on a standard cryptographic hashing function, designed along cryptographic principles, and able to guarantee the uniformity condition.

  It is possible to identify a number of robust hashing security scenarios parallel to the ones studied in data hiding (originally proposed in [20], and in turn inspired in Diffie-Hellman settings for cryptography). Two of these scenarios, which assume that the secret key K stays constant, can be meaningfully translated into robust hashing contexts:
  - Known original attack (KOA), in which the attacker observes N_o hash values and input signal pairs. In this situation it is possible to study the average information leakage about the key due to the availability of that data, which using the conditional entropy of K, or equivocation, is

    $$H(K|\mathbf{H}^{(1)}, \cdots, \mathbf{H}^{(N_o)}, \mathbf{X}^{(1)}, \cdots, \mathbf{X}^{(N_o)},). \tag{20}$$

    The problem can also be formulated as obtaining the mutual information between the key and these N_o pairs of signals.
  - Hash only attack (HOA), in which the attacker stands in the less favourable case of having only has access to N_o robust hash values. Security may be assessed in a similar way as in (20).

  These types of problems have been considered already in robust hashing scenarios in [21].
- Oracle attacks. Robust hashing systems may also be subject to forgeries when the whole hashing function is available as a black box, that is, without

the key being explicit. An attacker may exploit this situation to forge a signal which is not perceptually similar to another one, but that yields the same hash. Equivalently, the attacker wants to find $\mathbf{x}'$ such that $h(\mathbf{x}) = h(\mathbf{x}')$ while roughly achieving $d(\mathbf{x}, \mathbf{x}') > D$. This is a sort of rephrasing of the sensitivity attack (oracle attack) from data hiding or cryptography. Algorithms like the Blind Newton Sensitivity Attack [22] from data hiding could perhaps be adapted to this scenario, allowing to evaluate the practical security of different methods. Notice that a system will be clearly weak to oracle attacks if its probability of collision is high (or if its preimage resistance is low). However a low P_c does not necessarily imply that it is difficult for an intentional attacker to forge a suitable $\mathbf{x}'$.

4 Robust Hashing and Desynchronization

Differently to the ideal setting considered in Section 3.1, input signals not synchronized with those initially enrolled are quite common in robust hashing settings. By desynchronization we mean a lack of alignment between the original signal used in the enrollment stage and the one present at the matching stage. For instance, if the signal originally enrolled was $\mathbf{x} = \{x_1, x_2, \cdots, x_n\}$, desynchronized versions may be for instance $\mathbf{y}_1 = \{0, x_1, x_2, \cdots, x_{n-1}\}$ or $\mathbf{y}_2 = \{x_i, x_{i+1}, \cdots x_{i+\Delta}\}$. The fundamental question of robust hashing synchronization was first posed in [7].

In communications problems —for instance, in data hiding— synchronization issues can be addressed by means of pilot signals or pilot symbols interspersed or overlapped with the information-carrying signals. In robust hashing this possibility has to be ruled out, since the system has no control whatsoever on the input signals.

A naive way to achieve robust hashing synchronization could be to have a database of enrolled signals —that is, not only of enrolled hashes— against which to attempt registration. This approach is however not scalable, and moreover it defeats one of the main purposes of robust hashing which is to attempt identification solely by means of low-dimensional descriptors.

In practice robust hashing synchronisation issues are tackled in two different ways, depending on two different synchronization scenarios:

- Global desynchronization. A previously enrolled signal may be presented again in full to the system at a later time, but desynchronized with respect to the original. An example of this case is the signal $\mathbf{y}_1$ above with respect to $\mathbf{x}$. A more complex case would for instance an image rotated with respect to the one initially enrolled. Depending on the scenario considered, global desynchronization can be dealt with by means of the use of transforms leading to invariant domains such as the Fourier-Mellin transform [23], the Radon transform [24], and others.
- Desynchronization due to partial matching. This important feature deals with the ability to cope with partial signal samples presented at the matching stage, such as the signal $\mathbf{y}_2$ above. A strategy to deal with this situation, first

proposed in [8] to identify full audio files by means of short snippets, consists in dividing $\mathbf{x}$ into T heavily overlapped short subsignals $\mathbf{x}_j^s$ of length $l < k$, which are then individually hashed in the enrollment stage. If the degree of overlap is given by $0 < \alpha < 1$, then $T = \lceil k/(1-\alpha)l \rceil$. In this way, any randomly clipped sample will roughly correspond with high likelihood to one of the previously hashed samples if $\alpha \to 1$. As in the global desynchronization case, a transform or feature extraction stage is still required to withstand the ensuing fine desynchronization. This strategy pays the price of generating much longer hashes.

Therefore the use of transforms as an initial stage before the application of a core robust hashing function —such as the one discussed in Section (10)— is a fact of life in practical robust hashing methods wanting to cope with desynchronization. There are two other benefits derived from the use of transforms or feature extraction stages: 1) they can be keyed; and 2) they can provide domains with good statistical models (for instance, in the wavelet domain used in [6]), which we have seen are the only way in which it is feasible to guarantee cryptographic properties.

5 Conclusion

A general overview of the robust hashing problem has been presented, discussing the core driving forces behind this area, namely, cryptographic hashing and source coding with distortion constraints, and proposing a general framework to design robust hashing methods based on combining concepts from these disciplines. We have argued that despite the fact that the core robust hashing problem is independent of transforms or of feature extraction stages, these are needed in practice both in order to cope with synchronisation issues and in order to provide domains with reliable statistical models. It is likely that future efforts in the area will focus on the security properties of robust hashes, which have been up to now mostly overlooked in favour of robustness evaluations. Also, the development of methods able to achieve optimal behaviour in the basic setting herein discussed is still open. Since this setting is independent of the nature of the multimedia signals and of the use of initial transforms, such optimal methods would find general application in robust hashing.

References

1. Oppliger, R.: Contemporary Cryptography. Artech House (2005)
2. Kilian, J., Leighton, F., Matheson, L., Shamoon, T., Tarjan, R., Zane, F.: Resistance of digital watermarks to collusive attacks. In: Proceedings of the IEEE International Symposium on Information Theory, p. 271 (1998)
3. Subramanya, S., Simha, R., Narahari, B., Youssef, A.: Transform-based indexing of audio data for multimedia databases. In: Proceedings of IEEE International Conference on Multimedia Computing and Systems 1997. pp. 211–218 (1997)
4. Cohen, S., Guibas, L.: Shape-based image retrieval using geometric hashing. In: Procs. of the ARPA Image Understanding Workshop (1997)

5. Fridrich, J.: Robust bit extraction from images. In: Proc. IEEE ICMCS, Florence, Italy, pp. 536–540 (1999)
6. Venkatesan, R., Koon, S., Jakubowski, M., Moulin, P.: Robust image hashing. In: Procs. of the IEEE International Conference on Image Processing, Vancouver, Canada (2000)
7. Mihçak, M.K., Venkatesan, R.: A perceptual audio hashing algorithm: A tool for robust audio identification and information hiding. In: Moskowitz, I.S. (ed.) IH 2001. LNCS, vol. 2137, pp. 51–65. Springer, Heidelberg (2001)
8. Haitsma, J., Kalker, T., Oostveen, J.: Robust audio hashing for content identification. In: Procs. of the International Workshop on Content-Based Multimedia Indexing, Brescia, Italy, pp. 117–125 (2001)
9. Martinian, E., Wornell, G., Chen, B.: Authentication with distortion criteria. IEEE Transactions on Information Theory 51(7), 2523–2542 (2005)
10. Cano, P., Batlle, E., Kalker, T., Haitsma, J.: A review of audio fingerprinting. J. VLSI Signal Process. Syst. 41(3), 271–284 (2005)
11. Wang, A.: The Shazam music recognition service. Communications of the ACM 49(8), 44–48 (2006)
12. McCarthy, E.P., Balado, F., Silvestre, G.C.M., Hurley, N.J.: A framework for soft hashing and its application to robust image hashing. In: Procs. of the IEEE International Conference on Image Processing, Singapore (2004)
13. Voloshynovskiy, S., Koval, O., Beekhof, F., Pun, T.: Conception and limits of robust perceptual hashing: towards side information assisted hash functions. In: Media Forensics and Security, vol. 7254. SPIE, San Jose (2009)
14. Rivest, R.: The MD5 message-digest algorithm. Technical report, IETF Network Working Group (1992)
15. Federal Information Processing Standard: Secure hash standard. Technical report, National Institute of Standards and Technology, US Department of Commerce, Washington, USA (1995)
16. Contini, S., Steinfeld, R., Pieprzyk, J., Matusiewicz, K.: A critical look at cryptographic hash function literature. In: ECRYPT Hash Workshop (2007)
17. Shannon, C.E.: A mathematical theory of communication. Bell System Technical Journal 27, 379–423, 623–656 (1948)
18. Pradhan, S., Chou, J., Ramchandran, K.: Duality between source coding and channel coding and its extension to the side information case. IEEE Transactions on Information Theory 49(5), 1181–1203 (2003)
19. Johnson, M., Ramchandran, K.: Dither-based secure image hashing using distributed coding. In: Procs. of the IEEE International Conf. on Image Processing, Barcelona, Spain, vol. 2, pp. 751–754 (2003)
20. Cayre, F., Fontaine, C., Furon, T.: Watermarking security: theory and practice. IEEE Transactions on Signal Processing 53(10), 3976–3987 (2005)
21. Koval, O., Voloshynovskiy, S., Beekhof, F., Pun, T.: Security analysis of robust perceptual hashing. SPIE Conference Series, vol. 6819 (2008)
22. Comesaña, P., Pérez-Freire, L., Pérez-González, F.: Blind Newton sensitivity attack. IEE Proceedings of Information Security 153(3), 115–125 (2006)
23. Swaminathan, A., Mao, Y., Wu, M.: Robust and secure image hashing. IEEE Transactions on Information Forensics and Security 1(2), 215–230 (2006)
24. Lefebvre, F., Macq, B., Legat, J.D.: Rash: Radon soft hash algorithm. In: Eurasip (2002)

Content-Based Video Copy Detection – A Survey

Shiguo Lian[1], Nikolaos Nikolaidis[2], and Husrev Taha Sencar[3]

[1] France Telecom R&D (Orange Labs) Beijing, China
shiguo.lian@orange-ftgroup.com
[2] Department of Informatics
Aristotle University of Thessaloniki, Greece
nikolaid@aiia.csd.auth.gr
[3] Computer Engineering Department
TOBB University of Economics and Technology, Turkey
htsencar@etu.edu.tr

Abstract. In the presence of overwhelming amount of digital video data, the need for automated procedures to protect owners against unauthorized use of their content. to enable content publishers and distributors to monitor broadcast usage of videos and to help businesses manage storage of videos in large databases has become ever more critical Till now, two technical approaches have been proposed toward these goals, namely, watermarking-based systems and content-based copy detection (CBCD). Among them, the former one has been more thoroughly studied and explored, while the latter one is still in its early stages. This chapter reviews existing video copy detection systems, investigates some typical CBCD algorithms, compares the algorithms through appropriate performance metrics, and lists some challenges and open issues in this research field

1 Introduction

The widespread availability of video content and services on the Internet and the rapid diffusion of broadband platforms have brought new challenges to content owners, publishers and distributors [1]. Video copy detection (also known as fingerprinting, perceptual hashing and near-replica detection) is a recent research domain which has emerged largely in response to this technological trend. The goal of video copy detection is to develop automated video analysis procedures to identify the original and modified copies of a video among a large amount of video data for the purposes of copyright control, monitoring, and structuring large video databases.

Content-based video copy detection has primarily emerged as a key tool to cope with digital video piracy and illegal distribution of videos. For content owners and producers this problem has been a major concern over the years. Existing policies try to address this problem by punishing illegal content service providers (e.g., content sharing websites) or illegal content downloaders (e.g., users), and governments all around the world are adopting similar measures. For example China's State Administration of Radio, Film and Television (SARFT) [4] recently set out plans to crack down on unlicensed content distribution via Internet TV, where licensed content is defined as the content (music, audio broadcast, film, TV program, text report, etc) that passes the necessary legality checks (regarding privacy, copyright, rating, etc.).

H.T. Sencar et al. (Eds.): Intel. Multimedia Analysis for Security Appli., SCI 282, pp. 253–273.
springerlink.com

Similarly, the stance taken by Swedish government [2,3] against the peer-to-peer file sharing catalogue company, the Pirate Bay, and its newly adopted laws and the announcement by the UK government [5] for tackling unlawful peer-to-peer file sharing are all steps taken toward this direction. Considering the scale of the Internet and the dimensionality of the problem, success of all these measures essentially relies on the ability to track and identify illegal content through automated procedures which also involve analysis of video content.

Another application area where video copy detection is needed most is the tracking of commercials and media content (e.g., video clips, movies, etc.) over various broadcast channels, for applications like automated collection of royalties or verification of proper airtime allocation to commercials. Currently most such monitoring involves manual work and cannot be performed in real-time. In a similar manner, video copy detection is also needed for accomplishing various tasks involved in identifying, searching and retrieving videos from a database. With the increase in the scale of video databases, the ability to accurately and rapidly perform these tasks become increasingly crucial. For example, the number of videos in the video sharing site YouTube's database is in the order of tens of millions and it grows at a rapid rate. In such systems, copy detection techniques are needed for efficient indexing, copyright management and accurate retrieval of videos as well as detection and removal of duplicate videos to reduce storage costs.

Realizing the above tasks requires techniques that are capable of providing descriptive and distinguishing characteristics of videos that are robust to both intentional and non-intentional modifications [37]. Initially, to achieve the above goals watermarking systems have been considered. Essentially, watermarking approach relies on embedding a distinct pattern into the video content which can later be extracted to obtain information about the origin of the video and to link a copy of a video to its original. However, due to various issues concerning the deployment of the technology, watermarking was never widely adopted by the industry. Video copy detection is a complementary approach to watermarking but unlike watermarking it does not involve any form of embedding operation. Rather, it is based on the idea of extracting and matching content-based signatures through analysis of video content which can ideally serve as a fingerprint for the video. Today, a great deal of research work is being done in this field, and this chapter aims at providing a survey of content-based video copy detection techniques. All the key components and most prominent detection algorithms are investigated in detail.

The rest of the chapter is organized as follows. In Section 2, watermarking-based copyright detection systems are introduced together with their performance analysis. Content-based copy detection systems are investigated in Section 3, and their key components are presented in Sections 4, 5 and 6. In Section 7, performance comparisons of copy detection algorithms, and some open issues and potential challenging topics are discussed. Finally, the chapter is concluded by Section 8.

2 Watermarking-Based Systems

In watermarking-based systems [37][38][39], as shown in Figure 1, an information W (e.g., identifying information about content manager, producer, distributor, user or

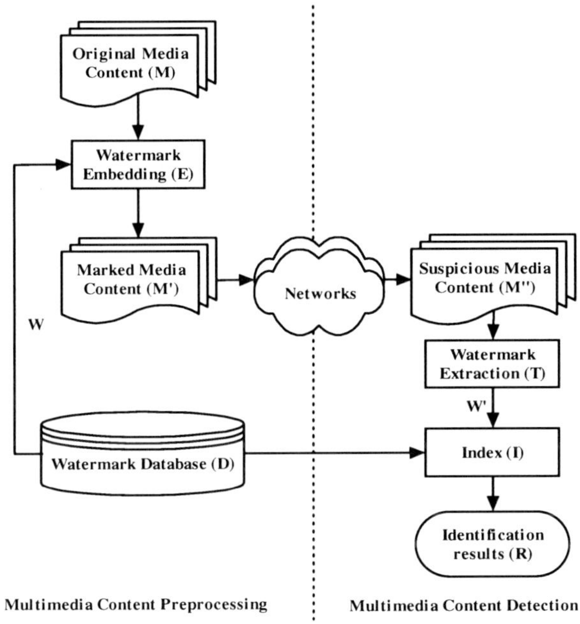

Fig. 1. Watermarking-based copyright identification

content order) is embedded into media content M imperceptibly before it is distributed. To detect the copyright of a suspicious media copy, the copyright information W' is extracted or detected from the media content M'', and compared with the watermark database D. The comparison results dictate, for example, whether free distribution of the media copy is permitted or not. In this system, there are four key components, i.e., watermark database, watermark generation, watermark embedding and extraction, and watermark index.

Watermark-based copyright detection has certain advantages:

- Important content-related information (e.g., producer, owner, distributor, user, etc.) can be embedded into the media content [40]. The information is combined with the media content during the entire distribution process, and can be used to trace the content's illegal distributors. For example, if the user's identity is detected from the content, then the copy's illegal distributor is identified.
- Only copyright information is stored, which may be assigned by content producers or service providers and can be short enough to save storage space.

However, there are also apparent disadvantages:

- The media content should be modified (by watermarking) before transmission and consumption. Content that has already been distributed without a watermark (e.g., legacy content) obviously cannot be protected or traced through watermarking.

- Generally, the modification induced by the watermark is imperceptible [41]. However even the smallest modification slightly degrades the content quality, which is not acceptable in some applications (such as those involving medical images).
- The watermark should survive various transformation operations on the content [41]. This robustness is necessary to identify the transformed media content. However, there is a tradeoff between the robustness and imperceptibility, and the existing methods robustness is still not adequate.

3 Content-Based Copy Detection Systems

In content-based copy detection (CBCD) [6][7][8], as shown in Figure 2, a content hash or fingerprint G is extracted from the visual/audio media M by extracting a feature vector that describes this content. The hash or fingerprint G is often expected to be invariant to content transformation operations, and is stored in the fingerprint database. Generally, the fingerprint extraction process is defined as

$$G = \mathrm{F}(M)\,. \tag{1}$$

Here, F is the fingerprint extraction operation that computes an appropriate feature vector from media content.

To decide whether the media M" is registered or not, the fingerprint G' is computed from the content according to

$$G' = \mathrm{F}(M''), \tag{2}$$

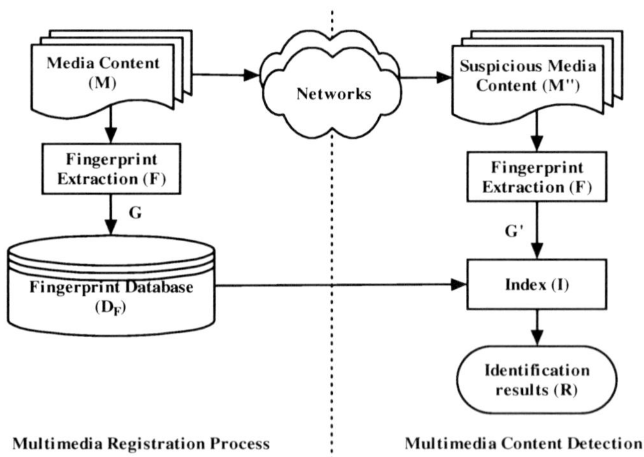

Fig. 2. Copyright identification based on CBCD

and compared with the ones stored in the fingerprint database D_F by an indexing operation (or a general matching or classification operation) defined as follows.

$$R = \mathrm{I}(G', \mathrm{D}_{\mathrm{F}}) . \tag{3}$$

The comparison results R indicate whether the media content is registered or not.

There are three key components in such a system, i.e., fingerprint database, fingerprint extraction, and fingerprint indexing and identification. These components will be described in detail in the following sections.

4 Fingerprint Database

The fingerprint database, as shown in Table 1, stores the fingerprints (feature vectors) and the corresponding media information, including the content name and content producer. In this database, each fingerprint uniquely corresponds to one multimedia document. That is, given a fingerprint, the corresponding multimedia content can be found in the database, and given a multimedia content, the corresponding fingerprint can be searched.

Table 1. Fingerprint Database

Order	Fingerprint (Feature Vector)	Content Information	Content Producer
0	G_0	Film 1	Film Company 1
1	G_1	TV program 1	CCTV
2	G_2	Film 2	Film Company 2
3	G_3	Film 3	Film Company 1
...	...	...	...

The fingerprint is extracted by computing a feature vector from the multimedia content, which can represent the content in a unique way. Generally, the extracted fingerprint should have certain properties:

Distinguishability denotes the ability to distinguish two different media contents. A good fingerprint extraction algorithm should generate different feature vectors for different multimedia items even if these items are similar (e.g. different photos of the same object). Otherwise, false positive errors might occur in the copy detection process.

Robustness refers to the ability of the fingerprint to survive various content transformations. In general, multimedia content is often processed and edited using operations such as the video manipulations, shown in Table 2. In some cases, these operations are performed with the intention to fool the copy detection process. A good fingerprint should remain nearly unchanged even after these content transformations. Otherwise, false negative errors will occur in the copy detection process.

Table 2. Typical video content transformations

Order	Transformations
0	Camera recording
1	Picture in picture
2	Insertion of patterns
3	Recompression (bit rate changes, frame frequency changes)
4	Pixel-level changes (blur, gamma, contrast, noise, filtering, etc.)
5	Geometric changes (resize, shift, rotation, etc.)
6	Temporal domain changes (frame dropping, insertion, resampling, etc)

Generally, the distinguishability and robustness are evaluated by certain metrics, i.e., *Recall* Rate, *Precision* Rate and *F1*-measure. In the case of copy detection, assume that we query a system with a video in order to retrieve all its copies. In such a setup, *Precision* rate is the number of retrieved videos that are indeed copies of the query video divided by the total videos that have been retrieved through this query ,whereas *Recall* rate is defined as the number of retrieved videos that are indeed copies of the query video divided by the total number of existing copies of the query video and *F1*-measure is the harmonic mean of *Recall* rate and *Precision* rate. These metrics are formally defined as

$$\begin{cases} Recall = \dfrac{N_{TruePositive}}{N_{AllTrue}} \\ Precision = \dfrac{N_{TruePositive}}{N_{AllPositive}} \\ F1 = \dfrac{2 \cdot Recall \cdot Precision}{Recall + Precision} \end{cases} . \tag{4}$$

Here, $N_{TruePositive}$ denotes the number of detected (retrieved) video sequences that are true copies of the query video, $N_{AllTrue}$ denotes the number of all existing copies of the query video (i.e. those that should have been detected as copies), and $N_{AllPositive}$ denotes the number of all video sequences detected by the algorithm as copies. Thus, *Recall* Rate quantifies robustness, while *Precision* Rate quantifies distinguishability.

Efficiency refers to the storage and computational cost of the copy detection system. The storage cost depends mainly on the size of the fingerprint database, which is determined by two aspects, i.e., the length of a fingerprint (feature vector), and the number of multimedia items. Thus, if L is the length of the fingerprint and N the number of multimedia items, the size of the fingerprint database (S) is

$$S = N \cdot L . \tag{5}$$

Additionally, the computational cost depends on the computational complexity of the fingerprint extraction and fingerprint indexing procedures. Thus, fingerprint extraction algorithms with low cost, and efficient fingerprint indexing methods are preferred since they increase the system's detection speed.

5 Fingerprint Extraction

5.1 Classification of Fingerprint Extraction Methods

Till now, various fingerprint extraction methods have been proposed. These can be partitioned into two classes, i.e., traditional hashing, and content-based hashing, as shown in Figure 3.

The first class adopts traditional hash functions defined in cryptography [36], such as SHA and MD5, to compute a hash value (a short string) from the media content. Since the hash value is very sensitive to modifications of the multimedia content, this kind of hash has good distinguishability, but very poor to non-existent robustness. A slight transformation on the media content, e.g., recompression, will change the hash value. Thus, this class of hash functions is not content-based, and is not appropriate for multimedia copy detection.

Content-based hashing methods compute the fingerprint by extracting a feature vector from the multimedia content [6][7][8]. Since these methods consider the multimedia content's meaningful information, they can survive certain content transformations. The tradeoff between the distinguishability and robustness can be controlled by choosing an appropriate feature extraction method. In the rest of this chapter, we will focus on the Content-Based Copy Detection (CBCD) algorithms for video data that adopt content-based hashing approaches.

5.2 Classification of Content-Based Fingerprint Extraction Methods

Till now, various CBCD algorithms have been reported. According to the extracted features, they can be classified into three types, i.e., those that utilize coarse features, local features and global features.

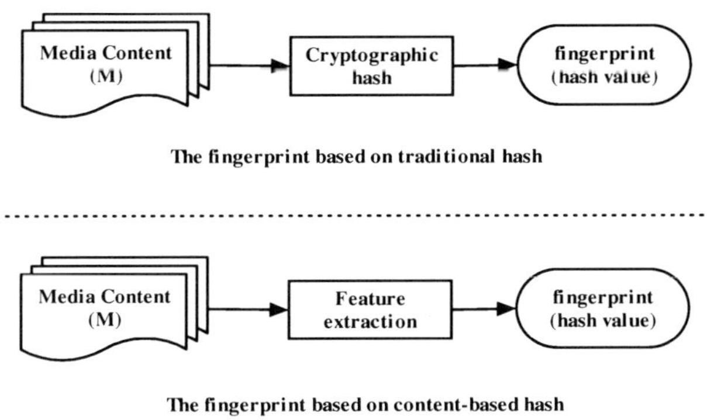

Fig. 3. Different fingerprint extraction methods

5.2.1 Algorithms Based on Coarse Features

In this class of algorithms, coarse features are extracted to represent the media content. For example, adopted features include shot boundaries of the video sequence [9], the luminance differential of partitioned grids in spatial and temporal regions [10], the Discrete Cosine Transform coefficients [11], the key frames extracted from a video sequence [12], the source device characteristics extracted from a video sequence [42], and the video representation model termed Bounded Coordinate System (BCS) that captures the dominant content and content-changing trends of a video [43]. Since these coarse features can only provide an approximate representation of the video content, they usually can not lead to good detection accuracy.

In the algorithm proposed in [9], the shot boundaries of a video sequence are detected, and the sequence of the shot boundaries' positions forms the feature vector or fingerprint, as shown in Figure 4. To detect whether a suspicious video clip is the subclip of the video sequence or not, the video clip's shot boundaries are detected, and the generated position sequence is compared with the original one. Since the algorithm is based on shot boundary detection (SBD) that is a basic component in multimedia analysis, it can use existing algorithms. Additionally, the SBD operation is often of low computational cost, which keeps the copy detection method efficient in terms of complexity. The disadvantages of this method are the following: 1) the detection accuracy depends on the SBD operation that, in most cases, is not accurate enough, 2) only the boundary frames are used, thus the method is not suitable for short video clips, and 3) the method depends on the relative positions of shot boundaries, and thus is not robust against such content transformations as frame removal and frame insertion.

Fig. 4. Feature extraction procedure in the CBCD algorithm [9].

In the algorithm proposed in [10], the luminance differential of partitioned grids in spatial and temporal regions is computed, the differential value is quantized into a binary bit, and the regions' bit sequence is used as fingerprint. In the spatial domain, each frame generates a sub-fingerprint. As shown in Figure 5(a), the frame is partitioned into MxN blocks, each block's average luminance is computed, and one bit ("0" or "1") is assigned to each block according to the comparison between the block's average luminance and the other adjacent blocks' average luminance. Thus, a sub-fingerprint composed of Mx(N-1) bits is generated for each frame. In the temporal domain, each frame generates only one fingerprint bit. As shown in Figure 5(b), the

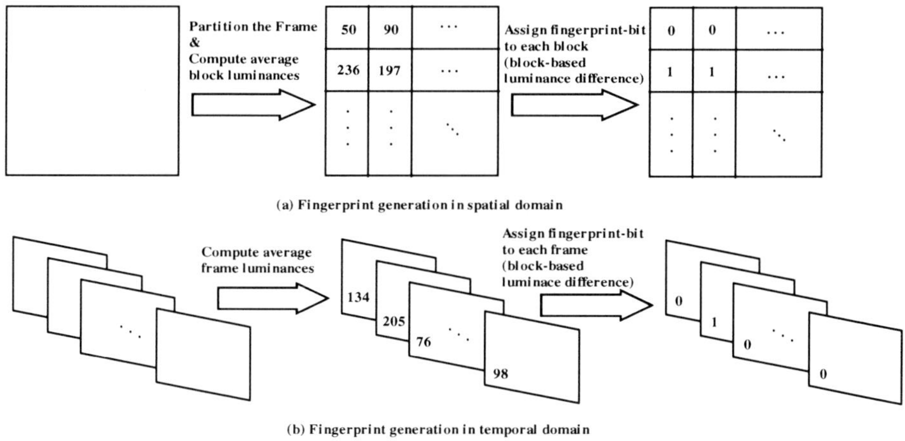

Fig. 5. Fingerprint evaluation in the CBCD algorithm based on differential luminance

average luminance of each frame is computed, and the bit ("0" or "1") is assigned to the frame according to the comparison between two adjacent frames' average luminance. The fact that the method utilizes the differential information between adjacent regions or frames makes it robust to such content transformations as noise, recompression, contrast changing and filtering. However, the method is not robust against some other transformations, such as region cropping, frame deletion or insertion, pattern insertion, picture in picture, etc.

In the algorithm proposed in [11], the video sequence's luminance component is transformed by 3D transformations after preprocessing that includes temporal or spatial normalization. The authors recommend the video sequence to be normalized (subsampled) into a 32x32x64 cube (width, height, frames) and the applied transformation may be 3D DCT or 3D Random Bases Transform. Then, the low-pass transform coefficients are ordered, and quantized using the median of the rank-ordered coefficients. The quantization operation produces a binary fingerprint-bit for each coefficient, and totally 4x4x4 bits are generated for each 3D cube. The process is shown in Figure 6. Since this method takes into account both the temporal and spatial video information simultaneously, it can resist some temporal transformations, such as frame dropping or frame rate change, while maintaining its robustness against certain spatial manipulations, such as noise, blurring, recompression and contrast changing. However, it is not robust against transformations that destroy the spatial/temporal information, such as frame insertion, insertion of patterns, picture in picture, etc.

In the algorithm proposed in [12], the video clip is partitioned into shots, the key frame is extracted from each shot, and the sequence of key frames is used to represent the video clip. Thus, the copy detection is based on key frames. That is, for each key frame in the query, a similarity value (using color, texture, and motion) is obtained with respect to the key frames in the database video. Consecutive key frames in the database video that are highly similar to the query key frames are then used to generate the set of replicated video clips. This method, as shown in Figure 7, reduces the number of video

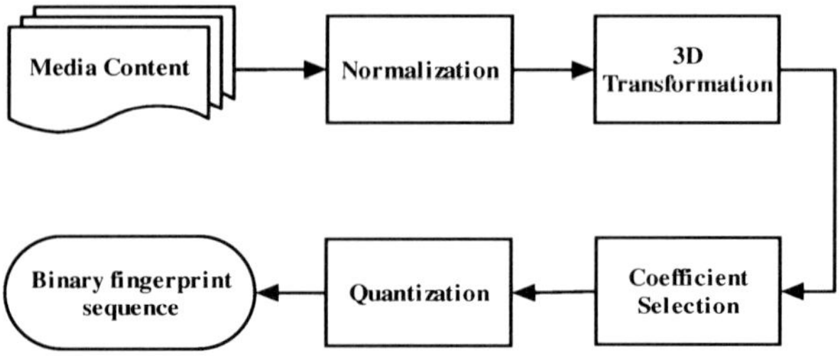

Fig. 6. The fingerprint method based on 3D transformations

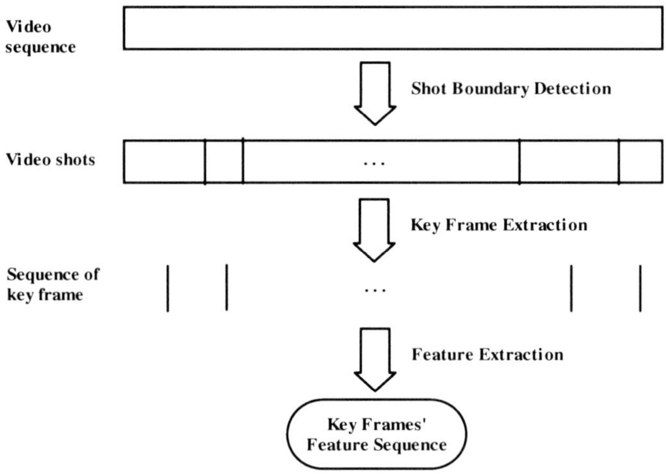

Fig. 7. Feature extraction procedure for the CBCD algorithm in [12]

frames to be processed and compared, and thus reduces the processing speed. However, it depends on the consecutive repeated shots, and is thus not robust to intentional editing that changes the shots' order.

The method proposed in [44] makes use of information about the appearance of faces of distinct individuals (e.g. actors), in order to characterize a video segment. Pulse-series-like signals that provide information on whether a certain actor appears or not in each frame of the video are used as feature vectors. The authors do not deal with the issue of how to obtain the feature vectors, i.e. they assume that efficient face detection, tracking and recognition algorithms are available. Since face-related information is of semantic nature, the proposed algorithm is robust to video noise and other types of video processing (e.g. color-related manipulations), as far as they do not drastically affect the face analysis modules that generate the feature vectors. A fast approach for the matching of the pulse-series feature vectors is proposed by the authors. This approach is convolution-based and thus robust to changes of query segment boundaries and (up to a point) to face detection and recognition errors. The computational complexity of the method is high and stems mainly from the face detection and recognition modules.

5.2.2 Algorithms Based on Local Features

In algorithms based on local features, features of local regions in the spatial or temporal domain are extracted and make up the fingerprint vector. For example, the Harris interest points are detected, and the features are extracted from the local region around these interest points [13][14][15]. In [16], trajectories are built by matching the interest points from frame to frame. These trajectories along with the points are used to construct the fingerprint. The method in [17] uses not only spatial interest points, but also temporal interest points. Since the local features are often robust against transformations, CBCD algorithms based on these features can often achieve good detection accuracy and robustness, especially for the transformations on small regions [6]. However, they often require considerable computational cost.

In the method proposed in [15], the interest points are detected by an improved version of the Harris interest point detector [14] and a differential description of the local region around each interest point is constructed, as shown in Figure 8. To reduce the number of features, the features are not extracted from every frame of the video but only from key frames [13]. The resulting local features are 20-dimensional vectors and the mean rate is about 17 local features per second of video. A local feature $\vec{F}$ is defined as

$$\vec{F} = \left(\frac{\vec{f}_0}{\left\|\vec{f}_0\right\|}, \frac{\vec{f}_1}{\left\|\vec{f}_1\right\|}, \frac{\vec{f}_2}{\left\|\vec{f}_2\right\|}, \frac{\vec{f}_3}{\left\|\vec{f}_3\right\|} \right), \tag{6}$$

where $\vec{f}_i$ (i=0,1,2,3) corresponds to a 5-dimensional sub-vector computed at four different spatiotemporal positions distributed around the interest point. For example, the typical $\vec{f}_i$ is defined as

$$\vec{f}_i = \left(\frac{\partial \vec{I}}{\partial x}, \frac{\partial \vec{I}}{\partial y}, \frac{\partial^2 \vec{I}}{\partial x \partial y}, \frac{\partial^2 \vec{I}}{\partial x^2}, \frac{\partial^2 \vec{I}}{\partial y^2} \right). \tag{7}$$

where $\vec{I}(x, y)$ is the gray level image. Thus, $\vec{f}_i$ is the differential decomposition of the gray level signal up to the second order. Since the relative positions of corner points are often kept unchanged even after transformations such as rotation, shifting and resizing, the fingerprints are robust against various geometric transformations. The disadvantage of this method is the high computational cost required for the evaluation of corner points.

The method proposed in [16] is similar with the one in [15]. The difference is that the corners' trajectories are formed and a label of behavior is assigned to the corresponding local description. Thus, as shown in Figure 9, the computed features include the following components:

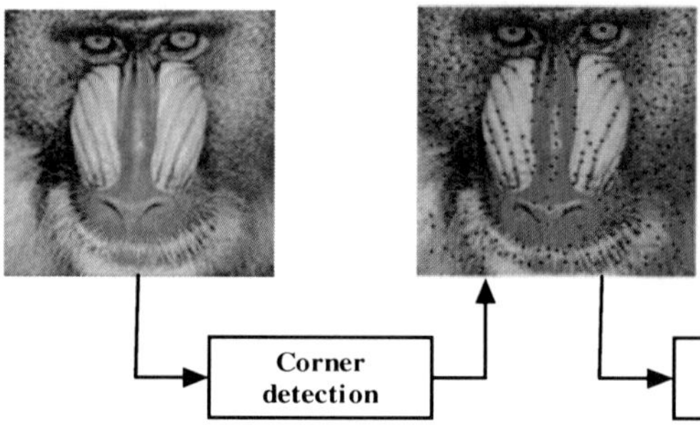

$$\vec{F}=\left(\frac{\vec{f}_0}{\|\vec{f}_0\|},\frac{\vec{f}_1}{\|\vec{f}_1\|},\frac{\vec{f}_2}{\|\vec{f}_2\|},\frac{\vec{f}_3}{\|\vec{f}_3\|}\right)$$

Fig. 8. The CBCD algorithm [15] based on corner points

- The 20-dimensional feature vectors of interest points
- The trajectory properties computed by associating the interest points from frame to frame with an algorithm such as KLT [26].
- A label (background or motion) of behavior that is decided based on the properties of each trajectory. Background denotes motionless and persistent points along frames, while motion denotes moving and persistent points.

Besides the corner points, the trajectory properties and the labels of behavior have also high robustness against various transformations. However this method incurs high computational cost.

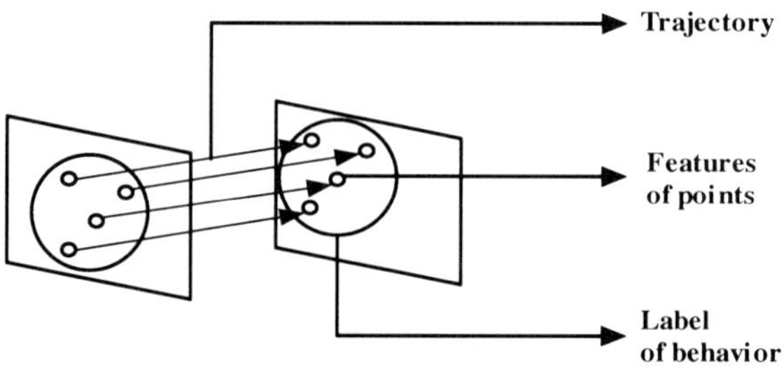

Fig. 9. Interest points and trajectories as utilized in [16]

Another method uses the space time interest points (STIP) [17][24] to construct the feature vector, as shown in Figure 10. Space time interest points correspond to points where the image values have significant local variation in both space and time. This detector was originally used for the classification of human actions and detection of periodic motion. Since interest points can represent a video sequence's meaningful content, they are robust against various transformations and thus suitable for CBCD algorithms. The typical space time interest points [24] are described by the spatiotemporal third order local jet leading to a 34-dimensional vector

$$\vec{F}=\left(I_x,I_y,I_t,I_{xx},\cdots,I_{tt},I_{xxx},\cdots,I_{ttt}\right) \tag{8}$$

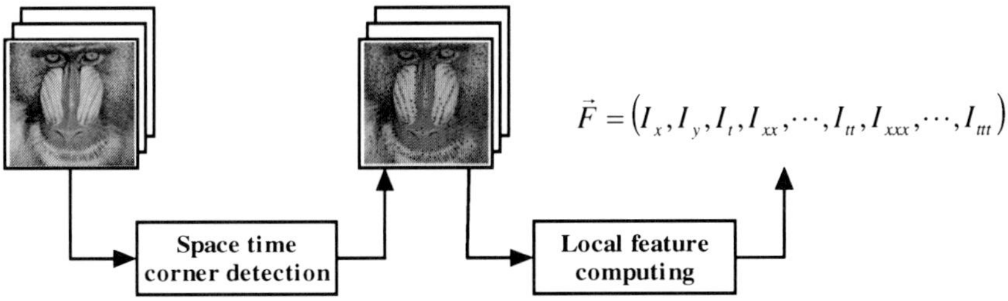

Fig. 10. Schematic representation of the CBCD algorithm based on space time corner points

where I(x,y,t) is the spatiotemporal pixel cube (the video), and I_{x^m, y^n, t^k} are spatiotemporal Gaussian derivatives normalized by the spatial detection scale and the temporal detection scale [17]. The spatiotemporal corner points can represent the video content in a robust manner, while the computational cost is the biggest drawback of this algorithm.

5.2.3 Algorithms Based on Global Features

In algorithms that are based on global features, the fingerprint vector is extracted from the entire video sequence by a frame-to-frame computation. For example, the ordinal measure which is composed of the orders (in terms of average intensity) of the blocks that make up the frame is computed for each frame [12][18][24][25]. The ordinal measure has been also extended so as to apply to the region's motion, color and spatio-temporal distribution of intensities [19][20]. Since the global features are computed from the whole video sequence, they can identify copies of the entire sequence, but they usually cannot survive some local operations, such as logo insertion and cropping [21][22][23]. An advantage of these algorithms is that they are of low complexity.

In [24], a global descriptor in the temporal domain is extracted and used for the fingerprint. In this method, the t-th frame's feature value is computed as the weighted sum of per-pixel squared differences of corresponding pixels in the t and t-1 frames. In more detail, the feature value F(t) is computed by

$$F(t) = \sum_{i=0}^{N-1} A(i)(I(i,t) - I(i,t-1))^2 . \tag{9}$$

Here, I(i,t) (i=0,1,…,N-1) is the t-th frame's i-th pixel's intensity, N is the number of pixels for each frame, A(i) is a weight function to enhance the importance of the central pixels. Finally, the fingerprint is computed around each maximum of the temporal feature F(t). Spectral analysis by FFT, leads to a 16-dimensional vector that is based on the phase of the feature value. This method uses only the content relation in the temporal domain. It is of low computational cost, but is not robust against certain local transformations, such as frame insertion, frame removal, region cropping, etc.

The ordinal measure has been proposed in [25] for computing the similarity between images, and then extended to video in [18]. The measure has also been used for copy detection [19][21][22]. The ordinal measure is computed by partitioning the image into N blocks, and sorting the blocks using their average gray level. For example, the ordinal measure R(t) of the t-th frame is defined as

$$R(t) = (r_0, r_1, \cdots, r_{N-1}) . \tag{10}$$

Here, r_i is the rank of the i-th block. Figure 11(a) shows an example for N=4. Generally, the ordinal measure is robust to transformations such as noise, recompression, filtering and resizing that are applied to the whole frame. Tests reported in [19][21] show that the ordinal measure based copy detection has superior performance compared to other methods based on motion and color features. However, it is difficult to survive such local operations as logo insertion, cropping, shifting, etc.

The ordinal measure has been extended from the spatial to the temporal domain [20][23] by ordering the blocks along the temporal axis. If each frame is divided in K blocks and if R^k is the ordinal measure of the block k in a temporal window with the length M, then the k-th block's temporal measure is defined as

$$R^k = (r_0^k, r_1^k, \cdots, r_{M-1}^k) . \tag{11}$$

An example for M=4, N=4 is shown in Figure 11(b). This measure is robust against certain transformations applied to the entire video clip, such as resizing, recompression, time shifting, etc, but cannot survive transformations that change a subset of the frames in the clip, such as frequent frame removal or insertion, frequent region cropping or insertion of large areas, etc.

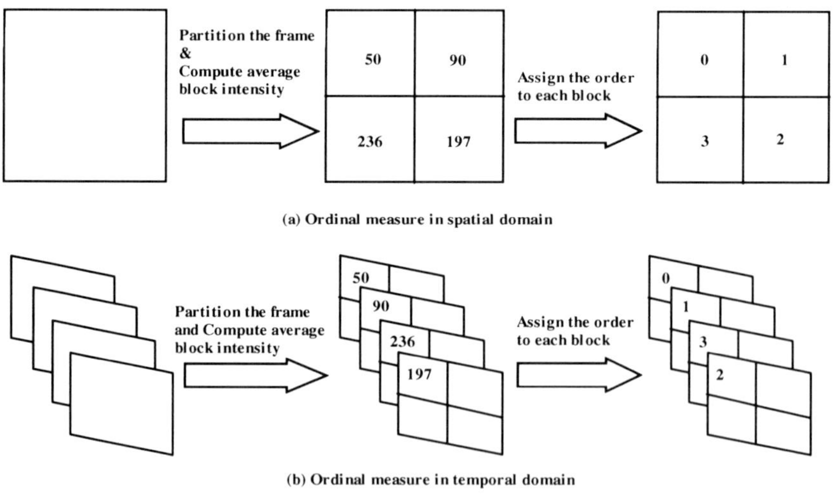

Fig. 11. The CBCD algorithm based on differential luminance

6 Fingerprint Indexing and Identification

After extracting the fingerprint from the video clip that will be used as a query to the copy detection system, fingerprint indexing and identification are applied to get the matching fingerprint in the database and the corresponding video sequence. Research often focuses on two issues, namely fingerprint/feature matching, and search acceleration. The former refers to methods for deciding whether two fingerprints are similar or not, and the latter one deals with the means to accelerate the search or indexing process.

6.1 Feature Matching

There exist various feature matching approaches that can be classified into five types, i.e., histogram intersection, l_1-distance, fuzzy distance, statistical measure, and model matching. They are often adopted together with the corresponding feature extraction methods, and have different properties.

Histogram intersection. The feature vectors are quantized and the corresponding histogram is calculated. Then, the similarity between the reference video sequence and the input one is defined as their histogram intersection [27][28][29]:

$$D = \frac{1}{N}\sum_{l=0}^{L-1} \min(h_{R,l}, h_{I,l}) \tag{12}$$

where $h_{R,l}$, $h_{I,l}$ are the values of the l-th bin of the two histograms, L is the total number of bins, and N is the total number of feature vectors. The bigger the D is, the more the two feature sequences are similar.

l_1-distance. The feature vectors are compared by using the sum of absolute differences as a metric. If $f_{R,l}$, $f_{I,l}$ are the l-th elements of the feature vectors that correspond to the original and the input video respectively, then the l_1-distance [30][31][32] is defined as

$$D = \frac{1}{L}\sum_{l=0}^{L-1} \left| f_{R,l} - f_{I,l} \right| \tag{13}$$

where L is the total number of feature vector elements. Smaller values of D denote stronger feature sequences similarity.

Fuzzy distance. The distance between two feature sequences is measured in a fuzzy manner, for example, by combining the notions of l_1-distance and hamming distance [33]. This kind of fuzzy distance measure is resilient to temporal transformations, such as random frame dropping or insertion.

Statistical measure. To match an input video clip against a video sequence involves two operations. First, to find each frame's matching. The best matches in the inverted video index to the frames from the input video are found, and the matches are accumulated in a hit table. Secondly, to get the matched clip according to temporal matching. If a video clip receives enough hits, then the temporal ordering consistence of frame matches is evaluated by a statistical measure like Spearman's rank order correlation coefficient [34].

Model matching. In video copy detection, a new, shorter version of a video clip is produced by removing some shots from the original one. The clip can be modeled by the left-right Hidden Markov Model (HMM). The codebook of the model can be obtained by quantizing the feature vectors of all the video clips, HMM parameters can be estimated for each clip [28], and the corresponding matches are then obtained by comparisons.

6.2 Search Acceleration

Considering the large scale of fingerprint databases, it is often necessary to accelerate the search process. An intuitive method is to reduce the length of fingerprint. For example, the fingerprint is extracted from the key frames only, and each fingerprint is compacted by fingerprint compression and distance bounding [29] that reduces the parameter space of fingerprint.

Another method is feature clustering [35][28] that can increase the search speed and improve the retrieval performance.

The skip strategy [27], which reduces the number of candidate segments for matching, is also a good means to accelerate the search. For example, the skip width w in terms of frame is defined as

$$w = \begin{cases} \lfloor N(\theta - D) \rfloor + 1 & if\, D < \theta \\ 1 & otherwise \end{cases} \tag{14}$$

where $\lfloor x \rfloor$ denotes the floor operator, θ is a given threshold, D is the similarity between the feature sequences, as defined in Eq. (12), and N is the number of feature vectors. Thus, the less the similarity is, the more segments will be skipped.

7 Methods Comparison and Open Issues

7.1 Performance Comparison

As it is obvious from the above analysis, different copy detection algorithms have different properties. Table 3 provides a comparison of some characteristic copy detection algorithms in terms of robustness, distinguishability, storage cost and computational cost. In general, methods based on traditional hashes can distinguish video copies exactly, while they can not survive general content transformations. Methods based on local features can often achieve high robustness and distinguishability but incur high computational cost. Methods based on coarse features and global features are difficult to reach a good tradeoff between robustness and distinguishability but have the advantage of low computational cost. Especially, the method based on corner points [13][14][15] has the potential to be a good solution to the problem if the computational cost can be reduced through a fast implementation. However, up to now, no algorithm can achieve a good tradeoff between robustness, distinguishability and efficiency.

Table 3. Comparison of various copy detection algorithms

Classification	Algorithm	Robustness	Distinguishability	Efficiency - storage cost	Efficiency - computational cost
Fingerprint based on traditional hash	Traditional hash [36]	Low	High	Low	High
Fingerprint based on coarse features	SBD based method [9]	Low	Low	Low	Low
	Method based on differential luminance [10]	Middle	Middle	Middle	Low
	Method based on 3D transformations [11]	Middle	High	High	High
	Method based on key frame matching [12]	Middle	Middle	Low	Low
Fingerprint based on local features	Method based on corner points [13][14][15]	High	High	Middle	High
	Method based on interest points and trajectories [16]	High	High	High	High
	Method based on space time corner points [17]	Middle	High	Middle	High
Fingerprint based on global features	Global descriptor in temporal domain [24]	Low	Low	Low	Low
	Ordinal measure [19][21][22]	Middle	Middle	Middle	Low
	Temporal ordinal measure [20][23]	Middle	Middle	Middle	Low

7.2 Properties of CBCD Algorithms

Based on the review and discussion above, one can claim that CBCD algorithms have the following advantages:

- The hash/fingerprint is computed from the media content without changing the content, which is important for numerous multimedia applications.
- The hash/fingerprint is constructed using robust features extracted from the media content. Thus the fingerprint is frequently more robust than a watermark against various content transformations.

- The hash/fingerprint can tell whether a suspicious piece of content matches a multimedia document (or part of it) registered in the fingerprint database. Thus, it can detect content repetition.
- Unlike watermarking, fingerprinting can be applied to "legacy" content, e.g. content that has already been distributed.

Of course, there are also some disadvantages:

- The hash/fingerprint is obtained by extracting robust features from the contents, which often incurs high computational cost.
- The hash/fingerprint can only characterize the content itself, but can not provide information regarding the content's other properties, such as ownership, producer, distributor, etc. This is possible only through matching the content to the fingerprint database.
- The hash/fingerprint is often composed of more bits than the watermark is. Thus, more storage space is needed for the fingerprint database.
- The number of hashes/fingerprints is proportional to the number of media contents. Thus, for large-scale applications, the fingerprint comparison or indexing process may cost a significant amount of time.
- Since fingerprinting does not alter the content, it is extremely difficult to achieve distinction between very similar multimedia items, e.g. two pictures of a monument taken by two photographers in similar lighting conditions and from the same viewpoint. This issue does not apply to watermarking.

7.3 Challenges for CBCD Algorithms

As can be seen, compared with watermarking-based copyright detection, content-based copy detection can often survive more content transformation operations. Considering that multimedia content is often transformed before being placed on the Web, content-based copy detection is preferable for the detection of illegal copies. The main challenges include:

- To achieve a good tradeoff between robustness and distinguishability
- To reduce the storage space required for each fingerprint.
- To improve the efficiency of fingerprint database search
- To reduce the computational cost of fingerprint extraction and matching, and get a good tradeoff between robustness, distinguishability and efficiency.
- To combine fingerprinting and watermarking in order to provide content and user identification

8 Conclusions

This chapter provided a review of existing pirate video copy detection methods, with emphasis on content-based copy detection (CBCD). The main elements of a CBCD system namely fingerprint extraction, fingerprint database and fingerprint index and identification have been introduced and described in detail. Comparison of existing content-based fingerprint extraction algorithms in terms of robustness, distinguishability, storage cost and computational cost has been also provided. Additionally, some open issues and challenges in content-based copy detection have been presented. It has been shown that, for CBCD methods, the biggest challenge is how to get a good tradeoff between robustness, distinguishability and efficiency. Future research is expected to contribute towards extraction of features with high robustness and distinguishability, increase of the feature extraction and database search speed, and decrease of the features storage space. Although the area is still in its early stages, the urgent need for practical anti-piracy applications makes it a hot topic, and numerous innovations are expected in the near future. We hope this chapter will provide useful information to readers interested in this research field.

Acknowledgments

This work is partially supported by France Telecom projects, i.e., Crypto and Invenio.

References

1. Lian, S., Zhang, Y.: Handbook of research on secure multimedia distribution. IGI Global (formerly Idea Group, Inc.) (2009)
2. Preel, M.: Swedish crackdown on piracy leads to fall in illegal filesharing, `http://www.physorg.com/news168596077.html` (accessed September 3, 2009)
3. Zee. The Pirate Bay Taken Offline by ISP, `http://thenextweb.com/2009/08/24/pirate-bay-offline-isp/` (accessed September 3, 2009)
4. SARFT crack down on unlicensed content distribution via Internet TV, `http://dtvinchina.blogspot.com/2009/08/sarft-to-crack-down-on-unlicensed.html` (accessed September 3, 2009)
5. Holton, K.: UK govt proposes cutting Web access to file sharers, `http://www.reuters.com/article/rbssTechMediaTelecomNews/idUSLP18813620090825` (accessed September 3, 2009)
6. Gengembre, N., Berrani, S.-A.: The Orange Labs Real Time Video Copy Detection System - TrecVid 2008 Results. In: TRECVID 2008 (2008)
7. Yang, X., Sun, Q., Tian, Q.: Content based video identification: a survey. In: Proceedings of 2003 International Conference on Information Technology: Research and Education (ITRE 2003), August 11-13, pp. 50–54 (2003)
8. Gengembre, N., Berrani, S.-A.: A probabilistic framework for fusing frame-based searches within a video copy detection system. In: Proc. of the ACM International Conference on Image and Video Retrieval, Niagara Falls, Canada (July 2008)
9. Indyk, P., Iyengar, G., Shivakumar, N.: Finding pirated video sequences on the internet. Technical report, Stanford University (1999)
10. Oostveen, J., Kalker, T., Haitsma, J.: Feature extraction and a database strategy for video fingerprinting. In: Chang, S.-K., Chen, Z., Lee, S.-Y. (eds.) VISUAL 2002. LNCS, vol. 2314, pp. 117–128. Springer, Heidelberg (2002)
11. Coskun, B., Sankur, B., Memon, N.: Spatio-temporal transform-based video hashing. IEEE Transactions on Multimedia 8(6), 1190–1208 (2006)

12. Jain, A.K., Vailaya, A., Xiong, W.: Query by video clip. Multimedia Syst. J. 7(5), 369–384 (1999)
13. Eickeler, S., Muller, S.: Content-based video indexing of TV broadcast news using hidden markov models. In: Proc. of Int. Conf. on Acoustics, Speech, and Signal Processing, pp. 2997–3000 (1999)
14. Harris, C., Stevens, M.: A combined corner and edge detector. In: 4th Alvey Vision Conference, pp. 153–158 (1988)
15. Joly, A., Buisson, O., Frelicot, C.: Content-based copy detection using distortion-based probabilistic similarity search. IEEE Transactions on Multimedia 9(2), 293–306 (2007)
16. Law-To, J., Buisson, O., Gouet-Brunet, V., Boujemaa, N.: Robust voting algorithm based on labels of behavior for video copy detection. In: ACM Multimedia, MM 2006, pp. 835–844 (2006)
17. Laptev, I., Lindeberg, T.: Space-time interest points. In: International Conference on Computer Vision, pp. 432–439 (2003)
18. Mohan, R.: Video sequence matching. In: 1998 Int. Conference on Audio, Speech and Signal Processing, Seattle, WA, USA, May 12-15, vol. 6, pp. 3697–3700 (1998)
19. Hampapur, A., Bolle, R.: Comparison of sequence matching techniques for video copy detection. In: Conference on Storage and Retrieval for Media Databases, pp. 194–201 (2002)
20. Chen, L., Stentiford, F.W.M.: Video sequence matching based on temporal ordinal measurement. Technical report no. 1, UCL Adastral (2006)
21. Hua, X.-S., Chen, X., Zhang, H.-J.: Robust video signature based on ordinal measure. In: International Conference on Image Processing, vol. 1, pp. 685–688 (2004)
22. Iwamoto, K., Kasutani, E., Hamada, A.: Image signature robust to caption superimposition for video sequence identification. In: 2006 International Conference on Image Processing, Atlanta, GA, October 8-11, pp. 3185–3188 (2006)
23. Kim, C., Vasudev, B.: Spatiotemporal sequence matching techniques for video copy detection. IEEE Transactions on Circuits and Systems for Video Technology 1(15), 127–132 (2005)
24. Law-To, J., Chen, L., Joly, A., Laptev, I., Buisson, O., Gouet-Brunet, V., Boujemaa, N., Stentiford, F.: Video copy detection: a comparative study. In: Proceedings of the 6th ACM international conference on Image and video retrieval (CIVR 2007), pp. 371–378 (2007)
25. Bhat, D., Nayar, S.: Ordinal measures for image correspondence. IEEE Transactions on Pattern Analysis and Machine Intelligence 20(4), 415–423 (1998)
26. Tomasi, C., Kanade, T.: Detection and tracking of point features. Technical report CMU-CS-91-132 (1991)
27. Kashioo, K., Smith, G., Murase, H.: Time-series active search for quick retrieval of audio and video. In: Proc. of ICASSP 1999, vol. 6, pp. 2993–2996 (1999)
28. Kuronuni, T., Kashino, K., Murase, H.: A method for robust and quick video searching using probabilistic dither-voting. In: Proc. of International Conference on Image Processing (ICIP), vol. 2, pp. 653–656 (2001)
29. Kimura, A., Kashino, K., Kurozumi, T., Murase, H.: A Quick Search Method for Multimedia Signals Using Feature Compression Based on Piecewise Linear Maps. In: Proc. of International Conference on Acoustics, Speech and Signal Processing (ICASSP), vol. 4, pp. 3656–3659 (2002)
30. Cheung, S.-C., Zakhor, A.: Estimation of web video multiplicity. In: Proc. SPIE, vol. 3964, pp. 34–36 (2000)
31. Oostveen, J.C., Kalker, A.A.C., Haitsma, J.A.: Visual hashing of digital Video: applications and techniques. In: SPIE applications of digital image processing XXIV, San Diego, USA (July/August 2001)

32. Pua, K.M.: Feature-based Video Sequence Identification. PhD Dissertation, The Univ. of Kansas (2002)
33. Shivakumar, N.: Detecting Digital Copyright Violation on the Internet. PhD Dissertation, Stanford University (1999)
34. Hampapur, A., Bolle, R.: VideoGREP: Video Copy Detection using Inverted File Indices. Technical Report, IBM Research (2002)
35. Cheung, S.-C., Zakhor, A.: Video similarity detection with video signature clustering. In: Proceedings of ICIP 2001, vol. 1, pp. 649–652 (2001)
36. Lian, S.: Multimedia Content Encryption: Techniques and Applications. Taylor & Francis Group, Abington (2008)
37. Lian, S., Kanellopoulos, D., Ruffo, G.: Recent Advances in Multimedia Information System Security. Informatica, Slovenian Society Informatika 33(1), 3–24 (2009)
38. Swanson, M.D., Kobayashi, M., Tewfik, A.H.: Multimedia Data-Embedding and Watermarking Technologies. Proc. IEEE 86(6), 1064–1087 (1998)
39. Kalker, T., Epema, D.H.J., Hartel, P.H., Lagendijk, R.L., Steen, M.: Music2Share-Copyright-Compliant Music Sharing in P2P Systems. Proceedings of the IEEE 92(6), 961–970 (2004)
40. Cox, I.J., Miller, M.L., Bloom, J.A.: Digital Watermarking. Morgan Kaufmann, San Francisco (2002)
41. Barni, M., Bartolini, F.: Watermark Systems Engineering. Marcel Dekker, New York (2004)
42. Bayram, S., Sencar, H.T., Memon, N.: Video copy detection based on source device characteristics - a complementary approach to content-based methods. In: 2008 ACM Conference on Multimedia Information Retrieval (MIR 2008), Vancouver, British Columbia, Canada, October 30-31, pp. 435–442 (2008)
43. Huang, Z., Shen, H.T., Shao, J., Zhou, X., Cui, B.: Bounded Coordinate System Indexing for Real-Time Video Clip Search. ACM Transactions on Information Systems 27(3), Article 17 (2009)
44. Cotsaces, C., Nikolaidis, N., Pitas, I.: Semantic video fingerprinting and retrieval using face information. Signal Processing: Image Communication 24(7), 598–613 (2009)

Image Steganalysis

Fangjun Huang[1,3], Bin Li[2], Yun Qing Shi[3], Jiwu Huang[1], and Guorong Xuan[4]

[1] Sun Yat-sen University, Guangzhou, China
{huangfj,isshjw}@mail.sysu.edu.cn
[2] Shenzhen University, Shenzhen, China
lib3366@gmail.com
[3] New Jersey Institute of Technology, NJ, USA
{huangfj,shi}@njit.edu
[4] Tongji University, Shanghai, China
grxuan@gmail.com

1 Introduction

No matter how unbreakable an encryption algorithm is, the encrypted messages will attract an adversary's awareness. Steganography can resolve this difficulty by embedding secret data into an innocuous cover medium, such as image, video, text, audio, etc. The transmission of the innocuous medium will not arouse much suspicion. Therefore, whereas cryptography only protects the contents of the secret message, steganography conceals the very existence of secret information, making the secret communications imperceptible [6].

Comparable to cryptanalysis applied to countering cryptography, the technique of steganalysis [6, 50] is applied to resisting steganography, in order to determine whether or not a suspected medium has a secret payload being embedded inside, and, if possible, determine the length of the payload and even discover the payload. The research on steganalysis is of importance. On the one hand, it can help to deter secret communications by illegal parties. On the other hand, it may stimulate the development of steganography.

An attacker (or the so called steganalyst), does not know before hand which of the files, if any, contains a payload. His only certainty is that the embedded secret message may change the statistical properties of the carrier medium, though its presence is often imperceptible to the human perception. Thus almost all the steganalytic tools are handled with statistical analysis. The media with and without secret data embedded inside are referred to as stego media and cover media, respectively. As images become ubiquitous in our daily life, we focus on discussing the steganalytic techniques for digital images here.

According to the objective of steganalyst, steganalysis can be divided into two kinds: passive and active. Passive steganalysis [e.g., 4, 9, 10, 11, 13, 14] involves only the detection of the presence of hidden information or a modified carrier, while active steganalysis [e.g., 12, 16, 17, 18, 19] involves estimating the message length, extracting the hidden information as well. Some passive steganalytic methods formulate the issue of steganalysis as a hypothesis testing problem, namely, null hypothesis for no message being embedded, and alternative hypothesis for message being embedded.

H.T. Sencar et al. (Eds.): Intel. Multimedia Analysis for Security Appli., SCI 282, pp. 275–303.
springerlink.com

Many other methods regard steganalysis as a pattern classification problem. In this perspective, some steganalytic features are extracted from a questionable medium and a trained classifier is used to determine which class the medium belongs to.

According to the embedding domain of the image steganographic algorithm, steganalysis can be divided into spatial domain steganalysis [e.g., 12, 17, 25, 26] and transform domain steganalysis [e.g., 4, 18, 19]. Spatial domain steganalysis mainly pays attention to the steganographic schemes that hide secret message in the spatial domain, while transform domain steganalysis tries to analyze the steganography designed in transform domain, such as discrete cosine transform (DCT) or discrete wavelet transform (DWT) domain.

Nowadays, many researchers prefer to divide the steganalytic methods into two classes, i.e., specific steganalysis [e.g., 4, 16, 18, 19, 25, 26, 27, 28, 29, 30, 33] and universal steganalysis [e.g., 9, 13, 14, 22, 35, 44, 46, 48, 56]. Specific methods address how to defeat one kind of steganographic scheme, while universal methods assume no priori knowledge of the embedding mechanism and try to detect multiple steganographic schemes. In the next two sections, some representative specific and universal steganalytic methods will be introduced with some details.

2 Specific Steganalysis

If the data embedding manner of a steganographic scheme is known, the steganalyst can fully utilize the knowledge to devise a steganalytic detector targeted to such a steganographic scheme. Many successful specific steganalytic methods have been proposed in the past years. We discuss the main idea of some specific attacking methods, whose targeted steganographic scheme is well known, such as least significant bit (LSB) substitution [5], LSB matching [43], F5 [52], Outguess [36], MB [40], and YASS [47].

2.1 Attacking LSB Substitution

Substituting the LSBs (least significant bits) of pixel/coefficient values of a cover medium with the secret message bits is a straightforward way to conduct data hiding [5]. The intended data receiver can retrieve the secret message by extracting the corresponding LSBs. Such a steganographic method is called LSB substitution or LSB replacement. It was broadly used early days, such as Steganos [1], S-tools [2], Jsteg [49], etc. However, the idea of simply overwriting the LSBs is not a secure way for concealing embedded data. Two pixel/coefficient values, which are only different in the LSB, form the so-called pair of value (PoV). Westfeld and Pfitzmann [53] observed that the frequencies of occurrence of each elements in PoV will become close in a stego medium if the secret message bits are uniformly distributed. This is because the two values in PoV can only be flipped into each other in LSB substitution, i.e., $v_{2i} \leftrightarrow v_{2i+1}$, i=1,2,⋯. Fig. 1 illustrates the PoV equaling phenomenon. A statistical attack, known as Chi-square attack, has been proposed to detect LSB substitution. The Chi-square statistic in the attack is defined as

$$\chi_{k-1}^2 = \sum_{i=1}^{k} \frac{v_{2i} - (v_{2i} + v_{2i+1})/2}{(v_{2i} + v_{2i+1})/2}, \tag{1}$$

where k is the number of PoVs, and the probability of embedding is computed by

$$p = 1 - \frac{1}{2^{\frac{k-1}{2}} \Gamma\left(\frac{k-1}{2}\right)} \int_0^{\chi_{k-1}^2} e^{\frac{x}{2}} x^{\frac{k-1}{2}-1} dx. \tag{2}$$

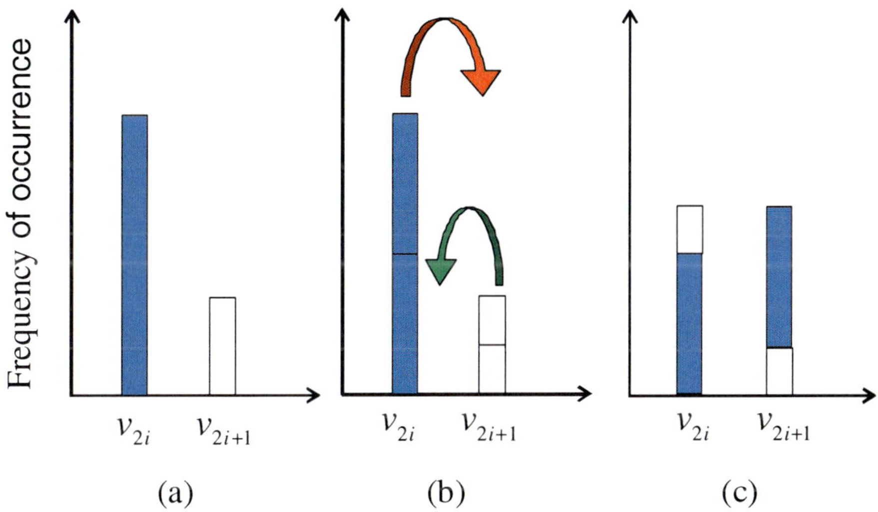

Fig. 1. Illustration of the PoV equaling phenomenon. (a) PoV before data embedding; (b) The two values in the PoV flip into each other when data are embedded; (c) PoV after data embedding.

The Chi-square attack is only effective in detecting the LSB substitution while the secret message bits are embedded sequentially. Many steganalytic methods have been proposed in the recent years to detect the phenomenon of LSB substitution with random embedding location. In [37], Provos and Honeyman noted that the Chi-square detector could still be applied to detecting secret messages in randomly scattered locations by applying the same idea to smaller portions of the image. This modified Chi-square attack is referred to as the extended Chi-square attack. Another two well-known specific steganalyzers for detecting LSB substitution of cover pixels are regular and singular group analysis (RS) [17] and sample pairs analysis (SPA) [12]. These two methods can not only detect stego images with high reliability, but also estimate the length of secret message accurately. More generalized forms of RS and SPA have been proposed by Ker [27, 28]. Based on a concept of weighted stego image, Fridrich [16] presented a novel method for estimating the length of the embedded message and the method was later improved by Ker [29, 30]. Due to the prolificacy of the methods for detecting LSB substitution, we do not intent to discuss all of them in details. Simply speaking, the effectiveness of all of these methods is built on the weakness of LSB substitution that the two elements in PoV can only be flipped into each other.

2.2 Attacking LSB Matching

LSB matching [43], a more advanced LSB embedding steganography, makes the detection of hidden message much more difficult. The LSB matching does not simply overwrite the LSBs. Instead, the selected sample value is increased or decreased by one randomly if its LSB does not match the secret message bit to be embedded. Different from LSB substitution, LSB matching does not cause equalized PoVs. Therefore, the methods designed for attacking LSB substitution [12, 16, 17, 27, 28, 29, 37, 38, 53] will fail in detecting LSB matching. Recently, a few attacking methods have been reported. We discuss Ker' method [26], which is mainly based on the work of Harmsen and Pearlman [23], and Hunag et al.'s method [25] by only considering the statistics in the least two significant bit-planes.

2.2.1 Ker's Method

In [23], the process of information hiding is modeled in the context of additive noise. A block diagram of this framework is shown in Fig. 2, where $h_c[n]$ and $h_s[n]$ represent the histogram of cover image and stego image respectively, and $f_\Delta[n]$ represents the probability mass function (PMF) of the additive noise.

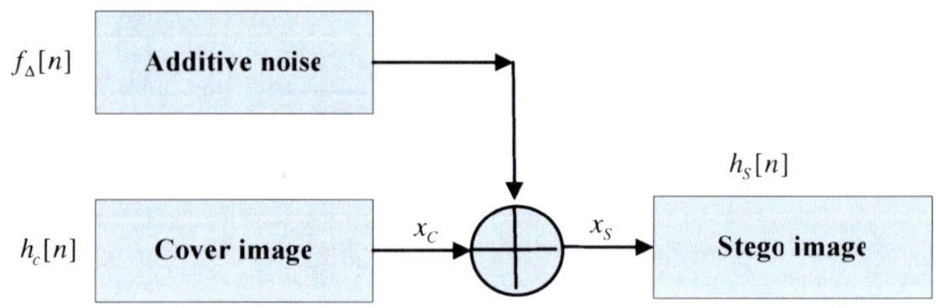

Fig. 2. Additive noise model

Under an independence assumption, the histogram of the stego image is a convolution of the noise PMF and the original histogram of the cover image [54], i.e,

$$h_s[n] = h_c[n] * f_\Delta[n]. \tag{3}$$

Thus, given knowledge of any hiding scheme in the form of $f_\Delta[n]$ as well as knowledge of $h_c[n]$, the histogram of the stego image can be known. According to the convolution theorem, Eq. (3) can be rewritten in the frequency domain as,

$$H_s[k] = H_c[k] F_\Delta[k], \tag{4}$$

where $H_s[k]$, $H_c[k]$, and $F_\Delta[k]$ are discrete Fourier transform (DFT) of $h_s[n]$, $h_c[n]$ and $f_\Delta[n]$, respectively. In particular, the DFT of a histogram is referred to as the histogram characteristic function (HCF). The center of mass (COM) of HCF is defined as

$$C(H[k]) = \frac{\sum_i i|H[i]|}{\sum_i |H[i]|}. \tag{5}$$

It is proved in [23] that, the COM of HCF of stego image is less than that of the corresponding cover image, i.e.,

$$C(H_s[k]) < C(H_c[k]) . \tag{6}$$

Harmsen and Pearlman directly used the COM of HCF to build a one-dimensional (1-D) feature space for grayscale images or a 3-D feature space for color images. Cover images and stego images can be differentiated in this feature space.

Ker [26] noticed that Harmsen and Pearlman's classifier can be revised to more effectively analyze the LSB matching in spatial domain. Two techniques were presented to remove the influence of the image content on the COM. Consider a grayscale cover image, made up of pixels with intensity in the range 0,…,N-1 (usually N = 256). Let $p_c(i, j)$ represents the intensity of the cover image at location (i, j), and $p_s(i, j)$ for the intensity of the stego image at location (i, j).

The first technique considers calibrated COM and it is based on downsampling an image by a factor of two in both horizontal and vertical dimensions using a straightforward averaging filter. Precisely, let $p_c'(i, j)$ be the pixel intensities of the downsampled cover image given by

$$p_c'(i, j) = \left\lfloor \sum_{u=0}^{1}\sum_{v=0}^{1} \frac{p_c(2i+u, 2j+v)}{4} \right\rfloor , \tag{7}$$

and $p_s'(i, j)$ the similarly downsampled version of the stego image. Ker observed that there exist

$$C(H_c'[k]) \approx C(H_c[k]) \tag{8}$$

and

$$C(H_c[k]) - C(H_s[k]) > C(H_c'[k]) - C(H_s'[k]) \tag{9}$$

In Eqs. (8) and (9), $C(H_c'[k])$ and $C(H_s'[k])$ are the COM of downsampled cover and stego images respectively. They are the so called calibrated COM. Combining Eqs. (8) and (9) gives a new detector for the presence of LSB matching steganography: $C(H[k]) < C(H'[k])$ in most images that have been subject to the data embedding. In view of the variation between the magnitudes of such values, $C(H[k]) / C(H'[k])$ is used as a dimensionless discriminator.

The second technique is based on considering the two-dimensional *adjacency histogram*. The adjacency histogram $h(m,n)$ is defined as the number of horizontal neighboring pixel pairs with grayscales m and n, i.e.,

$$h[m,n] = \left|\{(i, j) \mid p(i, j) = m, p(i, j+1) = n\}\right| . \tag{10}$$

Because adjacent pixels of cover image tend to have close intensities, this histogram is sparse off the diagonal. The two-dimensional COM corresponding to this histogram is computed as

$$C(H[k,l]) = \frac{\sum_{i,j}(i+j)\left|H[i, j]\right|}{\sum_{i,j}\left|H[i, j]\right|} , \tag{11}$$

where $H[k,l]$ is the two-dimensional DFT of the adjacency histogram $h[m,n]$. It is expected that

$$C(H_s[k,l]) < C(H_c[k,l]). \tag{12}$$

Thus $C(H[k, l])$ can be used as the discriminator.

It was also pointed out by Ker that these two methods can be combined together to form a more effective classifier for attacking LSB matching. By comparing the performance of these two schemes with Harmsen and Pearlman's original scheme, Ker showed that using the calibrated COM leads to a significant performance improvement.

2.2.2 Huang et al.'s Method

In [25], Huang et al. presented a new detector through analyzing the alteration rate of the number of gray levels in the least two significant bits. Suppose an $M \times N$ grayscale image $I(x, y)$ is composed of eight bit-planes $I_0 \sim I_7$, ranging from bit-plane 0 for the least significant bit to bit-plane 7 for the most significant bit. By combining the least two significant bit-planes, a new image $A(x, y)$ can be formulated as follows. That is,

$$A(x, y) = I_0(x, y) + I_1(x, y) \times 2 \quad (1 \le x \le M, 1 \le y \le N). \tag{13}$$

There are only four gray levels in this image, i.e., 0, 1, 2 and 3.

A 3x3 subimage is defined as the neighborhood of a point $A_{i,j}(2 \le i \le M-1, 2 \le j \le N-1)$ in $A(x, y)$. The center of the subimage is moved pixel by pixel starting at the top left corner. At each position, the number of gray levels in this 3x3 subimage is counted. There are in total four types of subimages.

T_1: There is only one gray level in this type of subimage, i.e., all pixels in the subimage have the same value.

T_2: There are two gray levels in this type of subimage.

T_3: There are three gray levels in this type of subimage.

T_4: There are four gray levels in this type of subimage.

It is pointed in [25] that while embedding a secret message, the probability that the subimages change from T_1 to $T_i (2 \le i \le 4)$ is much larger than the probability that the subimages change from $T_i (2 \le i \le 4)$ to T_1. A justification can be given below. For example, given a subimage belonging to T_1, if any one pixel is shifted, the probability that the subimage would belong to $T_i (2 \le i \le 4)$ is 100%. However, for any subimage belonging to $T_i (2 \le i \le 4)$ initially, if one pixel is modified, the probability that the subimage would belong to T_1 is much less than 100%. Thus the number of subimages belonging to T_1, denoted by $|T_1|$, would decrease after the embedding.

Since the values of $|T_1|$ have a wide distribution for different images, in [25], a new re-embedding methodology is presented to eliminate the influence of the image content. Suppose $|T_1^C|$ is the number of elements belonging to T_1 of the cover image

$I_C(x, y)$ and $\left|T_1^S\right|$ is the number of elements belonging to T_1 of the corresponding stego image $I_S(x, y)$. Embedding another random sequence into the cover and stego images by LSB matching simultaneously, $\left|T_1^{C*}\right|$ and $\left|T_1^{S*}\right|$ can be obtained, which are the count of elements belonging to T_1 in the obtained images. Denote the alteration rate as

$$k_c = \frac{\left|T_1^C\right| - \left|T_1^{C*}\right|}{\left|T_1^C\right|}, \tag{14}$$

and

$$k_s = \frac{\left|T_1^S\right| - \left|T_1^{S*}\right|}{\left|T_1^S\right|}. \tag{15}$$

Huang et al. derived that there exists

$$k_c > k_s \tag{16}$$

Thus in [25], the alteration rate of the test image is computed and used as a dimensionless discriminator.

The experimental results on images from UCID [42] were given. All the 1338 high resolution TIFF files were converted to grayscale images for testing. The re-embedding rate was chosen as 0.3 *bpp* (bit per pixel). Fig. 3 (a) shows the receiver operation characteristic (ROC) curves, where the horizontal axis represents the probability of false positive and vertical axis represents the true positive rate. The results corresponding to different stego images (the embedding rate p=0.25, 0.5, 0.75, 1 *bpp* respectively) are given. In [25], Huang et al. have compared this new approach with Ker's two methods [26] with embedding rate p = 1 *bpp*. The experimental results shown in Fig. 3 (b) demonstrate that this new method can detect LSB matching better than Ker's two methods.

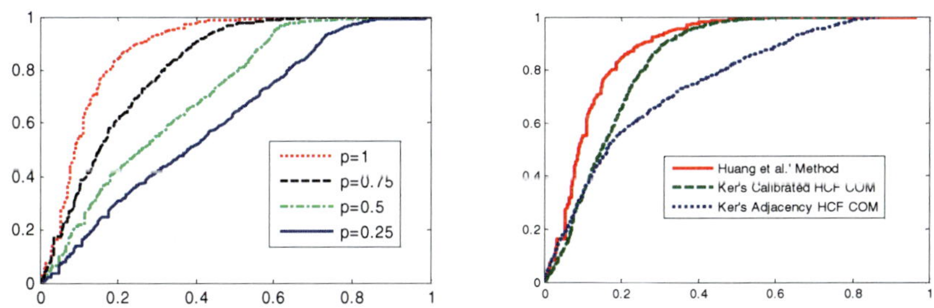

Fig. 3. The detection results of Huang et al.'s method. (a) The ROC curves about four different embedding rate. (b) The results compared with Ker's two methods.

2.3 Attacking F5

The JPEG coefficients, i.e., the quantized block DCT (BDCT) coefficients have some special properties. For example, the histogram of JPEG AC (alternate current) coefficients monotonously decreases as the absolute coefficient values increase, and the difference between two neighboring bars of the histogram decreases with the increasing absolute values. If LSB substitution is applied in JPEG domain, the JPEG coefficient histogram will show abnormality due to the equalization of PoV. Westfeld [52] presented a steganographic method, well known as F5, targeted to preserve the above mentioned two properties of JPEG coefficients. As in Jsteg [49], only the AC coefficients of the JPEG image are selected for embedding the secret message. The major feature of F5 is that the AC coefficient's absolute value will decrease if the coefficient's LSB does not match the secret message bit. Another important feature of F5 is that it employs matrix encoding technique to improve embedding efficiency, which is defined as the number of embedded bits per coefficient change.

F5 can preserve the aforementioned two properties of the histogram, but it causes another issue. Since the message can be extracted only from the non-zero coefficients at the receiving end, the same message bit must be embedded in the next coefficient again at the transmitting end if the subtraction leads to a zero coefficient. Thus the number of coefficients with value 0 would increase and the number of coefficients with value 1 or -1 would decrease after data hiding, which is called shrinkage effect. Through analyzing this shrinkage effect, Fridrich et al. [19] proposed a method which can effectively attack F5. Denote $h_{k,l}(d)$ and $H_{k,l}(d)$ as the total number of JPEG AC coefficients at frequency (k,l) with the absolute value d of a cover image and its corresponding stego image, where $1 \le k,l \le 8$. Let β be the probability that a non-zero AC coefficient will be modified in the data embedding procedure. The following relations are expected to be held between a cover image and its corresponding stego image.

$$\begin{cases} H_{k,l}(0) = h_{k,l}(0) + \beta h_{k,l}(1), & \text{for } d = 0 \\ H_{k,l}(d) = (1-\beta)h_{k,l}(d) + \beta h_{k,l}(d+1), & \text{for } d > 0 \end{cases}. \tag{17}$$

In the above equations, $H_{k,l}(d)$ are known by the attacker. If $h_{k,l}(d)$ can be estimated, β can be derived. To this end, Fridrich proposed a method, known as calibration, to obtain a calibrated image that can be used for estimating the cover image's statistics. According to [19], through decompressing the stego image, cropping by four pixels in each direction, and recompressing again using the same quality factor, a new calibrated image can be achieved. This newly obtained JPEG image has most macroscopic statistics similar to the original cover image. Denote $\hat{h}_{k,l}(d)$ as the absolute histogram of the calibrated image. The probability of modification of coefficients at each frequency (k,l) can be estimated from the first two values of the histogram ($d = 0$ and $d = 1$) by using minimal square error criteria

$$\beta_{k,l} = \arg\min_{\alpha} \left\{ \left[H_{k,l}(0) - \hat{h}_{k,l}(0) - \alpha \hat{h}_{k,l}(1) \right]^2 + \left[H_{k,l}(1) - (1-\alpha)\hat{h}_{k,l}(1) - \alpha \hat{h}_{k,l}(2) \right]^2 \right\}. \tag{18}$$

The least square approximation in Eq. (18) leads to

$$\beta_{k,l} = \frac{\hat{h}_{k,l}(1)\left[H_{k,l}(0)-\hat{h}_{k,l}(0)\right]+\left[H_{k,l}(1)-\hat{h}_{k,l}(1)\right]\left[\hat{h}_{k,l}(2)-\hat{h}_{k,l}(1)\right]}{\left[\hat{h}_{k,l}(1)\right]^2+\left[\hat{h}_{k,l}(2)-\hat{h}_{k,l}(1)\right]^2}. \tag{19}$$

Then probability of modification β can be computed by averaging $\beta_{k,l}$ from some low frequency subbands, i.e., $(k,l)\in\{(1,2),(2,1),(2,2)\}$.

2.4 Attacking OutGuess

OutGuess [36] is a JPEG steganographic tool that uses LSB substitution technique but is capable of evading the Chi-square attack. To achieve such capability, it divides the data embedding procedure into two phases. In the first phase, message bits are embedded into the LSBs of some usable JPEG AC coefficients. In the second phase, some of the remaining usable coefficients flip their LSBs in order to make the stego image histogram match the cover image histogram. Here the usable coefficients refer to the JPEG AC coefficients except 0's and 1's.

Fridrich et al. [18] presented a steganalytic method that can detect OutGuess quantitatively. Because OutGuess introduces random changes into the quantized coefficients, the spatial discontinuities at the boundaries of all 8×8 blocks will increase. The authors also observed that the blockiness is a linear function of the number of JPEG coefficients with flipped LSBs. In order to measure this discontinuities at the boundaries of 8×8 blocks, a blockiness measure is defined by the following formula,

$$B=\sum_{i=1}^{\lfloor (M-1)/8\rfloor}\sum_{j=1}^{N}\left|x_{8i,j}-x_{8i+1,j}\right|+\sum_{j=1}^{\lfloor (N-1)/8\rfloor}\sum_{i=1}^{M}\left|x_{i,8j}-x_{i,8j+1}\right|, \tag{20}$$

where $x_{i,j}$ is the pixel value at position (i, j) of an image with the size $M\times N$.

Since the embedding mechanism in OutGuess is overwriting the LSBs, embedding another message into the stego image may partially cancel out the previous embedding changes. Thus the increase of blockiness measure belonging to the stego image will be smaller than that of the cover image. This difference forms the basis of the steganalytic algorithm. Suppose the questionable image is *T*, four images need to be generated based on it. The first image, denoted as $\hat{S}^0$, is created by embedding data with maximum embedding rate to *T*. The second image, denoted as $\ddot{C}$ is the calibrated image of *T*. The third image is obtained by embedding data with maximum embedding rate to $\hat{C}$, and it is denoted as $\hat{S}^1$. The fourth image is formed by embedding data with maximum embedding to $\hat{S}^1$, and it is denoted as $\hat{S}^2$. The blockiness of the testing image and the four corresponding images are denoted by $B(T)$, $B(\hat{S}^0)$, $B(\hat{C})$, $B(\hat{S}^1)$, and $B(\hat{S}^2)$, respectively. The change rates of the blockiness of the three generated stego images, i.e., $\hat{S}^0$, $\hat{S}^1$, and $\hat{S}^2$, are computed as $R=B(\hat{S}^0)-B(T)$, $R_0=B(\hat{S}^1)-B(\hat{C})$, and $R_1=B(\hat{S}^2)-B(\hat{S}^1)$, respectively.

Assume p $(0 \le p \le 1)$ is embedding rate of the questionable image T. R, R_0, and R_1 has a linear relation as

$$R = R_0 - p(R_0 - R_1). \tag{21}$$

Therefore, the embedding rate can be computed as

$$p = \frac{R_0 - R}{R_0 - R_1}. \tag{22}$$

2.5 Attacking MB

Sallee [39] designed a JPEG steganographic method that can not only achieve a high embedding capacity but also maintain the shape of the coefficient histogram for individual JPEG AC coefficient subbands. It is based on the methodology of statistical modeling and information theory and thereafter named as model-based steganography (MB). In the method, the generalized Cauchy distribution is used to fit the low precision histograms of AC coefficients of each subband for a cover image, where the low precision histogram is set with the bin size larger than 1. Then an arithmetic decoder is used to convert the uniformly distributed message bits into symbols with the required distribution determined by the generalized Cauchy distribution model. Finally, the low precision histogram and the symbols are used to determine the new coefficients' values. In this way, the low precision histograms in the stego image remain unchanged, while the high precision histograms in the stego image fit the generalized Cauchy distribution. Chi-square attack [53] is unlikely to be effective in detecting the presence of MB. An improved version [40] was proposed with de-blocking measure in order to resist the steganalytic attack utilizing blockiness. These two methods are called MB1 and MB2, respectively.

Böhme and Westfeld [4] presented an illuminating steganalytic approach to attack MB1. It may also be effective in attacking MB2. The authors claimed that despite the low precision histograms follow the generalized Cauchy distribution in both cover image and stego image, but there are more outline bins in the high precision histograms for the cover image than the stego image. In other words, the histograms of the stego image fit the model better than the cover image and therefore show less non-conforming bins. In the attack, the steganalyst can obtain the low precision histograms to compute the model parameters. Then the expected high precision histogram can be derived. Through comparing the observed frequency and the expected frequency in the high precision histogram bins with a contingency test, the number of non-conforming bins can be counted. With a predefined threshold value, the type of the questionable image can be determined.

2.6 Attacking YASS

Yet Another Steganographic Scheme (YASS) [47] provides an novel perspective in implementing JPEG steganography. It does not embed data directly into the JPEG coefficients. Instead, data are embedded into the DCT coefficients of some blocks whose positions are randomly located in a given image. To this end, the cover image is firstly divided into consecutive and non-overlapping blocks with a larger size. Such blocks are called big-blocks (or B-blocks) and their sizes are set to $B \times B$ ($B > 8$).

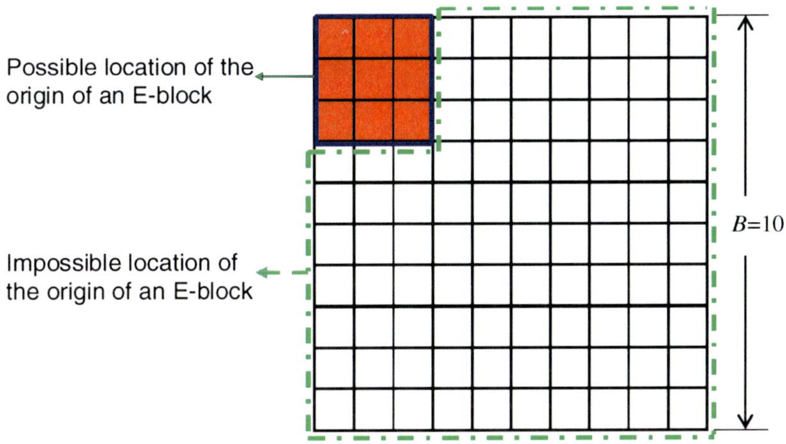

Fig. 4. Illustration of the possible locations and impossible locations of the origin of an E-block when the B-block size B=10.

Table 1. The detection results for YASS (in percentage). TP stands for true positive rate, TN for true negative rate, and AR for accuracy rate.

B		9	10	11	12	13	14	15
QF_h=50 QF_a=75	TP	99.24	99.18	98.60	97.51	96.77	94.93	91.38
	TN	100.00	100.00	100.00	100.00	99.81	98.49	95.87
	AR	99.62	99.59	99.30	98.76	98.29	96.71	93.62
QF_h=75 QF_a=75	TP	98.79	98.10	95.86	92.83	86.41	77.31	71.40
	TN	100.00	99.98	99.37	95.96	88.72	77.56	72.35
	AR	99.40	99.04	97.61	94.40	87.56	77.43	71.87

Then an 8×8 block, referred to as embedding host block (or E-block), is randomly selected in each B-block. In the third step, perform DCT on each E-block and divide the DCT coefficients by the quantization steps associated with a design quality factor QF_h. Data are embedded into the low frequency AC coefficients with quantization index modulation (QIM) embedding technique [8]. Next, multiply the DCT coefficients with the quantization steps associated with QF_h, and perform inverse DCT on the E-blocks. Finally, the whole image is JPEG-compressed with an advertised quality factor QF_a. In this way, YASS can disable the steganalyzers based on the powerful calibration technique [14, 35].

A specific steganalytic method that can detect the presence of YASS was proposed by Li et al. [33]. It is observed that the E-blocks are limited to be located entirely inside the B-blocks. Therefore, some locations are definitely not able to hold E-blocks. Besides, the QIM embedding strategy introduces more zero coefficients. The trace of embedding in the stego image can thus be exposed by comparing the statistical features extracted from the 8×8 blocks that are possible to be E-blocks and that from the blocks that are impossible to be E-blocks. More specifically, the input JPEG

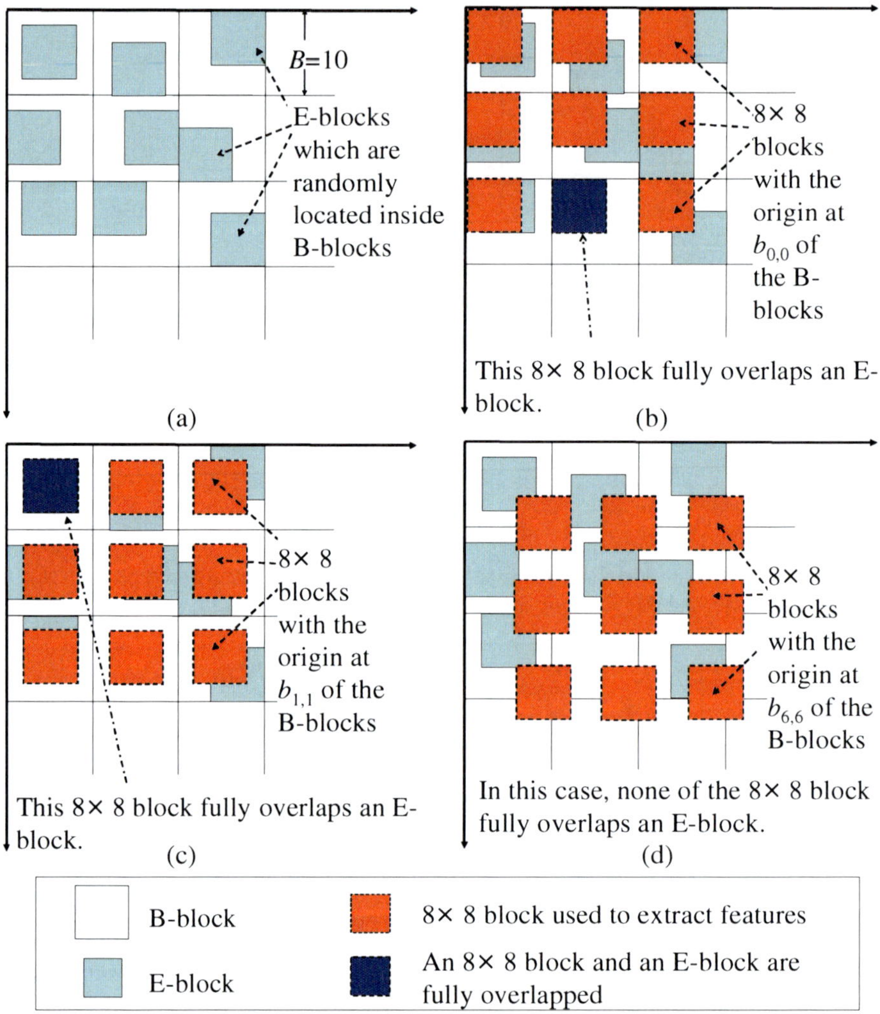

Fig. 5. Illustration of the relation between the location of the E-blocks and the location of 8×8 blocks for feature extraction. (a) Illustration of E-blocks which are randomly located inside B-blocks (B=10). (b) Illustration of 8×8 blocks with the origin at $b_{0,0}$ of the B-block. In this case, 1/9 of the total number of the 8×8 blocks will fully overlap H-blocks. (c) Illustration of 8×8 blocks with the origin at $b_{1,1}$ of the B-block. In this case, 1/9 of the total number of the 8×8 blocks will fully overlap E-blocks. (d) Illustration of 8×8 blocks with the origin at $b_{6,6}$ of the B-block. In this case, none of the 8×8 blocks will fully overlap E-blocks.

image of size $M \times N$ is decompressed into spatial domain and then divided into consecutive and non-overlapping B-blocks. Denote the B-blocks as $K_{m,n}$ ($m = 0,1,...\lfloor M/B \rfloor - 1$, $n = 0,1,...\lfloor N/B \rfloor - 1$) and denote the elements in $K_{m,n}$ as $b_{m,n}^{s,t}$ ($s,t = 0,1,...B-1$). Define the origin of a block as the element in the upper-left corner of the block. Then only the $(B-7)\times(B-7)$ elements in the upper-left corner of the B-block, i.e., $b_{m,n}^{s,t}$ ($s,t=0,1,...B-7$), can possibly overlap with the origin of the E-block, while other elements cannot. Fig. 4 shows the possible locations and impossible locations of the origin of the E-block in a B-block. Next, perform DCT on the 8×8 blocks whose origins are at $b_{m,n}^{i,i}$ ($i=0,1,...B-1$) and quantize them with the quantization steps associated with QF_a. The amount of zero DCT coefficients generated from the 8×8 blocks whose origins are at $b_{m,n}^{i,i}$ ($i=0,1,...B-7$) is expected to be greater than that at $b_{m,n}^{i,i}$ ($i=B-6,B-5,...B-1$), due to the fact that the blocks in the former case are possible to be coincide with some E-blocks which contain more zero DCT coefficients. For each possible B-block size, the frequency of occurrence of the zero DCT coefficients generated from these two different kinds of origins can be served as two features. Fig. 5 illustrates the relation between the E-blocks and the 8×8 blocks for feature extraction. For detecting the B-block size ranging from 9 to 15, a practical implementation algorithm was proposed in [33] to use the frequency of the zero DCT coefficients as the steganalytic features with 14 dimensions. With a Fisher linear discriminant (FLD) classifier, cover images and stego images can be differentiated. Table 1 demonstrates the detection results for the embedding parameter $QF_h=50$ with $QF_a=75$ and $QF_h=75$ with $QF_a=75$. It can be observed that the proposed method can detect YASS with a reliable accuracy rate.

3 Universal Steganalysis

An increasing number of steganographic schemes have posed great challenges to wardens. Therefore, universal steganalysis, aiming at detection of steganography without knowing the details of steganographic schemes, are called for. Normally, the universal methods are based on the methodology of pattern classification, which consists of three stages. In the first stage, steganalytic features are extracted from images. That is, each image is represented by a feature vector. In the second stage, the extracted feature vectors are inputted into the classifier for training purposes. Then in the third stage, the trained classifier is used to project the feature vector of a testing image into a decision region, in which the type of the testing image can be labeled. The classifier training is also called supervised learning. Nowadays, a number of supervised learning algorithms are available, such as Fisher linear discriminant (FLD), support vector machine (SVM), neural network (NN), etc. Since the frameworks of these classifiers are rather mature, the performance of a steganalytic method is primarily determined by the steganalytic features. In what follows, we mainly introduce the feature extraction methods of some typical universal steganalyzers.

3.1 Moment Based Methods

It is observed that the histograms of the pixels and DCT/DWT coefficients may be changed after data hiding. As a result, some statistics derived from the histogram can serve as the steganalytic features. In the literatures, mainly two kinds of moments, i.e., histogram moments and HCF (histogram characteristic function) moments have been proposed. Their definitions are introduced in the following.

Suppose the pixels or DCT/DWT coefficients of an image is represented by $S=\{s_1,s_2,\ldots,s_N\}$ and its corresponding histogram is $h(l), l\in\{0,1,\ldots,L-1\}$, where N is the number of pixels or DCT/DWT coefficients in the image, and L represents the number of disjoint categories (known as *bins*) in the histogram. The n-th histogram moment is defined as

$$H_n=\begin{cases}\frac{1}{N}\sum_{l=0}^{L-1} lh(l) \quad (n=1)\\ \frac{1}{N}\sum_{l=0}^{L-1}(l-\mu)^n h(l) \quad (n=2,3,\ldots)\end{cases}. \tag{23}$$

where μ is the mean value of S. The HCF is the discrete Fourier transform (DFT) of the histogram (defined in [23]). The K-point HCF is defined as

$$H(k)=\sum_{i=0}^{L-1} h(l)e^{-\frac{j2\pi lk}{K}}. \tag{24}$$

And its n-th HCF moment is defined as

$$C_n=\left(\sum_{k=0}^{K/2} k^n\left|H(k)\right|\right)\Big/\left(\sum_{k=0}^{K/2}\left|H(k)\right|\right). \tag{25}$$

Farid [13] proposed a method with the features based on the histogram moments of wavelet decomposition coefficients. In this method, firstly an image is decomposed by separable quadrature mirror filters (QMFs) into three scales. In each scale, there are four kinds of subbands, i.e., lowpass, vertical, horizontal, and diagonal subband. The i-th scale is created by using the QMFs to filter the lowpass subband in the (i-1)-th scale. Secondly, two sets of statistical features are extracted from the resulting subband coefficients. One set contains the mean, variance, skewness, and kurtosis of the subband coefficients at each high frequency subband and at each scale. They are corresponding to the first four histogram moments defined in Eq. (23) respectively. Another set of features contains these four histogram moments from linear prediction error of wavelet decomposition coefficients. The details of the prediction algorithm can be found in [13]. Finally, an FLD classifier is used to separate the images of cover class and that of stego class.

In [23], Harmsen and Pearlman derived that the HCF of image would change after data hiding, and proposed that the COM (center of mass) of HCF, i.e., the first order moment of HCF could be used as the feature. LSB matching[43], spread spectrum [34] and DCT hiding methods [3] are analyzed in this new framework.

Xuan et al. [56] proposed a method with the features based on the HCF moments of wavelet subband coefficients. To apply this method, an image is firstly decomposed by a three-scale 2-D Haar wavelet transform. There are four kinds of subbands, *LL*, *LH*, *HL* and *HH* in each scale. Then the first three HCF moments of the image itself, the wavelet subband coefficients at each level and each subband are served as steganalytic features. A Bayes classifier is applied afterwards for discriminating cover and stego images with the HCF moment based features. Based on theoretical analysis and various experiments, Xuan et al. [55, 56] pointed out that the HCF moments would more sensitive to data embedding than the histogram moments.

In [46], Shi et al. have improved Xuan et al.'s method by re-defining the HCF moments and extracting additional HCF moment features from the prediction-error image. The new defined n-th discrete HCF moment is in the form as

$$C_n' = \left(\sum_{k=1}^{K/2} k^n \left|H(k)\right|\right) \Big/ \left(\sum_{k=1}^{K/2} \left|H(k)\right|\right). \tag{26}$$

Compared to Eq. (25), $H(0)$ is excluded in Eq. (26) because it is the same for both a cover image and its stego image. The prediction-error image is generated by subtracting the predicted pixel values from the testing image, where the predicted value of a pixel x is computed from its neighboring pixels as [51]

$$\hat{x} = \begin{cases} \max(a,b) & c \le \min(a,b) \\ \min(a,b) & c \ge \max(a,b), \\ a+b-c & otherwise \end{cases} \tag{27}$$

where the locations of neighboring pixels are illustrated in Fig. 6. The employment of prediction-error aims to reduce the effect of image content and enhance the data hiding artifacts. Judging from the experimental results, Shi's method demonstrates further performance improvement over Xuan et al.'s method.

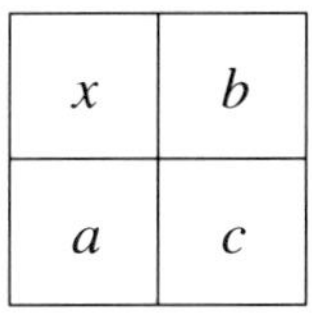

Fig. 6. Illustration of the neighboring pixel locations for prediction.

Based on Shi et al.'s work in [46], Chen et al. [10] presented another steganalytic method mainly targeted for JPEG steganography. In addition to compute HCF moments of image pixels, they also derived the HCF moments corresponding to the absolute values of JPEG coefficients. Moreover, the second order histogram is considered and the moments of 2-D HCF are also used for steganalysis. Thus this new method is more efficient than Farid's [13] and Shi et al.'s [46] methods in attacking some modern JPEG steganographic schemes such as OutGuess, F5 and MB. Inspired by this, Chen et al. [11] added some additional features to their steganalyzer targeted for texture images. These new features are generated from the BDCT (block discrete cosine transform) representations with a set of different block sizes, e.g., 2×2, 4×4, 8×8, 16×16, referred to as "rake transform". With various block sizes, the complicated frequency change can be perceived in BDCT coefficients with different block sizes, and hence the data embedded in texture images can be detected with HCF moments corresponding to these BDCT coefficients.

3.2 Calibration Based Methods

Fridrich [14] developed a universal steganalyzer designed for attacking JPEG steganography. In contrast to previous blind approaches, the features are calculated as an L_1 norm of the difference between a specific macroscopic functional calculated from the testing image and the same functional obtained from its corresponding calibrated image.

All features are constructed in the following manner. A vector functional F is applied to the testing JPEG image J_1 first, and then the same vector functional F is applied to the calibrated image J_2, where J_2 is obtained through de-compressing J_1 to the spatial domain, cropping 4 rows and 4 columns in both horizontal and vertical directions, and recompressing the cropped image again with the same quality factor. The final feature f is obtained as an L_1 norm of the difference

$$f = \|F(J_1) - F(J_2)\|_{L_1} . \tag{28}$$

The functional F could be the global JPEG coefficient histogram, a co-occurrence matrix, spatial blockiness, etc. Suppose the testing JPEG file is represented with a JPEG coefficient array $d_k(i, j)$, $i, j = 1,...,8$ and $k = 1,..., B$. The symbol $d_k(i, j)$ denotes the (i,j)-th JPEG coefficient in the k-th block (there are a total number of B blocks), and $L = \min_{k,i,j} d_k(i, j)$ and $R = \max_{k,i,j} d_k(i, j)$. The six different kinds of functionals F in Eq. (28) are defined by the following metrics.

(1) Global histogram functional

The first functional is the global histogram H of all 64×B JPEG coefficients

$$H = (H_L,..., H_R) , \tag{29}$$

where H_L and H_R represent the number of coefficients which are equal to L and R respectively.

(2) Individual histogram functional

For a fixed JPEG coefficient mode (i, j), let h_r^{ij} , $r = L, \ldots, R$, denotes the individual histogram of values $d_k(i,j)$, $k = 1, \ldots, B$. The next 5 functionals are individual histograms for 5 JPEG coefficient modes, i.e.,

$$h^{ij} = (h_L^{ij},..., h_R^{ij}) , \tag{30}$$

where $(i, j) \in \{(1,2),(2,1),(3,1),(2,2),(1,3)\}$.

(3) Dual histogram functional

For a fixed coefficient with value d, the dual histogram g_{ij}^d denotes the number of how many times the value d occurs at the (i,j)-th JPEG coefficient over all B blocks in the JPEG image. The following 11 functionals are dual histograms represented with

$$g_{ij}^d = \sum_{k=1}^{B} \delta(d, d_k(i, j)) , \tag{31}$$

where $d \in \{-5,-4,-3,-2,-1,0,1,2,3,4,5\}$ and the impulse function

$$\delta(x=i, y=j)=\begin{cases}1, & \text{if } x=i \text{ and } y=j \\ 0, & \text{otherwise}\end{cases}. \tag{32}$$

(4) Variation functional

The variation functional is used to explore the dependency existed in the inter-block sense, which is defined as

$$V=\frac{\sum_{i,j=1}^{8}\sum_{k=1}^{|I_r|-1}\left|d_{I_r(k)}(i,j)-d_{I_r(k+1)}(i,j)\right|+\sum_{i,j=1}^{8}\sum_{k=1}^{|I_c|-1}\left|d_{I_c(k)}(i,j)-d_{I_c(k+1)}(i,j)\right|}{|I_r|+|I_c|} \tag{33}$$

where I_r and I_c denote the vectors of block indices while scanning the image by rows and by columns, respectively.

(5) *B*lockiness functional

The blockiness functional is used to measure the block artifacts in the JPEG image. Two blockiness functionals are defined as

$$B_\alpha=\frac{\sum_{i=1}^{\lfloor (M-1)/8\rfloor}\sum_{j=1}^{N}\left|x_{8i,j}-x_{8i+1,j}\right|^\alpha+\sum_{j=1}^{\lfloor (N-1)/8\rfloor}\sum_{i=1}^{M}\left|x_{i,8j}-x_{i,8j+1}\right|^\alpha}{N\lfloor (M-1)/8\rfloor+M\lfloor (N-1)/8\rfloor}, \tag{34}$$

where M and N are the image size, $x_{i,j}$ are grayscale values of the decompressed JPEG image, and $\alpha\in\{1,2\}$.

(6) Co-occurrence functional

The co-occurrence functional is used to calculate the co-occurrence matrix of neighboring JPEG coefficients. These three functionals are defined as

$$\begin{aligned}N_{00}&=C_{0,0}(J_1)-C_{0,0}(J_2)\\ N_{01}&=C_{0,1}(J_1)-C_{0,1}(J_2)+C_{1,0}(J_1)-C_{1,0}(J_2)+C_{-1,0}(J_1)-C_{-1,0}(J_2)+C_{0,-1}(J_1)-C_{0,-1}(J_2)\\ N_{11}&=C_{1,1}(J_1)-C_{1,1}(J_2)+C_{1,-1}(J_1)-C_{1,-1}(J_2)+C_{-1,1}(J_1)-C_{-1,1}(J_2)+C_{-1,-1}(J_1)-C_{-1,-1}(J_2)\end{aligned} \tag{35}$$

where the co-occurrence matrix

$$C_{st}=\frac{\sum_{k=1}^{|I_r|-1}\sum_{i,j=1}^{8}\delta\left(s,d_{I_r(k)}(i,j)\right)\delta\left(t,d_{I_r(k+1)}(i,j)\right)+\sum_{k=1}^{|I_c|-1}\sum_{i,j=1}^{8}\delta\left(s,d_{I_c(k)}(i,j)\right)\delta\left(t,d_{I_c(k+1)}(i,j)\right)}{|I_r|+|I_c|} \tag{36}$$

It is noted that the results of these three co-occurrences N_{00}, N_{01} and N_{11} are used as the features directly.

An example of the performance of Fridrich's universal steganalyzer is in Table 2. The test image data set consists of 1814 images with the size 780×540, and quality factor 80. The FLD classifier was trained on 1314 cover and 1314 stego images, and the remaining 500 cover and 500 stego images were used for testing. The detection performance was evaluated using detection reliability ρ, which is defined as

$$\rho = 2A - 1 \quad (37)$$

where A is the area under the ROC curve. Five steganographic schemes such as F5, F5 without matrix encoding (F5_111), OutGuess, MB1 and MB2 were tested. In Table 2, the embedding rate is represented by *bpc* (bits per non-zero JPEG coefficients) values. It is noted that this is the first universal steganalyzer targeted to JPEG steganography, and it had performed best in defeating modern JPEG steganography such as Outguess, F5 and MB1 from 2004 to 2006.

Table 2. Detection reliability ρ for five steganographic schemes with different embedding rates (U = unachievable rate)

bpc	F5	F5_111	OutGuess	MB1	MB2
0.05	0.2410	0.6451	0.8789	0.2197	0.1631
0.1	0.5386	0.9224	0.9929	0.4146	0.3097
0.2	0.9557	0.9958	0.9991	0.7035	0.5703
0.4	0.9998	0.9999	U	0.9375	0.8243
0.6	1.0000	1.0000	U	0.9384	U
0.8	1.0000	1.0000	U	0.9916	U

3.3 Markov Process Based Methods

Some steganographic schemes, even some steganalytic methods, assume that the pixels/coefficients in a cover image are independent and identically distributed (i.i.d.). But in fact the image has strong dependency in local region, and thus correlation exists within adjacent pixels/coefficients. It is expected that the correlation will be weaken by the data hiding operation. Markov model can be exploited to provide a convenient way to measure such correlation.

3.3.1 Sullivan et al.'s 129-D

Sullivan et al. [48] proposed a steganalytic method by using the Markov chain to model the inter-pixel dependency. Firstly, a 2-D image is transformed into a 1-D sequence by vertical, horizontal, or zig-zag scanning the image pixels from the top-left to the bottom-right. Denote the sequence as $\mathbf{Y}$ and its n-th element as y_n ($n = 1, 2, \ldots L$), where L is the number of elements in $\mathbf{Y}$. Then an empirical transitional matrix $\mathbf{M}$ is obtained from the sequence by computing the element in the i-th column and j-th row as

$$\mathbf{M}_{i,j} = \sum_{n=2}^{L} \delta(y_n = i, y_{n-1} = j) \big/ (L-1) \quad i, j \in \{0, 1, \ldots, 255\} \quad (38)$$

where $\delta(x, y)$ is the impulse function, as described in Eq. (32).

Because the size of the empirical transition matrix is very large, e.g., the 65536 elements for a grey level image with bit depth of 8. It cannot be used as features directly. Some values in the empirical transitional matrix are selected as features, including six elements with the highest values on the main diagonal ($\mathbf{M}_{i,i}$), their ten nearest elements in the same row, i.e., $\{\mathbf{M}_{i,i-1}, \mathbf{M}_{i,i-2}, \ldots, \mathbf{M}_{i,i-10}\}$, and 63 elements that are sub-sampled from the remaining elements on the main diagonal by four, i.e.,

$\{\mathbf{M}_{1,1}, \mathbf{M}_{5,5}, \mathbf{M}_{9,9}, ..., \mathbf{M}_{253,253}\}$. In this way, a total number of 129 features have been extracted from an image. Experimental results show that the method is very effective to steganographic schemes whose stego noise is independent to the cover data.

However, as only part of the elements in the empirical transitional matrix is selected as features, some useful information might be ignored due to the random fashion of feature formulation. Finally, this deficiency may influence the detection accuracy rate that can be achieved.

3.3.2 Zou et al.'s 243-D

Inspired by [48], Zou et al. [57] presented a new universal steganalyzer through using Markov chain to model of the thresholded prediction-error image. In [57], the authors proposed to use neighboring pixels to predict the current pixel, and the predictions are made in three directions, namely, horizontal, vertical and diagonal. Suppose the questionable image is $f(x, y), 1 \le x, y \le N$. Three kinds of prediction-error images are described as follows.

$$\begin{aligned} e_h(x, y) &= f(x+1, y) - f(x, y) \\ e_v(x, y) &= f(x, y+1) - f(x, y) \\ e_d(x, y) &= f(x+1, y+1) - f(x, y) \end{aligned} \tag{39}$$

where $e_h(x, y)$, $e_v(x, y)$ and $e_d(x, y)$ represent the prediction-error images in horizontal, vertical and diagonal directions, respectively.

Although the range of the difference values in Eq. (39) is increased, the majority of the difference values are highly concentrated in a small range near zero owing to the high correlation between neighboring pixels in natural images. Considering the large values in the prediction-error image may mainly be caused by the image content rather than by the data hiding process, a certain threshold T_h is applied to remove the large values in the prediction-error images. That is, If the element in prediction-error image is larger than T_h or smaller than $-T_h$, it will be represented by T_h or $-T_h$, respectively. Then a 2-D Markov chain model is applied to the three kinds of thresholded prediction-error images, respectively. In [57], the authors selected T_h=4, and there are $(2T_h+1)\times(2T_h+1)\times3=243$ features in the final feature set. The experiments with the same data hiding methods on the same image data set demonstrated that Zou et al.'s method [57] outperforms Sullivan et al.'s method [48] generally.

3.3.3 Shi et al.'s 324-D

In [44], Shi et al. presented a JPEG steganalyzer by using the Markov process to model the difference JPEG 2-D arrays. For a given image, consider the 2-D array consisting of all the JPEG coefficients which have been quantized with a JPEG quantization table, and have not yet been zig-zag scanned and entropy coded. This 2-D array has the same size as the given image with each 8×8 block filled up with the corresponding JPEG coefficients. Take absolute value for each coefficient, resulting in a 2-D array as shown in Fig. 7. This resultant 2-D array is called JPEG 2-D array. The features proposed in this scheme are formed from the JPEG 2-D array.

Denote the JPEG 2-D array generated from a given test image by $F(u, v)(u \in [0, S_u-1], v \in [0, S_v-1])$, where S_u is the size of the JPEG 2-D array in horizontal direction and S_v in vertical direction. Then the difference arrays are generated by the following formulae:

$$\begin{aligned} F_h(u,v) &= F(u,v) - F(u+1,v) \\ F_v(u,v) &= F(u,v) - F(u,v+1) \\ F_d(u,v) &= F(u,v) - F(u+1,v+1) \\ F_m(u,v) &= F(u+1,v) - F(u,v+1) \end{aligned}, \tag{40}$$

where $u \in [0, Su-2]$, $v \in [0, Sv-2]$, and $F_h(u, v)$, $F_v(u, v)$, $F_d(u, v)$, and $F_m(u, v)$ denote the difference arrays in the horizontal, vertical, main diagonal, and minor diagonal directions, respectively.

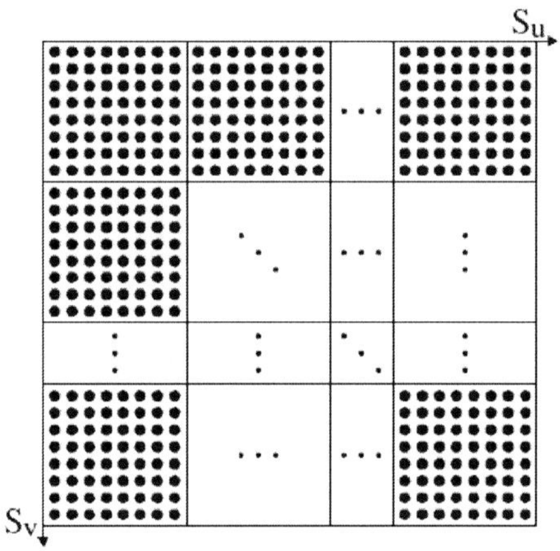

Fig. 7. A sketch of JPEG-2D array

Since the distribution of the elements of the difference 2-D arrays is Laplacian-like, most of the difference values are close to zero. The values of mean and standard deviation of percentage number of elements of horizontal difference 2-D arrays for the image set falling into [-*T*, *T*] when *T* = 1, 2, 3, 4, 5, 6, and 7 are shown in Table 3. The test image data set consists of 7,560 JPEG images with Q-factors ranging from 70 to 90, and the image size are 768×512 or 512×768. It is observed that more than 90% elements in the horizontal difference 2-D arrays fall into the interval [-3, 3]. The same is true for the difference 2-D arrays along other three directions.

Table 3. Mean and standard deviation of percentage numbers of elements of horizontal difference JPEG 2-D arrays falling within [-*T*, *T*] for *T* = 1, 2, 3, 4, 5, 6, and 7

	[-1, 1]	[-2, 2]	[-3, 3]	[-4,4]*	[-5, 5]	[-6, 6]	[-7, 7]
Mean	84.72	88.58	90.66	91.99	92.92	93.60	94.12
Deviation	5.657	4.243	3.464	2.836	2.421	2.104	1.850

* 91.99% is the mean, meaning that on statistic average 91.99% of all elements of horizontal difference JPEG 2-D arrays generated from the image set fall into the range [-4, 4]. The standard deviation is 2.836%.

The above-defined difference JPEG 2-D arrays are modeled by Markov random process. In order to further reduce computational complexity, the thresholding technique [57] is adopted. Only those elements in the difference JPEG 2-D arrays whose value falls into $\{-T_h, \cdots, -1, 0, 1, \cdots, T_h\}$ are considered. If an element's value is either larger than T_h or smaller than $-T_h$, it will be represented by T_h or $-T_h$ correspondingly. This procedure results in a transition probability matrix (TPM) of dimensionality $(2T_h+1)\times(2T_h+1)$. The elements of these four TPMs associated with the horizontal, vertical, main diagonal and minor diagonal difference 2-D arrays are given by

$$\begin{aligned}
p\{F_h(u+1,v)=n \mid F_h(u,v)=m\} &= \frac{\sum_{u,v}\delta(F_h(u,v)=m, F_h(u+1,v)=n)}{\sum_{u,v}\delta(F_h(u,v)=m)} \\
p\{F_v(u,v+)=n \mid F_v(u,v)=m\} &= \frac{\sum_{u,v}\delta(F_v(u,v)=m, F_v(u,v+1)=n)}{\sum_{u,v}\delta(F_v(u,v)=m)} \\
p\{F_d(u+1,v+1)=n \mid F_d(u,v)=m\} &= \frac{\sum_{u,v}\delta(F_d(u,v)=m, F_d(u+1,v+1)=n)}{\sum_{u,v}\delta(F_d(u,v)=m)} \qquad (41)\\
p\{F_m(u,v+1)=n \mid F_m(u+1,v)=m\} &= \frac{\sum_{u,v}\delta(F_m(u+1,v)=m, F_m(u,v+1)=n)}{\sum_{u,v}\delta(F_m(u+1,v)=m)}
\end{aligned}$$

where $m, n \in \{-T_h, \cdots, 0, \cdots, T_h\}$, the summation range for u is from 0 to S_u-2 and for v from 0 to $S_v - 2$, and δ is the impulse function as described in Eq. (32).

In summary, each of these four TPMs has $(2T_h+1)\times(2T_h+1)$ elements, and in total there are $(2T_h+1)\times(2T_h+1)\times4$ elements. All of them are serving as features for steganalysis. The threshold T_h is selected as 4 according to the statistical study shown in Table 3. Hence, there are $81\times4=324$ elements in total.

The 7,560 JPEG images mentioned previously are used in the experimental work to test the efficiency of this new JPEG steganalyzer. The 1/2 of the cover and stego

Table 4. Performance comparison using different methods (in the unit of %; TN stands for true negative rate, TP stands for true positive rate, and AR stands for accuracy rate; The four different steganalyzers are denoted as MW-72 [13], MC-78 [46], JF-23 [14] and MP-324 [44], where MW, MH, JF, and MP stands for moments of wavelet coefficient histogram, moments of HCF, JPEG features and Markov process, respectively, and 72, 78, 23 and 324 are the number of features.)

	bpnc	MW-72 [13]			MH-78 [46]			JF-23 [14]			MP-324 [44]		
		TN	TP	AR	TN	TP	AR	TN	TP	AR	TN	TP	AR
OG	0.05	59.0	57.6	58.3	55.6	58.5	57.0	49.8	75.4	62.6	87.6	90.1	88.9
	0.1	70.0	63.5	66.8	61.4	66.3	63.9	68.9	83.3	76.1	94.6	96.5	95.5
	0.2	81.9	75.3	78.6	72.4	77.5	75.0	90.0	93.6	91.8	97.2	98.3	97.8
F5	0.05	55.6	45.9	50.8	57.9	45.9	51.5	46.1	61.0	53.6	58.6	57.0	57.8
	0.1	55.5	48.4	52.0	54.6	54.6	54.6	58.4	63.3	60.8	68.1	70.2	69.1
	0.2	55.7	55.3	55.5	59.5	63.3	61.4	77.4	77.2	77.3	85.8	88.3	87.0
	0.4	62.7	65.0	63.9	71.5	77.1	74.3	92.6	93.0	92.8	95.9	97.6	96.8
MB1	0.05	48.5	53.2	50.8	57.0	49.2	53.1	39.7	66.9	53.3	79.4	82.0	80.7
	0.1	51.9	52.3	52.1	57.6	56.6	57.1	45.6	70.1	57.9	91.2	93.3	92.3
	0.2	52.3	56.7	54.5	63.2	66.7	65.0	58.3	77.5	67.9	96.7	97.8	97.3
	0.4	55.3	63.6	59.4	74.2	80.0	77.1	82.9	86.8	84.8	98.8	99.4	99.1

image pairs are randomly selected to train the SVM classifier [7] and the remaining 1/2 pairs to test the trained classifier. The test results shown in Table 4 are the arithmetic average of 20 random experiments. In Table 4, the embedding rate is represented as *bpnc* (bits per non-zero JPEG AC coefficients). It is observed that for three modern steganographic schemes OutGuess, F5 and MB1, this new JPEG steganalyzer has the most reliable detection performance compared with other three steganalyzers [13, 14, 46].

3.3.4 Chen et al.'s 486D

The features proposed in [44] mainly utilize intrablock correlation among JPEG coefficients. In [9], Chen and Shi proposed to use both the intrablock and interblock correlations among JPEG coefficients to enhance steganalysis capability. The interblock correlation is reflected among those JPEG modes, i.e., coefficients located in the same position within the 8×8 blocks, which capture the frequency characteristics of those blocks.

In this new steganalyzer, the intrablock features are the same as that in [44], and the interblock features are achieved as follows.

(1) Mode 2-D array

Mode 2-D arrays are formed by aligning each mode from all the 8×8 blocks of JPEG 2-D array. The size of each mode 2-D array is $S_s = S_u/8$ in horizontal direction and $S_t = S_v/8$ in vertical direction, where S_u is the size of the JPEG 2-D array in horizontal direction and S_v in vertical direction. For each given JPEG image, we can obtain 63 mode 2-D arrays but exclude Mode 1, i.e., the DC (direct current) component, which is not touched by modern JPEG steganography generally.

(2) Difference mode 2-D array

Difference mode 2-D arrays are formed the same way as the difference JPEG 2-D arrays were generated in [44]. Denoting a mode coefficient in a mode 2-D array by $M_h^{(z)}(s,t)$ ($s \in [0, S_s - 2]$, $t \in [0, S_t - 2]$, $z \in [2, 64]$), these difference mode 2-D arrays can be calculated by

$$\begin{aligned} M_h^{(z)}(s,t) &= M^{(z)}(s,t) - M^{(z)}(s+1,t) \\ M_v^{(z)}(s,t) &= M^{(z)}(s,t) - M^{(z)}(s,t+1) \\ M_d^{(z)}(s,t) &= M^{(z)}(s,t) - M^{(z)}(s+1,t+1) \\ M_m^{(z)}(s,t) &= M^{(z)}(s+1,t) - M^{(z)}(s,t+1) \end{aligned} \tag{42}$$

In above formulae, $M_h^{(z)}$, $M_v^{(z)}$, $M_d^{(z)}$ and $M_m^{(z)}$ denote the horizontal, vertical, main diagonal, and minor diagonal difference mode 2-D array, respectively.

(3) Markov process and TPM (transition probability matrix)

Difference mode 2-D arrays can be modeled by Markov process. Similarly, to reduce the dimension of feature vector, a threshold T_h is applied to each difference mode 2-D array. These TPMs along each direction are averaged and then four TPMs for all the 63 modes can be obtained. Elements of TPMs associated with the horizontal, vertical, main diagonal, and minor diagonal difference mode 2-D arrays are given by

$$p\{M_h(s+1,t)=n \mid M_h(s,t)=m\} = \frac{\sum \delta(M_h^{(z)}(s,t)=m, M_h^{(z)}(s+1,t)=n)}{\sum \delta(M_h^{(z)}(s,t)=m)}$$

$$p\{M_v(s,t+1)=n \mid M_v(s,t)=m\} = \frac{\sum \delta(M_v^{(z)}(s,t)=m, M_v^{(z)}(s,t+1)=n)}{\sum \delta(M_v^{(z)}(s,t)=m)}$$

$$p\{M_d(s+1,t+1)=n \mid M_d(s,t)=m\} = \frac{\sum \delta(M_d^{(z)}(s,t)=m, M_d^{(z)}(s+1,t+1)=n)}{\sum \delta(M_d^{(z)}(s,t)=m)}$$

$$p\{M_m(s,t+1)=n \mid M_m(s+1,t)=m\} = \frac{\sum \delta(M_m^{(z)}(s+1,t)=m, M_m^{(z)}(s,t+1)=n)}{\sum \delta(M_m^{(z)}(s+1,t)=m)} \tag{43}$$

In above equations, $m, n \in \{-T_h, \cdots, T_h\}$, the summation range for z is from 2 to 64, for s from 0 to $S_s - 2$, for t from 0 to $S_t - 2$, and δ is the impulse function.

Both the intrablock and the interblock correlations are combined for steganalysis. The TPM for the thresholded difference JPEG 2-D array along each direction gives $(2T_h+1)\times(2T_h+1)$ intrablock features, and the TPM for all the difference mode 2-D arrays along each direction gives $(2T_h + 1) \times (2T_h+ 1)$ interblock features. In [9], the features derived from four difference JPEG 2-D arrays and two difference mode 2-D arrays, i.e., horizontal and vertical difference mode 2-D arrays, are utilized. As a result, a 486-D feature vector is obtained for each JPEG image when T_h is selected as 4. According to the experimental results, these new approach can obtain a remarkable improvement over the original Markov process based JPEG steganalyzer [44].

3.4 Combining Calibration and Markov Process

In [35], Pevný and Fridrich constructed another JPEG steganalyzer. The feature set is created as a combination of the extended JPEG feature set [14] and calibrated Markov feature set [44].

As in [14], J_1 represents the testing JPEG image, and J_2 represents the corresponding calibrated image. The H_l, h^{ij}, g^d, and C_{st} represent the global histogram functional, individual histogram functional, dual histogram functional, and co-occurrence functional, respectively. If the functional F is a scalar, the calibrated feature is obtained as the difference $F(J_1)$-$F(J_2)$; otherwise if F is a vector or matrix, the calibrated feature is obtained as L_1 norm $\|F(J_1)-F(J_2)\|_{L_1}$. The features extended from the JPEG feature set are as follows.

(1) Global histogram features
The first 11 features are extracted from the global histogram H of all 64×B JPEG coefficients, i.e.,

$$H_l(J_1)-H_l(J_2), l\in\{-5,\ldots,5\} \tag{44}$$

(2) Individual histogram features
The next 55 individual histogram features are extracted from 5 different individual histograms, i.e.,

$$h_l^{ij}(J_1) - h_l^{ij}(J_2), l \in \{-5,\ldots,5\} \tag{45}$$

where $(i,j) \in \{(1,2),(2,1),(3,1),(2,2),(1,3)\}$.

(3) Dual histogram features

The next 99 dual histogram features are extracted from 11 different dual histograms, i.e.,

$$g_{ij}^d(J_1) - g_{ij}^d(J_2), d \in \{-5,\ldots,5\} \tag{46}$$

where $(i,j) \in \{(2,1),(3,1),(4,1),(1,2),(2,2)(3,2),(1,3),(2,3),(1,4)\}$.

(4) Co-occurrence features

The next 25 co-occurrence features are extracted from the central elements in of the co-occurrence matrix, i.e.,

$$C_{st}(J_1) - C_{st}(J_2), (s,t) \in [-2,+2] \times [-2,+2] \tag{47}$$

(5) Other features

The remaining 3 features, i.e., the one variation features and two blockiness features are the same as the features in [14].

In summary, there are 193 features extended from the JPEG features in [14]. The other features are derived from the Markov feature set [44]. Let **M** denote the TPM in a specific direction. The calibrated Markov features are formed by differences $\mathbf{M}^{(c)} = \mathbf{M}(J_1) - \mathbf{M}(J_2)$, where J_1 is the test image and J_2 is its corresponding calibrated version. The dimension of the calibrated Markov feature set remains the same as its original version.

A direct combination of both feature sets would produce a 517-dimensional feature vector. To reduce the dimensionality, the Markov features extracted in the horizontal, vertical, main diagonal, and minor diagonal directions are averaged. This feature vector has dimensionality 81. It is observed in [35] that the averaged features produced very similar performance as their full version. After merging the 193 extended JPEG features with the 81 averaged calibrated Markov features, the dimension of the resulting merged feature set became 193 + 81 = 274. Some experimental results given in [35] demonstrated that the new merged feature set provides significantly better results than previous JPEG feature set [14] and Markov feature set [44].

3.5 Universal Steganalyzers to Break YASS

As pointed in Section 2.6, YASS [47] is a newly developed JPEG steganographic scheme. It has been reported that YASS can effectively disable some major universal steganalyzers [13, 14, 35, 46, 56] to a great extent and is regarded as one of the most secure JPEG steganographic schemes today. However, as an entirely new steganographic scheme with complicated embedding strategy, the security performance of YASS is still a subject of much debate since then it has been proposed. In [24], Huang et al. pointed that with YASS' complicated embedding procedure, the intra- and inter-block dependency among the JPEG coefficients belonging to the original image would be disturbed after the secret message embedding. Thus some steganalyzers which utilizes intra- and/or inter-block correlation of JPEG coefficients can break YASS. In [24], the security performance of YASS against four state-of-the-art blind

Table 5. Detection rate against YASS achieved by the four steganalyzers (in the unit of %; The four different steganalyzers are denoted as JFMP-274 [35], MP-200 [22], MP-324 [44] and MP-486 [9], where JF stands for JPEG features and MP for Markov process, and 274, 200, 324 and 486 are the number of features.)

	$QF_h \rightarrow QF_a$	JFMP-274[35]	MP-200 [22]	MP-324 [44]	MP-486 [9]
YASS B9	50→50	62.8	68.3	77.2	**81.1**
	50→75	89.4	95.2	95.4	**95.8**
	75→75	62.8	68.1	74.0	**74.4**
YASS B10	50→50	65.6	73.9	83.0	**85.4**
	50→75	96.1	98.3	98.4	**98.5**
	75→75	67.5	75.2	**77.6**	77.1
YASS B12	50→50	59.7	64.4	73.5	**74.3**
	50→75	85.2	93.6	92.9	**94.2**
	75→75	60.3	64.0	66.0	**68.0**
YASS B14	50→50	55.3	56.2	63.9	**64.5**
	50→75	71.1	79.1	79.8	**81.1**
	75→75	54.0	57.0	**61.5**	60.7

JPEG steganalyzers [9, 22, 35, 44] are reported in detail. Below, the security performance of YASS with uncompressed input image is introduced.

There [24] the experiments are based on 5000 images, where the training set for every classifier contained 4000 cover images and 4000 stego images, and the remaining 1000 cover and 1000 stego images are used to test the trained classifier. The detection accuracy rates are shown in Table 5, where the B9, B10, B12 and B14 represent the B-blocks size in YASS, QF_h represents the quality factor for data hiding, and QF_a is the advertised quality factor of the output image of YASS.

The experimental results demonstrate YASS' embedding strategy is an efficient way to paralyze the calibration technique which is popularly utilized in JPEG steganalysis. However, YASS' complicated embedding procedure still leaves statistical artifacts. Consequently, the JPEG steganalyzers which do not rely on calibration can detect YASS.

4 Conclusion

In the last 10 years or so, great advances have been achieved in image steganalysis. Most of the existed steganographic schemes such as LSB substitution, F5, and MB1 etc. have been successfully analyzed. The rapid development in steganalysis casts new light on natural image models, pattern recognition methods, etc. It also triggers some new discovery in digital forensics. Some steganalytic strategies and means can even be applied directly in detecting image forgery, image splicing and so on [45].

However, steganalysis and steganography is just like a cat and mouse game, and the steganalyzers will always be chasing the development of steganography. Even though the state-of-the-art steganalyzers have progressed to such an extent that many steganographic schemes are broken or have to reduce the embedding rate for security, there are still a lot of challenges that need to be overcome.

Learning lessons from steganalysis, steganographic technique is heading towards the direction of being more secure. To this end, increasing the embedding efficiency [20, 21, 52] and reducing the embedding distortions [15, 31, 32] are two main approaches that the current steganographic schemes may adopt. To embed the same amount of message bits, increasing the embedding efficiency and lowering the embedding distortion can make the stego image more close to its cover image, thus lowering the possibility of being attacked by steganalytic methods. How to design more sensitive steganalyzers is a challenge issue. Steganalysts may need to use more advanced signal processing methods and statistical analysis approaches as well as more advanced image models to construct steganalytic classifiers.

The existing steganalyzers are still facing severe challenges, because there are so many steganographic schemes available nowadays due to the prosperous development of information hiding in the pass few years. For any given questionable image, we need to choose a trained classifier to detect it. However, which classifier should we select? Until now, we have not found an efficient way to classify all these existed steganographic schemes, not to mention those may appear in the future. The only way left for us is to try each possible steganalytic classifier that has been trained in turn. This may greatly increase the computational complexity and decrease the detection accuracy rate. Obviously, it is impractical.

Though steganalysis has been greatly progressed, our opinion is that steganalysis is still in its early stage. There still exist many open issues that have not been resolved or even untouched, such as classifying different steganographic schemes, identifying the positions of altered pixels/coefficients, extracting the embedded information, etc. We believe that with the rapid development of digital forensics, or some breakthrough in the field of pattern recognition and artificial intelligence, many problems perplexed us today in steganalysis will eventually be resolved.

References

[1] Steganos, `http://www.steganos.com/steganos-home/`

[2] Stools, `ftp://idea.sec.dsi.unimi.it/pub/security/crypt/code/s-tools4.zip`

[3] Alturki, F., Mersereau, R.: A novel approach for increasing security and data embedding capacity in images for data hiding applications. In: Information Technology: Coding and Computing, Las Vegas, NV, USA, pp. 228–233 (2001)

[4] Böhme, R., Westfeld, A.: Breaking cauchy model-based JPEG steganography with first order statistics. In: Samarati, P., Ryan, P.Y.A., Gollmann, D., Molva, R. (eds.) ESORICS 2004. LNCS, vol. 3193, pp. 125–140. Springer, Heidelberg (2004)

[5] Bender, W., Gruhl, D., Morimoto, N., Lu, A.: Techniques for data hiding. IBM System Journal 35, 313–336 (1996)

[6] Chandramouli, R., Kharrazi, M., Memon, N.: Image steganography and steganalysis concepts and practice. In: Kalker, T., Cox, I., Ro, Y.M. (eds.) IWDW 2003. LNCS, vol. 2939, pp. 35–49. Springer, Heidelberg (2004)

[7] Chang, C.C., Lin, C.J.: LIBSVM: a library for support vector machines (2001), Software available at `http://www.csie.ntu.edu.tw/~cjlin/libsvm`

[8] Chen, B., Wornell, G.W.: Quantization index modulation: A class of provably good methods for digital watermarking and information embedding. IEEE Transactions on Information Theory 47, 1423–1443 (2001)

[9] Chen, C., Shi, Y.Q.: JPEG image steganalysis utilizing both intra-block and inter-block correlations. In: IEEE International Symposium on Circuits and Systems, Seattle, Washington, USA, pp. 3029–3032 (2008)

[10] Chen, C., Shi, Y.Q., Chen, W., Xuan, G.: Statistical moments based universal steganalysis using JPEG 2-D array and 2-D characteristic function. In: IEEE International Comference on Image Processing, Atlanta, Georgia, USA (2006)

[11] Chen, C., Shi, Y.Q., Xuan, G.: Steganalyzing texture images. In: IEEE International Conference on Image Processing, Texas, USA (2007)

[12] Dumitrescu, S., Wu, X.L., Wang, Z.: Detection of LSB steganography via sample pair analysis. IEEE Transactions on Signal Processing 51, 1995–2007 (2003)

[13] Farid, H.: Detecting hidden messages using higher-order statistical models. In: International Conference on Image Processing, Rochester, NY, USA (2002)

[14] Fridrich, J.: Feature-based steganalysis for JPEG images and its implications for future design of steganographic schemes. In: Fridrich, J. (ed.) IH 2004. LNCS, vol. 3200, pp. 67–81. Springer, Heidelberg (2004)

[15] Fridrich, J.: Minimizing the embedding impact in steganography. In: 8th ACM workshop on Multimedia and Security, Geneva, Switzerland, pp. 2–10. ACM Press, New York (2006)

[16] Fridrich, J., Goljan, M.: On estimation of secret message length in LSB steganography in spatial domain. In: Delp, E.J., Wong, P.W. (eds.) IS&T/SPIE Electronic Imaging: Security, Steganography, and Watermarking of Multimedia Contents VI, pp. 23–34. SPIE, San Jose (2004)

[17] Fridrich, J., Goljan, M., Du, R.: Reliable detection of LSB steganography in color and grayscale images. In: ACM workshop on Multimedia and security: new challenges, Ottawa, Ontario, Canada, pp. 27–30. ACM Press, New York (2001)

[18] Fridrich, J., Goljan, M., Hogea, D.: Attacking the outguess. In: ACM Workshop on Multimedia and Security 2002, Juan-les-Pins, France, pp. 3–6. ACM Press, New York (2002)

[19] Fridrich, J., Goljan, M., Hogea, D.: Steganalysis of JPEG images: Breaking the F5 algorithm. In: 5th Information Hiding Workshop, pp. 310–323 (2002)

[20] Fridrich, J., Goljan, M., Soukal, D.: Wet paper codes with improved embedding efficiency. IEEE Transactions on Information Forensics and Security 1, 102–110 (2006)

[21] Fridrich, J., Lisonek, P., Soukal, D.: On Steganographic Embedding Efficiency. In: Camenisch, J.L., Collberg, C.S., Johnson, N.F., Sallee, P. (eds.) IH 2006. LNCS, vol. 4437, pp. 282–296. Springer, Heidelberg (2007)

[22] Fu, D., Shi, Y.Q., Zou, D., Xuan, G.: JPEG steganalysis using empirical transition matrix in block DCT domain. In: 2006 IEEE Workshop on Multimedia Signal Processing, Victoria, Canada, pp. 310–313 (2006)

[23] Harmsen, J., Pearlman, W.: Steganalysis of Additive Noise Modelable Information Hiding. In: Delp, E.J., Wong, P.W. (eds.) IS&T/SPIE Electronic Imaging: Security, Steganography, and Watermarking of Multimedia Contents V, pp. 131–142. SPIE, San Jose (2003)

[24] Huang, F., Shi, Y.Q., Huang, J.: A study on the security performance of YASS. In: IEEE Int. Conference on Image Processing, San Diego, California, USA (2008)

[25] Huang, F., Li, B., Huang, J.: Attack LSB matching steganography by counting alteration rate of the number of neighbourhood gray levels. In: IEEE International Conference on Image Processing 2007, San Antonio, Texas, USA, pp. 401–404 (2007)

[26] Ker, A.D.: Steganalysis of LSB matching in grayscale images. IEEE Signal Processing Letters 12, 441–444 (2005)

[27] Ker, A.D.: A general framework for the structural steganalysis of LSB replacement. In: Barni, M., Herrera-Joancomartí, J., Katzenbeisser, S., Pérez-González, F. (eds.) IH 2005. LNCS, vol. 3727, pp. 296–311. Springer, Heidelberg (2005)

[28] Ker, A.D.: Fourth-order structural steganalysis and analysis of cover assumptions. In: Delp, E.J., Wong, P.W. (eds.) IS&T/SPIE Electronic Imaging: Security, Steganography, and Watermarking of Multimedia Contents VIII, pp. 60720301–60720314. SPIE, San Jose (2006)
[29] Ker, A.D.: A weighted stego image detector for sequential LSB replacement. In: 3rd International Symposium on Information Assurance and Security, Manchester, United Kingdom, pp. 453–456 (2007)
[30] Ker, A.D.: Locating steganographic payload via ws residuals. In: 10th ACM workshop on Multimedia and security, Oxford, United Kingdom, pp. 27–31. ACM Press, New York (2008)
[31] Kim, Y., Duric, Z., Richards, D.: Modified matrix encoding for minimal distortion steganography. In: Camenisch, J.L., Collberg, C.S., Johnson, N.F., Sallee, P. (eds.) IH 2006. LNCS, vol. 4437, pp. 314–327. Springer, Heidelberg (2007)
[32] Kim, Y., Duric, Z., Richards, D.: Limited distortion in LSB steganography. In: Delp, E.J., Wong, P.W. (eds.) IS&T/SPIE Electronic Imaging: Security, Steganography, and Watermarking of Multimedia Contents VIII, 60720N01-60720N09, SPIE, San Jose (2006)
[33] Li, B., Shi, Y.Q., Huang, J.: Steganalysis of YASS. In: Proceedings of the 10th ACM workshop on Multimedia and security. ACM, Oxford (2008)
[34] Marvel, L.M., Boncelet Jr., C.G., Retter, C.T.: Spread spectrum image steganography. IEEE Trans. Image Processing 8, 1075–1083 (1999)
[35] Pevný, T., Fridrich, J.: Merging Markov and DCT features for multi-class JPEG steganalysis. In: Delp, E.J., Wong, P.W. (eds.) IS&T/SPIE Electronic Imaging: Security, Steganography, and Watermarking of Multimedia Contents IX, pp. 650503.03-650503.14. SPIE, San Jose (2007)
[36] Provos, N.: Defending against statistical steganalysis. In: 10th USENIX Security Symposium, Washington, DC, USA, pp. 323–325 (2001)
[37] Provos, N., Honeyman, P.: Detecting Steganographic content on the internet. University of Michigan, Ann Arbor (2001)
[38] Provos, N., Honeyman, P.: Hide and Seek: An Introduction to steganography. IEEE Security & Privacy Magazine 1, 32–44 (2003)
[39] Sallee, P.: Model-based steganography. In: Kalker, T., Cox, I.J., Ro, Y.M. (eds.) IWDW 2003. LNCS, vol. 2939, pp. 154–167. Springer, Heidelberg (2004)
[40] Sallee, P.: Model-based methods for steganography and steganalysis. International Journal of Image and Graphics 5, 167–189 (2005)
[41] Sarkar, A., Nataraj, L., Manjunath, B.S., Madhow, U.: Estimation of optimum coding redundancy and frequency domain analysis of attacks for YASS - A randomized block based hiding scheme. In: IEEE International Conference on Image Processing 2008, San Diego, California, USA, pp. 1292–1295 (2008)
[42] Schaefer, G., Stich, M.: UCID - An uncompressed colour image database. School of Computing and Mathematics. Nottingham Trent University, UK (2003)
[43] Sharp, T.: An implementation of key-based digital signal steganography. In: Moskowitz, I.S. (ed.) IH 2001. LNCS, vol. 2137, pp. 13–26. Springer, Heidelberg (2001)
[44] Shi, Y.Q., Chen, C., Chen, W.: A Markov process based approach to effective attacking JPEG steganography. In: Camenisch, J.L., Collberg, C.S., Johnson, N.F., Sallee, P. (eds.) IH 2006. LNCS, vol. 4437, pp. 249–264. Springer, Heidelberg (2007)
[45] Shi, Y.Q., Chen, C., Xuan, G., Su, W.: Steganalysis versus splicing detection. In: Shi, Y.Q., Kim, H.-J., Katzenbeisser, S. (eds.) IWDW 2007. LNCS, vol. 5041, pp. 158–172. Springer, Heidelberg (2008)
[46] Shi, Y.Q., Xuan, G., Zou, D., Gao, J., Yang, C., Zhang, Z., Chai, P., Chen, W., Chen, C.: Steganalysis based on moments of characteristic functions using wavelet decomposition, prediction-error image, and neural network. In: IEEE International Conference on Multimedia and Expo. 2005, Amsterdam, Netherlands, pp. 768–773 (2005)

[47] Solanki, K., Sarkar, A., Manjunath, B.S.: YASS: Yet another steganographic scheme that resists blind steganalysis. In: Furon, T., Cayre, F., Doërr, G., Bas, P. (eds.) IH 2007. LNCS, vol. 4567, pp. 16–31. Springer, Heidelberg (2008)
[48] Sullivan, K., Madhow, U., Chandrasekaran, S., Manjunath, B.S.: Steganalysis for Markov cover data with applications to images. IEEE Transactions on Information Forensics and Security 1, 275–287 (2006)
[49] Upham D Jsteg, `http://zooid.org/~paul/crypto/jsteg/`
[50] Wang, H.Q., Wang, S.Z.: Cyber warfare: Steganography vs. steganalysis. Communications of the ACM 47, 76–82 (2004)
[51] Weinberger, M., Seroussi, G., Sapiro, G.: LOCOI: A low complexity context-based lossless image compression algorithm. In: IEEE International Comference on Data Compression, Snowbird, Utah, pp. 140–149 (1996)
[52] Westfeld, A.: F5-A Steganographic algorithm: high capacity despite better steganalysis. In: Moskowitz, I.S. (ed.) IH 2001. LNCS, vol. 2137, pp. 289–302. Springer, Heidelberg (2001)
[53] Westfeld, A., Pfitzmann, A.: Attacks on steganographic systems - Breaking the steganographic utilities EzStego, Jsteg, Steganos, and S-Tools-and some lessons learned. In: Pfitzmann, A. (ed.) 3rd Information Hiding Workshop, Dresden, Germany. LNCS, pp. 61–76. Springer, Heidelberg (1999)
[54] Woods, J., Stark, H.: Probability and Random Processes With Applications to Signal Processing, 3rd edn. Prentice-Hall, Upper Saddle River (2001)
[55] Xuan, G., Gao, J., Shi, Y.Q., Zou, D.: Image Steganalysis Based on Statistical Moments of Wavelet Subband Histograms in DFT Domain. In: IEEE International Workshop on Multimedia Signal Processing, pp. 1–4 (2005)
[56] Xuan, G., Shi, Y.Q., Gao, J., Zou, D., Yang, C., Zhang, Z., Chai, P., Chen, C.H., Chen, W.: Steganalysis based on multiple features formed by statistical moments of wavelet characteristic functions. In: Barni, M., Herrera-Joancomartí, J., Katzenbeisser, S., Pérez-González, F. (eds.) IH 2005. LNCS, vol. 3727, pp. 262–277. Springer, Heidelberg (2005)
[57] Zou, D., Shi, Y.Q., Su, W., Xuan, G.: Steganalysis based on Markov model of thresholded prediction-error image. In: IEEE International Comference on Multimedia and Expo., Toronto, Ontario, Canada (2006)

WBE-Based Anti-collusion Fingerprints: Design and Detection

Zang Li and Wade Trappe

Wade Trappe is at the Wireless Information Network Laboratory, Rutgers University, 671 Rt. 1 South, North Brunswick, NJ 08902
trappe@winlab.rutgers.edu

Summary. Digital fingerprinting is a technique whereby a unique digital fingerprint associated with each legitimate user is hidden in the multimedia content so that consumers who redistribute the content illegally can be traced. Typically, robust watermarking techniques are used to embed digital fingerprints in order to provide resilience to a variety of processing attacks. However, one cost-effective strategy to attack digital fingerprints is collusion, where several colluders average their individual copies to disrupt the underlying fingerprints. Recently, a new class of fingerprinting codes, called anti-collusion codes (ACC), was proposed for use with code-modulated data embedding. Two important issues need to be considered when designing ACC: accommodating as many users as possible for a given fingerprint dimensionality, and identifying the colluders effectively from the colluded signal. In this chapter, we identify an underlying similarity between the colluder detection problem and the multiuser detection problem from *code division multiple access* (CDMA). We propose that fingerprints can be constructed using sequence sets satisfying the Welch Bound Equality (WBE). In order to identify the colluders when employing WBE-based ACC, we further propose a powerful detection algorithm utilizing sphere decoding to identify the colluders from the colluded signal. We compare our WBE-ACC against orthogonal fingerprints and BIBD-based anti-collusion codes via simulations, and show that the proposed WBE-based ACC and detection algorithm gives a better performance than the BIBD-based ACC under the same configuration.

1 Introduction

The increasing popularity of multimedia applications has made protecting the intellectual property associated with multimedia content an increasingly critical issue. Although traditional encryption methods can effectively restrict access to the digital multimedia during transmission and storage [9, 24, 33, 25, 26], the protection provided by encryption disappears prior to media rendering. Once a consumer accesses the decrypted multimedia content, he can easily duplicate, compress and redistribute the content. To protect media after it has been decrypted, the technique of digital fingerprinting is used. Digital fingerprinting involves hiding a unique digital signature associated with each legitimate user in the multimedia content so that consumers who redistribute the content illegally can be traced.

H.T. Sencar et al. (Eds.): Intel. Multimedia Analysis for Security Appli., SCI 282, pp. 305–335.
springerlink.com

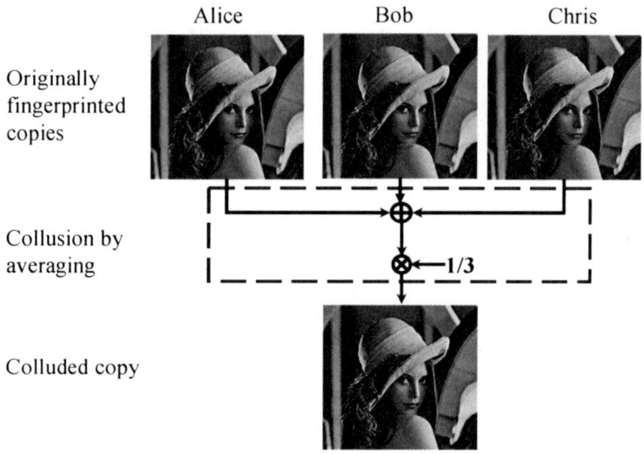

Fig. 1. Average collusion attack on fingerprinting.

In order for digital fingerprinting to be effective, it should satisfy the following basic requirements:

1. The fingerprint should be perceptually invisible to users, but be detectable by the producer using pre-designed algorithms.
2. The fingerprint should be unique to each user, so that after analysis, the original consumer who owned the media can be identified.
3. The fingerprint should be hard to remove or change without rendering the whole multimedia useless, i.e it must survive compression, filtering and attacks.

In particular, fingerprinting is not as simple as adding cipher data to a file header because that can be easily removed before illegal redistribution. Instead, fingerprints intended to have a security role should be embedded in the media content using robust watermarking techniques, such as spread spectrum methods[5, 17, 4, 34, 23], that are capable of withstanding signal processing attacks.

In spite of the well-documented resistance to a broad array of attacks, spread spectrum embedding methods are susceptible to a very simple, and cost-effective attack, known as collusion. A collusion attack is mounted by a coalition of users who have the same multimedia content with different watermarks. These colluders attempt to attenuate the underlying watermarks by combining their copies. An efficient and easy way that won't cause distortion of the host signal is linear collusion attacks, such as averaging several fingerprinted signals [21]. This collusion procedure is illustrated in Figure 1. With enough colluders, detection can be seriously disrupted. Therefore, it's important to design fingerprints that are resistant to collusion. Meanwhile, effective algorithms are needed to identify the colluders and in turn discourage the collusion attempts. The resistance of digital watermarks to such linear collusion attacks has been studied recently and will be reviewed in Section 2.

In this chapter, we propose to design anti-collusion fingerprints using WBE sequences, and devise an effective detection strategy for identifying colluders. Section 2 gives a review of previous work on this problem. Section 3 presents our fingerprint design and the basic detection statistics. We start from describing the fingerprinting and collusion problem in Section 3.1. Since our scheme is inspired from multiuser communication, we briefly show the similarity between collusion and multiuser communication problem in Section 3.2. Then, we review the optimality of the WBE sequence set for multiuser communication, and consequently why it is a sensible choice for our fingerprints. In Section 4, we propose our detection algorithm based on sphere decoding. The simulation results are presented in Section 5, where comparisons are made with BIBD ACC and orthogonal fingerprints. We conclude this chapter in Section 7.

2 Related Work

One of the first works on designing collusion resistant fingerprints was proposed by Boneh and Shaw [2] for generic digital data. In their model, colluders compare their copies to find the places where marks differ. These locations are then altered to mask the identities of the colluders, while positions where the marks are the same imply the mark is undetected. Their work assumed that the watermarks satisfy the *marking assumption*, that is, the users cannot change the state of an undetected mark without rendering the object useless. They constructed fingerprinting schemes that are able to trace the colluder with a high probability, as well as a corresponding detection algorithm.

It has been observed that Boneh and Shaw's construction of the fingerprinting schemes for generic data relies heavily on the validity of the marking assumption, which does not hold in the multimedia realm[27]. Unlike generic data, multimedia data is insensitive to minor changes of the data values, while the marks can be destroyed by these minor changes! For example, adding a low power noise signal to an image will not noticeably alter the visual quality, but it will alter a significant percentage of the pixel values available for representing the Boneh and Shaw fingerprints.

Another direction of work has focused on the theoretical limits of collusion-resistant watermarking. In [7], the authors modeled the fingerprinting of a document as a process that produced a marked document with a bounded distortion from the original signal. They proposed a collusion attack in which a small amount of noise is added to the average of fingerprinted documents, and derived a bound on the number of adversaries needed to defeat the underlying watermarks. In short, they showed that if the fingerprint length is n, then at most $O(\sqrt{n/\ln n})$ colluders can defeat any watermarking scheme with non-negligible probability. A similar resistance result is obtained in [14].

Further work on the resistance of digital watermarks to collusion attacks was presented by Su et al., who described a more general signal processing attack in which the colluders employ multiple-input single-output linear shift invariant (MISO-LSI) filtering plus additive Gaussian noise to thwart fingerprinting [22].

Subject to a distortion bound, the attackers aim to find the optimal filter to minimize the amount of information that can be recovered from the colluded signal. Because of the analogy between the collusion problem and the *Gaussian multiple access channel* (GMAC) problem, this information can be characterized by the rate capacity measure for GMAC. Under the assumption that the fingerprints are independent and have identical statistical characteristics, Su et al. showed that the optimal MISO-LSI attack involves each user weighting their marked documents equally prior to adding noise.

The collusion resistance of orthogonal fingerprints was thoroughly analyzed by Wang et al. in [31]. Wang et al. studied the maximum number of colluders K_{max} the system can handle given that the probability of catching at least one colluder (P_d) must be greater than a level β, and the probability of falsely accusing any innocent user (P_{fp}) is smaller than a level ϵ. They further derived the lower bound and upper bound of K_{max} as a function of n (number of users), N (fingerprint length), watermark-to-noise ratio (WNR) and the performance requirements $P_d \geq \beta$ and $P_{fp} \leq \epsilon$. Besides the averaging collusion attack, Wang et al. also studied several other collusion strategies based on median, minimum, maximum, and average of min and max. The authors showed that these nonlinear attacks can be approximated as an averaging collusion attack followed by additive noise. These attacks give similar performance as averaging collusion attacks as long as the overall mean squared error introduced to the host signal is the same. Therefore, from the detector's point of view, the exact underlying collusion model does not matter. A more detailed work on nonlinear collusion attack is done in [35].

In other work [30], Wang et al. proposed a group-oriented fingerprinting scheme where users likely to collude with each other are assigned correlated fingerprints. The users are divided into groups such that users are equally likely to collude with other members within the same group, but less likely to collude with members from other groups. The fingerprints assigned to different groups are statistically independent. Detection involves first identifying the groups containing the colluders, and then applying a correlator to identify colluders within each suspected group. Collusion-resistance increases with accurate knowledge of the collusion pattern, but decreases when the assumed collusion pattern deviates significantly from the true collusion pattern.

Celik et al. proposed a novel collusion-resistant fingerprinting scheme using random pre-warping [16]. In this scheme, the host signal is warped randomly prior to watermarking. Since the human perception is quite tolerant to geometric distortion, the warping does not significantly affect the perceived quality of the watermarked signal. However, due to independent warping, the average of several copies will produce significant blur in the resulting image. Visual quality does not improve when the number of copies is increased. This method can be used in conjunction with other watermarking schemes to prevent large-scale collusion.

One disadvantage of orthogonal fingerprinting is its capacity limitation. The total number of users cannot exceed the number of dimensions available. This could be a problem when the number of users is large or the number of basis functions is constrained by the dimension of the data or the application scenario. In [27], BIBD anti-collusion codes (ACC) were proposed to accommodate more users while providing collusion-resistance. The construction method used balanced incomplete block designs (BIBDs). The authors proposed three different detection schemes, with varying performance tradeoffs, for identifying the colluders.

The anti-collusion property of BIBD-ACC was based on the assumption that collusion can be modeled as an element-wise AND on the binary fingerprint code. In reality, the hypothesis testing involved in detection cannot be easily characterized with binary logical operations, and is instead better modeled via real-valued operations. Further, the BIBD-ACC construction requires knowledge of an upper bound for the amount of colluders during the design or selection of the fingerprint codes. If the actual size of the coalition exceeds the design limit, then the advantageous properties of the codes that was introduced in its design will not help in detection.

In this chapter, we propose a new class of ACC that allows for real-valued code vectors. This new family of codes, which we call Welch-Bound Equality Anti-collusion Codes (WBE-ACC), are motivated by similarities between the collusion detection problem and multi-user communication. We take the view that the collusion problem is similar with a multiuser communication problem where each user transmits 1 (colluding) or 0 (not colluding). Then it is desirable for the signatures of the users to be as uncorrelated as possible to each other from the detection point of view. The resulting WBE-ACC have both large capacity and good collusion-resistance. To facilitate improved detection of colluders, we propose a powerful decoder utilizing sphere decoding to identify the colluders. The collusion resistance is soft – it does not impose any constraint on the maximum number of colluders, except the fundamental limit that large-scale collusion naturally results in more severe watermark power attenuation due to averaging.

3 WBE Anti-collusion Fingerprint Design

3.1 Fingerprint and Collusion

In this chapter, we consider fingerprints embedded in the host signal using a spread spectrum technique. Suppose the host signal is represented by a vector $\mathbf{x}$, and the watermark signal is represented by a vector $\mathbf{w}$, the owner generates the watermarked signal by hiding the watermark in the host signal, via

$$\mathbf{y} = \mathbf{x} + \alpha \mathbf{w}, \tag{1}$$

where α is a scaling factor to control the energy of the embedded watermark $\mathbf{w}$. It's often set as the *just-noticeable-difference* from a human visual model.

Since there could be many users who purchase the right to access the host signals, the owner usually generates a family of watermarks, denoted as $\{\mathbf{w}_i\}$. Each sold copy contains a distinct watermark $\mathbf{w}_i$, and is associated with one user. This one-to-one correspondence allows tracing of pirated copies to the original

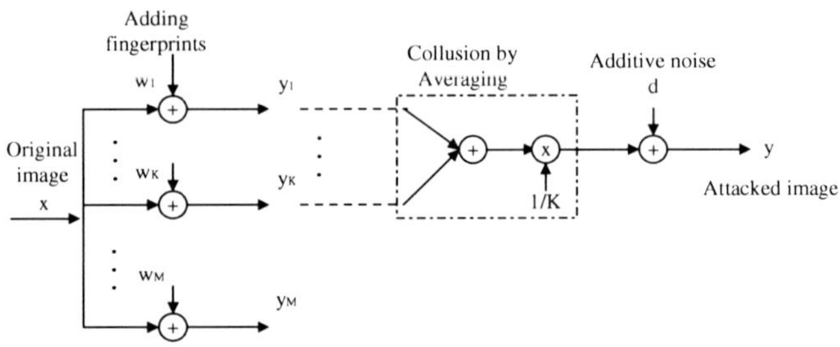

Fig. 2. An averaging collusion model

users. Usually the watermarks $\{\mathbf{w}_i\}$ are chosen to be noise-like signals, and can be represented as the linear combination of N orthonormal basis signals $\{\mathbf{u}_i\}$. In other words, the $\{\mathbf{u}_i\}$ are the basis signals spanning the watermark space so that each watermark can be written as

$$\mathbf{w}_j = \sum_{i=1}^{N} s_{ij}\mathbf{u}_i = \mathbf{U}\mathbf{s}_j, \tag{2}$$

where each column of $\mathbf{U}$ is an orthonormal basis vector, and $\mathbf{s}_j = [s_{1j}, s_{2j}, \cdots s_{Nj}]^\top$, is the vector representation of $\mathbf{w}_j$ in the space spanned by $\mathbf{U}$. Since the watermarks for every user should have the same power, we let $||\mathbf{s}_j|| = ||\mathbf{s}_k||$ for all users $j, k = 1 \cdots M$. It's convenient to put all $\{\mathbf{s}_j\}$ into one matrix $\mathbf{S}$, which will be referred to as the signature matrix from now on.

In the collusion scenario, several colluders linearly combine their copies, attempting to remove the underlying fingerprints. Since no one wants to take higher risk than the others, it's reasonable to assume that the weights for all users are equal. Assuming there are a total of K colluders, denoted by the set S_c, the colluded signal is

$$\mathbf{y} = \mathbf{x} + \frac{\alpha}{K}\sum_{j\in S_c}\mathbf{w}_j + \mathbf{d} \tag{3}$$

where $\mathbf{d}$ is an additive random noise vector representing noise from signal compression, transformation, etc. This collusion model is illustrated in Figure 2, where we assume user 1 to K collude without loss of generality. In the non-blind detection scenario, $\mathbf{x}$ is known and can be subtracted from the received signal. In the blind scenario, we treat $\mathbf{x}$ as part of the noise, and incorporate it into the overall noise vector $\mathbf{d}$. We note that the cut-and-paste attack, in which portions of the content signal are cut out from different users' copies and grafted together to form the colluded content signal has an effect that is comparable to linear averaging attacks when spread-spectrum embedding is employed[27]. Additionally, many families of non-linear collusion attacks have been shown to be well-approximated using averaging attacks[31].

At the detection stage, since we are only interested in detecting the colluders, we can project the received signal $\mathbf{y}$ onto the space spanned by $\mathbf{U}$. The resulting

$N \times 1$ coefficient vector, denoted as $\mathbf{T}$, is the sufficient statistics for detection. Similarly, denote the projection of $\mathbf{d}$ ($\mathbf{x}$ has been removed or incorporated into $\mathbf{d}$) onto $\mathbf{U}$ as $\mathbf{n}$, the noise in the watermark space. It's straightforward to show that

$$\mathbf{T} = \frac{\alpha}{K}\mathbf{S}\Phi + \mathbf{n} \tag{4}$$

where the column vector $\mathbf{\Phi} \in \{0,1\}^M$ indicates colluders via the location of its components whose values are 1. M is the total number of fingerprints or users.

One choice of fingerprint design is to make the $\{\mathbf{w}_i\}$, or equivalently the $\{\mathbf{s}_i\}$, orthogonal. For example, we can choose $\mathbf{S}$ to be the identity matrix. The advantage of an orthogonal signature matrix is that the fingerprints of different users do not interfere with each other. Thus, orthogonal fingerprinting has a simple encoder/decoder structure and good collusion resistance. However, when using N orthogonal basis signals, orthogonal fingerprinting can accommodate at most N users. If the number of users is large or the cardinality of the basis is constrained, this low capacity may not be satisfactory. To alleviate the conflict between the capacity and the anti-collusion ability, [27] proposed to use anti-collusion codes (ACC) which can handle more users and withstand collusions.

The ACC proposed in [27] were built using balanced incomplete block designs (BIBD), and had a hard collusion resistance limit. In particular, a BIBD-ACC was designed to resist up to K colluders and might not be able to handle more than K colluders. As an example, if we refer to the ACC code constructed from a $(16,4,1)$-BIBD, which is a 3-resistant ACC, there are several combinations of four colluders that yield the same colluded signal, and hence cannot be unambiguously distinguished. In this code, the averaging collusion of the $\{1,18,19,20\}$-th fingerprints yields the same result as the averaging collusion of $\{2,9,13,17\}$-th fingerprints, and it is not possible for the colluder-identification process to distinguish between these two outcomes. Further examination of this code reveals many more examples where there is ambiguity in distinguishing between sets of 4 colluders.

Hence, when using BIBD-ACC it is necessary to have a strict limit on the number of colluders. It would be more desirable, however, to have a scheme where the collusion resistance does not yield such ambiguity and does not require strict limits on the amount of colluders.

3.2 Collusion and Multiuser Communication

Since the fingerprints are embedded using spread spectrum, and the received colluded signal is simply a mixture of the watermarks, this should remind us of some similarity between the collusion problem and *Code-Division Multiple Access* (CDMA) in multiuser communication. In *Direct Sequence Spread Spectrum* (DSSS), the users embed their signal bits into signature sequences that have a broad spectrum. Spread spectrum embedding provides robustness against unknown channel distortion and enhances privacy. These properties are analogous to our requirements for fingerprints. The basic synchronous CDMA K-user channel model can be represented as

$$y(t) = \sum_{k=1}^{K} A_k b_k s_k(t) + n(t) \tag{5}$$

where A_k, $b_k \in \{-1, 1\}$ and $s_k(t)$ are the received amplitude, the signal bits transmitted and the signature sequence of the k-th user, respectively. $y(t)$ is the received signal. For simplicity, we let $A_k = A$. Assuming $\{s_k(t)\}$ is spanned by a set of N orthogonal functions, we can write the above model as

$$\mathbf{y} = A\mathbf{S}\mathbf{b} + \mathbf{n}, \tag{6}$$

where the i-th column of $\mathbf{S}$ is the signature sequence corresponding to the i-th user. $\mathbf{S}$ is of size $N \times K$. $\mathbf{b} \in \{-1, 1\}^K$ is a $K \times 1$ vector whose i-th entry is the signal bit of the i-th user. Note that the CDMA model (6) is in the same format as the collusion model (4). The only difference is that in CDMA $\mathbf{b} \in \{-1, 1\}^K$, while in the collusion model, $\mathbf{\Phi} \in \{0, 1\}^M$. In the extreme situation that all users send 1 in the CDMA case, and all users collude, the two equations are exactly the same. For best detection, it's desirable for both cases that the interference among users is minimum.

Extensive research has been done on optimal sequences for a synchronous CDMA system. Here we are mainly interested in a sequence set that has minimum total correlations and the largest user capacity. It was shown in [29] that the user capacity is maximized if the received powers of the users are identical and if the signature sequence set satisfies:

$$\mathbf{S}^\top \mathbf{S} = \mathbf{I_M} \qquad \text{when } M \leq N \tag{7}$$

and

$$\mathbf{S}\mathbf{S}^\top = (M/N)\mathbf{I_N} \qquad \text{when } M > N, \tag{8}$$

where M is the number of users and N is the number of dimensions. When this condition is satisfied, the MMSE (minimum mean squared error) receiver and the matched filter receiver are the same. Note that, when $M \leq N$, condition (7) implies that the codes are orthogonal.

Sequence sets that satisfy equations (7) and (8) are called the Welch Bound Equality (WBE) sequences. Welch [32] derived the following lower bound on the sum of the squared cross correlations, which is called total squared correlation(TSC) in [18],

$$TSC = Trace[(\mathbf{S}\mathbf{S}^\top)^2] = \sum_{i=1}^{M} \sum_{j=1}^{M} (\mathbf{s}_i^\top \mathbf{s}_j)^2 \geq M^2/N. \tag{9}$$

A WBE sequence set satisfies this bound with equality. In other words, a WBE sequence set provides the least total squared correlation among the signatures.

Since a WBE sequence set is simply an orthogonal set when $N \geq M$, we are more interested in the case of $N < M$ in terms of construction. There are several ways to obtain a WBE sequence set when $N < M$. For example, it can be constructed directly using *tight frames* from the wavelet literature [29].

When $M = 0(mod4)$ or $M = 1$ or 2, binary antipodal WBE sequence set can be constructed using the linear coding technique as proposed in [12] or the simple Hadamard matrix method mentioned in [13]. When M does not satisfy the above condition, the Welch bound cannot be achieved for binary antipodal sequence and alternative TSC bounds were provided in [13]. A detailed survey of WBE sequence sets and some extensions were given in [20]. A distributed and iterative approach to construct a WBE sequence set was proposed in [19]. We used the iterative method in our simulation since it can easily produce random WBE codes with real entries for any M.

Because of the similarity between the collusion problem and synchronous CDMA problem, we postulate that the optimal sequences for CDMA should be a good choice for fingerprints as well. If we assume that all users are equally likely to collude, then it's desirable to have the total squared correlation among users be minimum while the user capacity be maximum. Since WBE sequences are optimum in terms of least total squared correlation and user capacity, we suggest using WBE sequences as our fingerprints in a multimedia fingerprinting system.

4 Colluder Detection

The detection goal is to estimate the collusion vector $\mathbf{\Phi}$ from the noisy observation $\mathbf{T}$, while the signature matrix $\mathbf{S}$ and the scaling factor α are known. Assuming the noise is white Gaussian, the maximum likelihood solution of equation (4) is

$$\hat{\mathbf{\Phi}} = \arg\min_{\mathbf{\Phi}} ||\mathbf{T} - \frac{\alpha}{K}\mathbf{S\Phi}||, \tag{10}$$

i.e. the distance between $\mathbf{T}$ and $\frac{\alpha}{K}\mathbf{S\Phi}$ is minimum. Let's first assume that the number of colluders K is known. Then equation (10) is a typical integer least squares problem. When the fingerprints are orthogonal, the simple correlator is optimal. When the fingerprints are not orthogonal, the problem generally involves searching every possible input, which is NP-hard [1]. However, since the entries of $\mathbf{\Phi}$ are integer-valued, all possible inputs lie on a lattice. If the matrix $\mathbf{S}$ is full rank with $N \geq M$, multiplication by $\mathbf{S}$ simply skews the lattice. Then the sphere decoding algorithm can be used to find the maximum likelihood solution efficiently. The sphere decoding algorithm was initially proposed by Fincke and Pohst [8], and is widely used for maximum likelihood decoding in communication. We will briefly introduce the sphere decoding algorithm below and then discuss how it can be applied to solve the collusion problem.

4.1 Sphere Decoding Algorithm

The integer least squares problem can be formalized as

$$\min_{\mathbf{s} \in Z^m} ||\mathbf{x} - \mathbf{Hs}|| \tag{11}$$

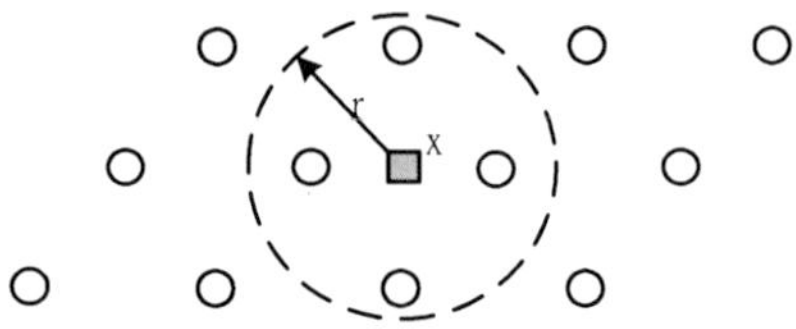

Fig. 3. Illustration of integer least squares problem and sphere decoding.

where $\mathbf{x} \in R^n$, $\mathbf{H} \in R^{n \times m}$, and Z^m denotes the m-dimensional integer lattice, i.e. $\mathbf{s}$ is an m-dimensional vector with integer entries. Usually the entries in $\mathbf{s}$ can only have some finite values, so the search will be on a finite subset of the integer lattice.

Assume $\mathbf{H}$ is full rank with $n \geq m$, then the lattice structure will be preserved by the transformation, i.e. the n-dimensional vector $\mathbf{Hs}$ will lie on a skewed lattice in R^n. Given the received n-dimensional vector $\mathbf{x}$ and the skewed lattice $\mathbf{Hs}$, the goal is to find the lattice point closest to $\mathbf{x}$ in Euclidean distance.

To find the closest lattice point, a brute-force method would calculate the distance to $\mathbf{x}$ for every point on the lattice and pick the one with the minimum distance. This method is computationally intensive, since the size of the lattice could be quite large. Instead, we can examine only the lattice points within a hypersphere of radius r centered on the received vector $\mathbf{x}$. Obviously the nearest point to $\mathbf{x}$ within this hypersphere is also the nearest point in the whole lattice. This is illustrated geometrically in Figure 3.

However, two problems exist: First, what's the appropriate choice of r? Clearly, if r is too large, we need to search many points which increases the computational cost; if r is too small, we may not have any valid lattice point within the hypersphere. There are many possible choices of r. For example, an adaptive r was chosen to reduce the complexity of the sphere decoder in [3].

Second, how to find the lattice points within the hypersphere? The sphere decoding algorithm provides an efficient way to solve this problem, as described below. For clarity, we will adopt the derivation given in [10]. We would like to find all possible $\mathbf{s}$ that satisfy

$$||\mathbf{x} - \mathbf{Hs}||^2 \leq r^2. \tag{12}$$

First, QR factorize the matrix $\mathbf{H}$ as

$$\mathbf{H} = \mathbf{Q} \begin{bmatrix} \mathbf{R} \\ \mathbf{0}_{(n-m)\times m} \end{bmatrix} = [\mathbf{Q}_1\ \mathbf{Q}_2] \begin{bmatrix} \mathbf{R} \\ \mathbf{0} \end{bmatrix} \tag{13}$$

where $\mathbf{R}$ is an $m \times m$ upper triangular matrix, $\mathbf{Q}$ is an $n \times n$ orthogonal matrix. $\mathbf{Q}_1, \mathbf{Q}_2$ represent the first m columns and the last $n - m$ columns of $\mathbf{Q}$, respectively. Then

$$||\mathbf{x} - \mathbf{Hs}||^2 = ||\mathbf{x} - [\mathbf{Q}_1\ \mathbf{Q}_2] \begin{bmatrix} \mathbf{R} \\ \mathbf{0} \end{bmatrix} \mathbf{s}||^2$$

$$= || \begin{bmatrix} \mathbf{Q}_1^\top \\ \mathbf{Q}_2^\top \end{bmatrix} \mathbf{x} - \begin{bmatrix} \mathbf{R} \\ \mathbf{0} \end{bmatrix} \mathbf{s}||^2$$
$$= ||\mathbf{Q}_1^\top \mathbf{x} - \mathbf{R}\mathbf{s}||^2 + ||\mathbf{Q}_2^\top \mathbf{x}||^2. \tag{14}$$

Now equation (12) is equivalent to

$$||\mathbf{Q}_1^\top \mathbf{x} - \mathbf{R}\mathbf{s}||^2 \leq r^2 - ||\mathbf{Q}_2^\top \mathbf{x}||^2. \tag{15}$$

Define $\mathbf{x}' = \mathbf{Q}_1^\top \mathbf{x}$ and $r'^2 = r^2 - ||\mathbf{Q}_2^\top \mathbf{x}||^2$. Both $\mathbf{x}'$ and $\mathbf{Rs}$ are $m \times 1$ vectors now. Since $\mathbf{R}$ is upper triangular, we can write the above condition as

$$r'^2 \geq ||\mathbf{x}' - \mathbf{Rs}||^2 = \sum_{i=1}^{m} \left(x'_i - \sum_{j=i}^{m} R_{i,j} s_j \right)^2, \tag{16}$$

which can be expanded as

$$r'^2 \geq (x'_m - R_{m,m} s_m)^2 + (x'_{m-1} - R_{m-1,m-1} s_{m-1} - R_{m-1,m} s_m)^2 + \cdots. \tag{17}$$

Therefore, a necessary condition for $\mathbf{Hs}$ to lie in the hypersphere is $r'^2 \geq (x'_m - R_{m,m} s_m)^2$, i.e. s_m lies within

$$\frac{r' - x'_m}{R_{m,m}} \leq s_m \leq \frac{r' + x'_m}{R_{m,m}}. \tag{18}$$

Since s_m is integer-valued, we can get all of its possible values from the interval defined above. For each possible s_m, after substituting it back into equation (17), we can get the interval for s_{m-1} with similar reasoning. Continuing the process until we reach s_1, we will get all possible $\mathbf{s}$ that lie within the sphere. The formal algorithm and analysis can be found in [10].

Since the sphere decoding algorithm decides the range of each input dimension sequentially, all points it searches form a tree structure with height m, the dimensionality of $\mathbf{s}$. Every path from the root to leaf represents one point within the hypersphere. Therefore, we can conveniently use a tree searching algorithm to get all points, calculate their distance to the received vector $\mathbf{x}$, and then pick the one with the minimum distance. The computational cost of the algorithm is directly affected by the number of leaves (equal to the number of points to search) of the tree. This will become more clear after we illustrate it with the collusion example in the following section.

The above procedure is valid when the matrix $\mathbf{H}$ is full rank with $n \geq m$, and the complexity of the algorithm is polynomial with m [10]. To solve the least integer squares problem when the matrix $\mathbf{H}$ is full rank with $n < m$, [15] proposed a generalized sphere decoding algorithm. The basic idea is to to do an exhaustive searching of the hypotheses in the subspace of $m - n$ dimensions, followed by the regular sphere decoding algorithm at the remaining n dimensions for each hypothesis. The complexity of the generalized sphere decoding algorithm is exponential in the difference $m-n$. Although a fast generalized sphere decoding algorithm was proposed in [6], the complexity order is still exponential.

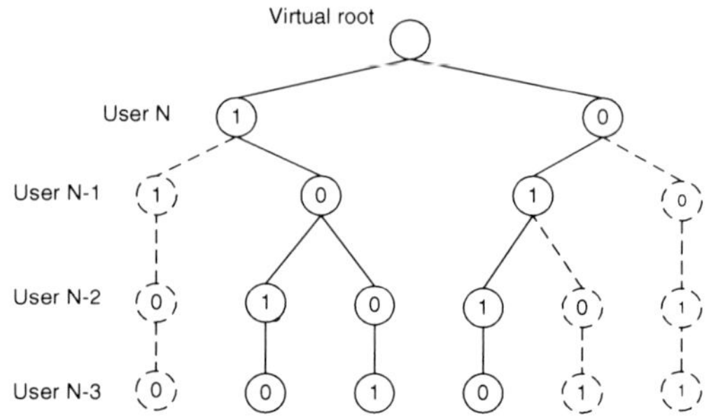

Fig. 4. Illustration of sphere decoding with a tree structure.

4.2 Colluder Detection with the Sphere Decoding Algorithm

Now let's come back to the collusion problem we are studying. When the number of colluders K is known, the optimal likelihood estimation of $\mathbf{\Phi}$, given by equation (10), is essentially the same as the integer least squares problem (11). So, the sphere decoding algorithm should help us get the maximum likelihood solution with reduced complexity. But in our collusion scenario, the signature matrix $\mathbf{S}$ is full rank with $N < M$. It can only be solved using generalized sphere decoding, and the computational complexity is large, as explained in last section. To alleviate this complexity burden, we examine a sub-optimal decoding algorithm. In particular, we will utilize the fact that the components of $\mathbf{\Phi}$ are either 1 or 0 to reduce complexity. Since innocent users do not contribute to the observed signal, we can simply remove the most likely innocent $M - N$ users from $\mathbf{S}$ so that the resulting matrix satisfies the condition for using the original sphere decoding algorithm. If the number of innocent users is much larger than $M - N$ and the noise power is not too large, as in most practical cases, we expect this removal to be relatively easy and reliable. A straightforward method is to use the correlator output as the decision variables – the users with the smallest $M - N$ correlator outputs are considered as the most likely innocent $M - N$ users. Their signatures are removed from the matrix $\mathbf{S}$. We note that this decision is final, and that these users will not be considered any further. So, if we miss a colluder in this stage, we can do nothing to alleviate this, and we will have ultimately harmed detection. Nevertheless, this probability is very low and, for our purposes, we consider it a negligible event. After we remove the most likely innocent $M - N$ users, equation (10) becomes

$$\hat{\mathbf{\Phi}}' = \arg\min_{\Phi'} ||\mathbf{T} - \frac{\alpha}{K}\mathbf{S}'\mathbf{\Phi}'||, \tag{19}$$

where $\mathbf{S}'$ is the tailored $N \times N$ full rank matrix. Now the minimum distance (maximum likelihood) $\mathbf{\Phi}'$ can be found using the sphere decoding algorithm.

To understand why sphere decoding can solve the problem efficiently, we can examine the tree structure shown in Figure 4. In this example, $N = 4$ and 2 users collude. For each user, a value of 1 means it colludes, while 0 means it does not participate in the collusion. A virtual root is plotted so that we have a single tree to accommodate all possible solutions. It's easy to see that every path from the root to the leaf is a possible realization of a collusion pattern (equivalent to $\mathbf{\Phi}'$). In general, for N users with K colluders, there are $\binom{N}{K}$ possible collusion patterns. Without sphere decoding, we need to check every possible pattern to find the most likely one. With sphere decoding, we only need to check a small portion, identified by the algorithm, of all possible patterns. In other words, sphere decoding identifies a subtree (indicated by the solid lines in Figure 4) so that the patterns in this subtree are within distance r to the received vector.

If K is known as we assumed, the number of 1's in $\mathbf{\Phi}$ should be strictly equal to K. Thus, as we traverse the tree on any path toward a leaf, once we have got K 1's (i.e. K suspects), we know that the values for the rest of the users must be 0. This observation helps us to speed the algorithm. To get K suspects faster, we order the users in $\mathbf{S}'$ in ascending order according to their likelihood of collusion (decided by the correlator outputs), so that these users lie at the higher layers close to the root of the tree. During the tree searching, the 1-branches are always searched earlier than the 0-branches if both of them are in the subtree specified by the sphere decoding.

In the above analysis, we assumed the number of colluders K is known. However, in practice, K is generally unknown at the detection side. To estimate K, we note that the minimum distance between the detection statistics T and $\frac{\alpha}{K'}\mathbf{S}'\mathbf{\Phi}'$ at each K' indicates how likely this K' equals the true number of colluders K. Hence, we can examine the minimum distance at every possible K', and choose the one with the minimum distance as our estimate of K. This is illustrated in Figure 5, where we run the algorithm for $K' = 1, 2, 3, \cdots, N$, and plot the change of the minimum distance with K'. The experiment setup is the same as what we used in simulation and will be explained in detail in Section 5. Let's first look at the case when 16 dimensions are used to handle 20 users with 5 colluders at watermark-to-noise ratio $WNR = -10$ dB (the solid curve in Figure 5a). The curve reaches the minimum when K' equals the the true number of colluders $K = 5$. Therefore, the proposed method will estimate K successfully. However, this solution has one obvious shortcoming: calculating the minimum distance solution for every possible K is computationally expensive and unnecessary. To reduce the computation, we utilized another phenomenon shown by the curve, that is, the minimum distance decreases monotonically with K' until K' reaches the true number of colluders $K = 5$, and then starts to increase. So, we can simply increase K' by 1 each time, and observe the change of minimum distance. Once it stops decreasing with increasing K', we stop our searching effort and consider the current minimum distance estimate as the final estimate.

It is true that at larger noise the valley may flatten out, which makes it more difficult to estimate K correctly. This is illustrated in Figure 5a as well, where the dashed curve shows the result at a lower $WNR = -15$ dB. The valley becomes flat as noise increases. Although not shown here, large noise can result in a mismatch between valley position and the true K, which in turn leads to estimation error. On the other hand, the number of colluders also affects the

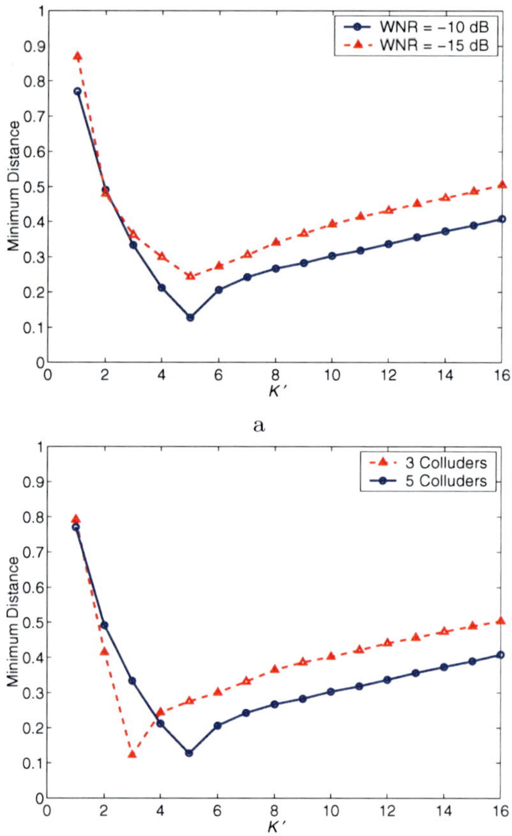

a

b

Fig. 5. Change of minimum distance with K' (a) under two WNR levels: $WNR = -10$ dB and $WNR = -15$ dB; (b) under two collusion scales: $K = 3$ and $K = 5$ when $WNR = -10$ dB.

sharpness of the valley, as shown in Figure 5b. The slope of the curve decreases as K' increases because it becomes harder to distinguish between the individual watermark amplitude $1/K'$ and $1/(K'+1)$ as K' gets larger. So a valley at a small K will be sharper than that at a large K.

The radius of the sphere also plays an important role in determining the computational cost of the sphere decoding algorithm. In our decoding scheme, we choose the minimum distance obtained at $K' - 1$ as the initial radius for the algorithm at K'. Meanwhile, we update the radius when we find a smaller distance during searching. These measures reduce the number of points to search and improve the detection efficiency. On the other hand, prior knowledge may help to further reduce the computational cost. For example, if we have knowledge of the upper and lower bounds of K, we can limit our search within this range.

Noise levels also affect the speed of the algorithm. Large WNR always results in faster convergence in our simulation. Nevertheless, these parameters only affect the computations needed for the algorithm and don't affect the detection result.

In summary, the algorithm can be formalized as below:

0. The signature matrix $\mathbf{S}$, scaling factor α and the observed statistics $\mathbf{T}$ are known.

1. Compute the correlator output for each user $y_i = \mathbf{s}_i^\top \mathbf{T}$, where $\mathbf{s}_i$ is the signature of user i.

2. Form a new signature matrix $\mathbf{S}'$ by removing those most likely innocent $M - N$ users that have the smallest y_i, and order the rest of the users in ascending order according to their y_i values.

3. Initialize $K' = 1, r_{min,1} = ||\alpha \mathbf{s}_i - T||$, where $i = \arg\max_i y_i$. This corresponds to the minimum distance with one colluder.

4. $K' = K' + 1$. Solve equation (19) and find the minimum distance $r_{min,K'}$ using the sphere decoding algorithm with the initial radius $r_{min,K'-1}$.

5. If $r_{min,K'} < r_{min,K'-1}$, repeat step 4; otherwise, stop.

6. The $\mathbf{\Phi}'$ with the minimum distance $r_{min,K'-1}$ is the final detection result. The users which correspond to the 1 entries in $\mathbf{\Phi}'$ are the users to be accused of colluding.

We now make some further comments regarding the complexity of this algorithm. The expected complexity of the sphere decoding algorithm is determined by the average number of lattice points searched, and thus is a function of cardinality, SNR, constellation size and also initial radius used in the algorithm. The exact forms of the expected complexity for various situations were studied thoroughly in [11, 28]. As a rule of thumb, for reasonable SNR/WNR, the complexity is approximately cubic in the amount of "pre-suspect" users that remain after removing those users who are most likely innocent. For fixed dimensionality, the complexity is decreased by increasing the WNR or decreasing the constellation size or decreasing the initial radius (though the initial radius must be large enough to ensure, with large probability, that there exists valid lattice points in the sphere).

Additionally, we note that it is possible to remove more than $M - N$ users in order to accelerate the operation of the sphere decoding algorithm. In such cases, the chances of missing one colluder (or a few colluders) might increase, but the strongest colluders will remain and these entities will very likely be captured. The system designer should consider whether it is sufficient to identify a few colluders, or whether it is necessary to identify more colluders. Ultimately, the amount of users we remove and hence the amount of "pre-suspect" users to consider is a design issue, and the system engineer should weigh the issues of algorithm complexity and the desire to capture more colluders.

4.3 Colluder Detection for General Linear Attacks

The above detection method is based on the averaging attack, where all colluders contribute equally to the colluded copy. This makes sense since no one user wants to take more risk than any other user. However, if the colluders intentionally change their ratio, will that disrupt the detection significantly? It turns out that a slight change in the algorithm will handle the more general case.

A general linear attack can be modeled as

$$\mathbf{y} = \alpha \sum_{i=1}^{M} \beta_i \mathbf{w}_i + \mathbf{d}, \tag{20}$$

where β_i is the weight for user i, and satisfies the condition $\sum_{i=1}^{M} \beta_i = 1$. An innocent user has weight 0. After projecting both sides onto the watermark space, we get the equivalent vector model

$$\mathbf{T} = \alpha \mathbf{S} \beta + \mathbf{n}, \tag{21}$$

where β is the weight vector $(\beta_1, \beta_2, \cdots, \beta_M)$.

Consider a simple example, three users collude via a weighted average of their copies. Without loss of generality, we assume the first three users collude. Also assume their weights are $\beta_1 = 0.5, \beta_2 = 0.25$ and $\beta_3 = 0.25$. Then

$$\beta = (0.5, 0.25, 0.25, 0, \cdots, 0) = \frac{1}{4}(2, 1, 1, 0, \cdots, 0).$$

Let us take $K = 4$ in the decoding algorithm, and $\mathbf{\Phi} = (2, 1, 1, 0 \cdots, 0)$, we can write

$$\beta = \frac{\mathbf{\Phi}}{K}. \tag{22}$$

Substitute it back to equation 21, we get

$$\mathbf{T} = \frac{\alpha}{K} \mathbf{S} \mathbf{\Phi} + \mathbf{n}. \tag{23}$$

This equation is exactly the same as our averaging collusion model (equation 4), except that the components of $\mathbf{\Phi}$ are no longer restricted to be 0 or 1 only. They can be any integer value between 0 and K, subject to $\sum_{i=1}^{M} \phi_i = K$. So the same integer least squares algorithm is applicable, given that we search over the lattice defined by this new $\mathbf{\Phi}$. We note that this will not increase the complexity significantly since we only search the lattice points within a small sphere. Although the lattice defined by the new $\mathbf{\Phi}$ has many more lattice points than before, most new lattice points are outside the sphere, and thus will not be examined at all. This is exactly the reason why the sphere decoding algorithm is more efficient than a brute-force searching – the complexity order of the sphere decoding algorithm does not increase significantly with alphabet size in each dimension.

In this simple example, β happens to be well represented in form of $\frac{\mathbf{\Phi}}{K}$ for some small K. If β does not have such a nice property, it can always be approximated to $\frac{\mathbf{\Phi}}{K}$ plus an error term, which can be considered as noise. Due to the constraint that the sum of β_i must be 1, the noise introduced by the approximation is bounded. If the contribution of a colluder is not too small, its detection shall not be hindered seriously. On the other hand, if the contribution of a colluder is very small, he likely will not be caught. Similarly, a colluder who does not contribute shall not be caught since he is actually innocent. We will further explore this property in Section 5 on performance analysis.

5 Performance Analysis

To test the proposed fingerprint design and detection scheme, we compared the performance of the WBE-based approach with the BIBD-based ACC codes proposed in [27] using the same abstract model and experimental conditions as in[27]. Both the distortion signal $\mathbf{d}$ and the orthogonal basis $\mathbf{u}_i$ were 10000 point vectors with i.i.d. Gaussian samples. The scaling factor α was used to control the watermark-to-noise ratio (WNR) defined as

$$\mathbf{WNR} = 10 \log_{10} \frac{||\alpha \mathbf{w}_i||^2}{||\mathbf{d}||^2}. \tag{24}$$

We used 16 orthogonal basis vectors to accommodate 20 users, 3 of which are colluders. For each trial, the three colluders were generated at random.

For each WNR, 2000 trials were performed to find the average detection performance measures. For the BIBD-ACC code, two detection methods were used: the adaptive sorting and the sequential algorithm. The following performance metrics were measured:

1) The average fraction of colluders that are successfully captured p_d.
2) The average fraction of innocent users that are falsely accused p_{fa}.
3) Probability of falsely accusing one or more innocent users in a trial.
4) Probability of missing one or more colluders in a trial.
5) Probability of not catching any colluders in a trial.
6) Probability of catching all colluders exactly without any false accusation in a trial.

In all results for the WBE-based fingerprinting scheme, we used sphere decoding where the amount of colluders K was estimated by the algorithm. The results are presented in Figure 6. Our WBE-based fingerprinting scheme shows improvement in all metrics.

To see whether the performance improvement over BIBD ACC is due to the code selection or the detection algorithm, we applied the modified sphere decoding algorithm to the BIBD ACC. The estimated p_d and p_{fa} from the simulation is shown in Figure 7. The BIBD ACC exhibited almost identical detection and false alarm performance when the modified sphere decoding algorithm and the sequential identification algorithm are used. This implies that the performance improvement between WBE codes and BIBD ACC is mainly due to the code construction. On the other hand, we emphasize that the modified sphere decoding is crucial for exploiting the collusion resistance of the WBE code since it provides nearly maximum likelihood estimation. As will be shown later in this section, when the number of users is large, or the collusion scale is large, the strong correlation between the fingerprints and the low power of the participating watermarks can easily cause the correlator outputs of the innocents to be

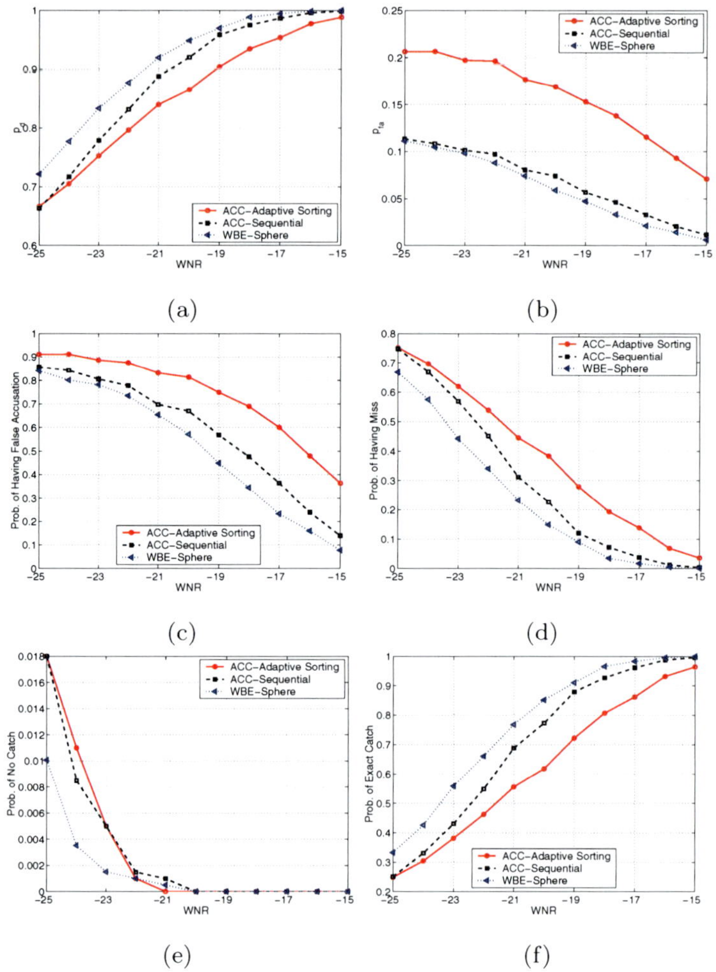

Fig. 6. The performance comparison of WBE-ACC and BIBD-ACC. 16 dimensions are used to handle 20 users with 3 colluders. (a) Average fraction of captured p_d. (b) Average fraction of false accusation p_{fa}. (c) Probability of falsely accusing one or more innocent users in a trial. (d) Probability of missing one or more colluders in a trial. (e)Probability of not catching any colluders in a trial. (f) Probability of catching all colluders exactly without any false accusation in a trial.

larger than the correlator outputs of some colluders. Therefore, a simple correlator is certainly not enough to detect the colluders effectively. Simulation shows that the modified sphere decoding algorithm performs well despite the strong correlations among the fingerprints. In particular, we conclude that the increase in the collusion capacity for WBE fingerprints is obtained at the price of a more complicated detection algorithm.

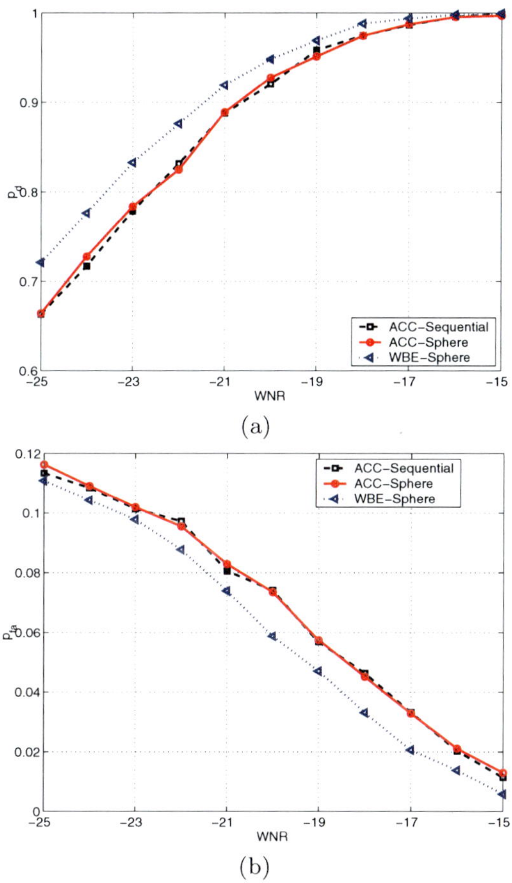

Fig. 7. The performance of the BIBD ACC with the modified sphere decoding algorithm. (a) p_d, (b) p_{fa}.

We then tested the performance of the proposed schemes with more colluders: 16 dimensions were used to accommodate 20 users, 5 of which acted as colluders. The results are shown in Figure 8 along with the 3-colluder case for comparison. As before, the sphere decoding algorithm estimated the amount of colluders when identifying colluders. Since more colluders causes a decrease in the fingerprint power for each colluder, it should degrade the detection performance. This is verified by the simulation results. However, the detection does not break as the number of colluders increases, and the performance of the 5-colluder case approaches that of the 3-colluder case as WNR increases.

We note that when the number of colluders increases from 3 to 5, due to the averaging effect, the power of each watermark is attenuated by $20\log_{10}(5/3) = 4.437$ dB. From the detection result shown in Figure 8, this is roughly equal to the gap between the two performance curves. In other words, the performance

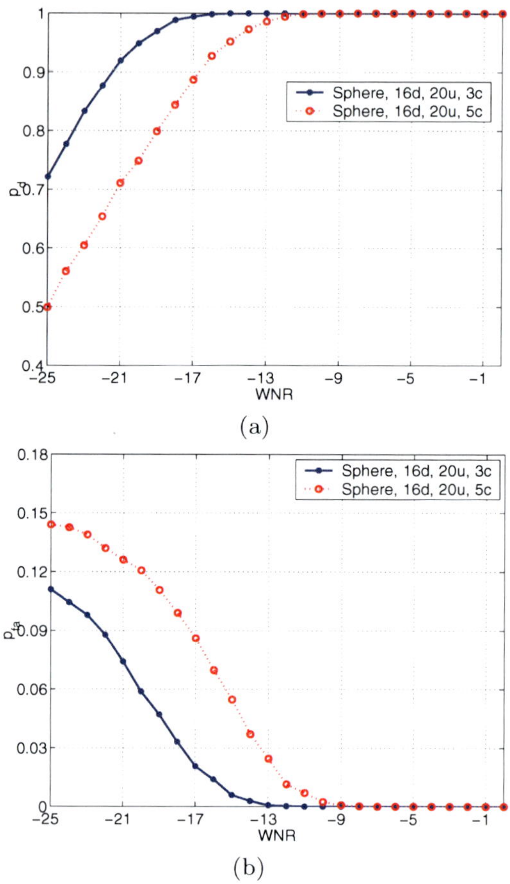

Fig. 8. The performance of the WBE fingerprints and the detection algorithm with 3 and 5 colluders. (a) p_d, (b) p_{fa}.

degradation is mainly due to the natural watermark power attenuation introduced by collusion. The detection algorithm does not set any upper limit on the possible collusion scale (except that it does not exceed the dimensionality N). This is contrary to the case for BIBD-ACC, where the algorithm breaks when the number of colluders exceeds the designed anti-collusion capability.

To illustrate the performance degradation with collusion scale, we fixed the WNR and varied the number of colluders from 3 to 10. Two WNR levels were tested: -10 dB and -15 dB. For all cases in this experiment, our sphere decoding algorithm estimated the amount of colluders prior to identifying colluders. The resulting detection performance degrades gracefully as we increase the collusion scale, as shown in Figure 9.

One natural question to ask is how much price we pay by using non-orthogonal fingerprints to save dimensionality. To answer this question, we also did simulations for orthogonal watermarks. 16 dimensions can only accommodate 16 users

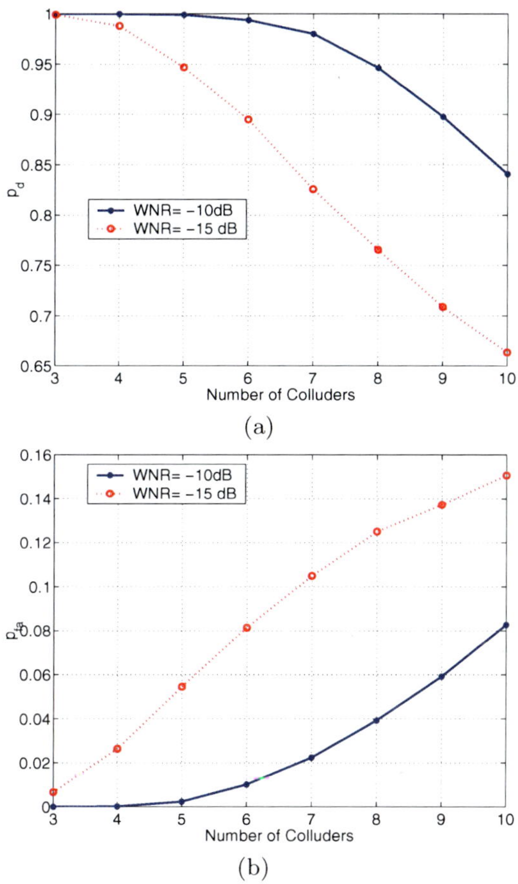

Fig. 9. The performance of the WBE fingerprints vs. the number of colluders at $WNR = -10$ dB and -15 dB, for a WBE-ACC using 16 basis vectors and supporting 20 users. (a) p_d, (b) p_{fa}.

for orthogonal fingerprinting. Since the watermarks are orthogonal, for a known K, a simple correlation detector is enough. In other words, the K users with the largest correlator outputs are the estimated colluders. However, because K is unknown for most practical cases, a similar maximum likelihood detection procedure as before is still needed to estimate K. The detection algorithm for the orthogonal watermarks is formalized below:

0. The signature matrix $\mathbf{S}$, scaling factor α and the observed statistics $\mathbf{T}$ are known.

1. Compute the correlator output for each user $y_i = \mathbf{s}_i^\top \mathbf{T}$, where $\mathbf{s}_i$ is the signature of user i. Sort the users according to their y_i.

2. Initialize $K' = 1, r = ||\alpha \mathbf{s_i} - T||$, where $i = \arg\max_i y_i$. This corresponds to the minimum distance with one colluder.

3. $K' = K' + 1$. $r' = ||\frac{\alpha}{K'} \sum_{i \in S_{K'}} \mathbf{s_i} - T||$, where the set $S_{K'}$ contains the users with K' largest correlator outputs y_i.
4. If $r' < r$, let $r = r'$ and repeat step 3; otherwise, stop.
5. The $K' - 1$ users with the largest y_i are the users to be accused of collusion.

The performance of WBE-ACC and the corresponding orthogonal fingerprinting scheme are plotted in Figure 10. WBE fingerprints with 20 users and 24 users are simulated for comparison with different amounts of colluders. The orthogonal scheme has a better performance than the non-orthogonal scheme – this is expected since it accommodates fewer users. However, the advantage is quite small, and is mainly in the probability of detection, and not in the probability of false accusation. The probability of false accusation even shows some improvement at low WNR. This is because the number of innocent users increases as the total users increases, as we fix the number of colluders. Therefore, the probability of false accusations, calculated as the total innocent users that are falsely accused divided by the total number of innocent users, might decrease if the former factor does not increase as fast as the latter one.

To show how the performance degrades as the total number of users increases, we tested our scheme with increasing users at a fixed $WNR = -10, -15$ dB, and fixed collusion scale $K = 5$. The results are given in Figure 11. Detection probability gradually decreases as the number of users increases, while the false accusation probability gradually increases. But the degradation speed is slow.

When the total number of users is large, the correlation between watermarks is also high. The nearly maximum likelihood detector we adopted here shows great advantage compared to the simple correlation detector to detect colluders. To illustrate this, we chose an example with highly correlated fingerprints at $WNR = -10$ dB, and 24 users are represented by 16 dimensions. The highest cross-correlation among the codes of the users is 0.4. Due to this high correlation among the watermarks, the correlator output of an innocent user will not be zero even without noise. However, for a legal copy with only one fingerprint, the correlator output for the watermark of the original user is still significantly higher than that of the others, as shown in Figure 12a. So the correlator is enough to test the fingerprint, and identify the owner. But for a colluded copy, the correlator responses of the colluding users are much weaker because of the attenuated power of their watermarks. Moreover, due to the correlation, the response of the innocent users can be even higher than that of some colluders. This phenomenon is shown in Figure 12b. The 2nd to 6th user are the five colluders, and the correlator outputs are indicated by the circles. The correlator response of the 24th user (innocent) is even higher than that of the 3nd and 6th user (colluders). So a thresholding-based method won't be able to identify all colluders correctly. But our detection algorithm based on sphere decoding correctly identifies the colluders, as indicated in stars (stars with non-zero heights show the index of the colluders) on the same figure.

Since it is desirable to support the distribution of content to a large amount of users, we explored the ability of WBE-based ACC in conjunction with sphere

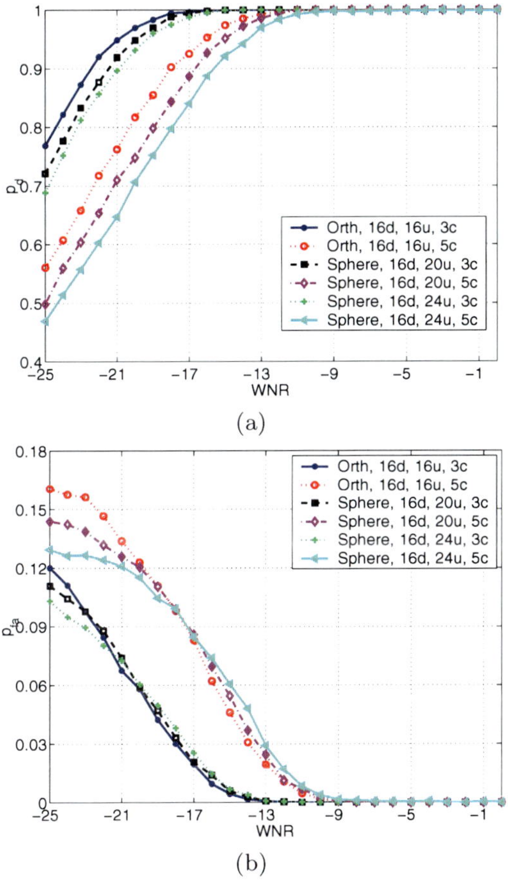

Fig. 10. The performance of the WBE fingerprints compared to the orthogonal scheme. (a) p_d, (b) p_{fa}.

decoding to identify colluders when we increased the size of the user base by an order of magnitude. We conducted an experiment where we increased the amount of users M from 100 to 500, while keeping the dimensionality of the basis N set at $N = 0.8M$. We had 10 colluders and examined a WNR of -10dB and -5dB. In our sphere decoding algorithm, rather than remove $M - N$ likely innocent users, we removed more so that we always had 20 pre-suspect users in the sphere decoding algorithm. The probability of detection and false alarm for this experiment is reported in Figure 13(a) and (b). From this figure, we see that the performance maintains a smooth characteristic with the p_d and p_{fa} comparable to the performance at smaller user sizes at the same WNR. At higher WNR cases, such as would correspond to non-blind detection cases, the performance is very good, with high probability of detection and low probability of false alarm.

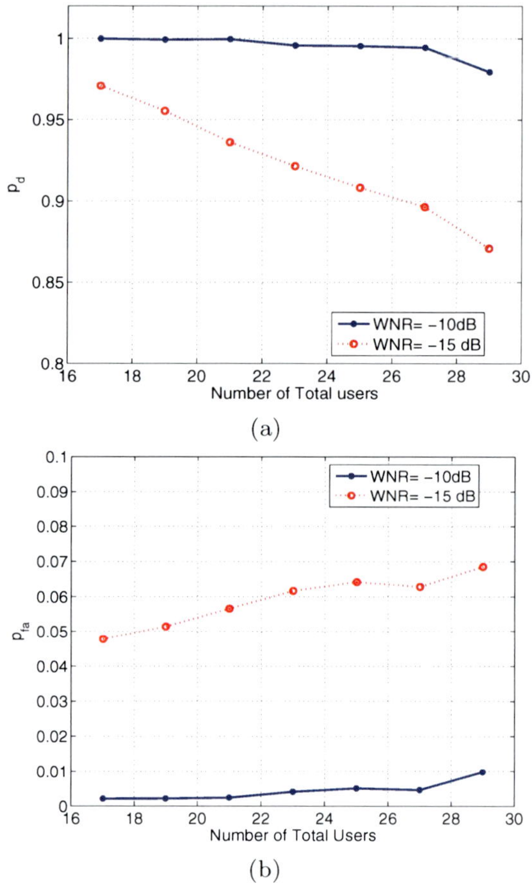

Fig. 11. The performance of the WBE fingerprints as the number of total users increases, at a fixed dimensionality $N = 16$, a fixed size of collusion $K = 5$, and a fixed $WNR = -10, -15$ dB. (a) p_d, (b) p_{fa}.

As a final comment, we would note that complexity of the sphere decoding algorithm is related to the operating WNR and the amount of pre-suspect users. In our experiment, we kept the amount of pre-suspect users fixed at 20, regardless of the value of M, so that we kept the computational complexity reasonable. However, if one would desire to have more pre-suspect users as M increases, then this would lead to increased complexity. In such cases, it might be desirable to employ more practical methods to associate users with fingerprints, as discussed in Section 6.

As discussed in Section 4, the algorithm also works for a general linear attack, where the colluded copy is a linear combination of the individual ones, i.e.

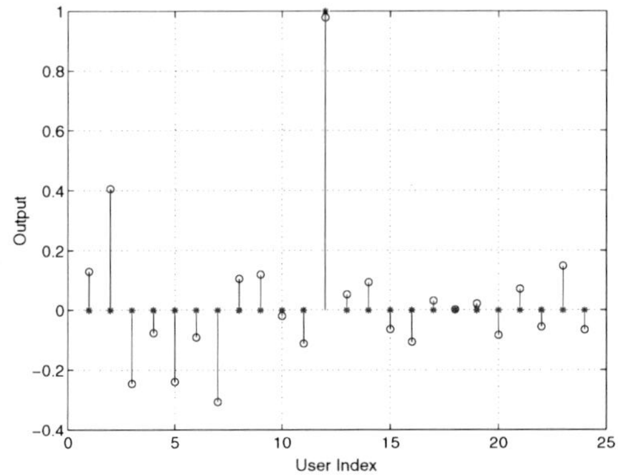

(a)

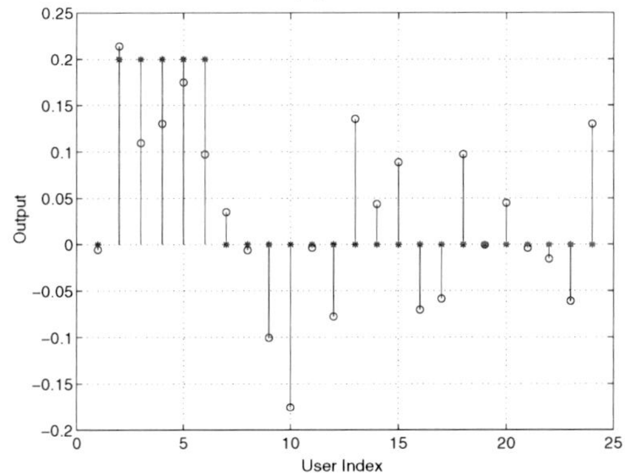

(b)

Fig. 12. The outputs of a simple correlator and the proposed sphere decoding detector. $N = 16, M = 24, WNR = -10$ dB. (a) Correlator output (circles) and the sphere decoding detector output (stars) when the signal contains only one watermark (the 12th user's). (b) Correlator output (circles) and the sphere decoding detector output (stars) when the signal is attacked by averaging collusion attack of size 5. Users of index 2 to 6 are the colluders.

$$\mathbf{y} = \alpha \sum_{i=1}^{M} \beta_i \mathbf{w}_i + \mathbf{d}.$$

We give two examples here. In the first one,

$$\beta = (0.5, 0.25, 0.25, 0, \cdots, 0) = \frac{1}{4}(2, 1, 1, 0, \cdots, 0).$$

The detection result is presented in Figure 14a. The output of the sphere decoding detector is the same as the real β. For the second example,

$$\beta = (0.18, 0.22, 0.34, 0.26, \cdots, 0),$$

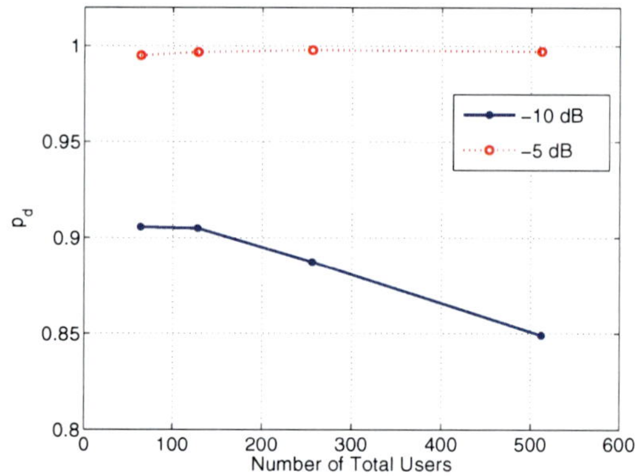

(a)

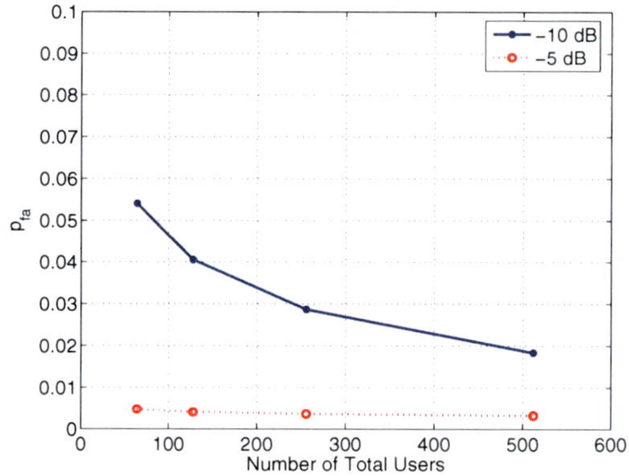

(b)

Fig. 13. The performance of WBE fingerprints as we increase the amount of users M while using a basis of size $N = 0.8M$. (a) p_d , (b) p_{fa}.

which can not be written as $\frac{\Phi}{K}$ for a small K. The output of the sphere decoder is shown in Figure 14b. Although it is different from the real β, it correctly identifies the colluders.

6 Practical Considerations

We also examined the performance of WBE-based fingerprints when embedded in a real image using a popular watermarking scheme. We used the scheme of Podilchuk and Zeng[17], which is a DCT-based embedding scheme that uses perceptual weighting. In this case, the scaling factor α varies across the feature vector according to the perceptual significance of the corresponding coefficient. Hence, we may represent the scaling as $\alpha(k)$, where k varies over the length of the feature vector. The fact that $\alpha(k)$ varies does not significantly affect the

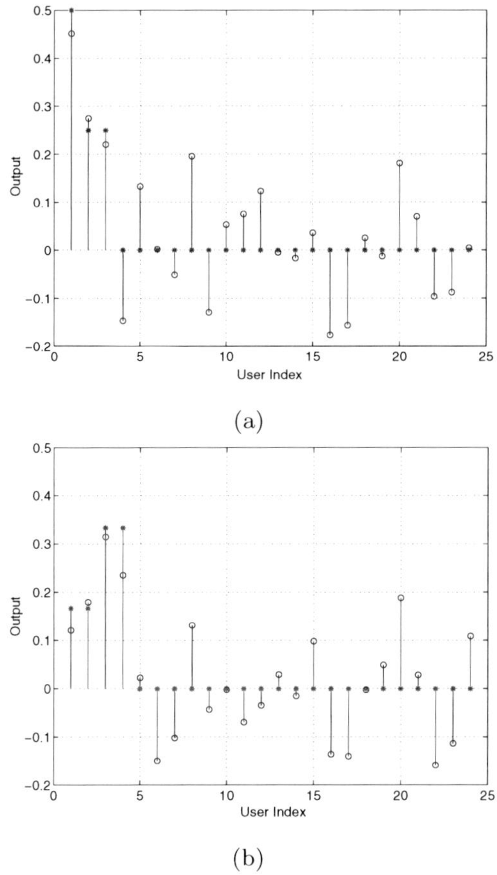

Fig. 14. The outputs of a simple correlator and the proposed sphere decoding detector for a general linear attack. $N = 16, M = 24, WNR = -10$ dB. (a) For $\beta = (0.5, 0.25, 0.25, 0, \cdots, 0)$, the sphere decoding detector (stars) gives a response that is the same as the real β. (b) For $\beta = (0.18, 0.22, 0.34, 0.26, 0, \cdots, 0)$, the sphere decoding detector (stars) has an output different than the real β, but it still correctly catches all colluders.

orthogonality of the feature vectors since, for images, the length of the basis vectors is large and the basis vectors are randomly generated. At the decoder, the perceptual scaling is either estimated using the original image (as in the case of non-blind detection), or must be estimated using the marked/colluded image. As long as the marked/colluded image is a faithful representation of the original content (i.e. the attacked media maintains its perceptual value), the estimated $\hat{\alpha}(k)$ will be very close to the $\alpha(k)$ used at the encoder. In practice, this discrepancy has negligible effect on performance.

In our experiment, we used the Lena image, and the Podilchuk embedding scheme identified $L = 37510$ embeddable coefficients. We used $N = 16$ basis

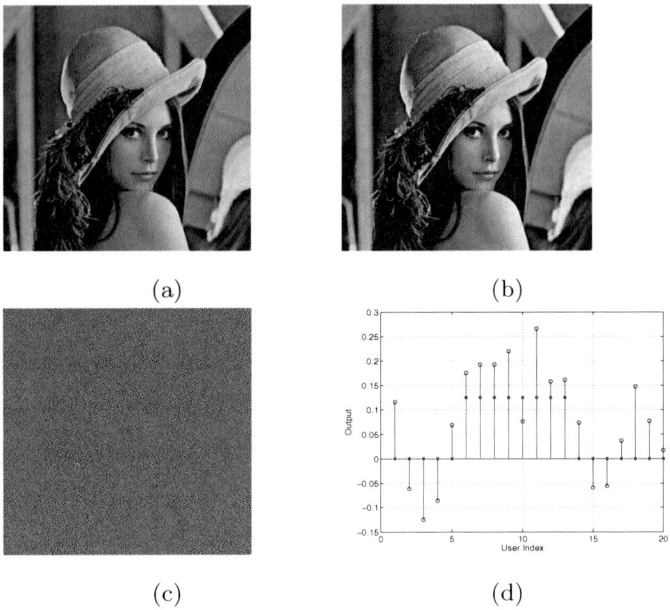

Fig. 15. (a) The original Lena image prior to fingerprinting. (b) The attacked content signal resulting from the collusion of 8 users. (c) The fingerprint contained in the attacked image. (d) The correlation output (circles) and the sphere decoding detector output (starts) in a blind detection scenario (the colluders are users 6 through 13).

vectors to support $M = 20$ users. In order to stress test the performance of WBE-based fingerprinting, we wanted the amount of colluders to be large relative to N, and thus chose $K = 8$ colluders. Using the 16 basis vectors, we constructed WBE-based fingerprints and embedded them in separate copies of the image. We selected the 8 colluders and formed a colluded image that was then used in the detection process. The results are presented in Figure 15. Here, we have presented the original image, the attacked image, and the remaining fingerprint signal after the collusion attack is performed. We conducted our detection and correlation in blind mode (without the aid of the original image), and the resulting individual correlation statistics are presented in Figure 15 (d). If only the individual correlation statistics are used for detecting colluders, we would capture 7 out of the 8 colluders, and would falsely accuse two users (user 1 and 18). Even in this blind detection mode, the sphere decoding algorithm, however, identified all 8 colluders and did not make any false accusations.

We now would like to discuss the important practical issue of scaling WBE-based fingerprinting to cope with more users. In Section 5 we saw that the sphere decoder was able to identify colluders as we increased the amount of users M and the size of the basis N. However, as we noted earlier, the computational complexity associated with identifying colluders grows cubically. This complexity can become prohibitive when we try to build a system that can handle a user base on the order of several thousand or more users, especially if we are performing blind detection where the WNR is low. In such cases, if we do not have the computational resources needed to conduct sphere decoding using a large basis of N basis vectors, it is possible to explore more compact representations of the users. We may batch our users into smaller groups, and use a different fingerprinting system for each group. For example, we could split our total M users into groups of 20 users, and each subgroup would have a different 16-dimensional basis that would be used to build 20 fingerprints. Identifying colluders is now done on separate batches of 20 fingerprints rather than using a single, significantly larger basis set. Consequently, the computational complexity can be reduced. Finally, we note that one further direction for reducing computational complexity could be to use a hierarchical naming scheme, similar to the group-oriented fingerprinting schemes described in [30].

7 Conclusion and Discussions

Digital fingerprinting is a popular technique to protect the copyright of digital data. A cost-effective way to attack fingerprinting is collusion, where several users average their copies. Although orthogonal fingerprints are known to be resistant to collusion attacks, the number of users the system can accommodate is greatly limited due to orthogonality constraints. In this work, we proposed a type of correlated WBE-based fingerprints that are resistant to collusion attacks, and also allow more users to be accommodated for a given fingerprint dimensionality than for orthogonal fingerprints. We then proposed a nearly maximum likelihood colluder detection algorithm based on sphere decoding. Our modified sphere decoding algorithm greatly reduces the computations needed for a nearly maximum likelihood detection. The performance of the proposed scheme was tested using an abstract model, and was shown to be better than BIBD-ACC in all performance metrics studied. The performance loss compared to orthogonal fingerprinting is small, especially at high WNRs, such as arise in non-blind detection scenarios.

Unlike schemes that utilize the combinatorial properties of the codes, we take the view that collusion problem is similar to a multiuser communication problem, as everyone transmits either 1 (collude), or 0 (innocent), and can be studied using similar techniques. To facilitate accurate detection, it is desirable to have as small of correlation as possible among users, which is achieved using WBE sequence sets. Several existing collusion-resistant fingerprinting schemes design for a specific level of collusion during the fingerprint construction phase, and then decide how many users can be accommodated. The design of the fingerprints is done independent of the WNR levels arising in the media distribution scenario. In contrast, our scheme does not place a hard limit on the size of collusion coalition, and the performance was shown to degrade gracefully as the number of colluders increases. Further, our construction is general, and allows for us to adjust the design of the fingerprints to accommodate varying amount of users based upon a target WNR level.

References

1. Ajtai, M.: The shortest vector problem in L_2 is NP-hard for randomized reductions. In: Proceedings of the 30rd ACM Symposium on Theory of Computing, May 1998, pp. 10–19 (1998)
2. Boneh, D., Shaw, J.: Collusion-secure fingerprinting for digital data. IEEE Tran. on Information Theory 44, 1897–1905 (1998)
3. Chan, A., Lee, I.: A new reduced-complexity sphere decoder for multiple antenna systems communications. In: IEEE International Conference on Communications, April 2002, vol. 1, pp. 460–464 (2002)
4. Cox, I., Bloom, J., Miller, M.: Digital Watermarking: Principles & Practice. Morgan Kaufmann, San Francisco (2001)
5. Cox, I., Kilian, J., Leighton, F., Shamoon, T.: Secure spread spectrum watermarking for multimedia. IEEE Tran. on Image Proc. 6(12), 1673–1687 (1997)
6. Dayal, P., Varanasi, M.K.: A fast generalized sphere decoder for optimum decoding of under-determined mimo systems. In: Proceedings of 41st Annual Allerton Conf. on Comm. Control, and Comput., Monticello, IL (October 2003)
7. Ergun, F., Kilian, J., Kumar, R.: A note on the limits of collusion-resistant watermarks. In: Stern, J. (ed.) EUROCRYPT 1999. LNCS, vol. 1592, pp. 140–149. Springer, Heidelberg (1999)
8. Fincke, U., Pohst, M.: Improved methods for calculating vectors of short length in a lattice, including a complexity analysis. Mathematics of Computation 44, 463–471 (1985)
9. H235 ITR: Security and encryption for H-series (H.323 and other H.245-based) multimedia terminals (November 2000)
10. Hassibi, B., Vikalo, H.: Maximum-likelihood decoding and integer least-squares: the expected complexity. In: Foschini, J., Verdu, S. (eds.) Multiantenna Channels: Capacity, Coding and Signal Processing. AMS
11. Hassibi, B., Vikalo, H.: On sphere decoding algorithm. i. expected complexity. IEEE Transactions on Signal Processing 53, 2806–2818 (2005)
12. Massey, J.L., Mittelholzer, T.: Welch's bound and sequence sets for code-division multiple access systems. In: Capocelli, R., DeSantis, A., Vaccaro, U. (eds.) Sequences II, Methods in Communication, Security and Computer Science, pp. 63–78. Springer, New York (1993)
13. Karystinos, G.N., Pados, D.A.: New bounds on the total squared correlation and optimum design of ds-cdma binary signature sets. IEEE Trans. on Communications 51, 48–51 (2003)
14. Kilian, J., Leighton, T., Matheson, L.R., Shamoon, T.G., Tarjan, R.E., Zane, F.: Resistance of digital watermarks to collusive attacks. Tech. Rep. TR-585-98, Department of Computer Science, Princeton University (1998)
15. Damen, M., Abed-Meraim, K., Belfiore, J.-C.: Generalised sphere decoder for asymmetrical space-time communication architecture. Electronics Letters 36, 166–167
16. Celik, M.U., Sharma, G., Tekalp, A.M.: Collusion-resilient fingerprinting using random prewarping. In: Proceedings, International Conference on Image Processing, vol. 1, pp. 509–512 (2003)
17. Podilchuk, C., Zeng, W.: Image adaptive watermarking using visual models. IEEE Journal on Selected Areas in Communications 16(4), 525–540 (1998)
18. Rose, C., Ulukus, S., Yates, R.: Wireless systems and interference avoidance. IEEE Trans. on Wireless Communications 1, 415–428 (2002)

19. Ulukus, S., Yates, R.D.: Iterative signature adaptation for capacity maximization of cdma systems. In: Proceedings of the 36th Allerton Conference on Communications, Control and Computing, Monticello, IL (September 1998)
20. Sarwate, D.: Meeting the welch bound with equality. In: Ding, C., Helleseth, T., Niederreiter, H. (eds.) Sequences and Their Applications: Proceedings of SETA 1998. DMTCS Series. Springer, Heidelberg (1999)
21. Stone, H.S.: Analysis of attacks on image watermarks with randomized coefficients. Tech. Rep. 96-045, NEC Research Institute (1996)
22. Su, J., Eggers, J., Girod, B.: Capacity of digital watermarks subjected to an optimal collusion attack. In: European Signal Processing Conference, EUSIPCO 2000 (2000)
23. Swanson, M., Kobayashi, M., Tewfik, A.H.: Multimedia data-embedding and watermarking technologies. Proceedings of the IEEE 87, 1064–1087 (1998)
24. Tang, L.: Methods for encrypting and decrypting MPEG video efficiently. In: Proceedings of the 4th ACM Inter. Conf. on Multimedia, November 1996, pp. 219–229 (1996)
25. Trappe, W., Song, J., Poovendran, R., Liu, K.: Key distribution for secure multimedia multicasts via data embedding. In: IEEE Int. Conference on Acoustics, Speech, and Signal Processing (2001)
26. Trappe, W., Song, J., Poovendran, R., Liu, K.: A dynamic key distribution scheme using data embedding for secure multimedia multicast. IEEE Trans. on Multimedia 5, 544–557 (2003)
27. Trappe, W., Wu, M., Wang, Z., Liu, K.: Anti-collusion fingerprinting for multimedia. IEEE Trans. on Signal Processing, 1069–1087 (April 2003)
28. Vikalo, H., Hassibi, B.: On sphere decoding algorithm. ii. generalizations, second-order statistics, and applications to communications. IEEE Transactions on Signal Processing 53, 2819–2834 (2005)
29. Viswanath, P., Anantharam, V., Tse, D.: Optimal sequences, power control, and user capacity of synchronous cdma systems with linear mmse multiuser receivers. IEEE Trans. on Information Theory 45, 1968–1983 (1999)
30. Wang, Z., Wu, M., Trappe, W., Liu, K.: Group-oriented fingerprinting for multimedia forensics. EURASIP Journal on Applied Signal Processing, Special Issue on Multimedia Security and Rights Management, 2153–2173 (2004)
31. Wang, Z., Wu, M., Zhao, H., Trappe, W., Liu, K.: Anti-collusion forensics of multimedia fingerprinting using orthogonal modulation. IEEE Trans. on Image Proc. 14, 804–821 (2005)
32. Welch, L.R.: Lower bounds on the maximum cross correlation of signals. IEEE Trans. on Information Theory 20, 397–399 (1974)
33. Wen, J., Muttrell, M., Severa, M.: Access control of standard video bitstreams. In: Proceedings of the Inter. Conf. on Media Future (May 2001)
34. Wu, M., Liu, B.: Multimedia Data Hiding. Springer, Heidelberg (2002)
35. Zhao, H., Wu, M., Wang, Z., Liu, K.: Nonlinear collusion attacks on independent multimedia fingerprints. IEEE Trans. on Image Proc. (May 2005)

An Introduction to Robust Transform Based Image Watermarking Techniques

Xi Zhao and Anthony T.S. Ho

University of Surrey, UK
{x.zhao,a.ho}@surrey.ac.uk

Abstract. One of the important applications of digital watermarking technology is copyright protection and ownership identification for digital images. To achieve this goal, robust watermarking has been rapidly developed in the past decade or so. Robust watermarking is designed to survive various non-geometric manipulations such as JPEG compression, additive noise and filtering as well as some geometric distortions such as rotation and scaling. In this chapter, the fundamental concept of digital watermarking, differences between visible and invisible watermarking, blind and non-blind watermark detection schemes, robust, fragile and semi-fragile watermarking algorithms, as well as four important properties for digital watermarking: imperceptibility, robustness, capacity and security will be described. A number of different transforms and algorithms used for robust image watermarking will be reviewed in detail. These include block based Discrete Cosine Transform (DCT), non-block based Discrete Wavelet Transform (DWT) and other state-of-the-art watermarking algorithms operating in the transform domain.

In contrast to conventional transform domain methods, a new adaptive robust watermarking algorithm using the non-redundant contourlet transform known as the wavelet-based contourlet transform (WBCT) is presented in this chapter. From experiments, we exploit the energy relations between parent and children coefficients, which are invariant before and after JPEG compression. Results show that even for QF set as low as 10, the percentages of invariant energy relations of all test images were above 75% after JPEG compression. This invariance feature is therefore very useful for robust image watermarking. The results of WBCT also are evaluated and compared with two other DWT based algorithms achieving high degree of robustness against most non-geometrical and geometrical attacks, while maintaining an excellent perceptual invisibility.

1 Introduction

Nowadays, the rapid development of technologies has led to the significant increase of digital information, particularly multimedia such as image, audio and video content. Such technological advances have led to the ease in which it is possible to illegally share, distribute, and copy Intellectual Property (IP). Subsequently, the copyright infringement issue has been identified as a "hot topic". According to a report from Oxford Economics in 2009, the UK film industry loses 531 million pounds per annum,

H.T. Sencar et al. (Eds.): Intel. Multimedia Analysis for Security Appli., SCI 282, pp. 337–364.
springerlink.com

as a direct result of copyright theft. Examples of this can include recording films at the cinema, the illegal sale or purchase of copyrighted DVDs, household copying, file-sharing and downloading, and streaming material from unauthorised web servers [1]. Furthermore, the music industry is also affected. According to the US copyright industry group International Intellectual Property Alliance, some 2 billion music tracks were illegally downloaded in Spain in 2008, compared to 2.2 million that were purchased legally [2]. Moreover, billions of digital images are widely available and can be accessed easily and quickly via almost any website containing graphics, or image search engines.

An obvious requirement, therefore, is the development of solutions for copyright protection and ownership identification for digital content. Whilst the traditional cryptographic solutions can encrypt digital content, however they require the receiver to use a key to decrypt the data successfully. Moreover, such solutions do not prevent or track the content against illegitimate reproduction after it has been decrypted [3]. As a result, digital watermarking techniques are often preferred as they can be used to protect the data content after decryption. Digital watermarking is the process of embedding relevant ownership information (such as a logo, fingerprint and serial number), into a media in order to protect the ownership of different media formats. This technique can be applied to different media types such as video, audio and image content. For the purpose of copyright protection and ownership identification, robust watermarking schemes are mainly used as they can tolerate a host of signal processing attacks that can be both unintentional and intentional.

The employing of digital watermarking technique is also recommended by the UK government agency, the Department for Culture, Media and Sport (DCMS), which expresses the following views:

"To develop and adopt pre-competitive standards and unique identifiers, which are open and interoperable, to cover hardware and software for secure delivery of music, including encryption, watermarking and usage rules of music on-line" [4].

"The UK Film Council's position is that an effective deterrence policy needs to be based on a blend of educative, technological and legislative interventions with the latter firmly enforced. This policy should be accompanied by detailed exploration by all parties of the potential of watermarking and other technologies to facilitate identification of illegal activity" [5].

The chapter begins with the fundamental concept and application of digital watermarking; follow by a more detailed description of different transform based watermarking techniques. A novel robust digital watermarking algorithm based on the non-redundant contourlet transform is also included in the chapter which should be of significant interest to experienced researchers as well as newcomers to the field.

The chapter is organised as follows:

- In Section 2, we briefly describe the background of digital watermarking in various forms, such as visible and invisible watermark, blind and non-blind watermarking schemes. We discuss the concept of digital image watermarking, classifications of digital watermarking techniques such as robust, fragile and semi-fragile watermarking, and also introduce and define some important terminologies. This section also introduces tools that measure the quality of watermarked images, as well as

describing several attacks that are frequently used to evaluate the robustness of watermarking schemes.
- Section 3 discusses various robust watermarking algorithms operating in the transform domain. We review algorithms applied in the spatial domain, and then extensively review transform-based robust watermarking methods for a comparative analysis. The algorithms described in this section range from traditional robust watermark embedding and detection techniques, to more state-of-the-art methodologies based on several image transform techniques. This section also gives a simple robust watermarking tutorial based on the Discrete Cosine Transform (DCT). In this tutorial, the watermark embedding and detection methods are demonstrated via a sample MATLAB program.
- Section 4 presents a novel robust watermarking algorithm using the non-redundant contourlet transform that exploits the energy relations between "parent" and "children" coefficients. The concepts and advantages of the proposed algorithms are described. The watermark embedding and detection processes are also discussed in detail with analysis and evaluation of results.
- Section 5 gives the conclusion and future work.

2 Background of Digital Watermarking

2.1 What Is Digital Watermarking?

The concept of watermarking has been used in many different forms and can be traced back to thousands of years ago. For instance, in the late 13th century in Italy, a thin, translucent layer was sewn with wire onto a paper mould to form a watermark [6]. Historically, postage stamps and currency were commonly watermarked. Indeed, the currency watermark is still used today when printing banknotes. A digital watermark can be either visible or invisible. An example of digital visible watermark is the translucent logos that are often seen embedded at the corner of videos or images, in an attempt to prevent copyright infringement. However, these visible watermarks can be targeted and removed rather simply by cropping the media, or overwriting the logos. Subsequently, the field of digital watermarking is primarily focused on embedding invisible watermarks, which operate by tweaking the content of the media imperceptibly. As the watermark cannot be seen, there must exist a robustness property that ensures the watermark data survives if the image is altered. Typical applications of digital watermarking can include broadcast monitoring, owner identification, proof of ownership, transaction tracking, content authentication, copy control, device control legacy enhancement and content description [3, 7].

Figure 1 illustrates a typical watermark embedding process. The watermarked work is produced from an embedding algorithm that is traditionally comprised of three inputs: the original work, the watermark and a key. A blind watermark detection process is shown in Figure 2. The watermark is extracted from the watermarked work by using a detection algorithm in conjunction with the same key that was originally used to embed the watermark. In contrast, Figure 3 illustrates a non-blind (or informed) watermark detection process that extracts the watermark. Here, the original work has to be provided

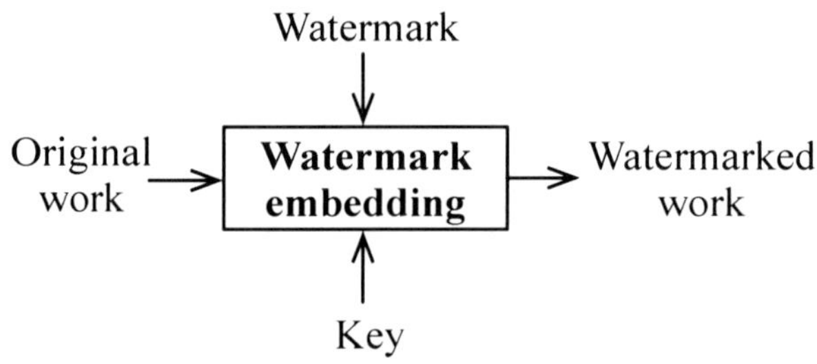

Fig. 1. A typical watermark embedding process.

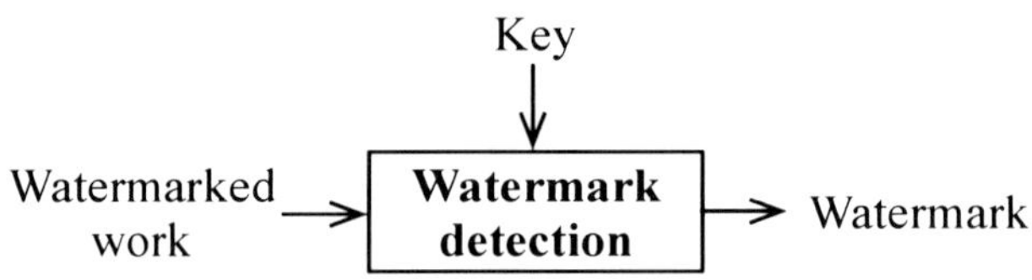

Fig. 2. A typical blind watermark detection process

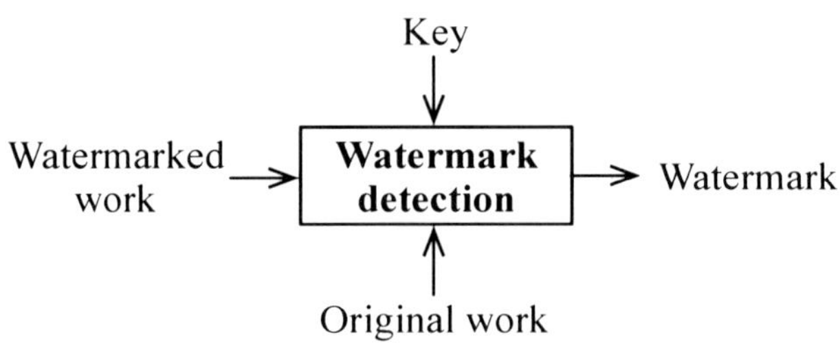

Fig. 3. A typical non-blind watermark detection process.

as a reference source in order for the detection algorithm to function. Therefore, the selection of a blind or non-blind watermarking detection system typically depends on whether the original work is accessible or not [7].

The original work is the host signal which is employed into diverse forms such as, video [8–10], audio [11, 12], image [13–15], halftone image [16–18], binary text [19, 20], 3D meshes [21–23], holography [24, 25], optical [26, 27] and network protocol [28, 29]. The watermark is a binary sequence of data produced from a logo image, fingerprint, serial number, owner's name or ID, or indeed anything that could identify the ownership of the media. The key is used to increase the security of the procedure; it prevents the possibility of a hacker modifying or removing the watermark as this can only be achieved if the key is known. Both of the watermark embedding and detection procedures from Figures 1 and 2 are demonstrated thoroughly in the tutorial part of Section 3.

2.2 Robust, Fragile and Semi-fragile Image Watermarking

There are three different classifications associated with digital watermarking, depending on the applications: *robust*, *fragile* and *semi-fragile*. Robust watermarking has been used extensively in the past decade, and is primarily designed to provide copyright protection and proof of ownership for digital images. The most important property of robust watermarking is its ability to tolerate certain signal processing operations that usually occur during the lifetime of a media object. A schematic diagram illustrating the main functions of robust watermarking is shown in Figure 4. The sender watermarks the original work via a watermark embedding process, and then sends the watermarked work to the receiver. The receiver extracts the watermark via a watermark detection process. During the transmission of the watermarked work, the image is open to attack, meaning the integrity of the watermark data is in jeopardy. Examples of common attacks include JPEG compression, additive noise, and filtering, and geometric distortions such as rotation and scaling. In this chapter, we will focus on robust transform based image watermarking schemes.

In contrast to the applications of robust watermarking, fragile and semi-fragile techniques are geared towards image authentication and localisation of tampered regions. Fragile watermarking can be used to detect any small manipulations made to the original image [30]. Hence, any attacks that ultimately alter the pixel values of an image can be detected, and the tampered regions can be located accurately when applying fragile watermarking schemes [31]. Many fragile watermarking algorithms are intentionally designed for use in the spatial domain (typically by altering the Least Significant Bits (LSB) of the image), as this domain is widely documented as being relatively fragile and sensitive to small changes [32–34].

Semi-fragile watermarking techniques for image content authentication have recently attracted much attention [35–37]. This is due to the fact that comparing to fragile watermarking, semi-fragile watermarking is not as sensitive as fragile watermarking. Semi-fragile schemes make it possible to verify the content of the original image, as well as permitting alterations caused by non-malicious (unintentional) modifications such as system processes. Moreover, semi-fragile watermarking is more focused on detecting intentional attacks than validating the originality of the image [38, 39]. During the image transmission, the mild signal processing errors caused by signal reconstruction and storage, such as transmission noise or JPEG compression, are permissible. However, the image content tampering such as copy and paste attack will be identified as a malicious attack.

2.3 Requirements for Digital Image Watermarking

In this subsection, four important properties for digital watermarking are discussed. These are imperceptibility, robustness, capacity and security.

Imperceptibility

The embedded watermark should be imperceptible from the watermarked work. The degradation from original work to watermarked work is permitted with the rule of maintaining image fidelity of the original work. Therefore, in order to evaluate the similarity

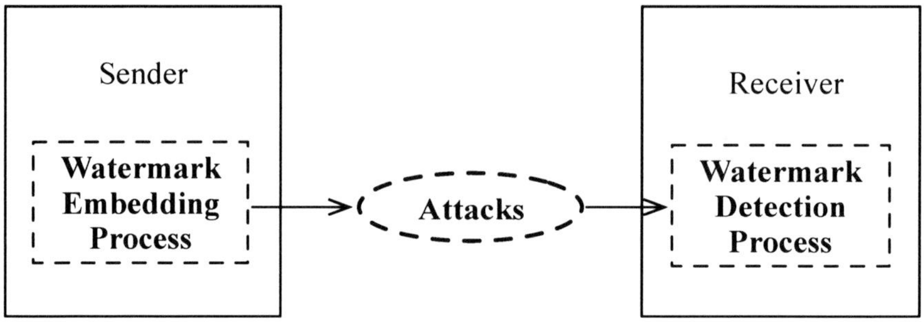

Fig. 4. Schematic diagram for robust watermarking

between original and watermarked image, objective and subjective evaluation methods will be needed. One of the standardised objective methods for image and video is the Double Stimulus Impairment Scale (DSIS) [40]. The watermarked images are visually reviewed by observers who have no background knowledge of image processing and are not experienced assessors. The watermarked images are compared against the original image and scored with a scale from one to five (*1 = very annoying, 2 = annoying, 3 = slightly annoying, 4 = perceptible, but not annoying, 5 = imperceptible*). To measure the quality of a watermarked image subjectively, a well-known statistical metric called the Peak Signal-to-Noise Ratio (PSNR), in Equation 1, is commonly used to evaluate the quality of the watermarked image by comparing it with the statistics of the original image. The quality of the watermarked image can be considered as acceptable if the PSNR value is above 30dB [41].

$$MSE = \frac{1}{mn} \sum_{i=0}^{m-1} \sum_{j=0}^{n-1} \left\| img\left(i,j\right) - img\_w\left(i,j\right) \right\|^2$$

$$PSNR = 10 \log_{10} \left(\frac{MAX^2}{MSE} \right) \qquad (1)$$

where m and n are the size of the image. (e.g. 512×512), img is the original image, img_w is the watermarked image, and MAX is the maximum possible pixel value of the image (e.g. 255 for a grayscale image).

Other alternative or supplementary methods of evaluating the similarity between the original image and the watermarked image are weighted PSNR (wPSNR) [42] and Mean Structural Similarity (SSIM) index [43]. wPSNR consists of the additional Noise Visibility Function (NVF) which is the difference between PSNR and wPSNR. NVF is exploited as a Gaussian model to estimate the degree of texture in the image by providing a value between 0 and 1 (*where 0 = high texture and 1 = smooth*) [44]. The wPSNR is defined in Equation 2.

$$wPSNR = 10 \log_{10} \left(\frac{MAX}{\sqrt{MSE} \times NVF} \right)^2 \qquad (2)$$

In contrast to PSNR and wPSNR, the MSSIM method, in Equation 3, separates the luminance, contrast and structure of the images for similarity measurement. This metric is based on the degradation of structural information and attempts to measure the attributes that reflect the key structure and objects of importance in an image. The MSSIM values exhibit much better consistency with the qualitative visual appearance [43].

$$SSIM(x,y) = [l(x,y)]^{\alpha} \cdot [c(x,y)]^{\beta} \cdot [s(x,y)]^{\gamma},$$
$$MSSIM(X,Y) = \frac{1}{M}\sum_{j=1}^{M} SSIM(x_j, y_j) \quad (3)$$

where the luminance comparison function is $l(x,y)$, the contrast comparison function is $c(x,y)$, the structure comparison function is $s(x,y)$, $\alpha > 0$, $\beta > 0$ and $\gamma > 0$ are parameters used to adjust the relative importance of the three components. X and Y are the original and the watermarked images, respectively, x_j and y_j are the image contents at the jth local window, and M is the number of local windows of the image [43]. However, due to its popularity and simplicity for calculation, the PSNR technique is used in this chapter for assessing the image quality between original and watermarked images.

Robustness

Robustness is an important property for robust watermarking schemes. The watermark that is embedded into the image should be robust (to varying degrees according to the application) to tolerate different forms of attack or image processing operations, when the watermarked image is transmitted. These image manipulations or attacks can be categorised into non-geometrical and geometrical groups. Non-geometrical distortion is derived from lossy compression algorithms such as JPEG or JPEG2000, as well as noise addition, image filtering and contrast stretching, while geometrical distortion includes rotation, scaling, cropping, translation, and shifting pixels. These distortions are often implemented to simulate possible attacks to analyse the performance trade-off of the proposed algorithms by the researchers in the community [3]. Maintaining the robustness of the watermark is much more difficult and challenging when considering geometrical attacks. This is due to the fact that each individual pixel location of the watermarked image is likely to be shifted or translated. A possible approach is to find an invariant property of an image that can be used in the watermark embedding process to enhance the robustness against different attacks. More details on the use of invariant properties for watermarking are described in Section 4.

Capacity

Capacity refers to the maximum amount of watermark bits that can be embedded into the original image. The number of watermark bits embedded into the image data can affect the overall perceptual quality of the image. Figure 5 illustrates the performance trade-offs concerned with watermarking; specifically, the imperceptibility of the watermarked image, the robustness of the watermark, and the capacity of the watermark

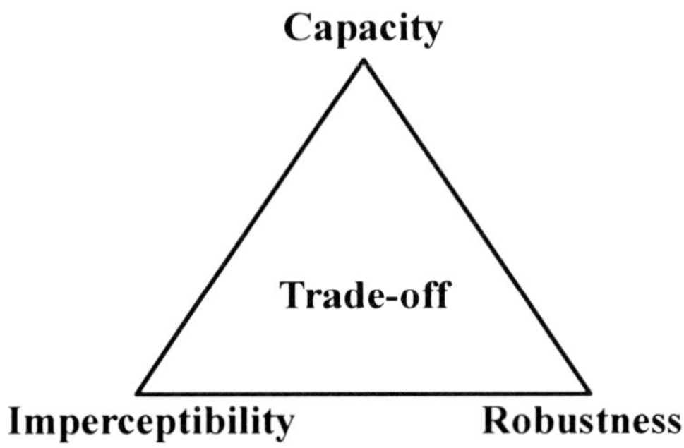

Fig. 5. Trade-off among the imperceptibility, robustness and capacity

data. If the quality of watermarked image is high, the robustness and capacity of the watermark data will be degraded.

Similarly, if the robustness is high, the quality of watermarked image is likely to be degraded, as a greater number of watermark bits will be used. Finally, if the capacity of watermark data is increased, the quality of the image and its robustness will decrease.

Security

The approach to security in digital watermarking is mainly focused on malicious removal or modification of the watermark bits. The watermark security can be defined as "the inability by unauthorised users to have access to the raw watermarking channel" [45]. The watermarking systems could be compromised if an attacker manages to obtain the secret key. In this case, the attacker will have access to parameters such as the watermark embedding locations, random frequency of the watermark bits, and the threshold for embedding the watermark bits. The secret key can be predicted by gathering the characteristics of a set of watermarked images and analysing their similarities, to evaluate whether the same secret key and watermark bits have been used repeatedly [46]. Some of the problems associated with secret key leakage have been studied by a number of researchers [47–49].

3 Digital Image Watermarking Schemes Survey

3.1 Spatial Domain Based Watermarking Schemes

Digital image watermarking techniques can be divided into two categories: spatial and transform domain. In the spatial domain, the watermark bits are embedded into the original image by modifying the image pixel values directly. These types of techniques are easy to implement and have higher embedding capacity than transform domain based watermarking techniques. However, the watermarks embedded in spatial domain based watermarking techniques are not robust against various manipulations, such as JPEG compression, additive noise, filtering and geometric distortions.

Least Significant Bits (LSB) Watermarking

The least significant bit (LSB) is the bit value 0 or 1 located in the most right bit of the binary integer sequence that is converted from a pixel value. LSB watermarking is one of the first watermarking schemes operating in the spatial domain [50]. As seen from Figure 6, the image pixel values 135, 98 and 221 are first converted into 8 bits binary sequences. The three watermark bits 0, 1 and 1 are then inserted by replacing the LSB of each binary sequence. Finally, the modified binary sequences are converted back to pixel values 134, 99 and 221, respectively. In essence, the original pixel values can only be changed either by ± 1 pixel value, or not at all (if the LSB already matches the watermark bit) when they are watermarked using this technique. As the changes are relatively minor, the quality of the watermarked image is high, especially if limited amount of changes occur. However, the watermark bits can be easily attacked by randomly overwriting the LSB values of the watermarked image, thereby destroying the watermark data.

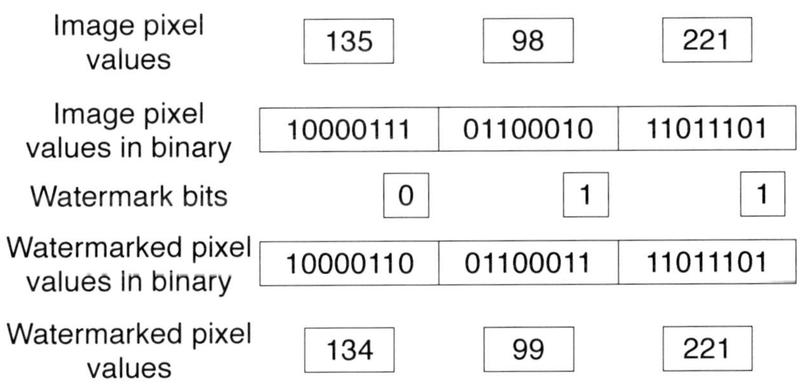

Fig. 6. An illustration of LSB watermark bits embedding.

3.2 Transform Domain Based Robust Watermarking Schemes

In contrast, transform domain schemes operate by embedding the watermark data into frequency coefficients of the image. Images in the transform domain have been comprehensively studied in the context of image coding and compression as well as digital watermarking [7, 52]. Image can be represented as frequency coefficients by mapping the pixel values. Low frequencies represent the overall shapes and outlines of features in the image, and its luminance and contrast characteristics, and high frequencies represent sharp edges and crispiness in the image, but contribute little spatial-frequency energy [53]. As some of the coefficients are not significantly distorted after some attacks (such as JPEG compression, additive noise, and filtering), robust watermarking schemes are commonly implemented in the transform domain [54–56]. One of first robust watermarking scheme based on spread spectrum was first proposed by Cox *et al.* [13]. In general, robust watermarking in transform domain can be classified into two groups: block based and non-block based algorithms. Figure 7 shows the difference between block and non-block based in watermark embedding process.

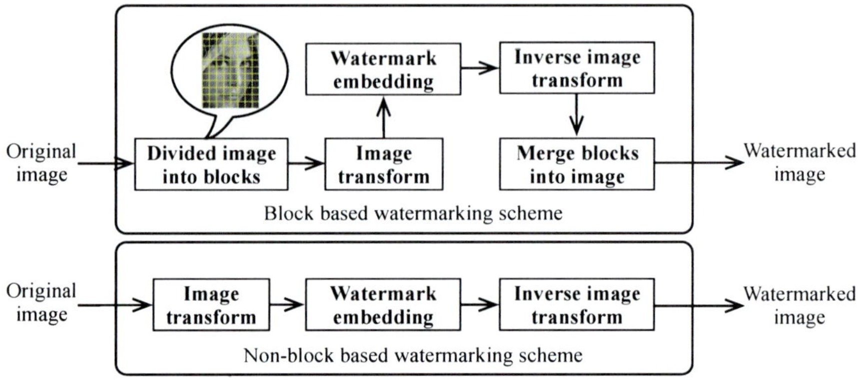

Fig. 7. Difference between block and non-block based in watermark embedding process.

Block Based Robust Watermarking

The most common image transform framework, the Discrete Cosine Transform (DCT), is frequently used for block based robust image watermarking [57–61]. Some researchers have also proposed adaptive watermarking schemes in the DCT domain based on image content. Perceptual models as supportive tools have been utilised to analyse the individual image content before the watermark bits are embedded into the DCT coefficients of each block, which could lead to an optimisation of the imperceptibility of the watermarked image [62].

Just Noticeable Distortion (JND) is designed to determine the maximum strength of a watermark signal that can be inserted into an image, as well as improving the quality of the watermarked image. Kay and Izquierdo [59] proposed a robust content based image watermarking scheme by estimating a JND mask from image characteristics such as texture, edges and smoothness, from both the spatial and DCT domains. The watermark embedding process is shown in Figure 8. The original image is first divided into non-overlapping blocks of size 8×8, then the JND mask is calculated from both spatial and DCT domains of each block. To embed the watermark bits, the selected DCT coefficients are modified according to the JND mask by using a key. An inverse DCT transformation is then applied to the modified DCT coefficients, and the blocks are merged back into a watermarked image.

In each block, the JND is derived in Equation 4.

$$D_T = log\left(\sum_{i=1}^{63} v_i^2 - v_0^2\right), D_E = \frac{64.P_E}{max\left(P_E\right)},$$
$$JND = \left(D_T - \frac{1}{2}\left(D_E + D_V\right)\right) + \left(128 - \bar{I}\right)^2 \tag{4}$$

where D_T is the texture information retrieved directly from DCT coefficients of the 8×8 block and v_i, $i = 0, ..., 63$ are the 64 DCT coefficients of the considered block. Edges and smooth areas D_E are extracted from the pixel domain, P_E is the cardinality

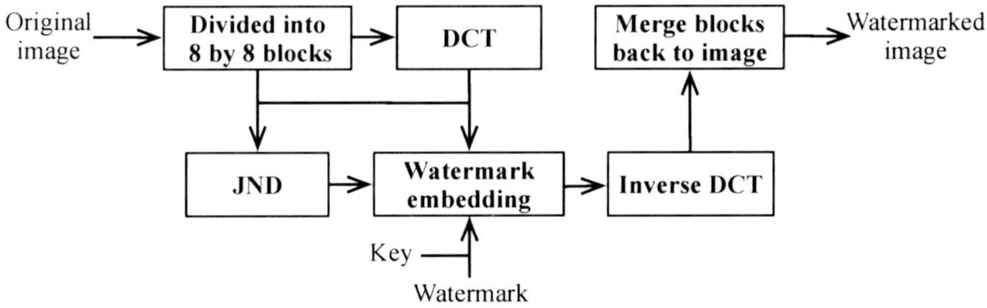

Fig. 8. Kay and Izquierdo watermark embedding process.

of the set of pixels within the block and at edge locations. The uniformity D_U in a block is defined as the number of pixels belonging to a uniform area that is extracted by the Moravec corner detection operator for each block. $\bar{I}$ is the mean of the luminance values of each block. In order to insert the watermark, the modified DCT coefficients v_i' are derived in Equation 5.

$$v_i' = v_i + \alpha.JND.\left|v_i\right| x_i \tag{5}$$

where α is a scaling parameter, and x is the watermark, consisting of a sequence of real pseudo-random numbers [59]. Another type of JND mask is generated from Watson's visual model which was utilised in Podilchuk and Zeng's image adaptive watermarking scheme [61], and then further adopted and extended by Hernandez *et al.* [60], Li and Cox [62] and Li *et al.* [63]. In Watson's visual model, the JND values are extracted by calculating the luminance and contrast masking of each block DCT coefficients of the image.

Wong *et al.* [64] proposed an iterative watermark embedding algorithm for JPEG compressed images capable of embedding multiple watermarks within the DCT domain with different keys. In addition, Dong *et al.* [65] also proposed two algorithms that embedded a multi-bit watermark in the DCT domain of the image. Their first algorithm utilised an image normalisation technique which was robust to general geometric transformation attacks. Their second algorithm utilised a resynchronisation scheme based on a mesh model to combat nonlinear geometric attacks. Moreover, in Yeo and Kim's scheme [66], a generalised patchwork algorithm (which is the combination of the additive patchwork algorithm and the multiplicative patchwork algorithm) was employed to embed the watermark bits in the DCT domain. Their experimental results showed that their method was robust against JPEG compression attacks and some signal processing attacks.

In order to increase the watermarked image quality as well as the watermark robustness, many researchers experimented with applying different image transform techniques to block based robust image watermarking schemes. Some examples include robust image watermarking in the Fast Hadamard Transform (FHT) which resulted in a much shorter processing time and simpler hardware implementation [67], Ho *et al.* [55] used Slant Transform (SLT) which provided significant advantage for watermark insertion and retrieval for images with complex textures such as satellite images,

Singular Value Decomposition (SVD) based scheme was found to be robust against typical attacks, such as low-pass and high-pass filtering [68]. A Curvelet Transform based scheme was proposed to overcome inherent limitations of traditional multiscale representations watermarking schemes in [69], and Xie *et al.* [70] chose the middle sub-band for embedding watermarks to achieve both good imperceptibility and robustness in Ridgelet Transform domain.

DCT Block Based Watermarking Tutorial

In this subsection a simple DCT block based watermarking tutorial is given with a sample MATLAB program provided. The watermark embedding and detection process are shown in Figures 9 and 10.

Watermark embedding process

Step 1: Read size of 512 by 512 grayscale image name "lena.bmp"

```
img=imread('lena.bmp');
```

Read Read size of 64 by 64 binary logo image name "logo.bmp" as the watermark.

```
logo_img=imread('logo.bmp');
```

Step 2: Divide into non-overlapping 8 by 8 blocks and then apply DCT to each block.

```
img_dct=blkproc(img,[8 8],@dct2);
```

Step 3: Watermark embedding

```
[r c]=size(img_dct);
k=1;
for i=0:r/8-1
    for j=0:c/8-1
        xx=img_dct(i*8+1:(i+1)*8,j*8+1:(j+1)*8);
        if logo_img(k)==1
                xx(5,5)=10;
            else
                xx(5,5)=-10;
        end
        y_dct(i*8+1:(i+1)*8,j*8+1:(j+1)*8)=xx; k=k+1;
    end
end
```

Step 4: Apply inverse DCT to each block to create watermarked image.

```
w_img=uint8(blkproc(y_dct,[8 8],@idct2));
imwrite(w_img, 'w_lena','bmp')
```

Watermark detection process

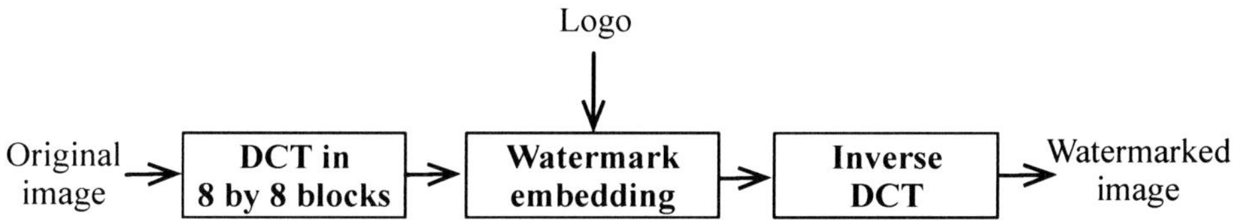

Fig. 9. A simple DCT block based watermark embedding process.

Step 1: Read size of 512 by 512 grayscale watermarked image name "w_lena.bmp"

```
img_w=imread('w_lena.bmp');
```

Step 2: Divide into non-overlapping 8 by 8 blocks and then apply DCT to each block.

```
img_w_dct=blkproc(img_w,[8 8],@dct2);
```

Step 3: Watermark Detection

```
[r c]=size(img_w_dct);
k=1;
for i=0:r/8-1
    for j=0:c/8-1
        yy=img_w_dct(i*8+1:(i+1)*8,j*8+1:(j+1)*8);
        if yy(5,5) > 0
            logo_img(k)=1;
        else
            logo_img(k)=0;
        end
    end
end
```

Step 4: Reconstruct the logo image

```
logo_new=reshape(logo_img,[64 64]); imwrite(logo_new,
'logo_new','bmp')
```

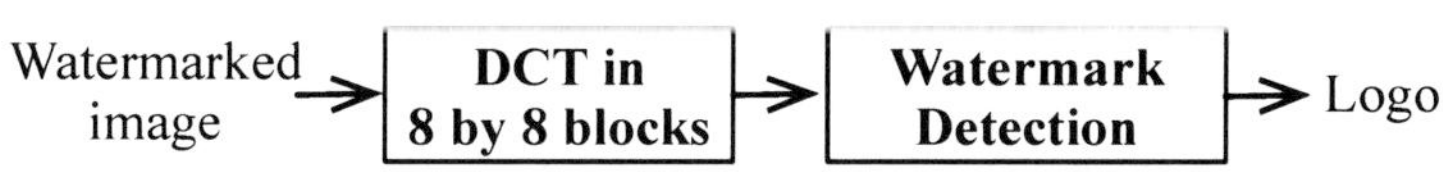

Fig. 10. A simple DCT block based watermark detection process.

Non-block based robust watermarking

One of the most popular image transform domains for non-block based robust watermarking is the discrete wavelet transform (DWT) [35, 54, 71–76]. In contrast to DCT, the original image is not divided into blocks in the DWT domain. DWT is one of the most computationally efficient frequency transforms that utilises the human visual

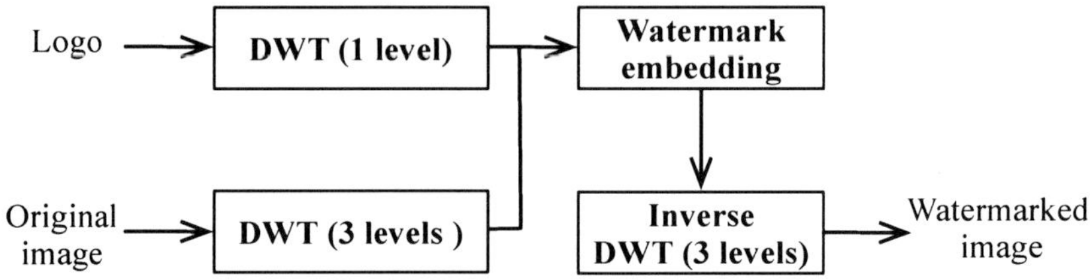

Fig. 11. Kundur and Hatzinakos's DWT watermark embedding process.

system (HVS). Moreover, in DWT based watermarking schemes, it is possible to embed watermarks with more energy thus significantly increasing their robustness [77]. A DWT domain based robust watermarking algorithm was proposed by Kundur and Hatzinakos [35], and their watermark embedding process is shown in Figure 11. In their scheme, both the original and logo images were first transformed into the wavelet domain and then decomposed into three levels (Figure 12(a)), and one level (Figure 12(b)), respectively. The DWT coefficients of the logo image were next embedded into the coefficients of the original image by using the multi-resolution fusion technique integrated with a model of HVS. Finally, the modified coefficients are transformed back into the spatial domain to create the watermarked image. In the experimental results, the authors claimed their proposed scheme was highly robust to compression and additive noise attacks, as well as being resilient to moderate linear mean filtering.

Two other schemes for DWT watermarking were proposed by Xia *et al.* [71] and Zhu *et al.* [78]. Both of these algorithms used a Gaussian sequence of pseudo-random real numbers as the watermark data, instead of a logo image. Xia *et al.* [71] proposed a watermarking scheme that decomposes the image into two levels in the DWT domain. The watermark was embedded into the middle and high-pass sub-bands in the wavelet domain of the image, denoted by HL_1, LH_1, HH_1, HL_2, LH_2 and HH_2, as shown in Figure 12(a). The authors claimed that their algorithm could tolerate additive noise, rescaling/stretching, compression attacks and that the algorithm was also more robust than the DCT approach. Zhu *et al.* [78] proposed a unified DWT watermarking approach that decomposed the original image into four levels in DWT domain. Next, the watermark was embedded into the high-pass sub-bands in the wavelet domain of the image, denoted by HL_1, LH_1 and HH_1 as shown in Figure 12(a).

In recent years, many researchers have attempted to develop watermarking algorithms based on the combination of two or more image transform techniques, with the aspiration of improving the schemes [56, 79–81]. For example, Yang and Zhang [80] improved the algorithm proposed by Lin *et al.* [82] in the distributed discrete wavelet transform (DDWT) domain. The DDWT technique is adapted largely from the DWT approach, along with SVD for robust watermarking. A fragile watermark is adaptively embedded into the spatial domain of the watermarked image. By embedding both robust and fragile watermarks into a single image, the method is capable not only of identifying the ownership, but also of authenticating the integrity of the image to deduce whether it has been tampered or not. Mabtoul *et al.* [81] implemented a robust watermarking

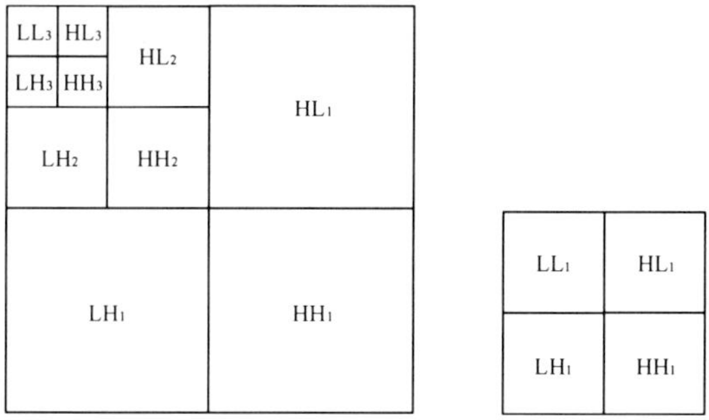

(a) 3 Levels (b) 1 Level

Fig. 12. Three levels and one level of DWT decomposition

scheme based on Kingsbury's Complex Dual Tree Wavelet Transform (DT-CWT) [83]. The aim of designing this scheme was to overcome the drawback of DWT caused by the lack of shift invariance and poor directional selectivity for diagonal features. Mabtoul et al claimed that the DT-CWT approach was more robust and effective than the DWT approach.

Wang *et al.* [84] proposed a novel feature-based watermarking scheme in the Discrete Fourier Transform (DFT), and their watermark embedding process is shown in Figure 13. In this scheme, the Local Characteristic Region (LCR) was first extracted from the original image. In the LCR extraction process, a set of feature points was obtained by employing the Harris-Laplace detector to the original image. Then LCRs were constructed from these characteristic scales of the feature points and their locations. The extracted LCRs were found to increase the robustness against various attacks such as signal processing and affine transformations. In order to embed the watermark bits into the DFT domain of the LCR, the zero-padding operation was applied to map the LCR circle areas into $n \times n$ blocks. Finally, after modifying the coefficients to embed the watermark bits in the DFT domain of these blocks, a zero-removing operation was applied to map these $n \times n$ blocks back into LCR circle areas to create the watermarked image. From the simulation results, the authors claimed that their proposed scheme was robust against common signal processing operators, such as median filtering, sharpening, noise adding, JPEG compression, rotation, scaling, translation, row or column removal, cropping and random bend attack.

4 Proposed Robust Watermarking Algorithm

In this section, we describe our proposed non-block based robust watermarking algorithm, adapted from the wavelet-based contourlet transform (WBCT) [14].

4.1 Wavelet-Based Contourlet Transform (WBCT)

The contourlet transform can be realised efficiently via a double-iterated filter bank structure. In the double filter bank, the Laplacian Pyramid (LP) [85] is first used to

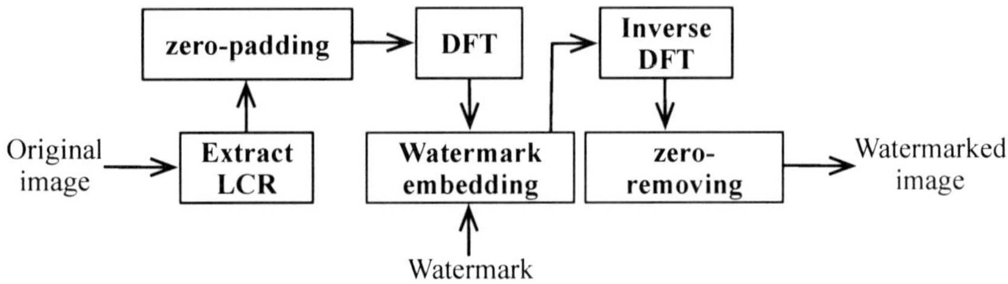

Fig. 13. Wang *et al.*'s watermark embedding process.

capture the point discontinuities. In the LP stage, the image is decomposed into a low-pass and a set of band-pass sub-bands. Each band-pass image is then further decomposed by a directional filter bank (DFB) [86] into a number of sub-bands to capture the directional information and link-point discontinuities into linear structures. Subsequently, the image is decomposed into several directional sub-bands at multiple scales.

Eslami and Radha [87] developed a WBCT, also as non-redundant contourlet transform, by replacing the LP with a wavelet, followed by implementing a directional filter bank (DFB) into the wavelet sub-bands to extract the directional information. As shown in Figure 14, at each level in the wavelet decomposition, the three high-pass bands corresponding to the LH, HL, and HH bands can be obtained. DFB is applied with the same number of directions to each band at a given level. WBCT was developed as an improvement to the wavelet transform that is inefficient when extracting smooth contours. It has the multiscale and time-frequency localisation property of wavelets, but it also provides a high degree of directionality and anisotropy [88]. The main advantage of WBCT is that a non-redundant multi-resolution and multidirectional expansion of images can be achieved. The transform has been successfully applied in image coding [87] and image fusion [89].

Eslami and Radha [87] stated that the WBCT parent-children relationship was different from the relationship that exists in conventional wavelet domains. In a conventional wavelet-domain, the parent-children links are always in the same direction among the three wavelet directions (15(a)). WBCT coefficients, on the other hand, comprise four children in two separate directional sub-bands for each LH, HL and HH sub-bands (15(b)). In Figure 15, the blank square is the parent coefficient and the four white squares (arrowed) are their children. This special relationship characteristic of WBCT is the basis for our novel robust watermarking algorithm.

4.2 Coefficient Relations after JPEG Compression of Original Images

Wang and Lin [90] proposed a watermarking method based on a wavelet tree quantisation and obtained a strong robustness to several different attacks. We first investigate the characteristics of the energy relations of the original images between the parent and the children coefficients before and after JPEG compression (the JPEG compression attack is one of the most common attacks in robust watermarking). For instance, if a relation between parent and the children coefficients is not changed after JPEG compression,

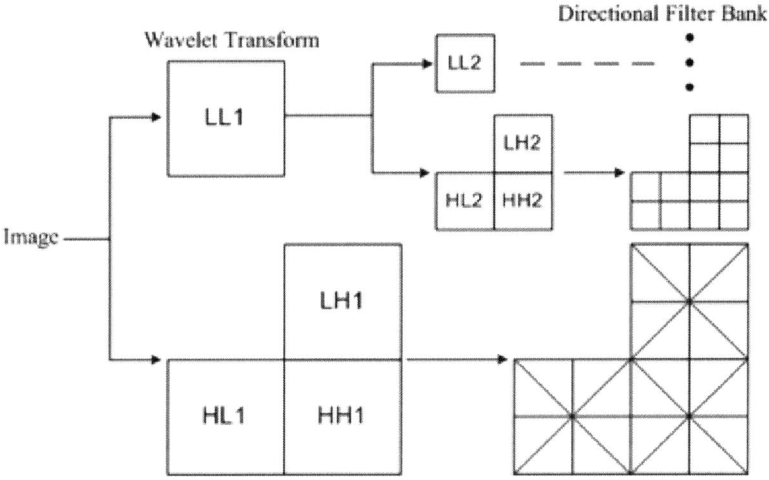

Fig. 14. The framework of the WBCT.

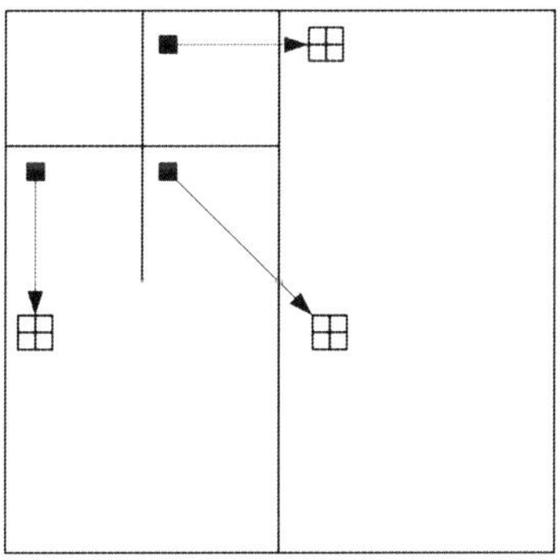

(a) DWT

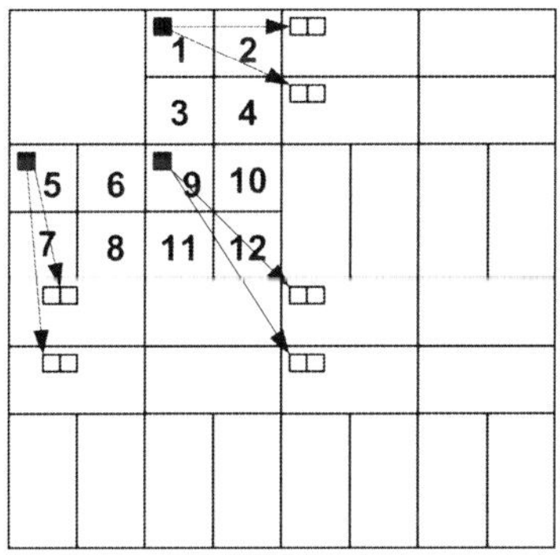

(b) WBCT

Fig. 15. Parent-children relationship for DWT and WBCT.

then we assume this energy relation is invariant. Six standard test images are used in this experiment to determine the invariant energy relationship as shown in Figure 16. As the quality factor (QF) decreases from 90 to 10, the average percentages of invariant relations also gradually decrease. For $QF = 90$, it reaches above 95%, and for $QF = 10$, although the image is distorted significantly, it still maintains above 75%. From Figure 16, it can also be observed that highly textured images such as 'San Diego', 'Bridge' and 'Baboon' all performed relatively better than the other images. Overall, an improved performance can be achieved for all images by exploiting the modulation of their energy relationship.

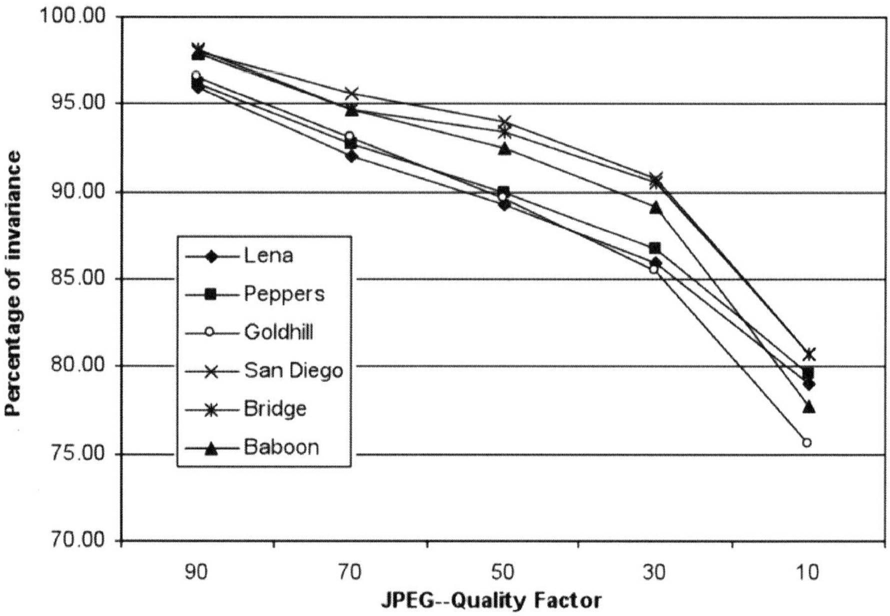

Fig. 16. Percentage of invariant energy relations after JPEG compression.

4.3 Watermark Embedding Process

In this subsection, we describe the watermark embedding process as shown in Figure 17. WBCT is first applied to the image, and then it is decomposed into three wavelet levels, then four, then eight, and finally sixteen directional sub-bands. We then randomly select a number of parent coefficients with their corresponding children coefficients using a key. The total number of these parent coefficients is equal to the length of the watermark, which is a pseudo-random binary sequence $\{-1, 1\}$. For each selected parent coefficient, we then embed the watermark bits by modulation as expressed in Equation 6.

$$P' = \begin{cases} P, & (|P| \geq |C_{avg}|) \wedge (w = 1) \\ P, & (|P| < |C_{avg}|) \wedge (w = -1) \\ P + K1 \times (|C_{avg}| - P), & (|P| < |C_{avg}|) \wedge (w = 1) \wedge (P \geq 0) \\ P - K1 \times (|C_{avg}| - |P|), & (|P| < |C_{avg}|) \wedge (w = 1) \wedge (P < 0) \\ P + K2 \times (P - |C_{avg}|), & (|P| \geq |C_{avg}|) \wedge (w = -1) \wedge (P \geq 0) \\ P - K2 \times (|P| - |C_{avg}|), & (|P| \geq |C_{avg}|) \wedge (w = -1) \wedge (P < 0) \end{cases} \quad (6)$$

where P' is the watermarked parent coefficient, P is the original parent coefficient, C_{avg} is the average of four children coefficients, w is the watermark bit, $K1$ and $K2$ are thresholds to determine the trade-off between imperceptibility and robustness. After the embedding steps, the watermarked image is reconstructed using the inverse WBCT transform.

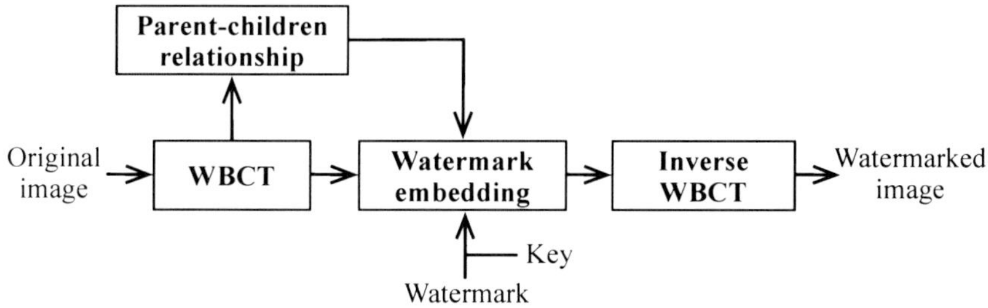

Fig. 17. The proposed WBCT watermark embedding process.

4.4 Watermark Detection Process

In Figure 20, we present our proposed algorithm for watermarking detection. First, the WBCT transform is performed on the watermarked image before the tree structures are selected using the key. The absolute value of the parent is compared with the absolute value of the average of the children, and if the former is greater or equal to the latter, then

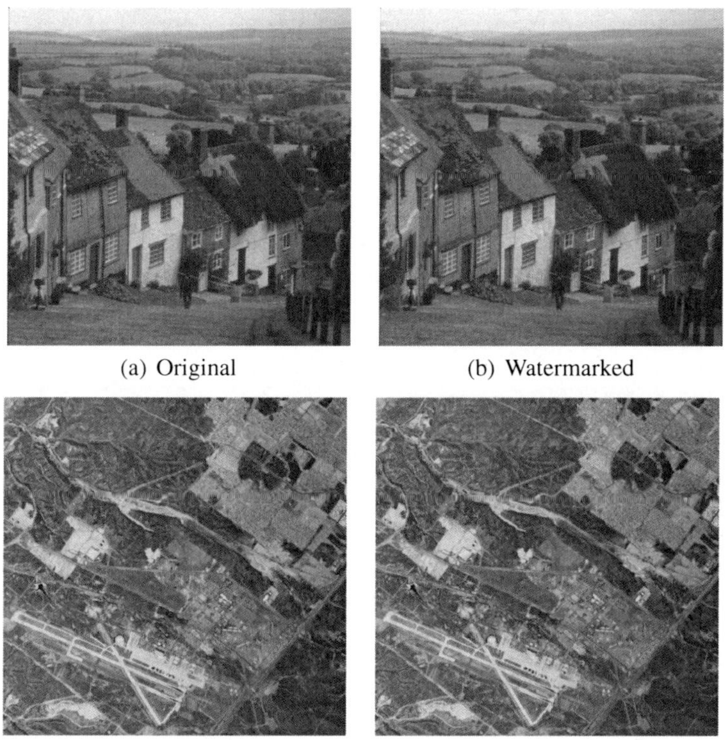

(a) Original (b) Watermarked

(c) Original (d) Watermarked

Fig. 18. 'Goldhill' - the original image and the watermarked image. 'San Diego' - the original image and the watermarked image.

Table 1. PSNR comparison between our and Wang' method.

Image	Our method (dB)	Wang's method (dB)
Lena	41.46	38.2
Peppers	39.24	38.7
Goldhill	42.22	39.8
San Diego	38.81	Not Available (NA)
Bridge	39.43	NA
Baboon	40.22	NA

the watermark bit '1' is obtained, otherwise, '-1' is obtained. The process is repeated for every tree structure to retrieve and construct the entire watermark data. A normalised correlation is used to determine whether a watermark is present or not, by comparing it to a pre-specified threshold. The normalized correlation is computed in Equation 7.

$$NC(w, \bar{w}) = \frac{\sum w(n)\, \bar{w}(n)}{\sqrt{\left(\sum w(n) \sum \bar{w}(n)\right)}} \tag{7}$$

where w is the given watermark, and $\bar{w}$ is the extracted watermark. If $NC \geq \tau$, then the watermark is present in the image. We adapt the threshold based on the false positive probability [91]. Based on empirical results for our algorithm, $N_w = 512$, τ is chosen to be 0.23 for a false positive probability of 1.03×10^{-7}.

4.5 Experimental Results

In this subsection, six grayscale images 'Lena', 'Peppers', 'Goldhill', 'San Diego', 'Bridge', and 'Baboon' (each of size 512×512) are used for our experiments to evaluate our proposed WBCT watermarking method.

Imperceptibility

In Table 1, the Peak signal-to-noise ratio (PSNR) is used to evaluate the perceptual distortion of these images before and after watermark embedding. High PSNR values indicate that the watermarked data is highly imperceptible. We compare these results with Wang and Lin [90]. Our proposed contourlet method achieves higher PSNR values

Table 2. Performance of our method under different attacks.

Image	JPEG ($QF = 10$)	Mean Filter (3×3)	Mean Filter(5×5)	Histogram Equalization
Lena	0.28	0.54	0.29	0.47
Peppers	0.26	0.60	0.23	0.45
Goldhill	0.34	0.60	0.32	0.55
San Diego	0.52	0.59	0.29	0.60
Bridge	0.48	0.63	0.33	0.67
Baboon	0.45	0.65	0.30	0.77

Table 3. Comparison of our method's performance with two other Methods

Attacks	Image	Wang's Method	Tsai's Method	Our Method
JPEG ($QF = 30$)	1	0.15	0.45	0.55
	2	0.23	0.44	0.54
	3	0.34	0.37	0.63
JPEG ($QF = 25$)	1	NA	0.37	0.45
	2	NA	0.29	0.56
	3	NA	0.23	0.50
JPEG($QF = 30$)	1	0.15	0.45	0.55
	2	0.23	0.44	0.54
	3	0.34	0.37	0.63
Median Filter (4×4)	1	0.23	0.38	0.28
	2	0.24	0.33	0.27
	3	0.25	0.36	0.30
Median Filter (5×5)	1	NA	0.43	0.29
	2	NA	0.32	0.25
	3	NA	0.41	0.29
Shifting A (9 pixels)	1	0.26	0.27	0.79
	2	0.29	0.35	0.86
	3	0.29	0.36	0.90
Shifting B (9 pixels)	1	0.25	0.29	0.54
	2	0.25	0.26	0.56
	3	0.28	0.31	0.54
Multiple Watermarking	1	0.11	0.24	0.93
	2	0.18	0.29	0.93
	3	0.22	0.25	0.91
Scale & Rotation ($1°$)	1	0.24	0.25	0.13
	2	0.15	0.25	0.04
	3	0.17	0.26	0.04
Scale & Rotation ($-0.75°$)	1	0.24	0.30	0.06
	2	0.25	0.38	0.03
	3	0.25	0.30	0
Gaussian Filtering	1	0.64	0.89	0.83
	2	0.56	0.91	0.91
	3	0.74	0.92	0.90
Sharpening	1	0.46	0.87	0.90
	2	0.39	0.63	0.71
	3	0.62	0.89	0.91

Where '1' means Lena image, '2' means Peppers image, '3' means Goldhill image.

than the other two wavelet methods, respectively. Figure 18 illustrates the original and watermarked images for 'Goldhill' and 'San Diego', with approximately $42dB$ and $39dB$, respectively.

Fig. 19. Attacked watermarked image 'Goldhill' - (a) JPEG Compression QF=20, (b)400 pixel random shifted, (c) 20 degree of contrast stretching, (d) Gaussian white noise, variance=0.01.

Robustness

The robustness of our proposed WBCT watermarking method has been tested against different attacks including non-geometrical and geometrical attacks. Furthermore, we compare these results with two conventional wavelet approaches based on Wang and Lin [90] and Tsai and Lin [92]. The experiments on WBCT against different attacks are summarised in Table 2, and comparative results between WBCT and the two wavelet transforms are shown in Table 3. The normalised correlation value below a threshold of approximately 0.23 means it has failed to detect the embedded watermark.

For JPEG and JPEG2000 compression attacks, different quality factors (QF) were used on watermarked images. Tables 2 and 3 show the effectiveness of our algorithm even when $QF = 10$, whereas the other two algorithms only given the results as $QF = 20$ and $QF = 30$. Tables 2 and 3 also highlight the results achieved for mean filtering, histogram equalisation, median filter, Gaussian filtering, and sharpening. From the results, our method outperformed the other two algorithms in all cases, with median filtering the single exception. Table 3 summarises the different non-geometrical and geometrical attacks used in comparing our method with the other two methods. In Table 3, "shifting A" indicates circular shifting, and "shifting B" indicates a deletion of lines followed by duplication of the adjacent lines. Figure 19 illustrates the watermark can still be detected from watermarked images 'Goldhill' after attacked with JPEG Compression $QF = 20$ (Figure 19(a)), 400 pixels random shifted (Figure 19(b)), 20 degree of contrast stretching (Figure 19(c)) and Gaussian white noise with $variance = 0.01$ (Figure 19(d)), respectively. Overall, our method achieved relatively better performance than Wang [90] and Tsai [92] except for the rotation and scaling attacks.

5 Conclusion and Future Work

In this chapter, the copyright infringement issues and the use of robust watermarking as a solution for copyright protection and ownership identification were discussed. The fundamental concept of digital watermarking was described. In spatial domain methods, the LSB watermarking algorithm was briefly discussed. For transform domain methods, DCT domain block based and DWT non-block based, robust watermarking algorithms were reviewed in detail and a discussion on other state-of-the-art transform based schemes was provided.

A novel robust image watermark embedding and detection algorithm in wavelet-based contourlet transform domain were presented. Through experiments, most of the energy relations between parent and children non-redundant contourlet coefficients maintained 75% of invariance before and after JPEG compression $QF = 10$, although the image was distorted significantly. Therefore, performance improvement was obtained by means of embedding a watermark exploiting the modulation of the energy relations. By comparing with two other wavelet methods, the experimental results showed that our non-redundant contourlet method was more robust to attacks such as JPEG, JPEG2000 compression, pixel shifting, mean filtering, histogram equalisation, median filter, Gaussian filtering, and sharpening.

For future work, we plan to improve our proposed algorithms further by improving the watermark robustness against different forms of mild signal processing attacks. The embedding method could be further improved by analyzing the parent and children WBCT coefficients before and after different attacks, such as, Gaussian white additive noise, different image filtering operations and some geometrical modifications. In addition, the watermark detection results could also be improved by having a more adaptive watermark detection threshold according to different forms of attacks.

Currently, many researchers are focusing on the analysis of the characteristics of the image to find the invariant areas for embedding the watermarks. This trend could lead to a better design and development of robust watermarking scheme against geometrical attacks such as print-and-scan process. Another potential area is the use of pattern recognition and neural network techniques to determine the type of attacks from the retrieved watermark patterns which could further enhance the development of robust image watermarking algorithms.

References

[1] Oxford Economics, Economic impact of legislative reform to reduce audio-visual piracy (2009),
http://www.bva.org.uk/files/images/AV_Piracy_Final_Report_-_FINAL.pdf (accessed June 28, 2009)

[2] The independent, P2P geek caught in $29m music copyright scrap (2009),
http://www.independent.co.uk/life-style/gadgets-and-tech/news/p2p-geek-caught-in-29m-music-copyright-scrap-1701449.html (accessed June 28, 2009)

[3] Zheng, D., Liu, Y., Zhao, J., et al.: A survey of RST invariant image watermarking algorithms. ACM Computing Surveys (2007), doi:10.1145/1242471.1242473

[4] Department for Culture, Media and Sport, The impact of new technologies on the music industry (2009), http://www.culture.gov.uk/images/publications/consumerscalltune.pdf (accessed June 28, 2009)

[5] Department for Culture, Media and Sport, UK film council response to digital Britain: The Interim Report (2009), http://www.culture.gov.uk/images/publications/UKFilmCouncil_DBIRResponse.pdf (accessed June 28, 2009)

[6] Hunter, D.: Papermaking: The History and Technique of an Ancient Craft. Gannon Distributing Co. (1978)

[7] Cox, I.J., Miller, M.L., Bloom, J.A., et al.: Digital watermarking and Steganography, 2nd edn. Morgan Kaufmann, USA (2008)

[8] Zhang, J., Ho, A.T.S., Qiu, G., et al.: Robust video watermarking of H.264/AVC. IEEE Trans. on Circuits and Systems (2007), doi:10.1109/TCSII.2006.886247

[9] Noorkami, M., Merserau, R.M.: A Framework for Robust Watermarking of H.264-Encoded Video With Controllable Detection Performance. IEEE Trans. on Information Forensics and Security (2007), doi:10.1109/TIFS.2006.890306

[10] Gong, X., Lu, H.M.: Towards Fast and Robust Watermarking Scheme for H.264 Video. In: 10th IEEE International Symposium on Multimedia (2008), doi:10.1109/ISM.2008.16

[11] Kirovski, D., Malvar, H.S.: Spread-Spectrum Watermarking of Audio Signals. IEEE Transactions on Signal Processing (2003), doi:10.1109/TSP.2003.809384

[12] Kumsawat, P., Attakitmongcol, K., Srikaew, A.: A robust audio watermarking based-on multiwavelet transform. In: International Symposium on Intelligent Signal Processing and Communications Systems, ISPACS 2008 (2009), doi:10.1109/ISPACS.2009.4806664

[13] Cox, I.J., Kilian, J., Leighton, T., et al.: Secure Spread Spectrum Watermarking for Multimedia. IEEE Trans. on Image Processing 6(12), 1673–1687 (1997)

[14] Duan, G., Ho, A.T.S., Zhao, X.: A novel non-redundant contourlet transform for robust image watermarking against non-geometrical and geometrical attacks. In: Proc. IET 5th International Conference on Visual Information Engineering (2008), doi:10.1049/cp:20080295

[15] Zhou, Y., Yang, E.H.: Joint robust watermarking and compression using variable-rate scalar quantization. In: Canadian Workshop on Information Theory, CWIT 2009 (2009), doi:10.1109/CWIT.2009.5069550

[16] Luo, H., Zhao, Z., Huang, J., et al.: Halftone Image Watermarking Based on the Binary Pseudo-Wavelet Transform. In: Proc. 3rd International Conference on International Information Hiding and Multimedia Signal Processing (2007), doi:10.1109/IIHMSP.2007.4457549

[17] Hsieh, C.T., Wu, Y.K., Chung, W.J.: Digital watermarking system for halftone images based on particle swarm optimization. In: 1st IEEE International Conference on Ubi-Media Computing (2008), doi:10.1109/UMEDIA.2008.4570890

[18] Jiang, W., Ho, A.T.S., Treharne, H.: A Novel Least Distortion Linear Gain Model for Halftone Image Watermarking Incorporating Perceptual Quality Metrics. In: Shi, Y.Q. (ed.) Transactions on DHMS IV. LNCS, vol. 5510, pp. 65–83. Springer, Heidelberg (2009)
[19] Qadir, M.A., Ahmad, I.: Digital Text Watermarking: Secure Content Delivery and Data Hiding in Digital Documents. In: 39th Annual 2005 International Carnahan Conference on Security Technology, CCST 2005 (2005), doi:10.1109/MAES.2006.284353
[20] Culnane, C., Treharne, H., Ho, A.T.S.: Authenticating Binary Text Documents Using a Localising OMAC Watermark Robust to Printing and Scanning. In: Shi, Y.Q., Kim, H.-J., Katzenbeisser, S. (eds.) IWDW 2007. LNCS, vol. 5041, pp. 173–187. Springer, Heidelberg (2008)
[21] Zafeiriou, S., Tefas, A., Pitas, I.: Blind robust watermarking schemes for copyright protection of 3D mesh objects. IEEE Trans. on Visualization and Computer (2005), doi:10.1109/TVCG.2005.71
[22] Motwani, R., Ambardekar, A., Motwani, M., et al.: Robust watermarking of 3D skinning mesh animations. In: IEEE International Conference on Acoustics, Speech and Signal Processing, ICASSP 2008 (2008), doi:10.1109/ICASSP.2008.4517969
[23] Hu, R., Rohao-Alface, P., Macq, B.: Constrained optimisation of 3D polygonal mesh watermarking by quadratic programming. In: IEEE International Conference Acoustics, Speech and Signal Processing, ICASSP 2009 (2009), doi:10.1109/ICASSP.2009.4959880
[24] Cheng, C.J., Lin, L.C., Dai, W.T.: Construction and detection of digital holographic watermarks. Optics Communications (2005), doi:10.1016/j.optcom.2004.12.010
[25] Lin, L.C., Chen, C.L.: Statistical detection of digital holographic watermarking system. Optics Communications (2008), doi:10.1016/j.optcom.2008.04.065
[26] Huang, S., Wu, J.K.: Optical watermarking for printed document authentication. IEEE Trans. on Information Forensics and Security (2007), doi:10.1109/TIFS.2007.897255
[27] Lu, Y., Li, X., Qi, W., et al.: A further study on an optical watermarking scheme. IEEE Signal Processing (2008), doi:10.1109/ICOSP.2008.4697588
[28] Zhang, J., Kou, W., Fan, K.: Secure buyer-seller watermarking protocol. IEE Proceedings Information Security (2006), doi:10.1049/ip-ifs:20055069
[29] Zander, S., Armitage, G., Branch, P.: A survey of covert channels and countermeasures in computer network protocols. IEEE Communications Surveys and Tutorials (2007), doi:10.1109/COMST.2007.4317620
[30] Ho, A.T.S., Zhu, X., Shen, J., et al.: Fragile Watermarking Based on Encoding of the Zeroes of the z-Transform. IEEE Transactions, Information Forensics and Security (2008), doi:10.1109/TIFS.2008.926994
[31] Izquierdo, E.: Fragile watermarking for image authentication. In: Furht, B. (ed.) Multimedia Security Handbook. CRC Press, Boca Raton (2005)
[32] Alomari, R., Al-Jaber, A.: A Fragile watermarking algorithm for content authentication. International Journal of Computing and Information Science 2(1) (2004)
[33] Byun, S., Lee, I., Shin, T.: A public key based watermarking for color image authentication. In: IEEE international conference on multimedia and expo. (2002), doi:10.1109/ICME.2002.1035851
[34] Fridrich, J.: Methods for tamper detection in digital images. In: The multimedia and security workshop at ACM multimedia (1999)
[35] Kundur, D., Hatzinakos, D.: A robust digital image watermarking method using wavelet-based fusion. In: Proc. International Conference Image Processing (1997), doi:10.1109/ICIP.1997.647970
[36] Ho, C.K., Li, C.T.: Semi-Fragile Watermarking Scheme for Authentication of JPEG Images. In: International Conference on Information Technology: Coding and Computing (2004), doi:10.1109/ITCC.2004.1286417

[37] Zhu, X., Ho, A.T.S., Marziliano, P.: A new semi-fragile image watermarking with robust tampering restoration using irregular sampling. Elsevier Signal Processing, Image Communication (2007), doi:10.1016/j.image.2007.03.004
[38] Rey, C., Dugelay, J.L.: A survey of watermarking algorithms for image authentication. EURASIP Journal on Applied Signal Processing (2002), doi:10.1155/S1110865702204047
[39] Bartolini, F., Tefas, A., Barni, M., et al.: Image authentication techniques for surveillance applications. Proc. of IEEE (2001), doi:10.1109/5.959338
[40] Wu, H.R., Rao, K.R.: Digital video image quality and perceptual coding. CRC Press, Boca Raton (2005)
[41] Gerhard, D.B., Kinsner, W.: Lossy compression of head and shoulder images using zerotrees of wavelet coefficients. In: Canadian Conference on Electrical and Computer Engineering (1996), doi:10.1109/CCECE.1996.548129
[42] Voloshynovskiy, S., Herrigel, A., Baumgaertner, N., et al.: A stochastic approach to content adaptive digital image watermarking. In: Pfitzmann, A. (ed.) IH 1999. LNCS, vol. 1768, pp. 211–236. Springer, Heidelberg (2000)
[43] Wang, Z., Bovik, A.C., Sheikh, H.R., et al.: Image quality assessment: from error visibility to structural similarity. IEEE Transactions on Image Processing (2004), doi:10.1109/TIP.2003.819861
[44] Abdel-Aziz, B., Chouinard, J.-Y.: On perceptual quality of watermarked images - an experimental approach. In: Kalker, T., Cox, I., Ro, Y.M. (eds.) IWDW 2003. LNCS, vol. 2939, pp. 277–288. Springer, Heidelberg (2004)
[45] Kalker, T.: Considerations on watermarking security. In: IEEE Fourth Workshop on Multimedia Signal Processing (2001), doi:10.1109/MMSP.2001.962734
[46] Perez-Freire, L., Perez-Gonzalez, F.: Spread-Spectrum Watermarking Security. IEEE Trans. on Information Forensics and Security (2009), doi:10.1109/TIFS.2008.2009603
[47] Barni, M., Bartolini, F., Furon, T.: A general framework for robust watermarking security. Elsevier, North-Holland, Inc., Amsterdam (2003)
[48] Cayre, F., Fontaine, C., Furon, T.: Watermarking security: theory and practice. IEEE Trans. on Signal Processing (2005), doi:10.1109/TSP.2005.855418
[49] Fei, C., Kundur, D., Kwong, R.H.: Analysis and design of secure watermark-based authentication systems. IEEE Trans. on Information Forensics and Security (2006), doi:0.1109/TIFS.2005.863505
[50] Van Schyndel, R.G., Tirkel, A.Z., Osborne, C.F.: A digital watermark. IEEE Int. Conf. on image processing 2, 86–90 (1994)
[51] Fridrich, J.: Security of fragile authentication watermarks with localization. In: Proc. SPIE, Security and Watermarking of Multimedia Contents IV, vol. 4675, pp. 691–700 (2002)
[52] Bow, S.T.: Pattern Recognition and Image Preprocessing. CRC Press, Boca Raton (2002)
[53] Muharemagic, E., Furht, B.: Survey of Watermarking Techniques and Applications. In: Multimedia Watermarking Techniques and Applications. CRC Press, Boca Raton (2006)
[54] Luo, K., Tian, X.: A New Robust Watermarking Scheme based on Wavelet Transform. In: 2008 Congress on Image and Signal Processing (2008), doi:10.1109/CISP.2008.739
[55] Ho, A.T.S., Zhu, X., Guan, Y., et al.: Slant transform watermarking for textured images. In: Proc. 2004 International Symposium on Circuits and Systems (2004), ISBN: 078038251X
[56] Kang, K., Huang, J., Shi, Y., et al.: A DWT-DFT Composite Watermarking Scheme Robust to Both Affine Transform and JPEG Compression. IEEE Trans. on circuits and systems for video technology (2003), doi:10.1109/TCSVT.2003.815957
[57] Wu, J., Sun, Z.: Robust Detection of DCT-domain Watermarks Using Alpha-stable Models. In: 4th IEEE Conference on Industrial Electronics and Applications (2009), doi:10.1109/ICIEA.2009.5138913

[58] Al-Gindy, A.N., Tawfik, A., Al Ahmad, H., et al.: A New Blind Image Watermarking Technique for Dual Watermarks Using Low-Frequency Band DCT Coefficients. In: 14th IEEE International Conference on Electronics, Circuits and Systems (2007), doi:10.1109/ICECS.2007.4511047

[59] Kay, S., Izquierdo, E.: Robust content based image watermarking. In: Proc. the Workshop on Image Analysis for Multimedia Interactive Services (2001)

[60] Hernandez, J.R., Perez-Gonzalez, F., Amado, M., et al.: DCT-Domain Watermarking Techniques for Still Images: Detector Performance Analysis and a New Structure. IEEE Trans. on Image Processing (2000), doi:10.1109/83.817598

[61] Podilchuk, C.I., Zeng, W.: Image-Adaptive Watermarking Using Visual Models. IEEE Journal on Selected Areas in Communications (1998), doi:10.1109/49.668975

[62] Li, Q., Cox, I.J.: Using Perceptual Models to Improve Fidelity and Provide Resistance to Valumetric Scaling for Quantization Index Modulation Watermarking. IEEE Trans. on Information Forensics and Security (2007), doi:10.1109/TIFS.2007.897266

[63] Li, Y., Zhu, H., Yu, R., et al.: An Adaptive Blind Watermarking Algorithm Based on DCT and Modified Watsons Visual Model. In: 2008 International Symposium on Electronic Commerce and Security (2008), doi:10.1109/ISECS.2008.83

[64] Wong, P.H.W., Au, O.C., Yeung, Y.M.: A Novel Blind Multiple Watermarking Technique for Images. IEEE Trans. on Circuits and Systems for Video Technology (2003), doi:10.1109/TCSVT.2003.815948

[65] Dong, P., Brankov, J.G., Galatsanos, N.P., et al.: Digital Watermarking Robust to Geometric Distortions. IEEE Trans. on Image Processing (2005), doi:10.1109/TIP.2005.857263

[66] Yeo, I.K., Kim, H.J.: Generalized patchwork algorithm for image watermarking. Journal of Multimedia Systems (2003), doi:10.1007/s00530-003-0097-0

[67] Ho, A.T.S., Shen, J., Chow, K.K., et al.: Robust Digital Image-in-Image Watermarking Algorithm Using the Fast Hadamard Transform. In: 2003 IEEE International Symposium on Circuits and Systems (2003), doi:10.1117/12.451250

[68] Calagna, M., Guo, H., Mancini, L., et al.: A robust watermarking system based on SVD compression. In: Proc. The Symposium on Applied Computing (2006), doi:10.1145/1141277.1141590

[69] Tao, P., Dexterb, S., Eskicioglu, A.M.: Robust Digital Image Watermarking in Curvelet Domain. In: Proc. SPIE, Security, Forensics, Steganography, and Watermarking of Multimedia Contents (2008), doi:10.1117/12.765895

[70] Xie, Z., Wang, S., Gan, L., et al.: Content Based Image Watermarking in the Ridgelet Domain. In: 2008 International Symposium on Electronic Commerce and Security (2008), doi:10.1109/ISECS.2008.32

[71] Xia, X., Boncelet, C., Arce, G.: Wavelet transform based watermark for digital images. Optics Express (1998), doi:10.1364/OE.3.000497

[72] Huang, D., Liu, J., Huang, J., et al.: A DWT-based image watermarking algorithm. In: IEEE International Conference on Multimedia and Expo. (2001), doi:10.1109/ICME.2001.1237719

[73] Yu, J., Chi, J., Zhuang, X.: A New Wavelet-based Robust Watermarking for Digital Image. In: IEEE International Conference on Networking, Sensing and Control (2008), doi:10.1109/ICNSC.2008.4525436

[74] Rawat, S., Bhatnagar, G., Raman, B.: A Robust Watermarking Scheme using Best Tree Wavelet Packet Transform. In: IEEE International Advance Computing Conference (2009), doi:10.1109/IADCC.2009.4809132

[75] Sun, D.: A Method to Estimate Watermark Invisibility before Embedding in DWT Domain. In: International Conference on Networking and Digital Society (2009), doi:10.1109/ICNDS.2009.135

[76] Taherinia, A., Jamzad, M.: A New Adaptive Watermarking Attack in Wavelet Domain. In: International Multimedia, Signal Processing and Communication Technologies (2009), doi:10.1109/MSPCT.2009.5164240
[77] Langelaar, G., Setyawan, I., Lagendijk, R.L.: Watermarking Digital Image and Video Data. IEEE Signal Processing Magazine (2000), doi:10.1109/79.879337
[78] Zhu, W., Xiong, Z., Zhang, Y.Q.: Multiresolution watermarking for images and video. IEEE Trans. on Circuits and Systems for Video Technology (1998), doi:10.1109/ICIP.1998.723530
[79] Santhi, V., Rekha, N., Tharini, S.: A hybrid block based watermarking algorithm using DWT-DCT-SVD techniques for color images. In: International Conference on Computing, Communication and Networking (2008), doi:10.1109/ICCCNET.2008.4907259
[80] Yang, H., Zhang, T.: A New Algorithm of Compound Image Watermarking Based on DDWT. In: International Conference on MultiMedia and Information Technology (2008), doi:10.1109/MMIT.2008.88
[81] Mabtoul, S., Ibn Elhaj, E., Aboutajdine, D.: A Robust Digital Image Watermarking Method Using Dual Tree Complex Wavelet Transform. In: IEEE Symposium on Computers and Communications (2008), doi:10.1109/ISCC.2008.4625651
[82] Lin, C.H., Jen, J.S., Kuo, L.C.: Distributed Discrete Wavelet Transformation for Copyright Protection. In: 7th International Workshop on Image Analysis for Multimedia (2006)
[83] Kingsbury, N.G.: The dual-tree complex wavelet transform: a new technique for shift invariance and directional filters. In: IEEE Digital Signal Processing Workshop (1998)
[84] Wang, X., Jun Wu, J., Niu, P.: A New Digital Image Watermarking Algorithm Resilient to Desynchronization Attacks. IEEE Trans. on Information Forensics and Security (2007), doi:10.1109/TIFS.2007.908233
[85] Burt, P.J., Adelson, E.H.: The Laplacian pyramid as a compact image code. IEEE Trans. Commun. 31(4), 532–540 (1983)
[86] Bamberger, R.H., Smith, M.J.T.: A filter bank for the directional decomposition of images: Theory and design. IEEE Trans. Signal Proc.1992), doi:10.1109/78.127960
[87] Eslami, R., Radha, H.: Wavelet-based Contourlet Transform and its Application to Image coding. In: Proc. IEEE International Conference on Image Processing (2004), doi:10.1109/ICIP.2004.1421791
[88] Do, M.N., Vetterli, M.: The contourlet transform: An efficient directional multiresolution image representation. IEEE Trans. Image Processing (2005), doi:10.1109/TIP.2005.859376
[89] Tang, L., Zhao, Z.: Multiresolution image fusion based on the wavelet-based contourlet transform. In: Proc. International Conference on Information Fusion (2007), doi:10.1109/ICIF.2007.4407989
[90] Wang, S.H., Lin, Y.P.: Wavelet tree-quantization for copyright protection watermarking. IEEE Trans. on Image Processing (2004), doi:10.1109/TIP.2004.823822
[91] Kundur, D., Hatzinakos, D.: Digital watermarking using multiresolution wavelet decomposition. In: Proc. of IEEE International Conference on Acoustics, Speech and Signal Processing (1998), doi:10.1109/ICASSP.1998.678149
[92] Tsai, M., Lin, C.: Constrained wavelet tree quantization for image watermarking. In: Proc. IEEE International Conference on Communication (2007), doi:10.1109/ICC.2007.227

Watermark-Based Authentication

Angela D'Angelo[1], Giacomo Cancelli[2], and Mauro Barni[3]

[1] Department of Information Engineering, University of Siena
angela.dangelo@unisi.it
[2] Department of Information Engineering, University of Siena
giacomo.cancelli@unisi.it
[3] Department of Information Engineering, University of Siena
barni@dii.unisi.it

1 Introduction

The large variety of signal processing tools available today permits to anyone to easily modify digital multimedia objects like audio, still images or video without leaving any perceptible trace of the modifications. Even though these tools are extremely useful in several scenarios, e.g. to increase the perceived quality of multimedia contents, there are cases in which we would like to be sure that a digital content is authentic, i.e. it corresponds to the original version. Stated in another way, we would like to find a way to avoid the loss of credibility of digital data. Authentication is the science that studies how to protect the integrity of digital media. Generally speaking, there are two basic forms of authentication: passive and active authentication. Passive authentication, also called forensic analysis, tries to understand whether a digital content has been tampered with by using statistical analysis without previously adding an authentication signal to the digital media. Passive authentication has the desirable characteristic of working with virtually any type of data, without requiring that they are modified at the time of creation. On the negative side, passive authentication is not always possible and doubts exist about its reliability and security. This is not the case with active authentication, whereby the integrity of a digital content is protected (and demonstrated) by embedding an authenticating signal within the digital content itself before sharing it with other users. Active authentication is also called watermark-based authentication, since it uses watermarking technology to embed the authenticating signal within the to-be-protected support.

The list of practical scenarios virtually interested by watermarking-based authentication is endless. The most obvious applications are related to forensics applications, wherein a strong insurance of image authenticity is needed due to the importance that images may have as evidence in a court. Similar to this, watermarking-based authentication has recently been proposed as a way to provide the originality of digital videos acquired within a video surveillance system [1].

Another possibility is to use the embedded watermark to guarantee the correct recovery of digital contents transmitted through an error-prone channel. As

H.T. Sencar et al. (Eds.): Intel. Multimedia Analysis for Security Appli., SCI 282, pp. 365–402.
springerlink.com

suggested in [2], the embedded watermark may also be used to recover part of the information lost because of transmission errors.

Watermarking-based authentication may also be used to prove the provenance and hence the originality of digital contents purchased or distributed through non-trusted channels, e.g. on the internet. In many cases, in fact, knowing the origin of a digital asset and proving its integrity is a good way to infer information about the quality of the asset itself, and the appropriateness of its cost or the lawfulness of a commercial transactions.

This chapter focuses on watermarking-based authentication, specifically on still image authentication. In particular in Section 2 we introduce the main ideas behind watermark-based authentication, by pausing to discuss the main differences between fragile, semi-fragile and robust authentication and by reviewing some of the most popular attacks the authentication systems should overcome. In Section 3, we review the relevant state of art. In Section 4 we describe a practical system for watermark-based authentication, whereby it is possible to verify the authenticity of still images and to detect tampered regions in which the analyzed image is different from the original one. In section 5, we draw some conclusions and highlight some trends and directions for future research.

2 Generic Framework

In this section we first review some basic digital watermarking concepts, then we show how watermarking technology can be used to authenticate multimedia contents.

2.1 Watermarking in a Nutshell

The general model of a data hiding system is given in figure 1. At the input of the system we find the to-be-hidden information and the original - cover - signal A. The cover signal may be an audio file, a still image, a piece of video or a combination of the above. The to-be-hidden information usually takes the

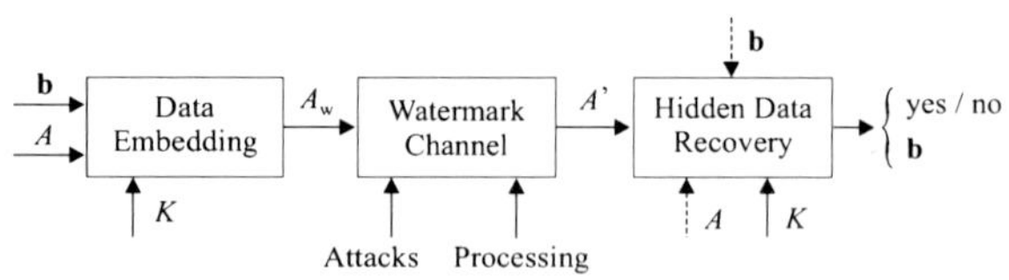

Fig. 1. Overall picture of a data hiding system. The watermark message **b** is embedded into the host signal A, producing the watermarked signal A_w. Due to possible attacks, A_w is transformed into A'. The hidden information is recovered from A', either by extracting the hidden message **b** or by deciding whether A' contains a known message **b** or not. Watermark embedding and recovery require the knowledge of a secret key K. Watermark recovery may benefit from the knowledge of the original, non-marked signal A.

form of a binary string $\mathbf{b} = (b_1, b_2 \dots b_k)$, with b_i taking values in $\{0, 1\}$. The data embedding module, or simply the embedder, mixes the cover signal A and $\mathbf{b}$ to produce a watermarked signal A_w. In order to increase the security of the system, the embedding function $\mathcal{E}$ usually depends on a secret key K:

$$A_w = \mathcal{E}(A, \mathbf{b}, K). \tag{1}$$

The definition of $\mathcal{E}$ goes through the extraction from A of a set of features $\mathbf{f} = (f_1, f_2 \dots f_n)$, called host features, that are modified according to the watermark message. Possible choices of $\mathbf{f}$ include audio or image samples, DFT or DCT coefficients, wavelet coefficients. In many cases it is useful to describe the embedding function by introducing a watermark signal $\mathbf{w} = (w_1, w_2 \dots w_n)$, that is added to the host feature set. In this case, by letting $\mathbf{f}_w = (f_{w,1}, f_{w,2} \dots f_{w,n})$ indicate the watermarked host features, we have:

$$f_{w,i} = f_i + w_i, \qquad i = 1 \dots n. \tag{2}$$

In the simplest case $\mathbf{w}$ depends only on $\mathbf{b}$ and K. For instance, according to the popular spread spectrum approach [3, 4], a pseudo random sequence $\mathbf{s} = (s_1, s_2 \dots s_n)$ is generated depending on K, then $\mathbf{s}$ is modulated by means of an antipodal version of $\mathbf{b}$, i.e. by letting $t_j = 1$ if $b_j = 1$ and $t_j = -1$ if $b_j = 0$, and by focusing on the j-th bit for simplicity, we have:

$$w_i = \gamma s_i t_j, \qquad (j-1)r + 1 \leq i \leq jr, \tag{3}$$

with γ determining the watermark strength and where $r = n/k$ represents the number of features used to embed a bit (we have neglected border effects for simplicity). A data hiding system for which $\mathbf{w}$ does not depend on A is called a blind embedding system, since embedding is carried out blindly without taking into account the particular host signal at hand. Better results can be obtained by tailoring $\mathbf{w}$ to the host asset A. This is indeed the case, as it has been shown in a number of seminal papers appeared since the late nineties [5, 6, 7] referring back to the theory of digital communications through channels with side information at the encoder [8, 9]. Data hiding systems obeying the strategy outlined in [8] are collectively termed informed watermarking systems. The most popular informed watermarking algorithm is Quantization Index Modulation (QIM [6]). In QIM schemes, data hiding is achieved through the quantization of the host feature vector, according to a set of predefined quantizers, where the particular quantizer used by the embedder depends on $\mathbf{b}$. Stated in another way, the to-be-hidden message modulates the quantizer index, hence justifying the QIM appellative. For further details on QIM watermarking readers may refer to [6, 7].

The second element in the scheme of figure 1 is the, so called, watermark channel. This accounts for all the manipulations the host signal may undergo after information embedding. Note that both intentional and non-intentional manipulations must be taken into account. The ability to survive intentional attacks is referred to as watermark security, whereas resilience against non-intentional manipulations is referred to as watermark robustness. After the host signal has passed the watermark channel, it enters the detector, whose scope is to retrieve the hidden information.

Extraction of the hidden information may follow two different approaches: according to the former, the detector looks for the presence of a specific message thus only answering yes or no, in the latter case the detector (which in this case is called decoder) reads the information conveyed by the host signal without knowing it in advance. These two approaches lead to a distinction between multibit watermarking and 1-bit watermarking. An additional distinction may be made between systems that need to know the original, non marked, signal A in order to retrieve the hidden information, and those that do not require it. In the latter case we say that the detector is blind (the term *oblivious* detection may also be used), whereas in the former case the detector is said to be *non-blind*. As it will be clear from the subsequent sections in authentication applications only blind detection/decoding makes sense.

In all the cases, the retrieval of $\mathbf{b}$ goes through the definition of a detection (decoding) function $\mathcal{D}$. In one-bit (oblivious detection) watermarking, $\mathcal{D}$ is a three-argument function accepting as input A', a watermark $\mathbf{b}$, and a secret key K. As an output $\mathcal{D}$ decides whether A' contains $\mathbf{b}$ or not, that is

$$\mathcal{D}(A', \mathbf{b}, K) = yes/no. \tag{4}$$

In multibit (oblivious decoding) watermarking, the decoder function takes as inputs A' a keyword K, and gives as output the string of bits $\mathbf{b}$ it reads from A':

$$\mathcal{D}(A', K) = \mathbf{b}. \tag{5}$$

Note that in multibit watermarking, the decoding process always results in a decoded bit stream, however, if the signal at hand is not marked the decoded bits are meaningless.

The exact form of $\mathcal{D}$ depends on the particular watermarking algorithm. The most common solution with spread spectrum systems relies on the analysis of the correlation between the spreading sequence $\mathbf{s}$ and the feature vector $\mathbf{f}'$ and its comparison against a detection threshold. Alternatively, the normalized correlation may be used.

In the case of QIM watermarking, decoding is straightforward, since it only requires that the host features are quantized again by considering a codebook which is the union of all the possible codebooks used by the embedder. In practice this corresponds to a minimum distance decoder, which in the case of a Gaussian noise addition attack, represents the optimum decoding strategy.

2.2 Authentication Systems Classification

As outlined above, it is common to classify watermark-based image authentication systems in terms of the robustness of the embedded watermark. Therefore, even if the concept of robustness is sometimes ambiguous and strictly depending on the application the authentication system is intended for, we can distinguish among fragile, semifragile and robust systems.

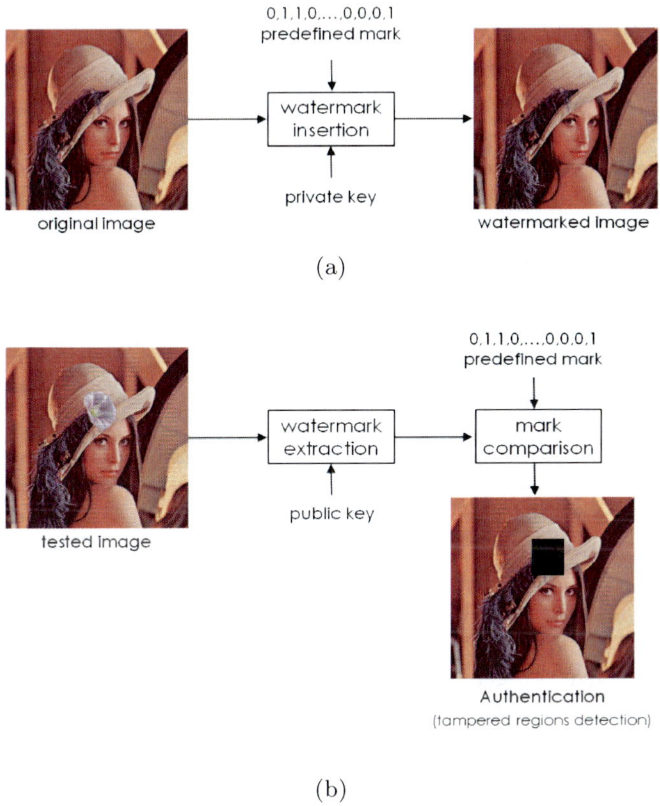

Fig. 2. Generic fragile watermarks scheme: (a) watermark embedding; (b) authenticity verification.

Early methods proposed for image authentication were based on a **fragile watermark**. The basic idea underlying these techniques is to insert a watermark in a such a way that any attempt to alter the content of an image will also alter the watermark itself. The fragile watermark, in fact, is very sensitive and it is designed to detect every possible change in the marked image. A so defined watermark can be used in all the applications for which no modification of the original content is allowed, like, for example, biomedical images.

A generic scheme based on fragile watermarking is described in Fig. 2. To verify the authenticity of the image, the watermark is extracted from the watermarked image and then compared to a predefined mark. If the sequences are equal the image can be considered authentic. The system depicted in Fig. 2, relies on multibit watermarking, authentication through fragile watermarking may be easily achieved by means of 1-bit watermarking as well.

While fragile watermarks can be used to verify the integrity of data, they are not suited for enforcing copyright ownership of digital images; an attacker would attempt to destroy the embedded mark and fragile marks are, by definition, easily destroyed. Many fragile watermarks have also the localization property, that is they are able to distinguish areas of the watermarked image that have been tampered with from areas where the image has not been modified.

The major drawback of this approach is that it is not able to distinguish between malicious and non-malicious modifications, for instance, a lossy compressed version of the original image is considered as a tampered image. Images, however, are generally transmitted and stored in a compressed form and the losses due to the compression process usually are not considered to affect the integrity of the image. Therefore it is necessary to introduce another type of image authentication systems, that are based on the concept of semi-fragile watermarking.

Semi-fragile watermarks are more robust than fragile watermarks and less sensitive to classical user modifications such as JPEG compression. The aim of these methods is to discriminate between malicious manipulations, such as the addition or removal of significant elements of the image, and non-malicious manipulations preserving the semantic content of the image, such as the addition of quantization noise due to lossy compression. In general, all the modifications which do not alter the perceptual content of the multimedia signal should be considered legitimate.

The main difficulty in the design of an authentication system relying on semi-fragile watermarking is the necessity of using a watermark that is fragile against non-allowed manipulations and robust against allowed ones. As it is easily guessed, the development of a watermarking system that has the above property is a very difficult task, hence a third class of authentication systems has been developed, relying on robust watermarking. Indeed, even if the primary applications of **robust watermarks** are copyright protection and content tracking, they can be also used to verify data integrity. Authentication based on robust watermarking relies on the idea that important components of the cover image are robust to channel noise, thus the watermark can be transmitted in a noiseless authentication channel. For example, a concise representation of the cover image can be communicated through the watermark channel. The form of a generic authentication system based on robust watermarking is described in Fig. 3. During the authentication step, both the watermark and the features of the image used for the watermark generation are extracted from the watermarked image. If the features are equal, the image can be considered authentic.

While fragile and semifragile systems can be based on a multibit or a 1-bit watermarking approach depending on the requirements and the particular application the system is intended for, authentication systems based on robust watermarking must rely on multibit watermarking since in all the cases the embedded watermark consists of a coarse representation of the host image that is not known in advance to the decoder.

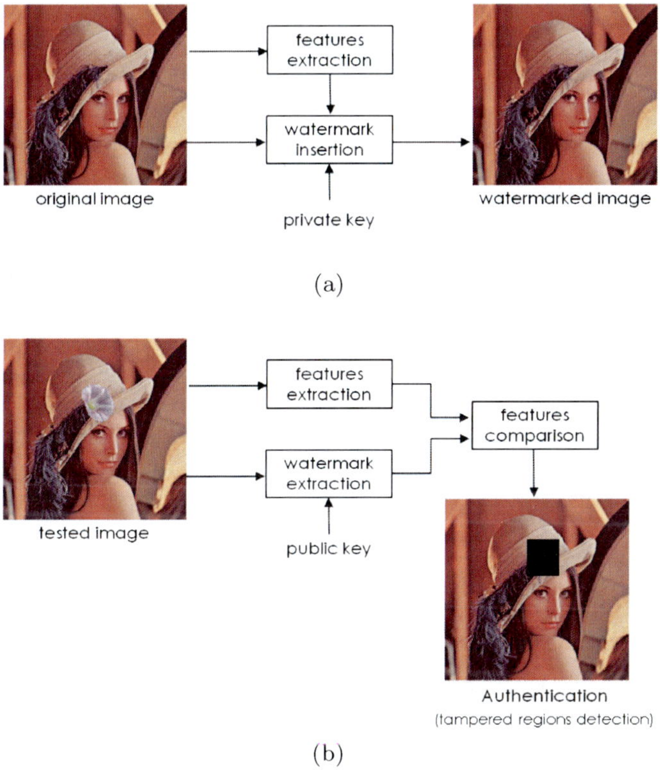

Fig. 3. Generic robust watermarks scheme: (a) watermark embedding; (b) authenticity verification.

2.3 Attacks on Authentication Watermarks

The goal of this section is to describe some of the attacks that an image watermark-based authentication system should overcome. The aim is not to present a list of all the possible attacks, but to describe some of the most frequent attacks. The common objective of these attacks is to fool the authentication system, i.e. to show that an image is authentic even if its content has been modified (or the other way round). Note that designing a system that is robust against all possible attacks proposed so far and those that will be invented in the future is likely to be an unfeasible task, nevertheless, being aware of the most popular attacks and finding a way to cope with them is an obvious requirement for the development any authentication algorithm.

It is possible to classify the attacks to a watermark-based authentication system in two categories: content-targeting attacks and scheme-targeting attacks. Content-targeting attacks aim at manipulating the content of the media without taking into account the protection measures provided by the algorithm. They include local tampering, such as removal of original objects/features and/or addition of new objects/features, and global tampering like scale, cropping, low-pass filtering, etc.

A variation of the local tampering attack is the *cover-up* [10] (or cut-and-paste) attack consisting in cutting one region/block of the image and pasting it somewhere else in the same image or in a different image.

Scheme-targeting attacks, instead, aim at defeating the security of the authentication scheme so that it fails to detect malicious content manipulations. Some examples of attacks belonging to this category are revised in the following.

One of the most common scheme-targeting attacks consists of trying to modify the protected image without altering the embedded watermark, or even more commonly, trying to create a new watermark that the authenticator will consider as authentic. Let us consider, for example, a system for which the integrity of the image is ensured by a watermark, generated independently on the image content, embedded in the LSB of the image pixels. In this case, if an attacker modifies the image without taking account which bits are affected by the manipulation, he will most likely degrade the watermark and therefore the attack will be detected. On the other hand, if the attacker alters the image without modifying the LSBs, the watermark will not change and the authentication process will fail to detect the tampering.

In general, when the integrity of an image is based on a mark that does not depend on the the image content, it is possible to develop an attack that copies a valid watermark from one original image to a fake image. By doing so, the fake image would be considered as authentic.

Based on this philosophy, Fridrich et al. [11] proposed the *collage attack* that creates a forgery by using a group of images protected by the same authenticator using the same mark and the same key. The idea of the attack is to forge a new watermarked image (collage) from a number of authenticated images watermarked with the same key and the same watermark, by combining portions of different authenticated images while preserving their relative positions in the final image. The authors show that counterfeiting is possible even when the watermark is unknown to the attacker provided that a large number of images is available. The main difficulty lies on obtaining a database of images rich enough to create a forgery of good visual quality.

Similarly to the collage attack, in the *vector quantization attack* [12] (also known as birthday attack), assuming the availability of a large library of images watermarked with the same key, the attacker builds a vector quantization codebook using blocks extracted from the authenticated images. The image to be faked is then approximated by using this codebook. Since each image block in the library is authenticated independently of the other images and blocks, the fake image is accepted as genuine by the detector. Obviously, the quality of the counterfeited image depends on two factors: the size of the database, as for the collage attack, and the size of the blocks used in the watermarking process.

Block-based methods, adopting a block-wise independent watermark, are also potentially vulnerable to the so-called *Holliman-Memon counterfeiting attack* [13]. In this paper the authors show that, given one or more images containing an owner's watermark W inserted using a key K, and a non-watermarked image Y, it is possible for an attacker to construct a watermarked image Y' such that $D_K(Y') = W$ (where D is the watermark detection function), without having any knowledge of the original key K, and, in the case of robust watermarking, without the knowledge of W.

Barreto et al. introduced the *transplantation attack* [10] which is an improved version of the cut-and-paste attack and works as follows. Let X' and X'' be two watermarked images. Let $X_A \rightarrow X_B$ denote the fact that the hashing of block X_B depends on the contents of block X_A. Suppose that images X' and X'' have blocks with the following dependence relationship:

$$... \rightarrow X'_A \rightarrow X'_D \rightarrow X'_B \rightarrow X'_C \rightarrow ...,$$

$$... \rightarrow X''_A \rightarrow X''_E \rightarrow X''_B \rightarrow X''_C \rightarrow ...,$$

where the block contents of X'_A are identical to those of X''_A, those of X'_B to those of X''_B and so on, and where the content of X'_D is different from that of X''_E. Then the pair of blocks (X'_D, X'_B) can be interchanged with the pair (X''_E, X''_B), without being detected by the scheme:

$$... \rightarrow X'_A \rightarrow X''_E \rightarrow X''_B \rightarrow X'_C \rightarrow ...,$$

$$... \rightarrow X''_A \rightarrow X'_D \rightarrow X'_B \rightarrow X''_C \rightarrow ....$$

Another classical attack tries to discover the secret key used to generate the watermark. This attack, that is well known in the security community, is called *Brute Force Attack*. Once the key has been found, it is easy for a hacker to falsify a watermark of an image that has been protected by this key. The only way to counter this attack is to use long keys to dissuade the attacker from trying to discover them, because of the high cost of the attack.

An effective watermarking scheme must have the essential capability of detecting content-targeting attacks and should have no security gaps to various kinds of scheme-targeting attacks. In general, as it will be more clear after the description of the state of the art provided in the next section, block-wise dependence is recognized as a key requirement to thwart vector quantization attacks [14, 13]. However, it is also proved that the dependency on deterministic context is susceptible to transplantation attack or even simple cover-up attack [10]. Nevertheless, Li and Yang [15] pointed out that even nondeterministic block-wise dependency (i.e. the neighboring relationship among individual pixels in the block is nondeterministic, as adopted in [14]) is still vulnerable to cropping attack.

The aforementioned attacks do not constitute an exhaustive list since more new attacks are expected to be devised in the future. Nevertheless, at present, an effective watermarking scheme used for authentication purpose should have the capability of thwarting at least these popular attacks.

3 State of the Art

In this section we give an overview of the watermarking algorithms for image authentication proposed in the scientific literature. The state of the art of image authentication systems is described according to the classification provided in the previous section.

3.1 Fragile Watermarks

Most methods proposed so far are based on a fragile watermark. We can further classify these techniques depending on the domain in which they work.

Pixel-domain schemes

By focusing on space domain techniques, several fragile watermarking methods that use the least significant bits of the image (LSB) have been developed. One of the first techniques used for tamper detection is based on inserting check-sums into the LSBsof the image data. The algorithm proposed by Walton [16] in 1995 consists in selecting, according to a secret key, a pseudorandom group of pixels. The check-sum value is obtained by summing the numbers determined by the 7 most significant bits of selected pixels. Then the check-sum bits are embedded in the LSB. The checking process consists in comparing, for each block, the check-sum determined by the MSB of the tested image with the original check-sum value recovered in the LSB.

The main advantages of the previous method are that it does not produce visible changes in the image and it provides a very high probability of tamper detection. One of the drawbacks of this method is that it is possible to attack it by swapping blocks of the same position from two authenticated images protected with the same key.

P. Wong describes another fragile marking technique based on LSB modification in [17], further improved by Wong et al. by using a public-key based scheme in [18]. The basic idea is that the LSB-zeroed target image and the binary watermark image are divided into blocks of the same size. The image size and each LSB-zeroed image block are then provided as inputs to a hash function, and the output together with the watermark block are subjected to XOR operation. The result of the XOR operation is then encrypted using a private key and embedded in the least significant bits of the original image. This scheme uses jointly cryptography and watermarking and works well in detecting cropping, and scaling. However, due to the lack of mutual dependence among neighbouring blocks during the watermarking process, this scheme is vulnerable to cover-up, vector quantisation and transplantation attacks, as decribed in the previous section. Moreover, the size of the block is limited from below by the length of the hash. Thus, the localization properties of such a scheme are also limited. In addition to that, it is possible to swap blocks within one image and among different images without introducing detectable changes.

Based on the previous method, Li et al. [19] suggested a slight variation so that neighboring blocks are related by blended data. The authors proposed a scheme that uses a binary feature map extracted from the underlying image as the watermark. The watermark is then divided into blocks of size 32×16 pixels. Blockwise dependence is established by blending the neighbouring blocks before encrypting and embedding into the LSBs of the image. On the decoding side, the feature map from the received image is extracted again and compared against the recovered watermark to verify the integrity and authenticity. This method is effectively resistant to vector quantisation and cover-up attacks and requires no a priori knowledge about the image to be watermarked. However, the accuracy of localisation is limited, like Wong's system, by the block size. Moreover, this scheme is vulnerable to transplantation attacks because the contextual dependence is established based on deterministic information.

To circumvent these drawbacks, the authors further proposed a scheme [15, 20] that is immune to transplantation attacks by introducing a 1-D neighborhood forming strategy. To watermark the underlying image, the gray scale of each pixel is adjusted by an imperceptible quantity in order to let a key-dependent binary watermark bit be equal to the parity of a bit stream obtained from the gray scale of a secret neighborhood of the pixel. To counter collage and look-up table attacks, the watermarking order of the pixels follows a zig-zag path and the secret neighborhood of a pixel is formed by picking previously watermarked pixels before the current pixels on the scanning path.

In [10], Barreto et al. show that Wong and Li's algorithms are insecure against attacks as simple as block cut and paste, the birthday attack or the Holliman and Memon's counterfeiting attack. Thus they introduce the HBC1 scheme, which counters these attacks by making the signature of each block depend on the contents of a neighboring block. Specifically, they proposed to generate the fingerprint of each image block through the calculation of a hash function taking the target block, a neighboring block and some random data as inputs. Since the random data is unique to each block, the fingerprint is unique and non-deterministic. However, the hash operation is relatively time-consuming and the accuracy of the tamper localization is limited by the size of the block. Finally, the HBC1 method is not secure against an improved form of cut and paste attacks, called by the authors transplantation attack.

Wolfang and Delp [21, 22] extended van Schyndel's work [23] to improve robustness and localization in their Variable-Watermark two Dimensional (VW2D) technique. This work is based on a block-based watermarking technique. To verify the authenticity of the image, the detector verifies the presence or the absence of the marks in all the blocks. The main drawback of this method is that for each block of the image it is necessary to store the result of the correlation between the watermark and the watermarked block, that is not feasible from a practical point of view.

In the popular technique proposed by Yeung and Mintzer [24], a watermark key is used to generate several pseudo-random look-up tables (one for each channel or color component), that control how subsequent modifications of the pixel data will occur. Then, after the insertion process is completed, a modified error diffusion process is used to spread pixel alterations, thus improving watermark invisibility. The security of this system depends on the difficulty of inferring the look-up tables. The search space for the table entries can be drastically reduced if knowledge of the bi-level watermark image is available. This scheme has very good localization properties but, unfortunately, it is not easy to develop a feasible key management infrastructure without introducing security gaps [25, 11].

Another drawback of Yeung-Mintzer's system is that the same bit is embedded in the same position of each watermarked image so that anyone who does not know the exact embedded bits can still forge an arbitrary watermarked image. To overcome this problem, a simple solution has been proposed by Fridrich et al. [14, 26]. The authors replaced the binary look-up tables for a single pixel by an encryption map defined on a local neighborhood rather than on a single pixel. To prevent the collage attack the authors embed an image index into each image repetitively into a set of non-overlapping 32×32 blocks. The authors claim that this scheme is secure from the viewpoint of computation load, however the bits embedded are still the same in corresponding positions of different watermarked images, which leaves some chances for the attacker. Another disadvantage of this scheme is that the localization property is not so good as that of the Yeung-Mintzer's scheme.

Several schemes have been proposed in order to eliminate the vulnerabilities of the previous methods to the quantization attack and the VQ counterfeiting attack of Holliman and Memon [13]. In [27] a multi-level hierarchy is introduced. While signatures of small blocks on the lowest level of the hierarchy ensure superior localization, higher level block signatures provide increasing resistance to VQ attack. The insertion of a binary watermark in the work by Lu et al. [28] is accomplished by replacing the least significant bits of the host image with a random binary image obtained by random permutations and suitable XOR operations among bitplanes. In [29], instead, similarity of image features from neighboring blocks are used for signature construction. This makes the method block-wise dependent, and thus not vulnerable to the VQ attack.

Another fragile, block-wise and content-based watermarking for image authentication and recovery is proposed in [30]. The watermark of each block is an encrypted form of its signature, which includes the block location, a content feature of another block, and a Cyclic Redundancy Checksum (CRC). An improved version of this system is described in [31] with the introduction of a hierarchical inspection of the image. While tamper detection is achieved through a three-level inspection of in-block information only (as for the previous work of [27]), the recovery of a tampered block relies on its feature information hidden in another block that can be determined by a one-dimensional transformation similar to Torus automorphism.

Several fragile watermarking methods based on the use of LSBs of image data were developed and discussed above. These methods treat every image pixel indiscriminately in the spatial domain with little consideration of human perception. Moreover, these techniques require changes of a large amount of image data to embed watermarks, which is quite inefficient and affects the quality of watermarked image.

Transformed-domain schemes

Several schemes working in the transformed domain have been proposed in the scientific literature. A brief review of the most popular among such schemes is given in the following.

Wu and Lu [32] in their paper introduce a system based on a modified JPEG encoder. The watermark is inserted by modifying the quantized DCT coefficients. To this purpose a look-up table is designed by ensuring mark invisibility and used to partition the space of all possible DCT coefficients values into two sets that are then used to encode the bi-level image describing the logo. At the receiver side, the decoding of the watermark is simply done by looking up the table. Although this scheme does not require the original image for watermark extraction, the same look-up table used in the embedding stage is necessary in the watermark extraction stage, thus the system is non-blind. Moreover, like Wong's scheme, this scheme is also blockwise independent and, therefore, vulnerable to cover-up, vector quantisation and transplantation attacks.

Kundur and Hatzinakos [33] and Xie and Arce [34] describe techniques based on the wavelet transform. Kundur et al. embeds a mark by modifying the quantization process of Haar wavelet transform coefficients while Xie et al. selectively insert watermark bits by processing the image after it has been compressed by using the SPIHT algorithm. Watermarks embedded in the wavelet domain are able to both locate and characterize the tampering.

Fridrich and Golian [35] propose an original method working on both the spatial and transform domain. The basic idea is to embed a compressed version of the image into the LSB of its pixels, in order to protect the image content. The algorithm consists in dividing the image into 8×8 blocks, setting the LSB of each pixel of the block to zero and then applying the DCT for each block. The matrix of the DCT coefficients is then quantified with the quantization matrix corresponding to a 50% JPEG quality, the result is encoded using 64 bits and the code is inserted into the LSB of another block. The quality of the recovered regions of the image is worse than a 50% JPEG quality but it is sufficient to identify the original content of these areas. As in all watermarking methods based on LSB embedding of the watermark, this method does not introduce visible artefacts, moreover this method also allows the region of the image that have been tampered with, cropped or replaced, to be partially repaired.

A common security gap inherent in many of the systems working in the transform domain is that they neither explicitly nor implicitly watermark all the transform coefficients. As a result, manipulation of those unwatermarked coefficients will go unnoticed. For example, in the wavelet transform domain approach proposed by Winne et al. [36], to minimise the embedding distortion and maintain high localisation accuracy, only the coefficients of the high-frequency subbands at the finest scale of the luminance component are watermarked. All the other coefficients and components are neither watermarked nor involved in the watermarking process.

In Fridrich et al. [37], all the coefficients are protected by taking all quantised DCT coefficients as input to the hash function and using the hash output as the signature, which is then embedded in the least significant bits of the lossless compressed version of some selected coefficients. However, the hash output conveys only global information about the image. When a local attack is launched against the coefficients that are not selected for embedding the hash, this algorithm can only tell that the image is not authentic without being able to locate the position where tampering occured.

A more recent method is the one described by Yuan and Zang in [38] and improved in [39]. The authors introduce a new system based on a Gaussian mixture model (GMM) to statistically characterize the images in the wavelet domain. This method embeds watermarks at image edges, resulting in fewer modified image data and imperceptible alteration of the original image.

3.2 Semifragile Watermarks

As already explained in the previous section, fragile watermarking methods are very sensitive to even the slightest deterioration of the image, but they offer only a strict integrity service, often far from user's need. In most of the applications, is preferable to use semifragile techniques. These systems are more tolerant with respect of non-malicious manipulations, such as good quality JPEG compression.

Recent approaches for providing localized alteration detection, while allowing some robustness to image compression, include: embedding a summary (or heavily quantized) version of the original image into the watermarked image, embedding key-dependent random patterns in blocks of the image, wavelet embedding to take advantage of the spatial and frequency information provided by the wavelet transform and embedding multiple watermarks. Some examples of systems based on these considerations are discussed in more detail below.

An example of embedding a highly compressed version of the original image is provided in [40]. In this paper, the authors introduce two techniques for self-embedding an image in itself as a means for protecting the image content. After embedding, it is also possible to recover portions of the image that have been cropped out, replaced, damaged, or otherwise tampered, without accessing the original image. The first method is based on transforming small 8x8 blocks using a DCT, quantizing the coefficients, and carefully encoding them in the least significant bits of other blocks. The quality of the reconstructed image areas with this method is roughly equivalent to a 50% quality JPEG compressed original. The second method uses a principle similar to differential encoding to embed a circular shift of the original image with decreased color depth into the original image. In this case, the quality of the reconstructed image gradually degrades with increasing amount of noise in the tampered image. Both the methods can detect intentional tamper, such as feature removal, adding, or replacement, and are robust against high quality PEG compression typically used in digital cameras.

Embedding multiple watermarks [41] has also been described as a method of authenticating an image with some degree of robustness. For example, a robust watermark can be embedded into an image to establish ownership followed by the embedding of a fragile watermark for authentication. The method embeds block-based patterns as the watermarks and detects the alterations by verifying the correlations on these patterns. This method still has some problems such as failure to detect alterations to DC coefficients only or substitutions of blocks with same address generated from the same key. It also presumes that most natural images have smooth features, thus, false alarms near edges can occur due to low correlations.

The idea of multiple embedding is also exploited in [42] and, later, in [43]. In [42], Fridrich proposes a hybrid image authentication watermark obtained as a combination of a fragile and a robust watermark. The fragile watermark has good localization and security properties and has been tested in presence of a wide variety of image distortions, but cannot be used to distinguish malicious changes, such as feature adding or removal, from common image processing operations, such as filtering, contrast/brightness adjustment or lossy compression. Hu et al. [43], instead, present a semi-fragile watermarking algorithm for image authentication which extracts image features from the low frequency domain to generate two watermarks, one for classifying the malicious content modification and the other for indicating the modified location. Experimental results show that the algorithm can identify intentional content modification, such as objects addition/removal, and incidental tampering, and also indicate the location where a modification takes place.

Sun et al. [44, 45, 46] propose a system in which robust signatures are cryptographically generated based on invariant features extracted from image content and are then signed and embedded back into the image again as watermarks, all in the wavelet domain. Error correcting coding is employed to further bridge between generated signatures and watermarks in a novel way: message bits are formed based on the same features, and parity check bits are taken as the seeds of watermarks. The generated signature is hashable and can be incorporated into a Public Key Infrastructure (PKI) framework. The use of error correcting codes addresses incidental distortions caused by acceptable manipulations such as lossy compression and the integration of PKI security infrastructure together with a hashing mechanism achieves system security and signature length reduction.

Lin and Chang [47, 48, 49] propose a semifragile watermarking algorithm that accepts JPEG compression and reject malicious attacks. They use two invariance properties of the DCT coefficients with respect to JPEG compression. The first property shows that if we modify a DCT coefficient to an integer multiple of a quantization step, larger than the steps used in later JPEG compression, then this coefficient can be exactly reconstructed after JPEG compression. This property is used for watermark embedding to guarantee robustness up to a certain level of JPEG compression. The second property is an invariant relationship between coefficients before and after JPEG compression. Because all DCT coefficient matrices are divided by the same quantization table in the JPEG compression process, the relationship between two DCT coefficients of the same coordinate position from two blocks will not be changed after the quantization process. The second property is used to generate the authentication bits. This technique allows to reconstruct an approximation of the original block and to locate precisely the zones of the images which were faded.

To prevent unauthorized removal or intentional watermark distortion, Fridrich [50, 51], proposes a technique based on a watermark that depends on the image in which it is embedded. Specifically, the binary watermark corresponds to a pseudo-random signal generated from a secret key, the block number and the content of the block represented by M bits. This system can be classified as a semifragile watermarking scheme since the watermark is robust with respect to brightness and contrast adjustment, noise addition, histogram manipulation, cropping and moderate JPEG compression.

In [52] and [53] the authors introduce a watermark-based authentication framework, which is robust to compression and coding, focusing on the JPEG 2000 image format. They decompose the image into phase and magnitude values. A signature is then generated from the phase values. In particular, binary phase-only filters (BPOF) and their variants are utilized for this. This signature is subsequently hidden into the magnitude part by a bit-plane embedding technique. The disjoint operations of signature generation and signature embedding minimize the embedding artifacts of the authentication signal. Moreover, they use wavelet decomposition, whereby, the signature can be generated from one subband, and then it can be embedded in other subbands, or the same subband. In addition, the phase quantization in the filter has built-in tolerance to minor changes to the image, which is a desirable feature for a semi-fragile image authentication watermarks. In the experimental results, the authors test the robustness of the system to compression and coding while no results are provided for other kinds of tampering thus the efficiency of the proposed system in an authentication framework can not be evaluated.

The above system has some additional limitations. For example, embedding the BPOF information in the bit plane of the rounded Fourier magnitude spectrum results in a very limited embedding strength. Moreover, to obtain a tradeoff between detection reliability and robustness to JPEG compression, the BPOF is usually embedded in the mid-level bit planes, which decreases the watermark imperceptibility more than embedding BPOF in the low-level bit planes. A possibility to overcome these problems is to use the identical ratio (IR) between the extracted watermark and the computed BPOF, which is simpler and more effective than correlation. Moreover, they embed the BPOF watermark in the low frequency ranges to improve the required fragility and robustness of a semi-fragile system. In the experimental results the authors show that the system is fragile in case of spatial-domain tampering, but robust against most common signal manipulation.

An original work is proposed in [54, 55] in which an approach for combined authentication and compression of color images by making use of a digital watermarking framework is proposed. The watermark, in fact, consists of two components: a soft-authenticator watermark for authentication and tamper assessment of a given image, and a chrominance watermark employed to improve the efficiency of compression. The authors propose the use of linear orthogonal separable transforms that work in orthogonal domains of the image for watermark generation and embedding. Specifically, the approach is implemented as a DCT-DWT dual domain algorithm. The proposed framework is semi-fragile to any form of acceptable distortions, both common signal processing operations and malicious manipulations. The capability of resistance to JPEG compressions is limited to larger quality factors up to 70%.

Maeno et al. proposed two semi-fragile techniques in [56]. The first method explores the correlation among coefficients in a local window. In fact, a manipulation tends to cause similar change patterns to coefficients in a local window and such similar patterns result in a clustered distribution in the plane. The fixed encoding boundary used in the technique in [47] has a potential issue of missing all the pairs of coefficients for such a malicious manipulation. In this method, the authors introduce a novel component which adds a random bias factor to the decision boundary. Such a randomization factor spreads out to each signature bit for catching the malicious manipulations. In the second method, the authors propose a nonuniform quantization scheme which uses a multibit nonuniform quantizer to encode the transform coefficient difference in each pair, and uses the different quantizers at the signature verification site. They use multiple bits to improve the accuracy in encoding the relationships between paired transform coefficients, and non-uniform quantizers to explore the nonlinear mapping between the coefficient differences in the original image and the compressed image. The coefficient pair is formed by selecting two coefficients from the same sub-band but at different locations. The authors proved that they are able to detect image manipulations even after JPEG2000 lossy compression with different filters and can detect various types of un-allowable manipulations (including object cropping), even for images with a very homogeneous background such as a document image.

A more recent study [57] investigates the resistance of a semi-fragile watermarking method to incidental manipulations with focus on compressions at low bit rates. The authors examine the characteristics of DCT coefficient distributions under compression-free and compression-imposed environments, respectively, and exploit them to design the system. Specifically, it is known that the distribution of DCT coefficients deviates from the true Laplacian distribution up to a certain degree. In view of this, they proposed to embed watermarks to enforce the approximated Laplacian distribution to become the true one. In other words, the pdf of DCT coefficients in an image is equal to a true Laplacian distribution, but when incidental manipulations (e.g., compressions) are encountered, the pdf of DCT coefficients deviates from the true Laplacian distribution. The goal is to reconstruct the deviated pdf back to the true form, from which the hidden watermarks can be successfully extracted. This is accomplished by performing a piecewise reconstruction of DCT coefficients: one value of the discrete quantized DCT distribution is taken at time and the corresponding quantization interval is approximate with a Laplacian. The authors verify the performance of proposed method in locating maliciously tampered regions under purely malicious modification and malicious plus incidental modifications. Specifically, it is able to distinguish incidental modifications from malicious modifications, even when JPEG compression is applied with quality factor as low as 50%.

In [58] a simple watermarking authentication algorithm based on ridgelet transform is proposed. The ridgelet coefficients of the host image are quantized so to form a binary sequence representing the image watermark. The sequence is stored in a sparse matrix. To preserve the quality of the original image, the weighted watermark is embedded into the host image. The system can distinguish the distortions caused by common image processing from those due to malicious tempering, even if the experimental results given in [58] refer only to a simple malicious attack. Moreover this technique is simple and fast, thus it can be applied, as the authors themselves suggest, to E-Government, i.e. for legal authentication of images routed outside the Internet.

A recent semi-fragile watermarking scheme for image authentication is proposed in [59]. The proposed scheme extracts the image content features in the low frequency domain to generate the watermark. Then it inserts the feature into the image by modulating the wavelet domain. To enhance robustness and invisibility, adaptive quantization is used, and the quantization step is calculated according to the local image characteristics. Integrity authentication and tamper detection are implemented by comparing the extracted watermark and the extracted content feature. Experimental result shows that the proposed scheme is robust to common image processing (such as JPEG compression, noise-adding, filtering), but malicious changes of the image will result in alarms of the watermark detection. In addition, proposed scheme can detect the exact locations-the illegal modified blocks. Even if the authors describe it belonging to the semi-fragile systems class, it can be considered a robust watermark based system described in the next sections.

3.3 Robust Watermarks

A different way to provide image authentication is through the use of robust watermarking, that is the embedded watermark is robust to any form of alteration of the original image but is able to localize malicious tampering actions that alter the semantic content of the image.

Zhu et al. [60] propose two techniques based on spatial and frequency masking. Their watermark is guaranteed to be perceptually invisible, yet it can detect errors up to one half of the maximal allowable change in each pixel or frequency bin depending on whether frequency or spatial masking is used. The image is divided into blocks and in each block a secret random signature (a pseudo-random sequence uniformly distributed in $[O, l]$) is multiplied by the masking values of that block. The resulting signal depends on the image block and is added to the original block quantized by using the same masking values. Errors smaller than one half of the maximal allowable change are readily detected by this scheme. The error estimates are fairly accurate for small distortions.

Lin and Chang [61] proposed wavelet-based image authentication. The image is decomposed into four sub-band (LL, LH, HL and HH) and the HH band is substituted by a pseudo-noise pattern selected according to a secret key. The watermarked image is obtained after applying the inverse wavelet transformation. In the authentication process the HH subband is first extracted and then convolved with the pseudo-noise pattern. If the image was not manipulated, the convolution result should be like a dot matrix, otherwise the distribution will loose its uniform character in the tampered areas. This system has been classified like a robust authentication technique since the authors show the robustness of the method against some filtering operations such as blurring and edge enhancing, and with soft JPEG compression. However, no results have been provide against specific attacks such as the substitution of the HH sub-band.

In [62] a simple and robust watermarking scheme based on the theory of torus automorphism is described. The authors test the performances of the system against classical attacks like blurring, JPEG compression and shearing while they do not evaluate the system in presence of malicious attacks typical of an authentication framework, like addition or removal of objects from the scenes. Thus the validity of the system in authenticating images can not be checked.

Another way to provide robust authentication, is by using methods based on feature-based watermark. This technique consists in first extracting features from the original image and then hiding them by means of an invisible watermark. In order to check if an image has been altered, it is sufficient to compare its features with those of the original image recovered from the watermark. If the features are identical, then the image has not been tampered with, otherwise the differences indicate the altered area. In case an attack is able to destroy the embedded watermark, the image will not be considered authentic and the robust watermark will act in this case like a fragile watermark. The choice of the image features is strictly related to the type of alterations that we wish to detect. Usually they are selected so that invariant properties are maintained under weak image alterations like JPEG compression and broken for malicious manipulations. Typical features used to provide image authentication are edges, colors, gradient, luminance. One of the problems of these methods is that the image undergoes slight modifications while inserting the watermark that can affect the image properties resulting in false positive detection.

Following the previous approach, Dugelay et al. [63, 64] have implemented an iterative watermarking algorithm working in the following way. The idea is to sign the image, to extract features from the newly obtained image, and then to repeat the watermarking process on the original image (in order to avoid cumulating distortions) using the newly computed features. Thanks to this iterative process, hidden features will perfectly coincide with the protected image features. The drawback of the described method is that the choice of the features to select from each image is not automatic. This makes the system difficult to use in a practical implementation.

In [65] the auhtors propose to use Zernike moments magnitudes (ZMMs) of the image for authentication. They demonstrated the good robustness and discriminative capability of this feature vector and propose a structure embedding method for the watermark generated from this feature vector to locate the tampered area. An improved version of this system is described in [66]. The improvements are obtained in the following ways: 1) embedding watermark in an image will affect its feature vector, thus the authors suggest to split the original image into several spaces randomly, one for generating feature vector, one for embedding ZMM based watermark; 2) presenting a two-stage authentication method to improve the discriminating capability of the authentication.

3.4 Merits and Drawbacks of Different Approaches

We now summarize and compare from different points of view the various classes of watermark-based authentication methods described so far. The merits and drawbacks of the various approaches are summarized in Table 1, where: sensitivity is the sensitivity of the system to malicious manipulations; tolerance is the tolerance of the system to some loss of information and to non-malicious manipulations; localisation and reconstruction are the ability of the system to locate the tampers in the image and to restore, even partially, the altered regions; complexity is the cost to implement the system and security means that there is no information leakage from the scheme and all the modifications are detected with very high probability derived from a secure cryptographic element. Note that those given in the table are generic indications of the pros and cons of the methods belonging to a given category. It goes without saying that the indications given in the table may not hold for some specific schemes. For instance, it is surely true that usually it is not possible to recover the original image content by means of techniques based on fragile watermarking. Nevertheless, some specific technique of this class exist that try to achieve this goal.

By looking at the table, we can notice that generally fragile watermarking methods allow the system to be very sensitive to any for of tampering but provide only a strict integrity control, whereas the semifragile and the robust watermarking methods guarantee content-based authentication. Moreover, fragile watermarking methods are the simplest to implement even. In addition they are the only one capable of ensuring cryptographic security.

Table 1. Properties of the various approaches to watermark-based image authentication

Property	Fragile	Semifragile	Robust
Sensitivity	very high	high	high
Tolerance	no	medium	very high
Localisation	yes	yes	yes
Reconstruction	no	no	yes
Complexity	very low	medium/high	high/very high
Security	very high	medium	medium

4 Watermark Authentication: A Practical Case

In this section we describe an authentication system based on a robust, multibit watermarking scheme that in addition to verify the authenticity of the image at hand, is also able to detect tampered regions and give an indication of the original image content. The proposed scheme is robust against classical, non malicious, image processing operations like color adjustments and jpeg compression, and, at the same time, is able to detect semantic tampering whereby the meaning of the image is modified, i.e. objects are added or removed from the scene. The main idea is the same of most systems based on robust watermarking: to embed within the to-be-authenticated image a summary of the original image describing its original content. Upon retrieval of the watermark, the actual content of the image can be compared, either automatically or manually, with the content of the watermark to detect a potential tampering of the image. Note that if the comparison is performed manually it is up to the human operator to decide whether the observed manipulation is an allowed one or not. Should the alteration be so strong to prevent watermark decoding the whole image would be declared as tampered. It is clear that the above procedure relies on a twofold assumption: the watermark is able to survive any kind of manipulations (including non allowed ones) and has a payload that is large enough to contain a significant summary of the host image. These are conflicting requirements since robustness is more easily achieved with low payloads. In fact, we can consider watermarking as a channel coding task in which the interference of the host image (for non-informed schemes), post processing and malevolent attacks can be assimilated to channel noise: by enlarging the admitted distortion, the watermark capacity goes down. The payload requirement can be worked out also by acting on the creation of the summary. The summary in fact, should be as small as possible while at the same time describing the semantic content of the image as closely as possible.

Summarizing, a trade off has to be found between the robustness of the watermarking scheme and the amount of information that we need for accurate tampering detection and localization. The next two sections investigate this trade off first by describing a robust high-payload watermarking algorithm which is able to robustly embed a 4096 bits message within a color image of size 1024×1024, then by presenting a way to create a meaningful image summary with such an extremely small number of bits.

4.1 A Robust High-Payload Watermarking Scheme

The watermarking technique the proposed authentication scheme relies on is a modification of the technique described in [67]. This watermark technique is a dirty paper coding based on orthogonal codes which assure a fixed maximum distortion level during the embedding phase.

In the next we briefly describe the watermarking algorithm and we show the implementation details that we have used to adapt it at the authentication scenario of interest here.

The watermarking algorithm is based on a random binning strategy. ¿From a very general (and theoretical) point of view, watermarking by random binning works as follows. Let U be an auxiliary source of randomness, let B indicate the set with all the possible to-be-hidden messages, and let 2^{nR} be the number of messages contained in it. Finally, let C be the source emitting the cover feature sequence. The embedder first generates a codebook $\mathcal{U}$ consisting of 2^{nR_t} entries (call them $\mathbf{u}$'s) which are randomly generated so to span uniformly the set of typical sequences of $\mathcal{U}$ [1].

Then $\mathcal{U}$ is randomly (and uniformly) split into 2^{nR} bins (sub-codebooks) each containing $2^{n(R_t-R)}$ codewords. It is then possible to associate each message $\mathbf{b} \in B$ to a bin of $\mathcal{U}$. In order to transmit a message $\mathbf{b}$, the embedder looks at the host feature sequence $\mathbf{c}$ that is going to host the message, then an entry in the bin indexed by $\mathbf{b}$ is looked for which is jointly typical with $\mathbf{c}$. Next it maps the cover features $\mathbf{c}$ into a marked feature sequence $\mathbf{c}_w$ which is jointly typical with $\mathbf{u}$ and $\mathbf{c}$. At the other side, the decoder receivers a sequence $\mathbf{r}$. In order to estimate the transmitted message, the decoder looks for an unique sequence $\mathbf{u}^* \in \mathcal{U}$ which is jointly typical with $\mathbf{r}$ and outputs the message corresponding to the bin $\mathbf{u}^*$ belongs to. The decoder declares an error if more than one, or not such typical sequence exists. If R is lower than watermarking capacity then it is possible to choose R_t so that the error probability averaged over all possible codes $\mathcal{U}$ tends to 0 as the length n of the transmitted sequence tends to infinity. The major problem with the above approach is that when n increases the dimension of the codebook becomes unmanageable, thus calling for the construction of structured codebooks allowing for an efficient search. The concept of typicality must also be replaced with a more manageable notion.

The most popular solution to put the random binning approach into practice is through the use of lattice codebooks, but in general this approach is weak against constant value-metric scaling the host feature. The method proposed in [67] and adopted here uses equi-energetic codebooks, in particular orthogonal codes, and correlation-based decoding, so that invariance to the presence of a constant gain is automatically achieved. Of course, since equi-energetic codes lie on the surface of a hypersphere, they are not able to cover uniformly the host feature space, hence it is necessary to devise a particular embedding strategy which permits to move the host features sequence into a point within the decoding region associated to the to-be-transmitted message. This can be obtained by fixing the watermark robustness and trying to minimize the embedding distortion, or by fixing the embedding distortion while maximizing the watermark robustness. Unfortunately, the standard fixed-robustness approach does not allow to take into account perceptual factors. As matter of fact, in order to diminish the visibility of the watermark, it is desirable that some features are marked less heavily

[1] Shortly a typical sequence is a sequence in which the symbols of the source alphabet appear in proportions equal to the corresponding probabilities of being emitted by the source. For example, a typical sequence of length n for a Bernoulli source with parameter p, is a binary sequence with approximately $n \times p$ ones and $n \times (1-p)$ zeros. For a more rigorous definition of typical sequences readers may refer to [68].

than others, leading to a constraint on the maximum allowable distortion. For this reason we used the reverse strategy of fixing the distortion and maximizing robustness. As a first step we need to define a robustness measure. As in [67], we use the maximum pairwise error probability function, that is the maximum error probability (under additive Gaussian noise) between the transmitted codeworks and all the codewords of $\mathcal{U}$ belonging to a bin Q_j with $j \neq l$, where by l we indicated the index associated to the transmitted information sequence. Even if such probability does not coincide with the true error probability of the system, it can be shown [69] that if the attack noise is not too strong, the maximum pairwise probability is a good approximation of the true error probability. Moreover, by using orthogonal codes, we can reformulate the embedding problem by fixing the distortion Δ and maximizing the watermark robustness, i.e. minimize the maximum pairwise error probability.

In the current authentication scheme, the fixed maximum distortion watermarking algorithm is applied to the amplitude of the 2D Fourier transform applied to the image, by using a full-frame perceptual Watson mask [70] and a tuned energy gain. Besides, we enhance the robustness of the scheme by embedding, instead of the message, its coded version by using a turbo-code scheme. The overall watermark system, calibrated to work with color images of 1024×1024 of size and a message of 4096 bits, is built as follows:

1. The 2D Fourier transform is applied to each color band[2] of the image by obtaining three amplitude coefficient matrices.
2. For each coefficient matrix the Watson's perceptual mask is calculated [70]. Watson's mask values are used as the maximum admissible distortion $\Delta(i, j)$ that the magnitude of each Fourier coefficient $|f(i, j)|$ may undergo (i and j are the coordinates of the mask and Fourier matrices). In this way we associate a distortion value to each coefficient by obtaining a mask-coefficient.
3. Given the 4096 bit of the message, the algorithm adds two additional bits and then it codes the message by using a turbo-code with puncture 1/3. The two added bits are used as tail of the code. At the end of the coding step we obtain an antipodal $\{+1; -1\}$ string of 12294 elements.
4. Each element of the antipodal string is encoded by using an appropriate orthogonal code of 32 bits and is embedded within 32 mask-coefficient pairs selected from the Fourier and mask matrices. Each coefficient is modified according to the distortion value of the mask. This embedding step is repeated for every element of the antipodal string. At the end 393408 mask-coefficient pairs are used to embed the antipodal string. The pairs are taken from the middle Fourier frequencies[3] through a key-dependent pseudo-random shuffling.
5. The watermarked image is finally obtained by applying the inverse Fourier transform.

[2] Red, green, and blue color bands.

[3] The highest frequencies are too weak for the watermark embedding, while by using the lowest frequencies the watermark visibility increases to much.

On the receiver side, the watermark decoder extracts the embedded message as follows:

1. The 2D Fourier transform is applied to each color band of the image by obtaining three matrices with the magnitude of Fourier coefficients.
2. 393408 coefficients are extracted from the middle Fourier frequencies by using the same embedding key to generate the pseudo-random shuffling of the coefficient matrices.
3. Each group of 32 coefficients is decoded by using the orthogonal decoder thus obtaining 12294 antipodal elements.
4. The 12294 bits stream is decoded by using the turbo-code algorithm. At the end 4098 bits are obtained.
5. The 4096 bits of the embedded message are extracted by removing the last two bits of the results of the turbo decoding step.

Implementation settings like the message length, the puncture of the turbo-code, and the length of the orthogonal code have been chosen in the attempt to reach an acceptable trade-of between the visibility and the robustness of the watermark. Experimental results reveal that the algorithm is robust to common color adjustment like contrast and luminance modifications thanks to the embedding rule and coding procedures. Moreover, it is robust to image cropping whereby only a part of the image is retained. Cropping has two effects: the first is that part of the watermark signal is lost, the second is that the watermark signal is shifted. With regard to the first effect, by working in the full-frame Fourier domain, each Fourier coefficient depends on all the pixels, thus, even if some of the pixels are lost, the remaining ones can be sufficient to adequately recover the embedded message. In this case, cropping can be modeled as the product of the signal by a rectangular window: this corresponds in the Fourier domain to the circular convolution of the Fourier transform of the signal with a sinc-like sequence. In practice, cropping introduces a correlation among the Fourier coefficients that, thus, can no longer be considered independent. However the watermark is robust to these kind of alterations. As to translation, thanks to the properties of the Fourier domain, the coefficients magnitude in which the message is embedded are invariant to this attack that results only in a modification of the coefficients phase. Finally we remind that in our scenario we assume that the original image size is known, so, after cropping we have to add a zero-padding in the pixel domain to re-obtain an 1024×1024 image, in order to avoid a wrong image resampling of the Fourier transform [4]

The watermarking scheme is also robust to JPEG compression at quality factors above 80, but it is not able to recover the embedded message in the presence of geometric attacks such as rotation and zooms applied to either the whole image or to small parts of it.

The next paragraph describes how the watermarking scheme is used in the overall authentication scheme.

[4] If the size of the original image is not 1024×1024, the image is resampled so to bring it to this size.

4.2 Image Content Extraction

In the previous paragraph we have introduced the watermarking algorithm used to embed a message within the image in a robust way. In this paragraph we study how we can extract a very short summary of the image which is able to describe the semantic of the image that we want to authenticate. Among the several approaches available in the literature we decided to use a perceptual description of a low resolution version of the image.

To obtain a good trade off between the size of the summary and its capability of describing the image content, we decided to consider only the image luminance, in addition we decreased the image size by zooming it out by a factor of 8 obtaining a 128×128 image. These steps alone are not sufficient to reduce the size of the summary to 4096 bits. We then had to apply to the gray-scale image a compression algorithm that avoids to spend too many bits to store the content of image regions that are not semantically significative. To do so, we used the points of interest (POI) detection and a foveation algorithm proposed in [71]. The system described in [71] is based on the observation that when looking at an image the human eye acquires the object by using several resolution levels depending on the so called "fixation point". The fixation point indicates the part of the image that has to be acquired at the highest resolution and corresponds to the center of the eye's retina, the fovea.

The motivation behind the use of foveation image processing is that there exists considerable high frequency information redundancy in the peripheral regions, thus a much more efficient representation of images can be obtained by removing or reducing such information redundancy. To this aim, we used the wavelet pyramid method explained in [72] that merges several image resolution due to a pyramid decomposition in according to the human eye model.

Specifically, psychological experiments have been conducted to measure the contrast sensitivity as function of retinal eccentricity [73, 74]. In [73], a model that fits physical data was given by:

$$CT(f, e) = CT_0 e^{\left(\alpha f \frac{e+e_2}{2_2}\right)} \tag{6}$$

where f is the spatial frequency (cycles/degree), e is the retinal eccentricity (degree), CT_0 is the minimal contrast threshold, α is the spatial frequency decay constant, e_2 is the half resolution eccentricity constant, and $CT(f, e)$ is the visible contrast threshold as function of f and e. The best fitting parameter values given in [73] are $\alpha = 0.106$, $e_2 = 2.3$, and $CT_0 = 1/64$. Next, a weighting function is computed using eq. (6) by calculating the distance from the observer to each point of the image (used to determine the maximum spatial frequency) and by determining the maximum spatial frequency resolvable by the eye at each point in the image (assuming a known, fixed point of gaze). The maximum frequency of the monitor is divided by the maximum frequency resolvable by the visual system at each pixel by obtaining the (fractional) pyramid level which is required to be sent at each point in the image. This will produce a matrix containing floating-point values between 1 and the number of pyramid levels which have

(a) Original image

(b) 64 × 64 thumbnail (c) 128 × 128 thumbnail (d) 256 × 256 thumbnail

Fig. 4. The original image (a) and the extracted thumbnail by using different sizes.

been computed. In order to use this matrix, each pyramid level is blurred and upsampled in order to make each pyramid image the same size as the original. These images form a 3-D dataset. The value at each point in the foveated image is computed by linear interpolation between images in the 3-D dataset based on the floating-point pyramid level value.

Just for completeness, the foveation process is applied to the gray-scale image. Then the foveated image is resized and a 128 × 128 image is obtained.

As the last step, we use a compression algorithm which works at very high compression rate to compress the 128 × 128 image. The algorithm that we used is the Set Partitioning in Hierarchical Trees (SPIHT) [75]. SPIHT is an image compression algorithm that exploits the inherent similarities across the subbands in a wavelet decomposition of the image. The algorithm codes the most important wavelet coefficients first, and transmits the bits so that an increasingly refined

Fig. 5. In the first row are showed the original images, while in the second row there are the watermark-authenticated images.

copy of the original image can be obtained progressively. To do so, SPIHT save the compressed image by using a progressive transmission scheme. In a progressive transmission scheme, the decoder can obtain a reconstructed image after receiving the value (approximate or exact) of some coefficients and considering zero the coefficients that it does not know. A major objective in a progressive transmission scheme is to select the most important information-which yields the largest distortion reduction-to be transmitted first. Specifically, SPIHT uses the mean squared-error (MSE) distortion measure to give a relevance rate to the to-be-compressed information.

In the proposed authentication scheme, we retain only the first 4096 bits of the SPIHT bitstream and we reject the other - negligible - information. Experimental analysis shows that 4096 bits are usually enough to represent the image content. Fig. 4 shows a comparison between the image that we want to authenticate and three versions of its 4096 bits summary by using different thumbnail sizes. Specifically, Fig. 4(a) is the original image that we want to summarize, while in Fig. 4(b), Fig. 4(c), and Fig. 4(d) represent, respectively, the summaries obtained by reducing the original image to 64 × 64, 128 × 128, and 256 × 256 pixels. We can notice that the combined effect of size reduction and SPIHT compression produces the best results when resizing to 128 × 128 is used.

The authentication scheme works by embedding the image summary built as described above into the cover image, by using the robust watermarking algorithm described in the previous section. During the validation phase a new image summary is derived from the image at hand and compared with the summary

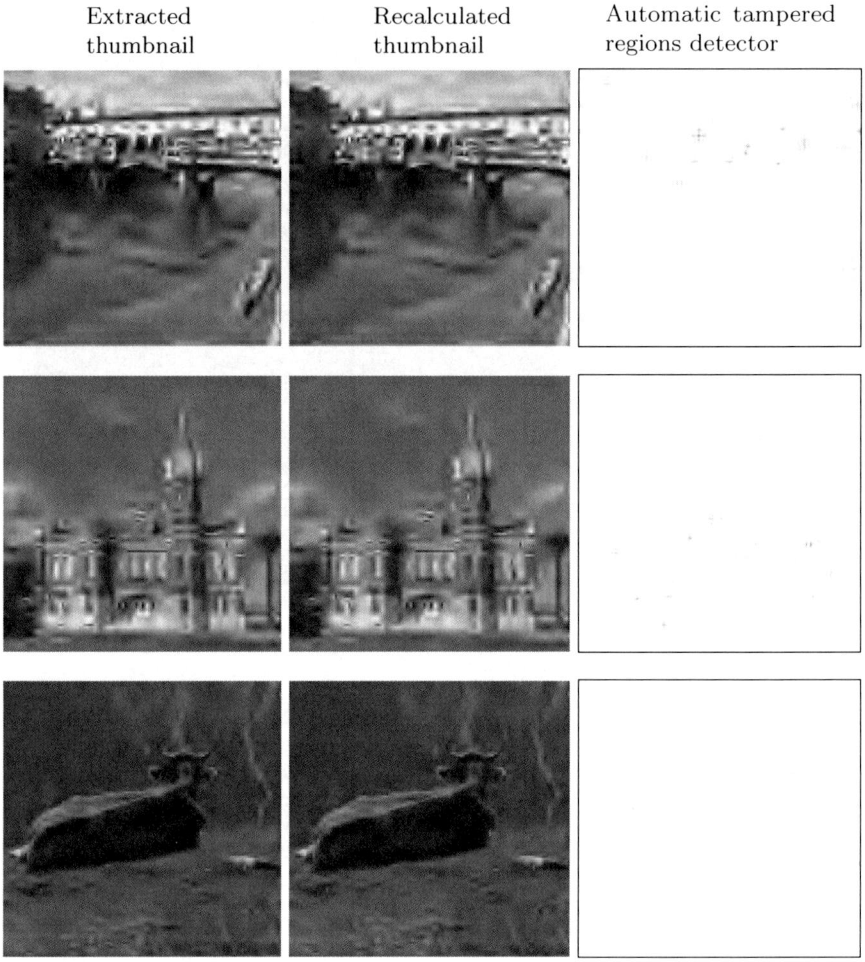

Fig. 6. Impact of JPEG compression at quality 80 on the extracted thumbnail and recalculated thumbnails.

extracted by the watermark decoder. Image authentication then reduces to the comparison between the extracted and actual image summaries. Of course, even if the image has not been tampered with the two summaries will be slightly different because the embedded one was calculated on the original-non-marked image while the recalculated one is derived from the watermarked version of the image. However in the case of non-tampered images the differences are negligible and easily filtered out by means of visual inspection[5]

[5] The automatic distinction between differences due to non-malevolent changes and to non-allowed manipulations may raise some difficulties.

(a) (b) (c)

Fig. 7. Tampered images by adding objects

4.3 Experimental Results

In this section we show some experimental results demonstrating the validity of the authentication scheme described so far.

To test the authentication scheme we need to take a subset of images and to define a set of attacks that measure the performance of the authentication algorithm. In Fig. 5 we can see in the first row three test images. First of all, in the second row of 5 we can verify that the presence of the watermark does not degrade the quality of the images.

Then we consider how the embedded summary changes as a consequence of allowed image manipulations. For simplicity we refer to the common case in which the watermarked image undergoes a lossy compression. Specifically, in Fig. 6 we show the results obtained by applying a JPEG compression with quality factor equal to 80%. The first column shows the image summary extracted from the JPEG-compressed image, while the second column shows the summary recalculated from the compressed images. As it can be seen by visual inspection, no significant difference exist between the two summaries hence allowing the authentication of the JPEG compressed image. In the third column in Fig. 6, the difference between the extracted thumbnail and the recalculated thumbnail is given: only very small differences randomly spread overall all the thumbnail are obtained hence permitting the automatic authentication of the image, the few existing differences, in fact, are easily removed by thresholding and the subsequent morphological operations.

As a next experimental setup the images have been tampered with. Several tampering actions may be considered:

1. post-processing like filters and color adjustments;
2. addition of objects to the scene;
3. removal of objects from the scene.

The first kind of manipulations are usually considered as allowed ones hence they mainly concern the robustness of the watermarking technique. If the attack does not degrade the watermark and the message can be completely extracted, then we can compare the two summaries and the image can be authenticated.

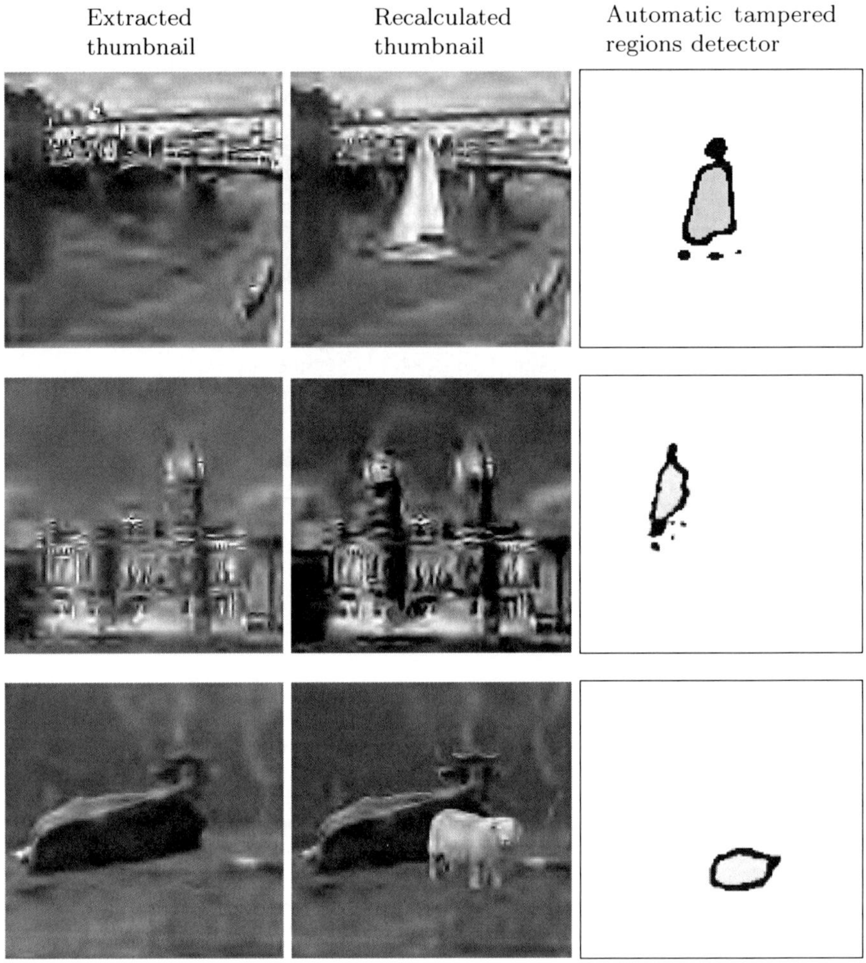

Fig. 8. Tampering adding detection

Otherwise, if the extracted message is corrupted or it is a random sequence of bits, its image-reconstructed version will be just a noise image that is completely different from the calculated summary. This case underlines that the attack was too strong and the attacker/user was not allowed to apply it without compromising image authenticity.

The second kind of tampering is applied by users that want to add one or more objects to the image.

In Fig. 7 three tampered images are shown. Specifically, in Fig. 7(a) a sailboat was added in the river, just in front of the bridge, in Fig. 7(b) a tower is added in the middle left of the building, and in Fig. 7(c) a sheep is added closed to the cow.

(a) (b) (c)

Fig. 9. Tampered images by removing objects

The results of the authentication procedure are illustrated in Fig. 8. In the first column of Fig. 8 the summaries extracted by the watermark decoder are shown, while second column contains the summaries re-calculated by starting from the image under analysis. We can notice that it is quite simple to detect the added objects even though the resolution of the summaries is extremely low. We have also implemented a simple detector that is able to recognise the tampered regions. The detector works by subtracting the two summaries and applying a threshold to the obtained value. The pixels for which the differences are over the threshold are classified as tampered pixels. Moreover an erosion-dilatation operator is applied to the thresholded image in order to remove isolated pixels. The results of this automatic tampered regions detector are shown in the third column of Fig. 8.

As a last kind of attacks we considered object removal.

In Fig. 9 the attacked images are shown. Specifically, in Fig. 9(a) the canoe in the bank is removed, in Fig. 9(b) the clock tower is removed from the building and in Fig. 9(c) the mountain face on the background is removed.

Like for the previous case, in Fig. 10 we show the results of the proposed authentication scheme.

Interestingly, we can notice that the luminance of the sky in the third row of Fig. 10 is sometimes similar to the mountain of the original image. Even though the color of the tampered background is extremely different from its original state, from a luminance point of view there are small regions in which they are almost the same and the automatic tampered regions detector is not able to distinguish between them. However, overall the system is able to detect a semantic alteration of the image as we can see in the first and the second row of Fig. 10.

As last consideration, we would like to underline the robustness of the proposed watermark technique. In fact, in the above scenarios in which objects are added or removed from the image, the watermark algorithm was always able to extract the correct message. Of course, if a very large portion of the image is removed or substituted, the watermark would be erased, however in this case

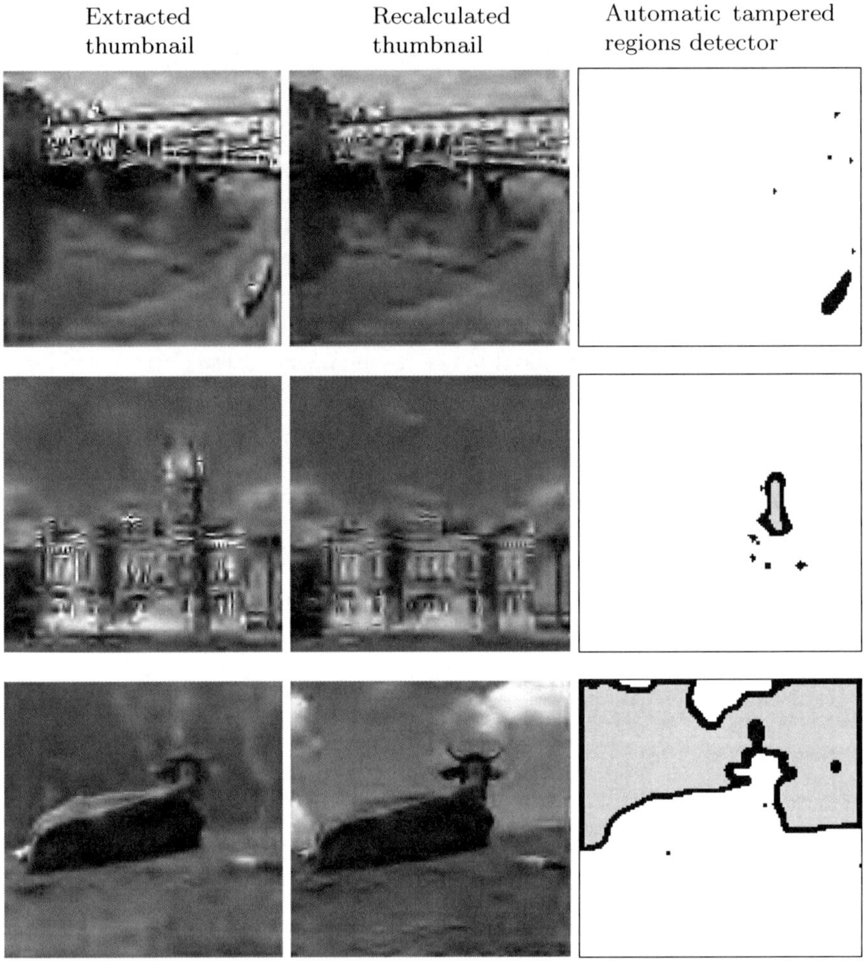

Fig. 10. Tampering subtracting detection

the authentication algorithm would no more validate the image, i.e. tampering is still detected though not localized.

5 Conclusions

The advent of digital era has brought many advantages in the creation and distribution of multimedia content but the ease of copying and editing facilitates unauthorized use, misappropriation and misrepresentation. Moreover, the wide availability of powerful image processing tools has also made imperceptible content modifications possible. As a result, multimedia content authenticity becomes greatly threatened. Content providers are naturally concerned about these issues and digital watermarking have been proposed as a tool to protect owner rights. While the main goal of a watermarking system has been for long enforcing copyright ownership, in the past few years many new techniques based on watermarking have been introduced as a means for tamper detection in multimedia content.

A review of watermark-based authentication systems have been proposed in this chapter together with a classification of the techniques into three categories: (1) fragile watermarking, which detects any modification of the image; (2) semi-fragile watermarking, which detects and localize malicious modifications while being tolerant to classical user modifications such as JPEG compression, (3) robust watermarking, which detects only significant changes in the content while permitting content-preserving processing.

In the last section of the chapter a new technique for robust watermark-based authentication has been described which is able to detect both non-malicious manipulations like filtering and color adjustments and malicious tampering such as addition or removal of objects from the scene.

It is important to point out that according to the current state of research, it is difficult to affirm which approach seems most suitable to ensure an integrity service adapted to multimedia documents. A solution perfectly matching all the constraints set by practical scenarios do not exist, since such a solution is going to be strictly depending on the particular application the system is intended for. Generally speaking, fragile watermarking methods are very sensitive to the slightest deterioration of the image, but they offer only a strict integrity service, relatively far from users' needs. Thus, the current tendency is more and more towards the use of semifragile or robust methods. The main obstacle to the wide deployment of such methods, however, is the definition of a method to extract a semantically meaningful summary of the image to be authenticated. In the current literature such a task is often referred to as robust hashing, a problem that has received considerable attention in recent years and that goes beyond the particular case of image authentication [76, 77, 78, 79].

In particular the security aspects of robust hashing have been studied only for a few years [80, 81] and still present several unknown aspects that still needs to be studied.

As a last aspect, we mention that the watermarking approach is not the only possible way to tackle the authentication problem. Other solutions are in fact available including, for instance, cryptographic solutions and passive authentication via forensics analysis. The identification of the exact merits and limits of such approaches is still an open problem calling for further research.

References

1. Bartolini, F., Tefas, A., Barni, M., Pitas, I.: Image authentication techniques for surveillance applications. Proceedings of the IEEE 89(10), 1403–1418 (2001)
2. Barni, M., Bartolini, F., Manetti, A., Piva, A.: A data hiding approach for correcting errors in h.263 video transmitted over a noisy channel. In: Proceedings of MMSP 2001, 2001 IEEE Workshop on Multimedia Signal Processing, October 3-5, pp. 65–70 (2001)

3. Cox, I., Kilian, J., Leighton, F., Shamoon, T.: Secure spread spectrum watermarking for multimedia. IEEE Transactions on Image Processing 6(12), 1673–1687 (1997)
4. Cox, I., Miller, M., Bloom, J., Fridrich, J., Kalker, T.: Digital watermarking and steganography. Morgan Kaufmann, San Francisco (2007)
5. Cox, I., Miller, M., McKellips, A.: Watermarking as communications with side information. Proceedings of the IEEE 87(7), 1127–1141 (1999)
6. Chen, B., Wornell, G.: Quantization index modulation: a class of provably good methods fordigital watermarking and information embedding. IEEE Transactions on Information Theory 47(4), 1423–1443 (2001)
7. Eggers, J., Girod, B.: Informed watermarking. Kluwer Academic Publishers, Dordrecht (2002)
8. Costa, M.: Writing on dirty paper. IEEE Trans. Info. Theory 29(3), 439–441 (1983)
9. Gel'Fand, S., Pinsker, M.: Coding for channel with random parameters. Problems of control and Information Theory 9(1), 19–31 (1980)
10. Barreto, P., Kim, V., Rijmen, H.Y.: Towards secure public-key blockwise fragile authentication watermarking. IEE Proceedings-Vision, Image and Signal Processing 149(2), 57–62 (2002)
11. Fridrich, J., Goljan, M., Memon, N.: Further attacks on Yeung-Mintzer fragile watermarking scheme. In: Proc. SPIE Int. Soc. Opt. Eng., vol. 3971, pp. 428–437 (2000)
12. Memon, N., Shende, S., Wong, P.: On the security of the yueng-mintzer authentication watermark. Proceedings of the IS&T Pics 99, 301–306 (1999)
13. Holliman, M., Memon, N.: Counterfeiting attacks on oblivious block-wise independent invisible watermarking shemes. IEEE Transactions on Image Processing 9(3), 432–441 (2000)
14. Fridrich, J., Goljan, M., Baldoza, A.: New fragile authentication watermark for images. In: Proc. IEEE Int. Conf. Image Processing, vol. 1, pp. 446–449 (2000)
15. Li, C., Yang, F.: One-dimensional neighborhood forming strategy for fragile watermarking. Journal of Electronic Imaging 12, 284–291 (2003)
16. Walton, S.: Information authentication for a slippery new age. Dr. Dobbs Journal 20(4), 18–26 (1995)
17. Wong, P.: A watermark for image integrity and ownership verification. In: IS&TS Pics Conference. Society for Imaging Science and Technology, pp. 374–379 (1998)
18. Wong, P., et al.: A public key watermark for image verification and authentication. In: Proceedings of the IEEE International Conference on Image Processing, vol. 1, pp. 455–459 (1998)
19. Li, C., Lou, D., Chen, T.: Image authentication and integrity verification via content-based watermarks and a public key cryptosystem. In: Proceedings of the IEEE International Conference on Image Processing, vol. 3, pp. 694–697 (2000)
20. Li, C., Yang, F., Lee, C.: Oblivious fragile watermarking scheme for image authentication. In: Proceedings of IEEE International Conference on Acoustics, Speech, and Signal Processing (ICASSP 2002), vol. 4, pp. 3445–3448 (2002)
21. Wolfgang, R., Delp, E.: A watermark for digital images. In: International Conference on Image Processing (ICIP 1996), vol. 3, pp. 219–222 (1996)
22. Wolfgang, R., Delp, E.: Fragile watermarking using the VW2D watermark. In: Proc. SPIE, Security and Watermarking of Multimedia Contents, vol. 3657, pp. 204–213 (1999)
23. van Schyndel, R., Tirkel, A., Osborne, C.: A digital watermark. In: International Conference on Image Processing (ICIP 1994), vol. 2, pp. 86–90 (1994)

24. Yeung, M., Mintzer, F., Center, I., Heights, Y.: An invisible watermarking technique for image verification. In: International Conference on Image Processing (ICIP 1997), vol. 2, pp. 680–683 (1997)
25. Memon, N., Shende, S., Wong, P.: On the security of the Yeung-Mintzer authentication watermark. In: IST Annual Conference, The Society for Imaging Science and Technology, pp. 301–306 (1999)
26. Fridrich, J.: Security of fragile authentication watermarks with localization. Proc. SPIE 4675(75), 691–700 (2002)
27. Celik, M., Sharma, G., Saber, E., Tekalp, A.: A hierarchical image authentication watermark with improved localization and security. In: International Conference on Image Processing (ICIP 2001), vol. 2, pp. 502–505 (2001)
28. Lu, H., Shen, R., Chung, F.: Fragile watermarking scheme for image authentication. Electronics Letters 39(12), 898–900 (2003)
29. Chotikakamthorn, N., Sangiamkun, W.: Digital watermarking technique for image authentication byneighbouring block similarity measure. In: TENCON. Proceedings of IEEE Region 10 International Conference on Electrical and Electronic Technology, vol. 2, pp. 743–747 (2001)
30. Lin, P., Huang, P., Peng, A.: A fragile watermarking scheme for image authentication with localization and recovery. In: Proceedings of IEEE Sixth International Symposium on Multimedia Software Engineering, pp. 146–153 (2004)
31. Lin, P., Hsieh, C., Huang, P.: A hierarchical digital watermarking method for image tamper detection and recovery. Pattern Recognition 38(12), 2519–2529 (2005)
32. Wu, M., Liu, B.: Watermarking for image authentication. In: Proc. of the IEEE Int. Conf. on Image Processing, pp. 437–441 (1998)
33. Kundur, D., Hatzinakos, D.: Towards a telltale watermarking technique for tamperproofing. In: Proceedings of the IEEE International Conference on Image Processing, vol. 2, pp. 409–413 (1998)
34. Xie, L., Arce, G.: Joint wavelet compression and authentication watermarking. In: Proceedings of the IEEE International Conference on Image Processing, vol. 2, pp. 427–431 (1998), 1998
35. Fridrich, J., Goljan, M.: Protection of digital images using self embedding. In: Symposium on Content Security and Data Hiding in Digital Media, Newark, NJ, USA (1999)
36. Winne, D., Knowles, H., Bull, D., Canagarajah, C.: Digital watermarking in wavelet domain with predistortion for authenticity verification and localization. In: Proceedings of SPIE, vol. 4675, pp. 349–356 (2002)
37. Fridrich, J., Goljan, M., Du, R.: Invertible authentication watermark for JPEG images. In: Proceedings of International Conference on Information Technology: Coding and Computing, pp. 223–227 (2001)
38. Yuan, H., Zhang, X.: Fragile watermarks based on the Gaussian mixture model in the wavelet domain for image authentication. In: Proceedings of the IEEE International Conference on Image Processing, vol. 1, pp. 505–508 (2003)
39. Yuan, H., Zhang, X.: Multiscale fragile watermarking based on the Gaussian mixture model. IEEE Transactions on Image Processing 15(10), 3189–3200 (2006)
40. Fridrich, J., Goljan, M.: Images with self-correcting capabilities. In: IEEE Int. Conf. Image Processing, vol. 3, pp. 792–796 (1999)
41. Mintzer, F., Braudaway, G., Center, I., Heights, Y.: If one watermark is good, are more better? In: Proceedings of IEEE International Conference on Acoustics, Speech, and Signal Processing (ICASSP 1999), vol. 4, pp. 2067–2070 (1999)

42. Fridrich, J.: A hybrid watermark for tamper detection in digital images. In: Proceedings of the Fifth International Symposium on Signal Processing and Its Applications, ISSPA 1999, vol. 1, pp. 22–25 (1999)
43. Hu, Y., Han, D.: Using two semi-fragile watermark for image authentication. In: Proceedings of International Conference on Machine Learning and Cybernetics, vol. 9, pp. 153–168 (2005)
44. Qibin, S., Shih-Fu, C.: Semi-fragile image authentication using generic wavelet domain features and ECC. In: Proceedings of the IEEE International Conference on Image Processing, Rochester, USA, vol. 2, pp. 901–904 (2002)
45. Sun, Q., Chang, S., Kurato, M., Suto, M.: A new semi-fragile image authentication framework combining ECC and PKI Infrastructures. In: IEEE International Symposium on Circuits and Systems, pp. 440–443 (2002)
46. Sun, Q., Chang, S., Kurato, M., Suto, M.: A quantitive semi-fragile JPEG2000 image authentication system. In: International Conference on Image Processing (ICIP 2002), vol. 2, pp. 1592–1595 (2002)
47. Lin, C., Chang, S.: Semi-fragile watermarking for authenticating JPEG visual content. In: Proc. SPIE Int. Soc. Opt. Eng., vol. 3971, pp. 140–151 (2000)
48. Lin, C., Chang, S.: SARI: self-authentication-and-recovery image watermarking system. In: Proceedings of the ninth ACM international conference on Multimedia, pp. 628–629. ACM, New York (2001)
49. Lin, C., Sow, D., Chang, S.: Using self-authentication-and-recovery images for error concealment in wireless environments. In: SPIE ITCom/OptiComm, vol. 4518, pp. 267–274 (2001)
50. Fridrich, J.: Image watermarking for tamper detection. In: Proceedings of the IEEE International Conference on Image Processing, vol. 2, pp. 404–408 (1998)
51. Fridrich, J.: Methods for detecting changes in digital images. In: IEEE Workshop on Intelligent Signal Processing and Communication Systems, Melbourne, Australia (1998)
52. Ahmed, F., Moskowitz, I.: Phase Signature-based Image Authentication watermark robust to compression and coding. In: Proceedings of SPIE, vol. 5561, pp. 133–144 (2004)
53. Ahmed, F., Moskowitz, I.: Correlation-based watermarking method for image authentication applications. Optical Engineering 43, 1833–1838 (2004)
54. Zhao, Y., Campisi, P., Kundur, D.: Dual domain watermarking for authentication and compression of cultural heritage images. IEEE Transactions on Image Processing 13(3), 430–448 (2004)
55. Kundur, D., Zhao, Y., Campisi, P.: A stenographic framework for dual authentication and compression of high resolution imagery. In: Proceedings of ISCAS 2004, International Symposium on Circuits and Systems, vol. 2, pp. 1–4 (2004)
56. Maeno, K., Sun, Q., Chang, S., Suto, M., Ltd, O., Saitama, J.: New semi-fragile image authentication watermarking techniques using random bias and nonuniform quantization. IEEE Transactions on Multimedia 8(1), 32–45 (2006)
57. Hsu, C., Lu, C.: Informed Authentication Watermarking Via Stego Data Reconstruction. In: Proceedings of IEEE International Conference on Acoustics, Speech and Signal Processing, vol. 5, pp. 743–747 (2006)
58. Huang, J.: A fast watermarking algorithm for image authentication. In: 2008 International Conference on Cyberworlds, pp. 511–514 (2008)
59. Jin, C., Chao, Y., Zhang, X.-L.: Semi-Fragile Watermark Based on Adaptive Quantization for Image Content Authentication. In: International Conference on E-Business and Information System Security, pp. 1–5 (2009)

60. Zhu, B., Swanson, M., Tewfik, A.: Transparent robust authentication and distortion measurementtechnique for images. In: Proc. of IEEE Digital Signal Processing Workshop, pp. 45–48 (1996)
61. Lin, C., Chang, S.: A robust image authentication method distinguishing JPEG-compression from malicious manipulation. IEEE Transactions on Circuits and Systems for Video Technology 11(2), 153–168 (2001)
62. Chang, C., Hwang, K., Hwang, M.: Robust authentication scheme for protecting copyrights of imagesand graphics. IEE Proceedings-Vision, Image and Signal Processing 149(1), 43–50 (2002)
63. Dugelay, J., Roche, S., It, P.: Process for marking a multimedia document, such an image, by generating a mark. Pending patent EP, 99480075 (1999)
64. Rey, C., Dugelay, J.: Blind detection of malicious alterations on still images using robust watermarks. In: Colloquium Digest-IEE, p. 13 (2000)
65. Liu, H., Lin, J., Huang, J.: Image authentication using content based watermark. In: IEEE International Symposium on Circuits and Systems, pp. 4014–4017 (2005)
66. Liu, H., Rao, J., Yao, X.: Feature based watermarking scheme for image authentication. In: Proc. IEEE International Conference on Multimedia and Expo., pp. 229–232 (2008)
67. Abrardo, A., Barni, M.: Informed watermarking by means of orthogonal and quasi-orthogonal dirty paper coding. IEEE Transactions on Signal Processing 53(2), 824–833 (2005)
68. Cover, T., Thomas, J.: Elements of Information heory. Wiley-Interscience, Hoboken (2006)
69. Proakis, J., Salehi, M.: Digital Communications. McGraw-Hill, New York (1995)
70. Watson, A.: DCT quantization matrices visually optimized for individual images. Proc. SPIE 1913, 202–216 (1993)
71. Wang, Z., Bovik, A.: Embedded foveation image coding. IEEE Transactions on image processing 10(10), 1397–1410 (2001)
72. Overall, B.: Foveated image: Applications to image and video compress (March 1999), `http://scien.stanford.edu/class/psych221/projects/99/wro/`
73. Geisler, W., Perry, J.: A real-time foveated multiresolution system for low-bandwidth video communication. Proc. SPIE 3299(1), 294–305 (1998)
74. Banks, M., Sekuler, A., Anderson, S.: Peripheral spatial vision: limits imposed by optics, photoreceptors, and receptor pooling. Journal of the Optical Society of America A 8(11), 1775–1787 (1991)
75. Said, A., Pearlman, W.: A new, fast, and efficient image codec based on set partitioning inhierarchical trees. IEEE Transactions on circuits and systems for video technology 6(3), 243–250 (1996)
76. Venkatesan, R., Koon, S.M., Jakubowski, M.H., Moulin, P.: Robust image hashing. In: Proceedings IEEE International Conference on Image Processing (ICIP), September 2000, pp. 664–666 (2000)
77. Mihcak, M.H., Venkatesan, R.: A tool for robust audio information hiding: a perceptual audio hashing algorithm. In: Proceedings of 4th International Information Hiding Workshop (IHW), PA (April 2001)
78. Kailasanathan, C., Naini, R.: Image authentication surviving acceptable modifications using statistical measures and k-mean segmentation. In: IEEE-EURASIP Work. Nonlinear Sig. and Image Processing, vol. 1 (2001)

79. Martinian, E., Wornell, G.: Multimedia content authentication: fundamental limits. In: Proceeding of International Conference on Image Processing, Rochester, NY, vol. 20 (2002)
80. Swaminathan, A., Mao, Y., Wu, M.: Robust and secure image hashing. IEEE Transactions on Information Forensics and Security 1(2), 215–230 (2006)
81. Monga, V., Mhcak, M.: Robust and secure image hashing via non-negative matrix factorizations. IEEE Transactions on Information Forensics and Security 2(3) Part 1, 376–390 (2007)

Author Index